GROUP THEORETICAL METHODS IN PHYSICS

Volume I

GROUP THEORETICAL METHODS IN PHYSICS

Proceedings of the Third Yurmala Seminar

Yurmala, USSR, 22–24 May 1985

Edited by
M. A. Markov, V. I. Man'ko and V. V. Dodonov

VNU SCIENCE PRESS
Utrecht, The Netherlands
1986

VNU Science Press BV
P.O. Box 2073
3500 GB Utrecht
The Netherlands

First published 1986

CIP-DATA KONINKLIJKE BIBLIOTHEEK, DEN HAAG

Group theoretical methods in physics: proceedings of the third Yurmala seminar: Yurmala, USSR, 22–24 May 1985/ ed. by M. A. Markov, V. I Man'ko and V. V. Dodonov – Utrecht: VNU Science Press
ISBN 90–6764–072–7 (Compl.: Vol. I and II)
Vol. I.
ISBN 90–6764–070–0 bound
SISO 531 UDC 512:53
Subject headings: physics; group-theoretical methods.

Printed in Great Britain by J. W. Arrowsmith Ltd., Bristol

PREFACE

In recent years, group-theoretical methods have been widely developed in various branches of physics; particularly in quantum field theory, elementary particle physics, nuclear physics, quantum optics and electronics, solid-state physics, and mathematical physics. Therefore, the subdiscipline "Group theoretical methods in physics" emerged. In 1972, the colloquium "Group theoretical methods in physics" was organized. This became traditional and meets annually. To date 14 meetings have been held, of which the proceedings have been published. The colloquium has an International Standing Committee which meets each year.

In addition to colloquia, seminars on group theoretical methods in physics are held in various countries in coordination with the Standing Committee.

In November 1979, the Nuclear Physics Department and the P. N. Lebedev Physical Institute of the USSR Academy of Sciences organized the first international seminar on group theoretical methods in physics which was held in Zvenigorod. Proceedings of the seminar were published in two volumes (Nauka publishers, Moscow, 1980). The second Zvenigorod seminar was held in November 1982. The proceedings of that seminar were published in two volumes by Nauka publishers (1983) and in one volume by Gordon & Breach (1985). The third such seminar was held in Yurmala in 1985. The organizers of the seminar were the Nuclear Physics Department and the P. N. Lebedev Physical Institute of the USSR Academy of Sciences and the Institute of Physics of the Latvian Academy of Sciences. This seminar was attended by more than 200 participants. The seminars in Zvenigorod and Yurmala may be considered to be the Soviet counterpart of the International Colloquia project.

The material of the Yurmala seminar is published in two volumes. The first volume contains the chapters: Gravitation, Cosmology, Supersymmetry and Supergravity; Quantum Field Theory; Superalgebras and their Applications; Nonlinear Integrable Equations; Group Representations and Special Functions; Quantum and Classical Mechanics. The second volume contains the chapters: Dynamical Symmetries and Symmetries of Equations; Representation Theory; Gauge Theories; Space Groups and Solid State Physics; Symmetries in Optics, Atomic, Molecular, and Nuclear Physics. As the reader will see, the range of subjects covered is diverse.

The editors thank the USSR Translation Centre and Dr D. A. Leites for their help in preparing English texts of the papers for the publication in the Proceedings of the Yurmala seminar.

THE EDITORS

SPONSORS

Nuclear Physics Department,
USSR Academy of Sciences

P. N. Lebedev Physical Institute,
USSR Academy of Sciences

Institute of Physics,
Latvian SSR Academy of Sciences

ORGANISING COMMITTEE

E. M. Anderson
Latvian State University, Riga

I. Ya Bersons
Insitute of Physics of Latvian Academy of Sciences, Riga

R. Ya Damburg
Institute of Physics of Latvian Academy of Sciences, Riga

V. V. Dodonov (*Scientific Secretary*)
Physical-Technical Institute, Moscow

E. S. Fradkin
P. N. Lebedev Physical Institute, Moscow

V. A. Kopstik
M. V. Lomonosov State University, Moscow

Yu. I. Manin
V. A. Steklov Mathematical Institute, Moscow

V. I. Man'ko (Vice-Chairman)
P. N. Lebedev Physical Institute, Moscow

M. A. Markov (Chairman)
P. N. Lebedev Physical Institute, Moscow

V. I. Ogievetsky
Joint Institute for Nuclear Research, Dubna

K. K. Zlobina
Nuclear Physics Division of the Academy of Science, Moscow

INTERNATIONAL STANDING COMMITTEE

A. Janner *Nijmegen (Chairman)*
H. Bacry *Marseille*
L. C. Biedenharn *Duke*
J. L. Birman *New York*
K. Bleuler *Bonn*
A. Bohm *Texas*
L. L. Boyle *Kent*
A. P. Cracknell *Dundee*
H. J. Elliott *Oxford*
F. H. Doebner *Clausthal Zellerveld*
M. Hamermesh *Minnesota*
P. Kramer *Tuebingen*
H. J. Lipkin *Rehovot*
V. I. Man'ko *Moscow*
M. Moshinsky *Mexico*
C. Piron *Genève*
S. Sternberg *Tel Aviv*
P. Winternitz *Montréal*
J. A. Wolf *Berkeley*

CONTENTS

Part V
Group representations and special functions

Part VI
Quantum and classical mechanics

Part I

GRAVITATION, COSMOLOGY, SUPERSYMMETRY AND SUPERGRAVITY

MICRO-MACRO SYMMETRIC UNIVERSE

M.A.Markov

Institute for Nuclear Research of the USSR Academy of Sciences, Moscow

In his Marchon lecture (1971) "Beyond the end of time" J.A. Wheeler considered the theory of gravitational collapse as one of the greatest crisises in the history of physics.

In the specific Wheeler's style, if we may say so, of the modern natural phylosophy he compares the collapse of the Universe predicted by the classical theory and classically predicted collapse of atom. Wheeler recalls the case of classically predicted collapse of atom, that is, the case where in a classical atom the electron must inevitably fall onto the nucleus ("collapse of atom"); recalls the unexpected solution of this problem (the solution that overstops the limits of the classical theory) in quantum theory, emphasizes the analogy between this great past crisis of the classical physica and another great crisis - collapse of the universe in the framework of the classical Einstein theory of gravitation. In this comparison he discusses different outcomes for this new "greatest crisis of the classical theory".

It is instructive to cite this page of the Wheeler's paper in the original [1].

In our opinion Wheeler's ideas concerning the universe in the limiting collapse phase, the character of this state ("black box") and peculiarity of reprocessing of the universe after collapse are very close to the truth.

In what follows we gradually approach discussing Wheeler's interpretation of the final collapse stage, the quantum stage.

COLLAPSE OF UNIVERSE PREDICTED BY CLASSICAL THEORY, COMPARED AND CONTRASTED WITH CLASSICALLY PREDICTED COLLAPSE OF ATOM

System	Atom (I9II)	Universe (I970's)
Dynamic entity	System of electrons	Geometry of space
Nature of classically predicted collapse	Electron headed toward point-center of attraction is driven in a finite time to infinite energy	Not only matter but space itself arrives in a finite proper time at a condition of infinite compaction
One rejected "way out"	Give up Coulomb law of force	Give up Einstein's field equation
Another proposal for a "cheap way out" that has to be rejected	"Accelerated charge need not radiate"	"Matter cannot be compressed beyond a certain density by any pressure, however high
How this proposal violates principle of causality	Coulomb field of point-charge cannot readjust itself with infinite speed out to indefinitely great distances to sudden changes in velocity of charge	Speed of sound cannot exceed speed of light; pressure cannot exceed density of mass-energy
A major new consideration introduced by recognizing quantum principle as overarching organizing principle of physics	Uncertainty principle; binding too close to center of attraction makes zero-point kinetic energy outbalance potential energy; consequent existence of a lowest quantum state; can't radiate because no lower available to drop to	Uncertainty principle; propagation of representative wave packet in superspace does not lead deterministically to a singular configuration for the geometry of space; expect rather a probability distribution of outcomes, each outcome describing a universe with a different size, a different set of particle masses, a different number of particles, and a different lengt of time required for its expansion and recontraction

We begin our presentation with discussing, by Wheeler's terminology, a "cheap way out", which suggests that "matter cannot be compressed beyond a certain density". This "cheap way out", according to Wheeler, should actually be rejected if, following Wheeler, one imposes the condition: "by any pressure, however high". This condition, naturally is not needed if one can formulate such a version of the theory where in the course of achieving some critical density (maximum density), for example, a Friedmann universe would collapse transforming into a De Sitter universe characterized by sign reversal of pressure (which is less than the Λ-term). In this case energy dominance is violated. This final De Sitter-like state of collapse is not singular. Hereafter we illustrate the possibility of such a classical scenario, i.e. "Friedmann-De Sitter universe". We emphasize that this is a "classical scenario" because the limiting density in this scenario is much less than Planck's density

$$\rho_0 \lesssim < \frac{c^5}{\hbar\kappa^2} \tag{I}$$

Constructing and discussing this classical scenario is, in our opinion, instructive, but mainly from the methodical point of view. The point is that in the framework of the classical theory one cannot determine the numerical value of a possible limiting matter density. In quantum theory the world constants (c, the speed of light; κ, the gravitational constant; $\hbar$, Planck's constant) constitute the quantity which may serve as the limiting density, namely

$$\rho_0 = \rho_{pl} = \frac{c^S}{\hbar\kappa^2} \sim 10^{94}\ gr/cm^3 \tag{2}$$

But a theory where the limiting density (2) plays the main role must inevitably be a quantum theory for which above-mentioned Wheeler's ideas can be decisive. But considering the consistent mathematical apparatus of the classical scenario of the Friedmann-De Sitter universe and assuming $\rho \rightarrow \rho_0$ one can obtain a number of results useful in the discussion of such a specific difficulty of an oscillating universe as a permanent entropy increase from oscillation to oscillation, the formation of daughter universes in the collapse process within a "mother" universe, etc.

The history of the idea of the "Friedmann-De Sitter universe". In recent years the majority of papers on cosmology have been devoted to the problems of the so-called "very early universe", where the initial "very early universe" is taken in an unstable De Sitter-like form a rapid matter creation of which transforms it into a Friedmann universe.

The way to understand the fact that De Sitter universe may be an essential element in cosmological constructions was not very easy. One had to overcome the psychological barrier which was an obstacle for considering a vacuum-like energy as a generally equal-right form of energy. The energy tensor density, usually written in the right-hand side of the Einstein equation, is characterized by the existence at each point of the medium of a single local co-moving frame of reference. Since De Sitter's solution appeared in the gravity theory, it has been though of as describing an empty universe, that is, a universe deprived of matter. Although the so-called Λ-term, typica of the De Sitter universe, stems, like the energy density, from the rank two symmetric tensor Λg_{ik} it is a peculiar one since for it any frame of reference is a co-moving one. The tensor $\Lambda\delta_{ik}$ is diagonal in any frame of reference.

Recall that in the book by R. Tolman [2] it is said that in the De Sitter model the proper pressure and density are related as

$$\rho_{oo} + \rho_o = 0$$

In its physical nature proper matter density ρ_{oo} can be only positive or zero. The pressure can, in principle, be negative. But even if we take into account the internal cohesive force, none of the existing forms of matter which we think of as an idealized liquid cannot create such a negative pressure.

This physical interpretation of the model of the De Sitter universe proposed half a century ago remained practically unchanged in the book by S. Hawking and G. Ellis [3] published in 1979:

These (De Sitter) spaces may be regarded as solutions of field equations in an empty space with $\Lambda = R/4$ (R is Ricci scalar) or as solutions for an ideal liquid with constant density $R/32\pi$ and with constant pressure $- R/32\pi$. But the latter interpretation is, evidently, invalid because in this case either pressure or density must be negative.

But as far back as 1965 there appeared a paper by E.B. Gliner [4] where "vacuum-like" energy of De Sitter's world was considered as a full and equal member of the family of different forms of matter. This equal footing is achieved in the framework of the classification of the properties of the energy tensor (T_{ik}), proposed by the author, according to whom "uniqueness of a co-moving frame of reference cannot be a priori considered necessary for any physical medium".

"The possibility for a relation between motion and matter other than that for an ordinary substance is seen from the fact that, for instance, a free electromagnetic field ($E^2 - H^2 = 0$; $EH = 0$) has no co-moving frame of reference".

Concerning equal footing of the energy forms, one may naturally assume also possibility of their mutual conversion.

"Negative pressures imply that internal volume forces in a substance are not repulsive, but attractive forces...

this makes us think that in a superdense state of matter, when baryons are so compressed that the meson fields which provide interaction between them (repulsion!) cannot appear, a continual medium is formed, the conditions in which correspond to attraction among elements of the substance and are described phenomenologically by a negative pressure. Such a state could occur, for instance, in a gravitational collapse" (underlined by M.M.). And five years later (in 1971) Gliner published paper titled "A vacuum-like state of a medium and Friedmann type cosmology" [5].

"The aim of the present paper", Gliner writes "is to show that if a vacuum-like state of a medium may become an initial one in any of the three Friedmann models, we are dealing with a vacuum-like state of the medium as some form of pramatter".x) But for realization of such a program of history of the universe there must necessarily exist "a starting point", in other words, an initial universe with a vacuum-like state of matter. There must also exist possibility for energy conversion from a vacuum-like form to one characterized by unique co-moving frame of reference already in Friedmann's world.

x) In Gliner's interpretaion, the vacuum-like state of a medium is very close to the concept of the primary state of the world according to Anaxagoras. The primary matter of the world, according to Anaxagoras, is some fragmentary mixture of all substances which are further formed in this world. We are not so far from the truth when we state that, in fact, Anaxagoras expressed the hypothesis about the origin of matter in the form of General Relativity. This form of matter, according to Anaxagoras, "is deprived of any moving capacity". Before space formation, infinitely small particles of this medium were initially at rest, each in its own place: they neither changed, nor were displaced because there was nothing to force them to change and to be displaced" [6]. The question of whether or not that was a brilliant prediction of Anaxagoras had better remain open. But in any case, from the modern standpoint the opinion of Anaxagoras of the primary state of matter in the universe is much more closer to us than the concept of matter in the form of water (Thales), air (Anaximenas), fire (Heraclitus).

Great progress in this direction was made by A. Starobinsky [7] and in very similar papers by V. Mostepanenko [8], S. Mamayev and V. Mostepanenko [9].

A.A. Starobinsky [7] gave an example of a conformally covariant field of matter where up to two-loop accuracy self-consistent solutions exist describing a De Sitter's world (with a total vacuum energy close to Planck's energy) which transforms into a Friedmann universe in the process of exponential expansion. The indicated solutions were derived under condition that $R_{iklm}R^{iklm} << l_{Pl}^{-4}$, where l_{Pl} is Planck's length, i;e; that higher loop corrections are disregarded. Such an approximation makes it possible to consider gravitational field as a classical field, to eliminate quantization of gravitational field and all the difficulties connected with the still unsolved problem of quantum gravity.

Recall that by the time discussed here from the historical point of view many essentially new things had appeared in the theory of field as such. Possibility of existence of vacua with different levels of vacuum-like energy, understanding of possibility of matter tunelling "between vacua" had become trivial in some sense. The unified theory of weak and electromagnetic interactions proposed by Weinberg-Gleshaw-Salam has substantially enriched physics with new concepts of properties of matter.

We mean, in particular, spontaneous symmetry breaking in gauge fields. The fundamental contribution for understanding properties of superdense matter has been made by D. Kirzhnits [10] who suggests that in a superdense substance, in a substance with a sufficiently high temperature, a broken symmetry must be reconstructed, the classical scalar field ϕ, which leads to symmetry breaking, disappears, in particular, vector bosons ($W_{\mu}^{\pm}$ and Z_{μ}^{o}) become massless. The idea of phase transitions was used to describe the behaviour of matter in early universe and in subsequent variations of its density (or temperature, which is the same). We mean the so-called "inflationary" versions of the universe. The term "inflationary universe" was coined by Guth [11]. This paper by Guth was pioneering in the discussion of various scenarios of early universe for explaning such properties of our universe as its "planeness" which imposes a rigorous condition on the matter density in very early universe and the presence of horizon - the realization of homogeneity in regions which are "not casually connected". This was the subject of fruitful and concrete studies, which, after [11], were carried out by A. Linde and others. These authors, however, make attempts also to overcome many difficulties that appeared in the problem of early universe due to the development of ideas of Grand Unification, due to supersymmetric theories and even to supergravity, which are still far from being completed. For example, in his review [12] A. Linde gives a list of twelve problems to be solved within the theory of hot universe[x)].

The versions of all the scenarios of an inflationary universe assume that at early stages of its evolution the universe could be in an unstable vacuum-like state with high energy density. As a result of an exponential expansion (inflation) of the universe in the process of this decay, as a result of the decay of a vacuum-like state there arises a high energy density characterized already by a co-moving reference frame in a Friedmann universe. In other words, the version of an inflationary universe is a realization of the Gliner's idea of an early history of Friedmann universes.

In the detailed review [12] A. Linde, who has made a substantial contribution to the development of the for-

---

[x)] It is not at all still excluded, for example, that even such a problem as baryon asymmetry can find its solution in properties of black holes, as was conjectured by Hawking and Zel'dovich some time ago. Neither is it excluded that monopoles will be a real problem of the future theory, i,e; in the future theory monopoles may be absent,etc.

malism of an inflationary universe, summarizes the three-year's activity in the development of this type of scenarios of early universe.

"As is seen from what has been said above, the scenario of an inflationary universe has been greatly modified for the three years of its existence. According to the first version of this scenario, inflation must have proceeded before the phase transition from a greatly overcooled symmetric state" (A. Guth).

"It was further understood that this version of the scenario of an inflationary universe does not work, but a new, a better scenario can be proposed according to which inflation proceeds <u>during</u> phase transition" [14] (see also the paper by Albrecht A, Steinhardt P.J.) [15]. "Finally, it has lately been understood that everything is much simpler and more natural in the scenario of chaotic inflation, which is <u>not at all based on the high-temperature phase transition theory</u> (underlined by M.M.) in early universe" [16], [17].

In the latter version of an inflating universe, some initial vacuum-like state and some "effective potential" are required. This potential could exist locally, e.g. in an infinite universe. For a chaotic scenario it is of importance that the existence of regions of a "required type" in unit volume in the initial universe was not equal to zero.

"In an infinite universe there will be infinitely many such regions, each of which, after inflation (and a subsequent expansion) becomes a kind of separate mini-universe with dimensions exceeding the dimensions of the observed part of the universe, $l \sim 10^{28}$ cm". Moreover, these necessary "nuclei" of a future concrete universe, the discussed space regions, could in principle, occur as quantum fluctuations "from nothing". The appearance of matter "from nothing" is, generally, not a new idea. The model of a Steady State Universe, proposed by Bondi and Gold [18] is based on the assumption of continuous matter creation, and Pirani [19] and Hoyle and Narlikar [20] consider also the mechanism of such creation if in addition to the ordinary matter one introduces a scalar field with a negative energy density (the so-called C-field). In such a universe, the same as in the De Sitter universe, the energy dominance principle is violated. It is of importance that in a closed universe matter creation does not contradict the energy conservation law if matter creation is compensated by gravitational mass defect [21].

For our further presentation the following two facts are essential: 1. If the energy conservation law is not violated, creation "from nothing" implies creation of only closed universes. In a closed universe the collapse problem arises. 2. Quantum fluctuations, as a possible initial state, require the existence of at least a vacuum medium - an arena for possible quantum fluctuations, i.e., in fact, an already existing universe[x)] if we do not mask the existence of such a primary universe by a number of now vague considerations. But if we mean the formation of "mini-universes" in the framework of some already existing original universe, we can also mention a more natural mechanism of "decoupling" of "daughter" universes, which is not at all connected with quantum fluctuations. As is clear from what follows, such a mechanism of the appearance of daughter universes is possible, in principle, also in the framework of classical (not quantum) physics.

<u>The classical model of a closed Friedmann - De Sitter universe</u>. The classical model of a closed Friedmann - De Sitter universe is a scenario of an oscillating universe, where collapsing from the state of its maximum expansion and in the course of achieving some minimal (but mean-

x) This remark concerns in the same measure also the model, proposed by Starobinsky, to which the developing scenarios of inflationary universes has come fairly close.

while an ultra-macroscopical) radius a closed Friedmann universe transforms into a macroscopic De Sitter universe. After this there starts a new cycle of a rather short (exponential) expansion in the course of which a De Sitter universe transforms into a Friedmann universe. The minimum radius of the universe is provided by the requirement that the tensor density of matter should not exceed some limiting density ρ_o. The mathematical apparatus of the theory is constructed so that a maximum but finite value of the matter density achieved in the process of collapse is not some additional condition imposed arbitrarily up on a given model, but is realized automatically in modified Einstein equations. Modification of Einstein equations consists, for instance, in writing the right-hand side as

$$R^{\nu}_{\mu}-1/_2 R\delta^{\nu}_{\mu}=\frac{8\pi\kappa_o}{c^4}\left[F\left(\frac{T^{\nu}_{\mu}\ T^{\mu}_{\nu}}{c^4\rho^2_o}\right)+\Lambda'\Theta\left(\frac{T^{\nu}_{\mu}\ T^{\mu}_{\nu}}{c^4\rho^2_o}\right)\delta^{\nu}_{\mu}\ c^2\right] \qquad (1)$$

where κ_o is a gravitational constant, Λ' is a constant, the functions F and Θ are such that $F\rightarrow 0$ as $T^{\nu}_{r}\ T^{r}_{\nu}\rightarrow\rho^2_o c^4$ and $\Theta\rightarrow I$ as $T^{\nu}_{r}\ T^{\mu}_{\nu}\rightarrow\rho^2_o c^4$ and $F\rightarrow I$ and $\Theta\rightarrow 0$ for $T^{\nu}_{\mu}\ T^{\mu}_{\nu}<<\rho^2_o c^4$.

Thus, for small energy densities the second term in (1) vanishes, $\Theta\rightarrow 0$, $F\rightarrow I$, and (1) describes a closed Friedmann universe. For $\rho^{\nu}_{\mu}T^{\mu}_{\nu}\rightarrow\rho^2_{\nu}c^4$ the first term in the right-hand side of (1) vanishes, the second term takes the form $\Lambda'\rho^{\nu}_{\mu}c^2$. That is, it becomes a De Sitter universe.

A mathematical apparatus was developed in a number of papers, where first a model of a dust-like universe with pressure p = 0 [22] and then p ≠ 0 [23] were considered. First we wrote ad hoc a modified Einstein equation in the form (1), using the variational principle we have derived, then equations of a similar class from the respective expression for the action [24]. Since the covariant divergence of the left-hand side of (1) vanishes, so does that of its right-hand side. Denoting the right-hand side of equation (1) by K^{ν}_{μ}, we get

$$K^{\nu}_{\mu;\nu} = 0 \quad (2)$$

In full $K^{o}_{o;o} = 0$ may be written as

$$\frac{1}{\sqrt{-g}} \frac{\partial}{\partial t} (T^{o}_{o} F \sqrt{-g}) + \Lambda' \frac{\partial}{\partial t} \Theta \quad (3)$$

Thus, in the theory there appears a peculiar law of "energy conservation".

The first term in the left-hand side of (3) is the energy in a co-moving frame reference. The energy decreases as the density increases and vanishes as the density becomes equal to the critical one. But this process is accompanied by increase of the vacuum-like energy and growth of Θ. Conversion of the energy density of a co-moving reference frame into the vacuum-like energy obays the "conservation law" (3).

Peculiarity of the mathematical apparatus of the theory is most easy to illustrate with a closed dust-like Friedmann universe. In the case of a closed isotropic and homogeneous Friedmann universe

$$ds^2 = c^2dt^2 - a^2(t)[d\chi^2 + \sin^2\chi \; d\theta^2 + \sin^2\theta d\varphi^2) \quad (4)$$

where a(t) is the radius of the universe, more precisely, the scale factor.

Let $F = (1 - \rho^2/\rho_o^2)$ and $\Theta = \Lambda' \rho^2/\rho_o^2$. Then, Einstein's equation is

$$\left(\frac{\dot{a}}{ca}\right)^2 + \frac{1}{a^2} = \frac{8\pi\kappa_0}{3c^2}\left[\rho\left(1 - \frac{\rho^2}{\rho_o^2}\right) + \Lambda' \frac{\rho^2}{\rho_o^2}\right] \quad (5)$$

The "conservation law" (2) now changes radically $K^{\nu}_{\mu,\nu} = 0$ and integration of the equation corresponding to (2) yields

$$\rho(\rho_o^2 - \rho^2) \left| \frac{\rho_o + \rho}{\rho_o - \rho} \right|^{\Lambda'/\rho_0} = \frac{B}{a^3} \quad (6)$$

The integration constant B equals $\frac{M_o \rho_o^2}{2\pi^2}$, where M_o is the total mass of the Universe.

Consider the model in the case $\Lambda'/\rho_o = 2$. Equation (6) now takes the form $\rho(\rho_o+\rho)^3/|(\rho_o-\rho)| = M_o\rho_o^2/2\pi^2a^3$ (7)

The equations for determinant a_{min} are as follows. For $\dot{a} = 0$ it follows from (5) that

$$1 = \frac{8\pi\kappa a^2}{8c^2}\left[\rho\left(1 - \frac{\rho^2}{\rho_o^2}\right) + \Lambda' \frac{\rho^2}{\rho_o^2}\right] \quad (8)$$

Equations (7) and (8) coincide provided that $\rho_{max} = \rho_c - \Delta$, where $\Delta <<< \rho_o$. It follows from (7) that

$$\rho_{max} \sim \frac{M \cdot \rho_o}{8\rho_o a^3_{min} + M_o} \quad (9)$$

and from (8) that

$$1 \sim 2a^2 \Lambda_o; \quad \Lambda_c = \Lambda \frac{3c^2}{8\pi\kappa_o} \quad (10)$$

In this case, e.g. for $\rho_o \sim 10^{21}$ gr/cm^3

$$\Lambda_o \sim 10^{-6} \text{ cm}^{-2}; \; a_{min} \sim 10^{-3} \text{ cm} \quad (11)$$

and

$$\rho_{max} \sim \frac{M_o \rho_o}{m_o + M_o}, \text{ where } m_o \sim \rho_\theta 8a^3_{min} \quad (12)$$

or

$$\rho_{max} \sim \rho_o (1 - m_o/M_o) \quad (13)$$

Thus, for $\rho_o \sim 10^{21}$ gr/cm^3 when the collapse stops, i.e. at the moment of bounce we estimate the radius of the universe $a_{min} \sim 10^{-3}$ cm for all the Universes of the masses $M_o >> m_o$ the Freidmann mass density

$$\rho_F = \rho_{max}(1 - \rho^2_{max}/\rho_o^2) \sim 10^{-21} \text{gr/cm}^3 \text{ and } \Lambda' \sim 10^{+21} \text{gr/cm}^3$$

The Friedmann mass density at this moment may be neglected as compared with the Λ'-term. This means that by this moment the universe had for a long time been in the De Sitter form:

$$(\dot{a}/ca)^2 + 1/a^2 \simeq \Lambda^o \quad (14)$$

$$a_{F \cdot S} \sim a_{min} \exp(c\sqrt{\Lambda_o}\, t) \quad (15)$$

In the Friedmann phase of collapse ρ approaches the critical value at $a_F \; (M_o/\rho_o)^3$. In this model the De Sitter phase ($\rho_o \sim 10^{21}$gr/cm^3) lasts during the collapse from $a_{F \cdot S}$ to a_{min} 10^{-3} cm. We have $a_F \sim 10^{14}$cm, $a_{min} \sim 10^{-3}$cm for $M_o \sim 10^{55}$ gr. Since the universe remains closed all the time, its total mass remains zero.

At the moment of maximum expansion of the universe:

$\rho_{min} \sim 10^{-29}gr/cm^3$; the Λ-like term is

$$8\pi\kappa_0\Lambda' /3c^2 \cdot \rho^2_{min}/\rho^2_0 \sim 10^{-100}cm^{-2} \quad (16)$$

Thus, at the moment of maximum expansion, the universe becomes strictly Friedmann's one, more precisely, at this moment Einstein's one. Such a universe cannot be stable [25] and must collapse. Neither can the universe be stable after collapse has stopped when it is in the De Sitter phase. As soon as collapse stops, the universe must start expanding, which is a classical analogue of the well-known inflation scenarios of universes.

As it is known, in the proposed inflationary universes there remain two fundamental difficulties:

1. The presence of singularity at the initial moment in all the versions of the universes (open, closed, plane).

2. A large value of the Λ-term in the Friedmann phase.

These two fundamental difficulties are absent from the classical model proposed here.

Formation of daughter universes, entropy increase in oscillating universe. This "demonstrative" model of the universe remains classical also for densities much higher than the one in the considered case. For this it suffices that

$$\rho_0 \ll \rho_{pl.} = \frac{c^5}{\hbar\kappa^2} \sim 10^{94}gr/cm^3$$

In the process of collapse, when, however it may be, very high densities are reached, such density fluctuation may appear in some region of the universe, that the total mass, say, of many particles (specks of dust) is extinguished by their gravitational defect [26,27]. If this is the case, this region of the universe, more precisely, a given number of dust specks, will get closed and form a closed universe with a certain internal radius determined by the amount of matter (the number of dust specks), i.e. these dust specks (say, baryons) will vanish for an observer external with respect to this universe. Such a universe will be called a daughter universe. For simplicity we have considered above a dust-filled universe, i.e. the one with a pressure $p = 0$. Aman and I proposed in our paper [23] a version of the universe for $p \neq 0$ too.

Tolman [25] was the first to pay attention to the fact that in an oscillating universe the entropy must increase during expansion or contraction. This process is accompanied by heating of the universe, by increase of its energy, i.e. mass, which leads to an increase of the maximal radius from oscillation to oscillation. Considering the history of an oscillating universe, the authors conclude that an oscillating universe must have its starting point. If the radius of our universe is finite at present and is close to $R \sim 10^{28}$cm then the universe must have once started its existence. The possibility of formation of daughter universes may be the solution of this problem, i.e. the problem of entropy increase, even in the framework of the classical model.

Interpreting parameters of our universe one should apply the "antropic principle": We find ourselves in such a daughter universe, namely, with such parameters, specifically, with such conditions under which organized matter can exist at its highest level. The formation of daughter universes and the daughter universes themselves are likely to be numerous.

O. Klein [26, 27] noted possible existence of semi-closed worlds. Indeed, if in the presence of a large number of particles the state of the system is such that the gravitational defect almost equals the bare mass of all the particles, then the system of particles turns out to be within the Schwartzschield sphere. This means that there appears an object with an arbitrarily small mass for an external Schwartzschield observer. Thus there appears so-called black hole of the second kind [28].

An impressive example of such a universe can be given:

Suppose that our universe with its bare mass M_0 is closed. This implies that its total mass equals zero. Its total electric charge equals zero too. If only a single extra electron is added into such a universe, it will no longer be closed. For an external observer such a universe has a Reissner-Nordstrom continuation and manifests itself as a charged micro-particle with Planck's mass and Planck's dimensions. Such a system is characterized by relativization of the concepts of macro and micro. Inside it is a universe with all its ultra-macroscopic dimensions, galaxies, possible civilizations, and outside it is a micro-particle. The title of this paper reflects that inside this universe there may exist similar particles and the universe in which our universe is, in turn, a micro-particle. The limit case is the one where all these universes enclosed in one another are closed. By the way, it is not at all easy to have a certain opinion of the fate of the discussed system, an "almost closed" universe with a single electric charge, because for an external observer such a universe remains a micro-object of Planck's dimensions with all the ensuing difficulties. Several cases are possible here:

a) such a universe eventuelly emits an electric charge, becomes closed and stops being observable in the space where the electric charge is emitted. Note that in this case the universe is a source of particle emission with about Planck's energy;

b) the system may be stable;

c) the system may decay with a substantial half-life. Return to discussing properties of systems with peculiarities of limiting (cold) elementary black holes whose total mass equals their charge or is entirely determined by their rotational energy. Such black holes have zero temperature and are not subject to the Hawking radiation.

The choice of ρ_0 in a realistic scenario (Quantum character of the scenario). In the classical (not quantum) physics there are no indications of any possible existence of a universal quantity as a maximum matter density in a co-moving reference frame, i.e. in a rest frame. In quantum theory, however, such world constants as c - the speed of light, $\hbar$ - the Planck constant, and κ - the gravitational constant may constitute a quantity with the dimension of mass density, namely, Planck's desity

$$\rho_{pl.} = c^5/\hbar\kappa^2 \sim 10^{94} gr/cm^3. \qquad (18)$$

Our attention is attracted by the fact, which is evidently important, that in modified Einstein equations (1), (5) ρ enters the ratio $(\rho/\rho_0)^n$. The terms of the type $(\rho/\rho_0)^n$ take the form

$$(\rho\kappa^2\hbar/c^5)^n \qquad (19)$$

and vanish as $\hbar \to 0$.

In other words, as $\hbar \to 0$, a modified Einstein equation (more precisely, its right-hand side) takes the form of the classical Einstein equation.

In our opinion, the essential role of the Planck's constant in the considered version raises the heuristic value of the developed viewpoint concerning the possible mechanism of collapse cessation. At first glance it may seem that the modification of the Einstein equations proposed here is some formal technique introduced ad hoc and deprived of any deep physical meaning. As a matter of fact, the modification of Einstein's equation is a mathematical realization of the idea of asymptotic freedom of gravitational interactions. In the scenarios considered it is just the weakening of gravitational interactions at small distances that is the cause of collapse cessation.

Asymptotical freedom of gravitational interactions may be the cause of collapse cessation. This feature of the discussed scenarios of the universe is proved most convincingly in the derivation of modified Einstein equations from the variational principle. Such a derivation is given by Mukhanov and the author in [24]. The initial action is expressed in the usual form

$$S = \frac{c^4}{16\pi G_0}\int (R + 2\kappa\varepsilon)\sqrt{-g}\, d^4x \qquad (20)$$

If G_0 and κ are constants, R is Ricci's curvature, g determines the metric, ε is energy density. But the situation changes radically if we assume that κ is some function of energy density. Varying the action with resput to g_{ik} we obtain an equation of the form

$$R^i_k - 1/2 R\delta^i_k = G(\varepsilon)T^i_k + \Lambda(\varepsilon)\delta^i_k \qquad (21)$$

Here $T^i_k + (\varepsilon + p)\, u^i u_k = p\delta^i_k$; p is the pressure, $G(\varepsilon)$ and $\Lambda(\varepsilon)$ are expressed, respectively, in terms of $\kappa(\varepsilon)$:

$$G(\varepsilon) = \varepsilon\frac{\partial\kappa}{\partial\varepsilon} + \kappa; \quad \Lambda(\varepsilon) = -\varepsilon^2\frac{\partial\kappa}{\partial\varepsilon} \qquad (22)$$

Notice two circumstances:

1. The consideration concerns the general case, that is for the pressure $p \neq 0$;

2. For an arbitrary function $\kappa(\varepsilon)$ the energy-momentum tensor obtained from the action (20) contains a Λ-like term.

THE Λ-LIKE TERM VANISHES IF AND ONLY IF THE ACTION COINCIDES WITH EINSTEIN'S ACTION, I.E. FOR κ = CONST. The dependence $\kappa(\varepsilon)$ on the energy density in eq. (20) is transformed into the dependence of $G(\varepsilon)$ on the energy density, which can be interpreted as the dependence of the gravitational constant on $(G(\varepsilon))$. And if $G(\varepsilon)$ decreases as the energy density increases, G may be regarded as a function that weakens gravitational interaction at small distances. Such a change in the energy density is connected with time (for instance, in a homogeneous and isotropic Friedmann universe),

$$t \sim \frac{\text{const}}{\rho^2} \qquad (23)$$

The dependence $G(\varepsilon)$ can be rewritten as time dependence on the gravitational "constant". The time dependence on the gravitational constant has been repeatedly discussed in literature (Dirac, Jordan, and others). Here we deal with the time dependence which is inverse to the dependence considered, for instance, by Dirac. The fact that the cause of the collapse cessation of the universe in this scenario is the asymptotical freedom of gravitational interactions raises the modification of the Einstein equation, proposed here ad hoc, to the rank of a physical theory.

The class of modified Einstein equations describing the scenario under discussion is formulated in a very general form (as a theorem) in [24]. "If $G(\varepsilon)$ in eq. (22) is positive for any ε and $G(\varepsilon)\varepsilon \to 0$ as $\varepsilon \to \infty$ then a homogeneous isotropic closed universe is described at high densities by a solution arbitrarily close to the De Sitter one".

Indeed, $G(\varepsilon)$ and $\Lambda(\varepsilon)$ are related via

$$\Lambda'(\varepsilon)=\int_0^{\varepsilon} G(\varepsilon)\,d\varepsilon - G(\varepsilon) \qquad (23')$$

Since $G(\varepsilon)\varepsilon \to 0$ as $\varepsilon\to\infty$ and $G(\varepsilon) \geq 0$, then (23) implies that as the density ε increases, the Λ'-like term tends to a constant equal

$$\Lambda_0' = \int_0^{t} G(\varepsilon)\,d\varepsilon \qquad (24)$$

For example, assuming in (20) $\kappa = (1 - \varepsilon^2/3\varepsilon_0^2)$, we obtain

$$G(\varepsilon) = (1 - \varepsilon^2/\varepsilon_0^2) \text{ and } \Lambda'(\varepsilon) = 2\varepsilon^3/3\varepsilon_0^2 \qquad (25)$$

Clearly $G(\varepsilon) \to 0$ and $\Lambda'(\varepsilon) \to 2\varepsilon_0/3$ as $\varepsilon\to\varepsilon_0$; (25) are just the values of F and Q which we chose to illustrate one of the scenarios [23]. We should emphasize once again that the introduced limiting maximum matter density in a co-moving frame of reference refers to what we all "Friedmann density", namely

$$\varepsilon' = \varepsilon G(\varepsilon) \qquad (26)$$

Just this density is associated with a co-moving frame of reference, just this density is a physically observed quantity. (In the case, e.g. [24], $G(\varepsilon) = (1+\varepsilon/\varepsilon_0)^{-2}$, ε' and $\Lambda'(\varepsilon)$ remain finite and positive as $\varepsilon \to\infty$.). The boundedness of this density we interprete just as a fundamental law of nature at the same level of generality as the law that bounds the speed of matter with the speed of light ($v<c$) and the law that bounds the action ($s>\hbar/2$). As is seen from the concrete versions of scenarios, ε' coincides with ε for small $\varepsilon/\varepsilon_0$ and ε differs from ε' only near the classical singularity, where ε has the character of a formal parameter which can, in principle, take arbitrarily large values subject to the value of $G(\varepsilon)$ and ε' remains smaller than that of ε_0.

As follows from [24], in varying the action (20) the Fock's methods [29] are used which exploit the expression for the particle density satisfying the continuity equation

$$(nu^i)_{;i} = 0 \qquad (27)$$

As follows from the above particular examples of scenarios in a region close to collapse cessation, "Friedmann" (observable) matter density ε' tends to zero, the radius, i.e. the volume of the universe decreases and, accordingly, the bare gravitational mass decreases too. With equation (27) for conservation of a number of particles this situation can be described in the following way: particles do not vanish in the process of collapse, but due to weakening of the gravitational interaction each of them becomes more and more "incorporeal" (of smaller mass), more precisely, of smaller weight. At the same time the vanishing gravitating (Friedmann-type) mass, as the whole ensemble of "incorporeal soules", transforms into a peculiar De Sitter-type vacuum-like energy. It is of interest that equation (5) written ad hoc does not follow from the variational principle and from the expression for the action presented in (21). Unlike (20), the variations of F and Θ are determined not only by asymptotical freedom of gravitational interactions, but also by the possibility for particles to vanish in a collapse and to be produced in an expanding universe. (Equations of the type (5) with a given constant Λ' are obtained from the action with the condition $(nu^i)_{;i} \neq 0$). By the Bianchi identity, this process is also connected with production of particles (Friedmann-type matter) from the De Sitter vacuum-like state and with vanishing of particles under transformation of a gravitating (Friedmann-type) energy into the De Sitter vacuum-like energy. Thus, not only the asymptotical freedom of gravitational interactions, but also vanishing of Friedmann-type matter may be the cause of collapse cessation. In the universe of the Hoyle-Narlikar type the energy dominance condition is also violated, this violation being required by the solution of the singularity problem. The metrics of these theories are known to belong to the class of De Sitter's metrics. As is known, the simplified x) picture of the classical Friedmann universe filled with a dust-like matter develops according to the law 30

$$R = (2/9M_{0}\varkappa)^{1/3}\, t^{2/3} , \qquad (28)$$

where M_0 is the total mass of the universe. More precisely, M_0 is the total mass of the dust particles that fill the universe at pressure p=0,   the gravitational interaction constant, t the time counted from t=0. The right-hand side of (28) determines the dimensions the universe at t. It is of interest that the at the initial moment the unjverse starts expanding at an infinitely high speed. As t 0, the time derivative

$$\dot{R} = 2/3((2/9M_{0}\varkappa)^{1/3}/t^{1/3}) \qquad (29)$$

tends to infinity.

It has long been known that such a picture of expansion universe leads to a "serious" difficulty.The point is

x) Simplification consists in the fact that in this case we consider the classical,in effect,Newtonian picture of the dynamics of a homogeneous gravitating medium in the so-called co-moving reference frame.A medium,in our case "dust",fills a sphere of finite radius.But the formulas appearing here in an apt way are similar to the expressions obtained in the solution of the relevant equations of General Relativity.But General Relativity deals with the dependence on time of the so-called scale factor which characterizes time variations of the distance between two points in space.

that on a large scale our universe is matter with a homogeneous and isotropic distribution. The question is what causal mechanism "lays" this homogeneity at the initial time moment if the spread velocities are such that they exclude causality between separate regions of the universe. Casuality is realized by signals propagating with a speed which does not exceed the speed of light. Thus, we deal with the long known difficulty of the Friedmann universe usually referred to as "horizon problem". The horizon problem would be solved if one could change the time dependence of the law of the expansion of a very early universe, i.e. if we had, for example, $R \sim t^n$, where $n > 1$. Then the speed of expansion of the universe $\dot{R} \sim t^{n-1}$ for $t > 0$ would be very small in contrast with the case (29). This could be achieved by assuming that for small times the factor $(2/9M_0\kappa)^{1/3}$ in the expression (28) is not a constant, but a function of time increasing from zero or from a small value. Such an assumption could imply that for small t (or, which is the same, for high matter densities) κ is an increasing function of time (or a decreasing function of density). Such an assumption would imply that at small distances between gravitating particles the so-called "asymptotical freedom" of gravitational interactions is realized. This possibility is realized by the version of modification of Einstein's equations (20), (21). But the same result could be obtained under the assumption that κ= const, but the mass of the universe, M_0, appears not instantly, but during some initial time t within which it reaches its maximum value M_0. It is of importance to mention that in present-day attempts to describe a very early universe the both possibilities are discussed in some form.

The case considered above (5), where the parameter Λ' is introduced as a new arbitrary constant, i.e. the Λ-like vacuum term appears not as a result of asymptotical freedom only (or ρ is not determined by the dependence

$\kappa(\varepsilon)$ only), is an example of realization of the second possibility to interpret F and Θ.

<u>From the classical illustrative model to a realistic scenario of early universe</u>. In the above-considered models illustrating the possibility of consistent scenarios of oscillating Freidmann-De Sitter universes all the versions have been developed in the framework of strictly classical theory. Both in the Friedmann and De Sitter phases universes possesse classical ultra-macroscopic dimensions. But the necessity to chose a physically meaningful parameter for the limiting value of the matter density, which plays the fundamental role of a world constant, has led us unambiguously to the expression for the Planck density ($\rho_{pl.}$).

According to (10) the expression for the limiting density (ρ_o) is connected with the minimal radius of an oscillating universe by the relations

$$a_{min} = \sqrt{1/2_{\Lambda o}};\quad \Lambda_o = \Lambda' \frac{8\pi\kappa}{3c^2};\quad \Lambda' = 2\rho_{pl}$$

Substituting $\rho_{pl.} = c^5/\hbar\kappa^2$, we obtain

$$a_{min} = (8\pi/3)^{1/2}(\hbar\kappa/c^3)^{1/2} \sim l_{pl.} \tag{30}$$

In this case, when the collapse stops or, which is the same, in the state of an "early" universe, the universe must have Planck's dimensions with all the difficulties that follow - the necessity of quantum consideration of micro-universe.

Previous illustrations of the classical models give instructive ideas about the character of changes in the collapse of a universe with an increase of the limiting density ρ. As is clear e.g. from the classical version of the model, namely, the case (5), the maximum density ρ_{max} reached in the process of collapse is smaller than the conventional limiting density ρ_o. But it is different for different universes with different values of their bare mass M_o.

By (13) the maximum density value reached in a collapse, ρ_{max} (for the parameters taken in the case (5), namely, ρ_0-limiting approximately equal to $10^{21}gr/cm^3$ and the bare mass of a universe, M_0, approximately equal to $10^{55}gr$) differs from ρ_0-limiting by

$$\Delta\rho \sim \rho_0 \frac{m_0}{M_0} \text{ or } \rho_{max} \simeq \rho_0 (1 - \frac{m_0}{M_0}) \qquad (31)$$

And only as $M_0 \to \infty$, does ρ_{max} reach ρ_0. If we dare consider the same version (5) for Planck's limiting density ($\rho_0 = \rho_{pl}$), then instead of (31) there appears the expression: $m_0 = m_{Pl}$ and

$$\rho_{max} = \rho_{pl}.(1 - \frac{m_{pl}}{M_0}) \sim \rho_{pl}(1-10^{-60}) \qquad (32)$$

On the basis of some of my recent results one can formulate a theorem with far-reaching corollaries*):

"For universes of any masses (any M_0) much larger than Planck's mass (m_{Pl}) in the process of collapse the maximum density (ρ_{max}) with a large accuracy reaches the limit density ρ_{Pl}. Equivalently collapse cessation occurs for the same minimal radius (in this case $a_{min} \sim l_{Pl}$, more precisely, $a_{min} \sim l_{Pl}(1-m_0/M_0)$ irrespective of $M_0 >> m_{Pl}$".

The indicated characteristic feature of the considered scenario is not typical of the limiting value ρ_0 which equals Planck's density - it is explicitly expressed also in the case $\rho_0 \sim 10^{21}gr/cm^3$, which is by 70 orders of magnitude smaller than Planck's density. In the latter case the collapse stops at $a_{min} \sim 10^{-3}cm$ for all universes with mass $M_0 >> m_0 \sim \rho_0 8a^3_{min} \sim 10^{13}$ gr when according to (10) and (13)

$$a_{min} \sim 1/\sqrt{2\Lambda_0}; \ \Lambda_0 = \Lambda' 3c^2/8\pi\kappa; \ \Lambda' = 2\rho_{max} \sim 2\rho_0(1-m_0/M_0)$$

$$a_{min} \sim 10^{-3} \ (1-m_0/M_0) \text{ or } a_{min} \sim 10^{-3}(1-10^{-42}) \text{ cm}, \qquad (33)$$

For our further discussion it is of great importance that

---

*) For the reasons clear from what follows this theorem will be hereafter referred to as the "reprocessing" (Wheeler's term) of universes theorem.

the state of a universe at the moment of collapse cessation (ρ_{max}, a_{min}) is at the same time a characteristic of the initial state which completely determines the fate of the expansion phase of the universe. If $a_{min} >> l_{Pl}$, $\rho_{max} << \rho_{Pl}$, the scenario completely, in all its phases falls in the framework of the classical theory and the replacement of t by -t repeats (in the simple case (p=0)) in an inverse order the whole process of collapse. But if $\rho_o \sim \rho_{Pl}$ and the dimension of the universe is of the order of Planck's dimensions, there arises the specifics of quantum nature of the universe in its early phase of anticollapse. In what follous we give a more detailed consideration (following Wheeler's ideas) to the quantum specifics of the state of such a universe at the moment of collapse and to its possible influence on the character of a subsequent expansion of the universe.

The place of the considered scenario of a "Friedmann-De Sitter universe" among other scenarios of a very early universes. In recent four or five years there has appeared a large number of papers devoted to the problems of a very early universe. Many of them are united by the same term "inflationary universes". Common for all these scenarios is the appearance at some early initial moment of the development of a universe of De Sitter-type phase which transforms into a Friedmann-type phase.

A. After collapse cessation the versions of Friedmann-De Sitter universe considered here are in the phase of inflationary universe.

B. But as distinguished from all the other versions of inflationary universes, the duration of the De Sitter phase of the classical universe is determined by the bare mass in the Friedmann phase.

C. In other versions of inflationary universe the conditions are usually chosen from the very start so as to provide the appearance in the Friedmann phase of a universe with parameters close to those of our universe. If such conditions were unique this would imply that the universe can only be long to the image of our universe.

D. The principle possibility of formation of daughter universes discussed above disproves such a viewpoint and supposes the existence of different universes with different dimensions and masses.

The versions of early universes are usually considered under the conditions that the gravitational fields at the initial moment of the history of the universe are limited to such an extent that their quantum effects can be neglected. The gravitational fields in this case are classical and are not subject to quantization. Only the material field in the one-loop approximation is quantized. The boundedness of the curvature is expressed via

$$R_{\mu\nu\lambda\delta} \cdot R^{\mu\nu\lambda\delta} \quad 1/l_{pl}^4 \tag{34}$$

E. If one analyzes the proposed condition on the boundedness of the maximum density

$$\rho_o < \rho_{pl}^2. \tag{35}$$

and multiplies by the constant κ/c^4, then the new (35) inequality

$$\kappa^2\rho_o^2/c^4 \quad c^6/\hbar^2\kappa^2 = 1/l_{pl}^4 \tag{36}$$

coincides in its expression and meaning with the estimate (34). But in our case this estimate is not simply justifying the estimate of the one-loop approximation, but is an expression of a strict and universal law of nature. The attempts to associate the essential features of a very early universe with supersymmetric Grand Unified Theories and, ultimately, with supergravity look very tempting. But these theories are not yet quite perfect and often introduce new essential features to the characteristics of a very early universe with their new interesting possibilities and sometimes with new cosmological difficulties. This kind of problems is considered in A.Linde's review among the twelve listed problems [12]. These are grounds to believe that some part of these problems is either connected with imperfection of the involved theories (for instance, the problem of domain walls) or is likely to find its possible solution in the framework of quite different physical concepts.

<u>A very early universe, a universe at the moment of bounce (for $\rho=\rho_{max}$).</u> When one speaks of a very early universe, one means a universe at $t>t_{Pl}$ 10^{-43}sec. At $t<t_{Pl}$ the universes which are usually discussed must have dimensions smaller than the Planck's length This restiction is only explained by the fact that a region of so small dimensions and high matter densities is beyond the framework of modern theories. In our versions of oscillating Friedmann-De Sitter universes the theorem discussed above is of substantial importance. By this theorem, for any value of the internal mass of a universe in its Friedmann phase, the universe has practically the same dimensions $l>l_{Pl}$ at the moment of collapse cessation. Since in a closed world the motion of matter is restricted by the dimensions of the world, the energy spectrum of particles (for instance, dust) that fill the universe is by necessity discrete. The corresponding Wheeler - De Witt equation, which is solved in the WKB approximation, gives a characteristic form of such a spectrum. For a Friedmann mini-universe filled with dust, an approximate solution is given in [31] . As is seen from numerical (computational) data, the energy spectrum of a Friedmann mini-universe is given with high accuracy by the expression

$$E_n = E_o(2n+1)^{1/2}, \quad n=0,1,2... \tag{37}$$

where

$$E_o = (3\pi^3\hbar c/\kappa)^{1/2} \tag{38}$$

Thus, a quantum universe cannot have dimensions smaller than Planck's ones. The lower values of the energy (mass) of the universe must also be Planck's ones. The discussed Friedmann-De Sitter universe at a state close to collapse cessation is characterized just by Planck's parameters.

For a specific dependence of the gravitational constant on the density of the type analyzed in [24], this case has also been investigated by Mal'tsev. Since in this case, too, the universe is closed and matter motion is finite, the energy spectrum, although differs from a purely Friedmann one, remains discrete. In the case of mini-universes, the quantization by Wheeler-De Witt takes into account in a finite motion of matter also zero oscillations of matter determining substantially the lower state of a universe. In their pioneering paper [32] V.L. Ginzburg, D.A. Kirzhnits and A.A. Lubushin do not take into account quantum vacuum corrections in the Wheeler-De Witt equations used in [31]. Their corrections modify to some extent the discrete energy spectrum of Friedmann-De Sitter mini-universes. It should be emphasized that in the framework of the Wheeler-De Witt equation, at a moment close to collapse cessation, the Friedmann-De Sitter universe discussed here is, in fact, an image of the initial state of the universe or, in terms of Hartle and Hawking, the ground states of the wave function of the universe [33].

"Beyond the end of time". Let us dare describe the case of Planck's value ρ_{max} in the framework of classical physics. We start describing a very early universe immediately after collapse cessation. To obtain in the Friedmann phase a universe with a mass of the order of our universe, we must in the initial data, according to (13), set the density ρ_{max} EXACTLY in the form

$$\rho_{max} = \rho_{pl}\left(1 - \frac{m_{pl}}{M_o}\right) \qquad (39)$$

It is natural to assume that quantum fluctuations of density can substantially violate such a small difference between ρ_{max} and the limiting Planck density. The value of the density ρ_{max} at the moment of classical bounce is in one-to-one correspondence with the radius of the universe at a_{max}. Wheeler estimated the change in the radius of our universe at the moment of maximum expansion

($R_{max} \sim 10^{28}$cm: [34]. The mean deviation from the classical radius of the universe R_o:

$$R = R_o + l_{pl}^{4/3} \ (24\pi^2)^{2/3} \ R_o^{1/3} = R_o + \Delta R; \ \Delta R \sim 10^{-55}\text{cm}.$$

Wheeler-De Witt's quantization of a Friedmann mini-universe for the lower state of Planck's dimensions gives the value comparable with the dimensions of the mini-universe. Deviations from the expected classical radius lie in the limits Δl_{pl}: [31]

$$l = l_{pl.} + \Delta l_{pl.}, \text{ where } \Delta l_{pl} \sim 0{,}4 \ l_{pl}$$

l_{pl} is the radius of a mini-universe in the lower energy state: n = 0. Quantum uncertainty in the dimensions of a mini-universe of Planck's dimensions ($\Delta a_{min}/a_{min} \sim 1$) can be interpreted as quantum uncertainty of the bounce radius. In the classical description the dimensions a_{min} are rigidly connected with the value of the mass of the universe at the moment of maximum expansion and, therefore, with entropy (its numerical value). From the above the only thing is now clear: the classical rigid connection between the bounce radius and the mass of the universe in the Friedmann phase in the models proposed by us is violated by Wheeler-De Witt's quantization. These considerations can be expressed in a more rigorous form. Here we again go back to Wheeler's ideas.

"In the real world of quantum physics, however, one cannot give both a dynamic variable and its time-rate of change. The principle of complementarity forbids. Given the precise 3-geometry at one instant, one cannot also know at that instant the time-rate of change of the 3-geometry. In other words, given the geometro-dynamic field coordinate, one cannot know the geometro-dynamic field momentum. If one assigns the intrinsic 3-geometry, one cannot also specify the extrinsic curvature.

The uncertainty principle thus deprives one of any way whatsoever to predict, or even to give meaning to, "the deterministic classical history of space evolving in time". [1] (p. 1182).

According to physical contents of the gravitational equations the collapse process in the classical description is a transformation of the initial 3-geometry (3G) i.e. all closed universes with any maximum radii at the bounce moment transform into mini-universes of small masses and small dimensions. In the case of the limiting density $\rho_{Pl} = c^5/\hbar\kappa^2 \sim 10^{94}$ gr/cm^3 the dimensions of all the universes at the moment of bounce ($\dot{a} = 0$) only within the Planck's length can exceed these of the micro-universe whose radius equals precisely Planck's length, i.e. $a_{min} = l_{Pl} + \Delta l_{Pl}$. According to formulas (32) and (33) only the universes which had in the past $a_{max} = \infty$ (or $M_o = \infty$) reach Planck's dimensions in the process of collapse. The radii of all other universes with $M_o \sim m_{Pl}$ in the state of bounce (a_{min}) are somewhere between l_{Pl} and $l_p + \Delta l_{Pl}$, and $\Delta l \sim l_{pl} \frac{m_{pl}}{M_o}$. In other words, if in the process of collapse a universe reached the "bounce" state, the place of bounce (a_{min}) cannot be precisely known. Or, which is the same, after a bounce the character of the universe, in Wheeler's language, "reprocessing" of the universe after collapse, in not determined. Our scenario of a collapsing Friedmann-De Sitter universe and "reprocessing" (Wheeler's term), after the collapse, of a new universe is close in its spirit to the Wheeler's "black box" model. In [1] Wheeler's "Beyond the end of time" there is Section entitled "The black box: the reprocessing of the universe". The reprocessing of the universe.

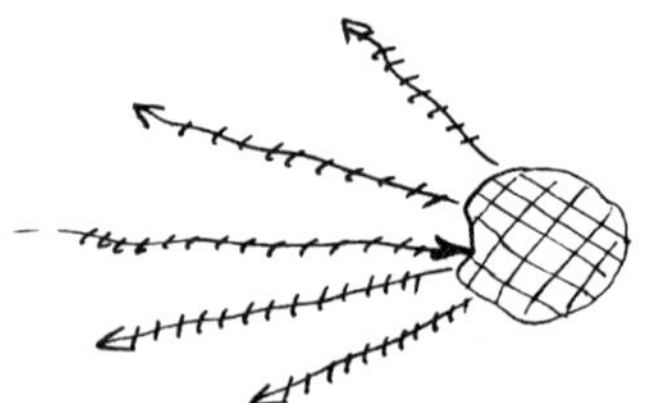

"The "black-box model" applied (1) to the scattering of an electron by a center of attraction and (2) to the collapse of the universe itself. The deterministic electron world line of classical theory is replaced in quantum theory by a probability amplitude, the wave crests of which are illustrated schematically in the diagram. The catastrophe of classical theory is replaced in quantum theory by a probability distribution of outputs. The same into a certain final 3-geometry (3G). Connection between adjoining 3-geometries (3G) is realized by the so-called extrinsic curvature. In this description (3G)'s are analogues of the coordinates of the classical mechanics of particles, and the extrinsic curvature is an analogue of momentum canonically conjugate with the coordinate. Collapse develops in the superspace of 3-geometries (3G). In quantum consideration the 3-geometry of bounce does not commute with the extrinsic curvature, with a "geometrodynamic momentum". Thus, precisely knowing the geometrodynamic momentum, we lose the knowledge of the canonically conjugate 3-geometry. In a simple case of homogeneous and isotropic universes, for instance, a Friedmann universe with

$$ds^2 = a^2(t)\left[d\chi^2 + \sin^2\chi(d\theta^2+\sin\theta d\phi^2)\right]$$

or a De Sitter universe with

$$ds^2 = a^2(t)\left[d\chi^2+\sinh^2\chi\ (d\theta^2 + \sin^2\theta d\phi^2)\right]$$

the expression for the extrinsic curvature involves a derivative with respect to the proper time of the scale factor (radius of the universe), that is, da/dt*).

In the classical case the place of bounce in the collapse is determined by equating a to zero. As follows from the classical picture of collapse, in our formalism the character of development of the universe after bounce is contained in the initial data.Namely,the place of the bounce "remembers" what maximum radius (or in the case p=0 what total mass) characterized the universe before its collapse.In other words,only the dimensions of the universe at the bounce moment determine the maximum radius of this universe at the moment of its maximum expansion. As we have seen above, all collapsing universes,

*) The extrinsic curvature tensor $\overline{K}$ is $-\dfrac{1}{2Na^2}\dfrac{da^2}{dt}I$, where N is the lapse fraction. For simplicity Wheeler wrote the expression for the tensor K disregarding the shift vector N^i ([1], p.513).

> diagram illustrates the "black-box account" of gravitational collapse mentioned in the text. The arena of the diagram is no longer spacetime, but superspace. The incident arrow marks no longer a classical world line of an electron through spacetime, but a classical "leaf of history of geometry" slicing through superspace (Chapter 43). The wave crests symbolize no longer the electron wave function propagating through spacetime, but the geometrodynamic wave function propagating through superspace. The cross-hatched region is no longer the region where the one-body potential goes to infinity, but the region of gravitational collapse where the curvature of space goes to infinity. The outgoing waves describe no longer alternative directions for the new course of the scattered electron, but the beginnings of alternative new histories for the universe itself after collapse and "reprocessing" end the present cycle ([1] , p. 1213).

Wheeler refers to the state of the universe in its last collapse phase as a black-box state. In early papers, where the formalism proposed by us was developed, this state was sometimes called "purgatory" in the sense that the universe is "reprocessed" from this state in some different quite new form, in particular, it can be "purified", for example, from the entropy "excess" accumulated during its past history. The possibility of such probabilistic reprocessing of the universe is commented by Wheeler in the following way:

> "Little as one knows the internal machinery of the black box, one sees no escape from this picture of what goes on: the universe transforms, or transmites, or transists, or is reprocessed probabilistically from one cycle of history to another in the era of collapse.
>
> However straightforwardly and inescapably this picture of the reprocessing of the universe would seem to follow from the leading features of general relativity and the quantum principle, the two overarching principles of twentieth-century physics, it is nevertheless fantastic to contemplate. How can the dynamics of a system so incredibly gigantic be switched, and at "the whim of probability, from one cycle that has lasted 10" years to one that will last only 10^6 years? At first, only the circumstance that the system gets squeezed down in the course of this dynamics to incredibly small distances reconciles one to a transformation otherwise so unbelievable" ([1] , p.1214).

The formalism we have proposed to maintain the "faith" with which Wheeler speeks, contains additional arguments of "smallness". Not only the dimensions, but also the masses of the universes in the bounce state (in a "black box") turn out to be small, and what is particularly important, in this state the universes become indistinguishable. This is the reason why we have called the theorem, which is the basis of this indistinguishability, "reprocessing theorem".But here one should also pay attention to some constraints to which the conditions imposed by a black box are reduced.If during the collapse a universe retains its total mass,if it is only density that increases infinitely in its small dimensions,then the existing peculear integral of motion (the total mass of the universe or,which is the same,its gravitational radius) will apparently not admit a rebirth of the universe with parameters other than those of the collapsing universe. The decrease of the total mass of the universe along with the decrease of its dimensions is therefore important for the realization in the collapse of the "black box" with the properties admitting the reprocessing of the universe with renewed parameters.

Moreover, the change in the total mass of the universe, which is only due to the appearance of asymptotical freedom, <u>with a retention of the number of particles</u>, i.e. equation (27) may also appear insufficient*) for reprocessing of the universe with parameters other than in the initial universe.

Only scenarios of the type (5) with violation of the particle conservation law (an arbitrary Λ and $(nu^i); i \neq 0$) are apparently adequate to Wheeler's idea of "black box". Another possibility is connected with a different interpretation of the wave function in quantum theory in the spirit of Everett. In other words, the question arises whether or not the usual Copenhagen interpretation of the

---

*) This place is vague: "beyond the end of time" equation (27) in ^4G does not make sense.

wave function is retained here, which is the solution of the corresponding Wheeler-De Witt equation. In the Copenhagen interpretation an important role is played by classical observed, a classical device for observation. A certain difficulty in the Copenhagen interpretation of the bounce phenomenon is the fact that for a closed universe there is no classical observer (the same refers to any version of a very early universe), that would fix the necessary initial conditions arising in a particular case of collapse cessation. The absence of a classical observer in quantum description of a closed universe has been repeatedly emphasized in the literature as a defect of the Copenhagen interpretation of quantum theory beginning from Everett who proposed his own interpretation of quantum theory. For the interpretation of the ψ-function in the description of bounce there are two possibilities mathematically equal but competing in the interpretation of the ψ-function:

I. An absolute chance[x)] ("God really dices"), and we found ourselves, fairly by chance, in a universe with conditions for life.

II. In the state of collapse cessation a universe is a set of various universes (Everett), and it is not by the absolute chance, but by the chance as it is understood

---

x) As has been stressed by De Witt [35] , the quantum theory (the Copenhagen interpretation - M.M.) describes in the mathemathical language the world where the chance is not a measure of our ignorance as in the case of statistical mechanics, i.e. not a "chance as an unknown necessity", but an absolute chance. From the experience in discussing interpretation of quantum theory, the interpretation by Everett and those holding the same views on the simultaneous existence of many universes hardly finds any supporters because of an unusual character of the assumption. But only the habit not to pay attention to the fact that one and the same term - chance - implies essentially different things helps accept the existence of a not less unusual possibility, the possibility of an absolute chance, which is so different from the concept of chance in statistical mechanics.

in statistical mechanics that we found ourselves in a universe where there were conditions for the appearance of organized matter. The "reprocessing" of the universe after collapse in a new form, in particular, with a lower entropy (less than R_{max}) solves the problem of an infinite entropy increase in closed universes.

In the framework of the Everett's interpretation of quantum mechanics, De Witt proposes a concrete possibility of reprocessing of a universe in such a state that its entropy does not increase, but on the contrary decreases with time. These considerations are based on symmetry with respect to the time functional which describes the universe [35] $(H + \kappa)$ *). We often say that we have no consistent quantum theory of gravitation. But can we state that we are not yet able to describe the quantum state of the initial universe or, which is the same, the state of a collapsing universe in the final state of its collapse, i.e. in the state of bounce and in the state of rebirth? Is there anything missing for an exhaustive description of these analogies in that obvious picture with which Wheeler illustrates the analogy between electron

---

*) "However, if the operator $H + \kappa$ is time-reversal invariant, and if its zero eigenvalue is nondegenerate, then the state functional of the universe is necessarily time-symmetric. This means that for every Everett branch in which entropy increases with time there must be another in which entropy decreases with time. To an observer in the second branch "time" in fact appears to be "flowing" in the opposite sense. Because of the extreme sensitivity of the state functional to slight changes in the operator $H + \kappa$ (see [56]) it is difficult to say how these conclusions must be modified if, as recent experiments suggest, the real world is not invariant under time reversal. However, the world being as complicated (and hence ergodic) as it is, it is still quite possible that there is no preferred direction in time. The ensemble of Everett branches in which time has a given direction of flow may very well be balanced by another ensemble in which time flows oppositely, so that reality as a whole possesses no over-all time orientation despite the absence of time-reversal invariance."

scattering on the centre of attraction and reprocessing of the universe in collapse? A single act of collapse and reprocessing of the universe, as well as a single act of the incidence of an electron onto the centre of attraction and its scattering at an uncertain angle are similarly undetermined.

We have every right to believe that none of the future theories can explain why in a single act of the scattering of an electron it was scattered at a certain angle. By analogy, the question why at the last stage of collapse (and a "black box") there arises a universe other than the primary one is evidently also irrelevant. But unlike the situation with the electron, in the case of the collapse the question is possible: In what form does matter arise in the process of conversion of the vacuum-like energy into the energy of the Friedmann universe? One may think that this question is similar to the following: In what form does the matter of the Friedmann universe transform in the process of collapse into the vacuum-like matter of the De Sitter world? It is quite obvious that in the final stage of collapse matter contains all its variable forms. All the forms of the so-called elementary particles are evidently produced under these extreme conditions.

The state of matter in the De Sitter phase of the universe. When the matter of a Friedmann universe is transformed into the Λ-like form of matter of a De Sitter universe, the whole variety of forms of matter appearing in this process - various particles arising under extreme conditions of almost Planck's density - must somehow disappear in a uniform Λ-like medium. The task of many theoretitian physicists is to construct in the framework of the group theory such a supermultiplet of elementary particles that would organically contain all the variety of elementary particles existing in nature. Let all the particles of such a multiplet turn out to be randomly localized in a small domain, namely, inside its common Schwartzschield sphere. For an external observer the appearing object is a black hole. In this case all the differences in the physical properties of the elements of the multiplet are buried in the common grave of the black hole. And again Wheeler: "Black holes do not have hair". In this case, in a co-moving reference frame the matter is characterized only by its density*).

Thus, there arises an attractive idea to suppose that the primary form of matter originating from the decay (transformation) of its vacuum form is also no other than elementary (of Planck's dimensions) black hole. Evaporation of elementary black holes must lead to the appearance of all the components of supermultiplets of particles, to the Big Bang with all the consequences of the hot phase of the universe. In other words, the set of all elementary particles could be limited to a decay of Planck's mass. In the known versions of early universe, the vacuum-like term in the Einstein equations is taken in the form of a homogeneous and isotropic scalar field. Here we consider only the transformation of De Sitter universes into Friedmann universes with a "transformation" of a vacuum-like scalar field ϕ, i.e. with a decay of the corresponding vacuum into an ordinary matter. If the problem were inversely formulated, i.e. if in the process of collapse Friedmann universes were transformed into De Sitter universes, one would face the transformation of the diverse matter of a Friedmann universe into the vacuum-like energy of a scalar field. By the way, A. Linde writes [12] that the field ϕ itself is not at all necesserily an elementary scalar field. It may be a

---

*) The gravitational waves, including relic and gravitational perturbations in general must not escape this transformation into elementary black holes. Otherwise, they alone form singular states.

compound field of the type of fermion (or vector) pairs $(\bar{\phi}\phi)$ $(\langle G^a_\mu G^a_\mu \rangle)$. It may also be just a chaotic distribution of the curvature tensor $R_{\alpha\beta\gamma\delta}$. In general, this is only one step to the generalization that all the existing fields contribute to the discussed scalar field. It seems simpler and more general to associate a vacuum-like energy only with the energy density, as is the case in our formalism. Paying tribute to fashion, Mukhanov and I [24] also expresse the energy density in terms of the scalar field ϕ. In this case a modified Einstein equation takes the form

$$R^i_k - \frac{1}{2}R\delta^i_k = 3\ell^2\left[(\phi^{;i}\phi_{;k} - \frac{1}{2}\delta^i_k\phi^{;\ell}\phi_{;\ell})F(\frac{\phi^2}{b^2}) + \frac{1}{2}\delta^i_k m^2\phi^2\psi(\frac{\phi^2}{b^2})\right]$$

where $\ell = (\frac{8\pi G}{3})^{1/2}$, $c = \hbar = 1$. Here b^2 plays the role of the limiting field value at which F vanishes, and ψ becomes 1, by analogy with F, G and θ, Λ in the above scenarios (5) and (21). Before that, in [22'] the action was given in the form

$$L = (-g)^{1/2}\left\{\left[g^{\alpha\beta}\frac{\partial\phi^*}{\partial x_\alpha}\frac{\partial\phi}{\partial x_\beta} - m^2\phi^*\phi\right](1 - m^2\frac{\phi^*\phi}{b}) + \Lambda m^4(\frac{\phi^*\phi}{b^2})^2\right\}$$

which, in accordance with the case (5), contains an arbitrary constant b. In our scenario we have only used the scalar field (ϕ). In [36] only the behaviour of a very early universe is described. The description of this phase of the Friedmann-De Sitter universe is very close to the Linde's scenario of a chaotic inflationary universe. At first the editors even returned the paper to us with the note of the refree that it was only a particular case of the chaotic inflationary universe*). If one takes into account that our scenario can use any field

*) Indeed, the paper described only the initial stage of the Friedmann-De Sitter universe, i.e. the previous stage, the collapse, was not considered. But the assumption concerning the existence of a limiting energy density played, as before, an important role. This is absent in the Linde's scenarios. But, in fact, any of the inflation schemes actually postulates a finite density of the vacuum-like energy.

and that it is not grounded to distinguish a certain field among other fields, then, in general, a theory that takes into account the existence of all possible fields forming the tensor T_{ik} is only permitted. Therefore, to use in the above scenarios the concept of energy density ε and the limiting density value ε_0 is, perhaps, to describe a situation not so far from the one probably existsing in reality.

In connection with a possible important role of elementary (Planck's) black holes in the history of the universe, it seems reasonable to mention some problems in the description of physical properties of these objects. The nature of elementary black holes ("maximons") [37]. Understanding of the nature of elementary black holes of Planck's dimensions in the framework of the existing possibilities of their theoretical description meets practically the same difficulties as those at the description of the lower state of the universe. Moreover, a macroscopic black hole decays as a result of Hawking radiation. As the mass of a black hole tends to Planck's mass in the course of this radiation, one should expect a substantial change in the radiation spectrum. The point is that the motion of matter inside the Schwartzschield sphere is finite. This leads to a discrete energy spectrum which should manifest itself in the Hawking radiation of a small black hole. Variations in the spectrum of the Hawking radiation have not yet been examined even approximately.

The description of small black holes by Wheeler-De Witt's equation, using the simplifications made by De Witt in the description of the Friedmann universe, leads, as it was said above, to the discrete spectrum of a black hole, $E_n = E_0\sqrt{2n+1}$, n=0,1. From this formula it would follow that in the final state (n=0) a black hole has Planck's mass and must evidently, be cold since its energy must clearly be determined only by zero oscillations of its matter. In other words, there appear some grounds to assume stability of the final state of a black hole. On the other hand, stability of elementary black holes is in contradiction with known Hawking's theorems [38]. According to these theorems, black holes which appear in the matter of a closed vessel have to vanish in some time. The necessity of such a state of a vessel's matter is required by the fundamental thermodynamics law.

Actually, joining N newly appeared black holes leads to radiation of matter and to disappearance of N - 1 black holes. But the state in which only one black hole does not disappear must necesserily remain. Although it is not clear whether such a quantitative violation of the law of thermodynamics [39] is in principle observable. On the other hand, an intense formation of stable elementary black holes (maximons) in an early universe contradicts the astrophysical data in the nucleosynthesis epoch [40]. In other words, there should be an intense process of disappearance of elementary black holes. The discussed situation becomes essentially different for a special class of extreme black holes when their mass is completely determined by their electric charge or angular momentum.

As is known due to Hawking, a black hole radiates as a black body heated to the termperature $T = \chi\hbar/2\pi k$, where k is the Boltsmann constant, $\chi = 4\pi(r_+ - M)/A$ is the surface gravity of the black hole, M is the black hole mass, $r_+ = M + (M^2 - Q^2)^{1/2}$ and $A = 4\pi(2M^2 - Q^2) + 2(M^4 - M^2Q^2)^{1/2}$, where Q is the electric charge of a non-rotating black hole, $c = \kappa = 1$.

If a black hole radiates energy so that $M \to Q$, $r_+ \to M$ then $\chi \to 0$ and $T \to 0$. In the limiting case ($T = 0$) the radiation of a black hole, as radiation of a black body, stops. In this case the remaining mass of the black hole is entirely the energy-mass of its electrostatic field. The same situation arises for an electrically neutral, but a rotating black hole if its mass is entirely due to its rotational energy. Thus, we deal with a class of absolutely cold black holes which cannot, of course, lose their energy for the Hawking radiation. Moreover, for a charge $Q \ll 137e$ such extreme black holes cannot lose their energy for emission of charged particles. It should be emphasized that existence of such holes in a closed vessel containing electrically neutral matter does not contradict the Hawking theorems. It is possible that such extreme black holes (maximons) are long-lived. Cosmology needs some matter hidden from observation, e.g. a neutrino with the corresponding nonzero rest mass [41] and other particles, in particular, the class of black holes ("maximons") [37].

Field singularities. Although we have discussed a description of universes, in particular, those filled with a scalar field ϕ, the field equation obtained from the corresponding Lagrangian is of interest regardless a description of the universe. The field equation contains the scale factor ought not be present in the equation. But as soon as we deal with field source singularities, one should take into account the appearance of all sorts of particles in this strong field and the appearance of a corresponding curvature of (3G). The assumption about the existence of a limiting density must, in principle, provide the absence of field singularities.

On can expect, of course, that many, even almost all, divergencies will probably be mutually compensated in the future theory of matter, but may be one of them, namely, the one connected with gravity, will require a formalism of the same type as ours in order the remaining "general" divergence to be finite.

References

1. Misner,Ch., Thorne,K., Wheeler,J.A. (1973). Gravitation, Freeman & Co., S,F.

2. Tolman,R. (1969). Relativity, thermodynamics and cosmology, Clarendon Press, Oxford.

3. Hawking,S.W., Ellis,G. (1979). The large scale structure of space-time. Cambridge Univ. Press.

4. Gliner,E.B. (1965). ZhETF 49, 543 (Russian).

5. Gliner,E.B. Vacuum-like state of the medium and Friedmann type cosmology. (1970). Doklady Akad. Nauk, 102, 771 (Russian).

6. Rozhansky,I.D. (1983). Anaxagoras, Mysl, p. 29 (Russian).

7. Starobinsky,A.A. (1980). Phys. Lett. B91, 99

. Starobinsky,A.A. (1979). Pis'ma v ZhETF (ZhETF Lett.) 30, 719 (Russian).

8. Mostepanenko,V.M. (1980). Nuclear Physics 31,1690.

9. Mamaev,S.G., Mostepanenko,V.M. (1980). ZhETF 78, 20 (Russian).

10. Kirzhnitz,D.A. (1972). Pis'ma v ZhETF (ZhETF Lett.). 15, 529 (Russian).

11. Guth,A. (1981). Phys. Rev. D23, 347.

12. Linde,A.D. (1984). A swelling Universe, UFN (Russian Physical Surveys) 144, 177 (Russian).

13. Linde,A.D. (1984). Reports on Progress in Physics, 47 8, 925.

14. Linde,A.D. (1982). Phys. Lett., ser. B, 108, 389.

15. Albrecht,A., Steinhardt,P.J. (1982). Phys. Rev. Lett. 48, 1220.

16. Linde,A.D. (1983). Pis'ma v ZhTEF (ZhETF Lett.), 38, 149 (Russian).

17. Linde,A.D. (1983). Phys. Lett. Ser. B, 129, 177.

18. Bondi,H., Gold,T. (1948). Mon. Not. Roy. Astr. Soc. 108, 252.

19. Pirani,F.A.E. (1955). On the energy-momentum tensor and the creatinn of scalar field. Proc. Roy. Soc. A228, 455.

20. Hoyle,F., Narlikar,J. (1964). A new theory of gravitation. Proc. Roy. Soc. , London, A282, 191.

21. Hayakawa,S. (1959). Progr. Theor. Phys. 21,324.

22. Markov,M.A. (1982). Pis'ma v ZhETF (ZhETF Lett.),36

214 (Russian).

22'. Markov,M.A. (1964). Ann. of Phys. 155, 33.

23. Aman,E.G., Markov,M.A. (1983). preprint p-0290 Inst. for Nucl. Research, Moscow.

24. Markov,M.A., Mukhanov,V.F, (1985). Il Nuovo Cim. 8613, 97.

25. Tolmal,R. (1969). Relativity, thermodynamics and Cosmology. Oxford and the Clarendon Press.

26. Klein,O.(1961).In: W.Heisenberg und die Physik unzer Zeit. Beiträge von Beek,G. et al. Hrsg. von F.Bopp. Braunschweig, Vieweg.

27. Zel'dovich,Ja.B. (1977). UFN (Russian Physical Surveys) 123, 487 (Russian).

28. Markov,M.A. (1973). Report at the Warsaw Gravitation Conference. In: DeWitt ed. On Gravitational Radiations and Gravitational Colapse, Morette (1974).

29. Fock,V.A. (1955). The theory of space, time and gravity, Gosizdat, Moscow (Russian).

30. Zel'dovich,Ja.B., Novikov,I.D. (1975). Structure and Evolution of the Universe, Nauka, Moscow (Russian).

31. Mal'tsev,V.K., Markov,M.A. (1980). Quantum Miniobjects in General Relativity, preprint p-0160. Inst. for Nucl. Res., Moscow.

32. Ginzburg,V.L., Kirzhnitz,D.A., Lubushin,A.A. (1976). ZhETF 60, 452 (Russian).

33. Hartle,J.B., Hawking,S.W. (1983). Phys. Rev. O28, 2960.

34. Wheeler,J.A. (1963). In: Relativity, Groups and Topology, Les Houches, p. 462.

35. Witt De,B. (1967. Phys. Rev. 160, 1140.

36. Mukhanov,V.F., Markov,M.A. (1983). Phys. Lett. 13

37. Markov,M.A. (1965). Progr. Theor. Phys. Suppl. Commemoration Issue for the 30th Anniversary of the meson theory by Dr. Yukawa. Extra number, 85-95.

38. Hawking,S.W. (1976). Phys. Rev. D14, 2460.

39. Markov,M.A. In: Quantum Gravity, ed. Markov,M., West,P. (1984). Plenum Press, N.Y.

40. Khlopov,M.Yu., Polnarev,A.G. (1981). Phys. Lett. B106, 307.

41. Markov,M.A. (1964). Phys. Lett., 10, 122.

HARMONIC SUPERSPACE AND EXTENDED SUPERSYMMETRIES

A.Galperin*, E.Ivanov, V.Ogievetsky, E.Sokatchev
Joint Institute for Nuclear Research, Dubna
*Institute of Nuclear Physics, Tashkent

1. At the first seminar on Quantum Gravity in 1978 a manifestly invariant formulation of the simplest (N=1) supergravity had been reported. Its main novelty was complexification of superspace. Manifestly invariant unconstrained off-shell descriptions become especially important for extended supersymmetries . The wonderful ultraviolet finiteness properties become transparent in these formulations and their proofs are getting both simple and compact [1,2]. Nevertheless an appreciable progress along this line was achieved only in 1984 despite rather deep an experience stored now in extended supersymmetry. We have succeeded in finding manifestly invariant formulations of N=2 matter and of N=2 supergravity and those of N=2 and 3 supersymmetric gauge theories in terms of off-shell unconstrained superfields. In doing so, a harmonization of superspace proved to be needed (instead of the complexification in the N=1 case). We introduce a harmonic superspace containing additional even coordinates. Its even part is

$$M^4 \times \frac{SU(2)}{U(1)} \qquad \text{(6 dimensions) for N=2} \tag{1}$$

$$M^4 \times \frac{SU(3)}{U(1)\times U(1)} \qquad \text{(10 dimensions for N=3)} \tag{2}$$

where M^4 is the Minkowsky space-time. On-shell this even part can be effectively reduced to M^4 (eliminating auxiliary fields with the help of the equations of motion).

A complete exposition of these topics may be found in our works [3,4] and forthcoming papers on harmonic Feynman rules. Here we shall confine ourselves to some comments and explanations.

2. Basic idea. The Minkowsky space-time is the quotient space of the Poincaré group over the Lorentz group

$$M^4 = \frac{\text{Poincaré}}{\text{Lorentz}}$$

having four standard coordinates, x_μ.

As a superspace of extended supersymmetry one considers usually a quotient space

$$\frac{\text{Super-Poincaré}}{\text{Lorentz}}$$

having both even, and odd (θ) coordinates

x_μ, θ_{α_i}, $\bar{\theta}^i_{\dot\alpha}$ ($i=1, 2, \ldots N$).

Superfields in such a superspace are known to contain a lot of superfluous field components. Such superfields have to be constrained just to kill these components. However, we wish to avoid constraints!

The way out. One has to include into the play the automorphism group SU(N) of N-extended supersymmetry. Instead of considering

$$\frac{\text{Super-Poincaré}}{\text{Lorentz}} = \frac{\text{Super-Poincaré} \ltimes SU(N)}{\text{Lorentz} \times SU(N)}$$

one has to replace SU(N) in the denominator by its certain subgroup H. This procedure leads to a harmonic superspace

$$\frac{\text{Super-Poincaré} \ltimes SU(N)}{\text{Lorentz} \times H}, \quad H \subset SU(N) \tag{3}$$

It has as its even part just (1) and (2) in particular cases N=2, H=U(1) and N=3, H=U(1) × U(1). The next step is to single out some analytic subspace of harmonic superspace. The usual constraints become now the analyticity conditions which are identically fulfilled if the corresponding superfield lives in the analytic subspace.

3. An example: N=2. An internal harmonic space is the sphere $S^2 \sim \frac{SU(2)}{U(1)}$. We take basic spherical harmonics $U_i^{\pm}$ as its coordinates. They are spinors with respect to SU(2) (index i) and have charges +1 or -1 with respect to U(1). The U(1) group is realized locally in S^2. Besides, on the sphere we demand

$$U^{+i}U_i^- = 1 \tag{4}$$

and as a result, the number of independent coordinates is just two. Now we pass from odd isospinor variables $\Theta_{\alpha i}$, $\bar{\Theta}$ to some new ones having definite U(1) charge

$$\Theta_\alpha^{\pm} = U^{\pm i}\Theta_{\alpha i}, \quad \bar{\Theta}_{\dot\alpha}^{\pm} = U^{\pm i}\bar{\Theta}_{\dot\alpha i} \tag{5}$$

In our extended superspace

$$(x_\mu,\ \Theta_\alpha^+,\ \bar{\Theta}_{\dot\alpha}^+,\ \Theta_\alpha^-,\ \bar{\Theta}_{\dot\alpha}^-,\ U) \tag{6}$$

we single out an analytic subspace

$$\{x_\mu^A = x_\mu - 2i(\Theta^+\delta_\mu\bar{\Theta}^- + \Theta^-\delta_\mu\bar{\Theta}^+),\ \Theta_\alpha^+\ \bar{\Theta}_{\dot\alpha}^+, U\}\ \{\zeta^A, U\} \tag{7}$$

It is remarkable that in this subspace the complete N=2 supersymmetry group acts (despite the absence of coordinates Θ_α^-, $\bar{\Theta}_{\dot\alpha}^-$). We shall refer to superfields defined on this subspace as to analytic superfields. They obey the Grassmann analyticity conditions [5] (the Causchy-Riemann conditions) having in the analytic basis (6) the simple form

$$D_\alpha^+ q = \frac{\partial}{\partial\Theta^{-\alpha}} q = 0, \quad \bar{D}_{\dot\alpha}^+ q = \frac{\partial}{\partial\bar{\Theta}^{-\dot\alpha}} q = 0 \tag{8}$$

Analytic superfields play the crucial role in our approach. They contain infinite sets of ordinary fields in the 4-dimensional space as the coefficients in decomposition in powers of $\Theta^+, \bar{\Theta}^+$ and symmetrized products of harmonic variables (higher harmonics).

For example, the matter Fayet-Sohnius hypermultiplet [6] is represented by analytic superfield q^+. The analyticity conditions

$$D_\alpha^+ q^+ = \bar{D}_{\dot\alpha}^+ q^+ = 0 \tag{9}$$

after substitutions

$$D^+_\alpha = U^+_i D^i_\alpha \text{ (a)}, \quad \bar{D}^+_{\dot\alpha} = U^+_i \bar{D}^i_{\dot\alpha} \text{ (b)}, \quad q^+ = U^+_i q^i \text{ (c)} \tag{10}$$

lead to the known constraints for the hypermultiplet (brackets mean symmetrization)

$$D_\alpha^{(i}q^{j)} = \bar{D}_{\dot\alpha}^{(i}q^{j)} = 0 \tag{11}$$

If superfield q^i is defined in the usual superspace, then constraints (11) contain the equations of motion and they cannot be obtained from the action principle. At the same time in our case they become simply the analyticity conditions and are satisfied identically if superfield q^+ lives in the analytic subspace (7). Now as the equations of motion (in the free case) we have a condition of independence of superfield q^i of harmonic variables. All the dependence on them in the free case must be reduced to the factor U^+_i in Eq.(10c):

$$D^{++}q^+ = 0 \tag{12}$$

where D^{++} is the derivative with respect to the harmonic U^- consistent with the equation (4),

$$D^{++} = U^{+i}\frac{\partial}{\partial U^{-i}} - 2i\theta^+\sigma^\mu\bar\theta^+\frac{\partial}{\partial x^\mu} \tag{13}$$

(in application to an analytic superfield).

It is worthwhile to stress that just Eq.(12) is now the equation of motion. For this equation it is rather simple to write the action principle and even to include interaction (this appeared inconceivable when working with superfields q^i or with the hypermultiplet field components which are scalars $\varphi_i(x)$ andtwo isosinglets fermions $\psi_\alpha(x)$, $\varkappa^\alpha(x)$). After introducing the integration measure

$$d\zeta_A^{(-4)} = d^4x d^2\theta^+ d^2\bar\theta^+ \tag{14}$$

we have

$$I = \int d\zeta_A^{(-4)} dU \left[\overset{*}{q}{}^+ D^{++}q^+ + \lambda(q^+)^2(\overset{*}{q}{}^+)^2\right] \tag{15}$$

Choice of self-interaction is rather restricted due to the U(1) - charge of the integration measure (14): instead of (15) one could choose $\lambda q^+(\bar{q}^+)^3$ + H.C., or

$\lambda(q^+)^4$ + H.C. and this is the end of the list. Note that if one is not afraid of breaking the SU(2) invariance, (and supersymmetry) then the possibilities become much richer. In the free case ($\lambda=0$) variation of the action (15) gives just the equation (12) while at $\lambda\neq 0$ one has

$$D^{++}q^+ = 2\lambda \overset{*}{q}{}^+ (q^+)^2 \qquad (16)$$

As was said before, the analytic superfield q^+ contains an infinite set of auxiliary fields which could be in principle eliminated with the help of the equations of motion (16). One may expect to obtain in such a way a nonlinear hyper-Kahler σ-model like those discussed e.g. by Witten and Bagger [7].

4. In the extended N=2 Yang-Mills theory a gauge superfield is again an analytic harmonic superfield but now it has U(1)-charge 2. It enters into the harmonic derivative (13) as a connection

$$\mathcal{D}^{++}=D^{++} + igV^{++}(\zeta_A, U) \qquad (17)$$

Correspondingly, it has a transformation law

$$V^{++\prime}=\frac{1}{ig}e^{i\lambda(\zeta^A,U)}(D^{++}+igV^{++}(\zeta_A,U)e^{-i\lambda(\zeta^A,U)} \qquad (18)$$

This law is of the same form as the one in the standard Yang-Mills theory. The only differences are that the localization is now performed in the analytic harmonic superspace and the parameter λ is an analytic superfunction not just a function and the space derivative turns into the harmonic derivative D^{++}. One can show further that special gauge exists (an analogue of the Wess-Zumino gauge) where V^{++} contains only the physical fields. In this gauge V^{++} has the following form:

$$\begin{aligned} V^{++}(x^A,\theta^+,\bar\theta^+,U) &= \theta^+\theta^+\left[M(x^A)+iN(x_A)\right] + \\ &+\bar\theta^+\bar\theta^+\left[\bar M(x_A)-iN(x_A)\right] + \theta^+\theta^+\bar\theta^+_{\dot\alpha}\psi^{\dot\alpha i}(x_A)U^-_i + \\ &+\bar\theta^+\bar\theta^+\theta^{+\alpha}\psi^i_\alpha(x_A)U^-_i + \theta^+\theta^+\bar\theta^+\bar\theta^+D^{(ij)}(x_A)U^-_iU^-_j \end{aligned} \qquad (19)$$

Fields entering into (19) are just the physical fields of

the known N=2 gauge supermultiplet. In this case the number of auxiliary fields is finite ($D^{(ij)}(x)$). However, the number of gauge degrees of freedom is already infinite.

5. We shall not give here the form of action and the equations of motion, these may be found in [3]. However, we shall write down propagators because formulas for them in [3] contain mistakes:

$$<q^{+}(1)\overset{*}{q}{}^{+}(2)> = \frac{1}{\Box}(D_1^{+})^4 (D_2^{+})^4 (D_1^{--})^2 \delta^{12}(z_1-z_2)\Delta_{2,1}^{(-2)}(u_1,u_2) \quad (20)$$

$$<V^{++}(1)V^{++}(2)> = \frac{1}{\Box}(D_1^{+})^4 \delta^{12}(z_1-z_2)\Delta_{0,2}^{(o)}(u_1,u_2) \quad (21)$$

(in the Feynman gauge), where

$$\Delta_{r,s}^{(-2t)}(u,v) = \sum_{n=o}^{\infty} (-)^{h+t+r} \frac{(2n+r+s+1)!}{(n+s)!(n+t+r)!} \times$$
$$\times\, u^{+}_{(i1} \dots u^{+}_{in+r} u^{-}_{in+r+1} \dots u^{-}_{i2n+r+s)} v^{+i1} \dots v^{+in+s-t} v^{-in+s-t+1} \dots v^{-i2n+z+s} \quad (22)$$

We know how also propagators in an external field needed for the background formalism, where counting of divergences becomes especially simple.

6. Without giving a full account of Feynman rules because of the lack of place we confine ourselves to demonstration of the absence of divergences in the multiloop Feynman graphs of the N=2 extended Yang-Mills theory in the framework of background formalism. In this formalism for multiloop diagrams one can guarantee (just as in the standard proofs of the nonrenormalization theorems [2,8]) that amplitudes contain spinor connections $A_\alpha(x,\Theta)$ (dimension $m^{\frac{1}{2}}$) instead of prepotentials V^{++} (dimension m^o)*. This fact simplifies the proof. Let us count powers. An amplitude for a process with n external lines is written down in the general case as

$$A_n = \int (d^4p)^{n-1} d^8\Theta A_{\alpha 1}(P,\Theta) \dots A_{\alpha n}(P,\Theta)\,\Gamma^{\alpha_1 \alpha_2 \dots \alpha_n}_{(P_1,P_2 \dots P_n,\Theta)} \quad (23)$$

\+ It can be shown [8] that in the improved background framework actually only vector connections (dimension m^{-1}) enter amplitudes.

where we indicate under each factor its dimension in mass units. Dimension of $A_\alpha(p,0)$ is $-\frac{7}{2}$ because $A_\alpha(p,0) = \int d^4x A_\alpha(x,\Theta) e^{ipx}$. We see that
-4 $\frac{1}{2}$

$$\Gamma^\alpha(p_1 \ldots p_n, 0) \sim m^{-\frac{n}{2}}$$

and, consequently, the diagram is convergent. It can be proved that it is true for each diagram and that subdiagrams with quantum external lines are inessential for analysis. Note that for one-loop contributions the situation is more delicate. One-loop amplitudes may contain the harmonic connections themselves. Moreover, logarithmic divergences are present in the pure gauge theory without matter. They disappear, however, if one includes couplings with matter hypermultiplets assigned to a given number of dedinite representations of the internal symmetry group (e.g. if there is one hypermultiplet in the adjoint representation and this is a remnant of the N=4 Yang-Mills theory.) Besides, one has to take appropriately into account the Faddeev-Popov and Kallosh-Nielsen ghosts. Without going into details we wish to note only that the latter are represented by analytic superfields with the U(1) - charge 4.

7. The extended supersymmetric N=3 gauge theory turns out to be especially interesting [4]. (As is known, this theory on-shell coincides with the N=4 theory. We expect, however, that off-shell there will appear differences in the composition and structure of auxiliary fields).

As was mentioned before, the even part of the corresponding harmonic superspace

$$M^4 \times \frac{SU(3)}{U(1) \times U(1)} \tag{24}$$

has 10 dimensions. Besides four coordinates x_μ the harmonic coordinates

$$U_i^{(1,1)}, \; U_i^{(-1,1)}, \; U_i^{(0,-2)} \quad \text{and} \quad \text{H.C.} \tag{25}$$

are introduced. They satisfy some unitarity and unimodularity conditions which are omitted here. The upper indices in (25) correspond to the charges with respect to $U(1)xU(1)$ while index i is the quark index of $SU(3)$. The analytic subspace contains two variables of the θ - type and two of the $\bar{\theta}$ - type

$$\{\zeta_A, U\} = \{ x_\mu^A, \; \theta_\alpha^{(1,-1)}, \; \theta_\alpha^{(0,-2)}, \; \bar{\theta}_{\dot{\alpha}}^{(1,1)}, \; \bar{\theta}_{\dot{\alpha}}^{(-1,1)}, U\} \tag{26}$$

The Yang-Mills potentials enter into harmonic derivatives as the analytic superfield connections

$$\begin{aligned} \mathcal{D}^{(1,3)} &= D^{(1,3)} + iV^{(1,3)}(\zeta_A, U) \\ \mathcal{D}^{(-1,3)} &= D^{(-1,3)} + iV^{(-1,3)}(\zeta_A, U) \\ \mathcal{D}^{(2,0)} &= D^{(2,0)} + iV^{(2,0)}(\zeta_A, U) \quad . \end{aligned} \tag{27}$$

The basic object in the standard approach is spinor connections. They were restricted by the constraints that led to the equations of motion [10-11].

A representation of these spinor connections in terms of the harmonic ones (27) identically resolves these constraints in our approach. These constraints admit an interpretation in the Ward-Penrose spirit [12] used in the supersymmetry context by Witten [11], and especially, by Rosly [13].

For harmonic potentials (27) constraints also appear and they also are equivalent to the equations of motion. However, in contradistinction to the constraints for spinor potentials, the equations for harmonic potentials are derivable from the action principle. Paradoxically, this action principle has very simple structure: the action is bi- and trilinear in superfields.

$$I = d\zeta_A^{(-2,-6)} dU \, \mathrm{Tr}\{ V^{(2,0)} (D^{(1,3)} V^{(-1,3)} -$$
$$- D^{(-1,3)} V^{(1,3)}) - V^{(-1,3)} (D^{(1,3)} V^{(2,0)} - D^{(2,0)} V^{(1,3)})$$
$$- V^{(1,3)} (D^{(-1,3)} V^{(2,0)} - D^{(2,0)} V^{(-1,3)}) -$$
$$- (V^{(1,3)})^2 + 2iV^{(1,3)} [V^{(-1,3)}, V^{(2,0)}]\} \qquad (28)$$

It is of the Chern-Simons type because the Lagrange density acquires harmonic divergence while the action is invariant and the equations of motion are completely covariant and have tensor nature.

The proof of finiteness of N=3 theory is easier than in the N=2 case. Even without passing to the background formalism the naive power counting from dimensional considerations gives

$$[V(\zeta_A, U)] = m^0 , \quad [V(p,\Theta)] = m^{-4}$$

$$\underset{0}{A_n} = \int \underset{4(n-1)}{(d^4p)^{h-1}} \underset{6}{d^{12}\Theta} \underset{-4n}{V(p_1,\Theta)\ldots V(p_n,\Theta)} \underset{-2}{\Gamma(p_1\ldots p_n,\Theta)} \qquad (29)$$

This leads us to the conclusion that

$$[\Gamma(p_1\ldots p_n\theta)] = m^{-2} .$$

Consequently, any diagram converges.

8. We finish this fluent review of theories related with the harmonic superspace by making a short mention of the N=2 supergravity theory in harmonic superspace. It is based on the general coordinate transformation group in analytic subspace. Besides, some auxiliary variable x_A^5 is added [3]:

$$\delta\zeta_A^M = \lambda^M(\zeta_A, U) , \quad \delta x_A^5 = \lambda^5(\zeta_A, U)$$
$$\delta\theta^{\mu-} = \lambda^{\mu-} \, (Z_A, U) \qquad (30)$$
$$\delta\theta^{\mu} = \bar{\lambda}^{\dot{\mu}-} \, (Z_A, U)$$

When covariantizing the harmonic derivative D^{++} there appear vielbeins V^{++M} and V^{++5}

$$\mathcal{D}^{++} = U^{+i} \frac{\partial}{\partial U^{-i}} + V^{++M}(\zeta, U) \frac{\partial}{\partial \zeta_A{}^M} + V^{++5}(\zeta, U) \frac{\partial}{\partial x_A^5} \tag{31}$$

These vielbeins serve as prepotentials of the N=2 supergravity. This theory corresponds exactly to the known component formulation [14] with (20+20) field components.

9. In the theory under consideration one can violate both supersymmetry and internal symmetry. Remarkably, these violations do not spoil wonderful finiteness properties of the theory provided these violations maintain N=2 (or N=3) gauge invariance *). This invariance will be maintained if the violated terms are constructed from the tensors of gauge groups multiplied by an appropriate number of Grassmann (Θ) and harmonic (U) coordinates. So, the future realistic theory will retain some relicts of exact extended gauge symmetries.

10. Therefore, if our world possesses an exact extended gauge symmetry, then it is four-dimensional only on-shell being multidimensional off-shell. (Stress that rigid supersymmetry may and will be broken). Correspondingly, the numbers of auxiliary and gauge degrees of freedom may be infinite. At the same time the number of physical degrees of freedom remains finite. It is the essence of difference between our theory and the Kaluza-Klein theory. In the latter a number of physical excitations is infinite. This is explained in the following way. Both theories contain additional coordinates. At the same time the Kaluza-Klein theory involves also ad-

---

*) Preservation of finiteness with the special choice of soft breaking terms has been discovered in 15 within the component approach.

ditional, "superfluous" symmetries while ours does not imply any other symmetries besides those which are really needed.

11. We only begin to develop the harmonic superspace approach. Besides the general matter outlined in this talk we have presented at the Seminar several reports on more special topics.

Here we give only their abstracts because the corresponding preprints are already out of print [16,17,18].

A. "Harmonic supergraphs" [16]

The quantization procedure in the harmonic superspace approach is worked out. Harmonic distributions are introduces and are used to construct the analytic superspace δ - functions and the Green functions for the hypermultiplet and the N=2 Yang-Mills superfields. The gauge fixing is described and the relevant Faddeev-Popov ghosts are defined. The corresponding BRST transformations are found. The harmonic superspace quantization of the N=2 gauge theory turns out to be rather simple and has many parallels with that of the standard (N=0) Yang-Mills theory. In particular, no ghosts for ghosts are needed. Based on these techniques, the Feynman rules for N=2 matter and Yang-Mills theories are derived and the various examples of harmonic supergraph calculations are given. Calculations do not appear to be more difficult than those in the N=1 case. The integration over harmonic variables does not lead to any trouble, non-locality disappears on-shell. An important property is that the quantum corrections are always written as integrals over the whole harmonic superspace. As a by-product our results imply a very simple proof of finiteness of a wide class of the N=4, D=2 nonlinear σ - models.

B. "Conformal invariance in harmonic superspace"

N=2 conformal supersymmetry is realized in harmonic superspace, its pecularities are analyzed. The coordinate group and the analytic prepotentials for N=2 conformal supergravity are found. Using the unconstrained hypermultiplet without central charges as a compensator we obtain a new version of N=2 Einstein supergravity with infinitely many auxiliary fields.

C. "Hyper-Kähler metrics and harmonic superspace" [18]

The most general unconstrained superfield action for self-interacting N=2 matter hypermultiplets in analytic N=2 superspace is argued to produce a most general N=2 hyper-Kähler σ - model after eliminating an infinite number of auxiliary fields. This suggests a new possibility of classifying hyper-Kähler metrics according to the N=2 analytic superfield self-interactions and provides an effective tool to compute these metrics explicitly. As a simple example the U(2)-invariant quartic self-coupling of a single q-hypermultiplet is analysed and is shown to yield the familiar Taub-NUT metric. To see the geometric pattern directly in terms of N=2 superfields we introduce a new on-shell representation of q-hypermultiplets in N=2 harmonic superspace similar to the τ - description of N=2 gauge theories. For U(2)-example this formulation is checked to coincide with that by Sierra and Townsend.

It is worthwhile to note that there have recently appeared several investigations of other authors concerning the harmonic superspace. R.Kallosh [19] have extended our work on N=3 Yang-Mills [4] by constructing a complete off-shell differential geometry of this theory in N=3 harmonic superspace. To know all irreducible tensors is important in view of the problem of soft breaking in N=3 gauge theory [15]. She also has applied the harmonic superspace approach to superstring theories [20]. In preprint [21] the theory of self-interacting hypermultiplets with central charges in harmonic superspace and its quantum content were analyzed. B.Zupnik (these Procedings) have applied our approach to D = 6 SUSY and suggested an elegant closed representation for $N = 2$ Yang-Mills action directly in analytic superspace (it is again of the Chern-Simons type, as in the $N = 3$ case). Some mathematical aspects of harmonic superspace are discussed by A.Rosly and A.Schwarz in 22 .

12. Finally, we would like to emphasize that a lot of things is yet to be understood and made within the present approach. Let us list some problems.

1) Background formalism for the $N = 2$ Yang-Mills theory in the harmonic superspace.

2) Quantization of the $N = 3$ Yang-Mills theory.

3) A complete description of soft violation in tthe harmonic superspace.

4) Infrared behaviour.

5) Anomalies.

6) Instantons and monopoles in the harmonic approach.

7) The most intriguing problem: how to construct the $N = 4$ Yang-Mills theory. We expect some principal differences in $N = 3$ case.

8) $N = 3$ supergravity.

9-100) $N = 4 \ldots 8$ supergravity theories.

101-200) A harmonic superspace approach to the superstring theory. A harmonic superstring?

201-1000) Finite models and superunified theories. The embedding problem of QED, QCD, etc. into them.

References

1. Howe,P.S., Stelle,K.S., West,P.C. (1983). Phys. Lett. 124B, 55

2. Howe,P.S., Stelle,K.S., Townsend,P.C. (1984). Nucl. Phys, B236, 125.

3. Galperin,A., Ivanov,E., Kalitzin,S., Ogievetsky,V., Sokatchev,E. (1984). Classical and Quantum Gravity,1, 469.

4. Galperin,A., Ivanov,E., Kalitzin,S., Ogievetsky,V., Sokatchev,E. (1984). Prepr. JINR E2-84-441.Dubna; (1985). Classical and Quantum Gravity, 2, 155; (1985). Phys. Lett. 151B, 215.

5. Galperin,A.S., Ivanov,E.A., Ogievetsky,V.I.(1981). JETP, Pisma, 33, 176.

6, Fayet,P. (1976). Nucl. Phys., B113, 135; Sohnius,M.F. (1978). Nucl. Phys., B138, 109.

7. Bagger,J., Witten,E. (1983). Nucl. Phys., B222, 1.

8. Grisaru,M.J., Zanon,D. (1984). Phys. Lett., 142B, 359.

9. Gates,S.J., Grisaru,M.T., Rocek,M., Siegel,W. (1983). Superspace or One thousand and one lessons in supersymmetry. Benjamin Cummings Reading.

10. Sohnius,M. (1978). Nucl. Phys. B136, 461.

11. Witten,E. (1978). Phys. Lett. 77B, 394.

12. Ward,R. (1977). Phys. Lett., 61A, 81.

13. Rosly,A.A. (1982). In: Proceedings of the International Seminar on Group Theoretical Methods in Physics. Moscow, Nauka, 1, 263.

14. Fradkin,E.S., Vasikiev,M.A. (1979). Phys. Lett., 85B, 47.; De Wit B., Van Holten W., Van Proeyen A. (1980). Nucl. Phys. B167, 186.

15. Parkes,A., West,P. (1983). Nucl. Phys. B222, 269.; Sazdovich,B.T., Tarasov,O.V. JETP Pisma (1983), 37, 602.; Namazie M.A., Salam A., Strathdee J. Phys. Rev. (1983), D28, 1481.

16. Galperin,A., Ivanov,E., Ogievetsky,V., Sokatchev,E. (1985). Prepr. JINR E2-85-127. 128, Dubna.

17. Galperin,A., Ivanov,E., Ogievetsky,V., Sokatchev,E. (1985). Prepr. JINR E2-85-363. Dubna.

18. Galperin,A., Ivanov,E., Ogievetsky,V., Sokatchev,E. (1985). Prepr. JINR E2-85-753. Dubna.

19. Kallosh,R. JETP, Pisma (1985). 41, 172.

20. Kallosh,R. (1985). Prepr. CERN-TH 4188/85.

21. Ohta,N., Sugata,H. , Yamaguchi,H. (1985). Prepr. OS-GE 85-02. Osaka University.

22. Rosly,A., Schwarz,A. (1985). Prepr. ITEP-39. Moscow October 1984.

SOLUTION OF THE SUPERFIELD CONSTRAINTS FOR SIX-DIMENSIONAL SUPERFIELD THEORIES IN HARMONIC SUPERSPACE

B.Zupnik

Research Institute of Applied Physics, Tashkent State University, Tashkent

1. Introduction. The method of harmonic superspace [1,2] gives the most elegant and complete description of the extended N=2, D=4 supersymmetric theories. From the formal point of view the harmonic method provides an effective solution of superfield constraints used to exclude nonphysical superfield components in the full N=2, D=4 superspace. The harmonic quantization is a constructive superfield method for calculating supergraphs with gauge and matter superfields [2] .

We will show that the harmonic method leads to an adequate superfield description of the theories with N=1, D=6 supersymmetry [3,4] . Basic quantities and equations of the harmonic 6-dimensional theories can be determined by a straightforward generalization of the harmonic N=2, D=4 theory. We use conventions and notations of Refs. [1,2] . The covariant derivatives are denoted ∇ , the flat harmonic derivative are $D^{(+2)}$, $D^{(-2)}$, D^{o} .

Analizing constraints of the supergauge theory SYM^1_6 we solve by iteration the equation for a harmonic connection $A^{(-2)}$. Note, that all the coefficients of the iterative expansion are calculated in the explicit form. These coefficients determine also the expansion of the SYM^1_6-action in terms of prepotentials. The obtained simple formula for the action of supergauge theory in the full superspace is of the universal form, which is valid for the dimensionally reduced $D< 6$ SYM-theories.

Using the harmonic formalism we consider a nontrivial dimensional reduction of SYM_6^1 to the 4D-supergauge theories with mass operators which play the role of central charges.

Constraints of the extended N=2, D=6 supergauge theory SYM_6^2 are equivalent to the equations of motion [3]. We propose a new interpretation of these constraints with the help of harmonization of the Lorentz group acting in (6-D)-dimensional space. The SYM_6^2-theory can be also formulated in terms of the N=1 harmonic superfields.

Constraints of the 6-dimensional supergravity SG_6^1 can be solved in the harmonic formalism by the operator method analogous to that in SYM_6^1.

2. Six-dimensional gauge theories in the harmonic superspace. The 6-dimensional analytic SU(2)/ U(1) superspace is determined by analogy with the corresponding 4-dimensional superspace [1]. Analytic coordinates ζ can be expressed in terms of the real 6-D superspace coordinates z and SU(2)/U(1) harmonics $u_i^{\pm}$:

$$\zeta^{AB} = x^{AB} + 2i\Theta_k^{[B}\Theta_l^{A]} u^{k+} u^{l-} ,$$
$$\Theta^{A(+)} = O^{Ak} u_k^{+} \tag{2.1}$$

where A,B are spinor indeces of $\overset{*}{SU}(4) \cong SO(5,1)$.

The superfield constraints of SYM_6^1 are considered in [3,4]

$$\{\nabla_A^i , \nabla_B^k\} + \{ \nabla_A^k , \nabla_B^i\} = 0,$$
$$\nabla_B^k = D_B^k + iA_B^k . \tag{2.2}$$

A transition to the analytic basis (AB) can be performed with the help of the matrix exp(iv) (bridge). SYM_6^1-constraints have the following form in AB:

$$\{ \nabla^{(+)}_A , \nabla^{(+)}_B \} = 0 , \qquad [\nabla^{(+)}_A , \nabla^{(+2)}] = 0 \tag{2.3}$$

$$\text{a)} \quad [\nabla^{(-2)}, \nabla^{(+)}_A] = \nabla^{(-)}_A , \qquad \text{b)} \quad [\nabla^{(+2)}, \nabla^{(-2)}] = D^0 \tag{2.4}$$

Let us introduce 6D-analytic prepotential $V^{(+2)}(\zeta,u)$ analogy with Ref. [1] :

$$\delta V^{(+2)} = -D^{(+2)}\lambda + i [\lambda, V^{(+2)}] , \tag{2.5}$$

where $\lambda(\zeta ,u)$ is an analytic parameter of the gauge transformation.

Nonanalytic components of connection $A^{(-2)}, A^{(-)}_B, A_{BC}$ obey the analogous transformation rules. To solve the constraints it suffices to express the bridge in terms of prepotentials [2] . We consider an alternative method for solving SYM-constraints, which reduces this problem to solving Eq. (2.4b) or the equivalent equation for $A^{(-2)}$:

$$D^{(+2)}A^{(-2)} - D^{(-2)}V^{(+2)} + i[V^{(+2)}, A^{(-2)}] = 0 . \tag{2.6}$$

This equation has a simple iterative solution $A^{(-2)} = \sum_{n=1}^{\infty} A^{(-2)}_n$:

$$A^{(-2)}_1(z,u) = \int du_1 \frac{V^{(+2)}(z,u_1)}{(u^+u^+_1)^2} , \tag{2.7}$$

$$A^{(-2)}_n(z,u) = (-i)^{n+1} \int du_1 \ldots du_n \frac{V^{(+2)}(z,u_1)V^{(+2)}(z,u_2)\ldots V^{(+2)}(z,u_n)}{(u^+u^+_1)(u^+_1u^+_2) \ldots (u^+_nu^+)} \tag{2.8}$$

where $(u^+u_1^+)^{-2}$ and $(u^+u_1^+)^{-1}$ are the harmonic distributions [2].

Other components of the connection can be written in terms of $A^{(-2)}$ using the AB-constraints.

The iterative expansion of an invariant SYM_6^1 action can be also written in the explicit form:

$$S = \frac{1}{g_6^2} \sum_{n=2}^{\infty} \frac{(-i)^n}{n} \int d^{14}z du_1 \ldots$$

$$\ldots du_n \frac{Sp(V^{(+2)}(z,u_1)\ldots V^{(+2)}(z,u_n))}{(u_1^+u_2^+)(u_2^+u_3^+) \ldots (u_n^+u_1^+)} = \tag{2.9}$$

$$= -\frac{1}{g_6^2}\int d^{14}z du_1 du_2\, \delta^{(1,-1)}(u_1,u_2) \cdot$$

$$\cdot\, Sp\left[\ln \frac{1}{D^{(+2)}} \nabla^{(+2)}\right](u_1,u_2) \; ,$$

where g_6 is a coupling constant and the following pseudodifferential operator is used:

$$\left(\frac{1}{D^{(+2)}} \nabla^{(+2)}\right)_{ab}^{(-1,1)}(u_1,u_2) =$$

$$= \left(1 + \frac{i}{D^{(+2)}} V^{(+2)}\right)_{ab}^{(-1,1)}(u_1,u_2) = \tag{2.10}$$

$$= \delta_{ab}\, \delta^{(-1,1)}(u_1,u_2) + \frac{f_{abc}V_c^{(+2)}(z,u_2)}{(u_1^+u_2^+)} \, .$$

The harmonic quantization [2] can be generalized to the case of D=6. All the 4-dimensional formulas for the effective quantum action and harmonic superfield Green functions have the corresponding 6-dimensional analogues.

The ultraviolet divergences of the 6D-supersymmetric theories have been studied in [3] using ordinary superfields.

The massless theory SYM_4^2 can be obtained from the SYM_6^1 by a trivial dimensional reduction: $x^5, x^6 \to 0$. We will consider a nontrivial dimensional reduction of SYM_6^1 to the massive gauge theory with central charges SYM_4^2 in the harmonic approach:

$$V_6^{(+2)} = \exp(ix^5 Z_5 + ix^6 Z_6)\, V_4^{(+2)}, \qquad (2.11)$$

where Z_5 and Z_6 are the commuting generators of the gauge group. The action of $SYM_4^2(Z)$ can be obtained from (2,9) by the simple replacements $V_6^{(+2)} \longrightarrow V_4^{(+2)}$, $g_6^{-2} d^{14} Z \longrightarrow g_4^{-2} d^{12} Z$.

3. Harmonic interpretation of the extended six-dimensional supergauge theory. The standard superfield treatment of D=6, N=2 supergauge theory SYM_6^2 has been discussed in [3] . In the framework of the harmonic D=6, N=1 superspace the SYM_6^2-theory can be treated as an interaction of $V^{(+2)}$ with ω-hypermultiplet in the adjoint representation of the gauge group. The corresponding $V^{(+2)} - \omega$ action is invariant under the following transformation of the on-shell D=6, N=2 supersymmetry:

$$\delta V^{(+2)} = \epsilon_A^i\, u_i^+\, \Theta^{A(+)} \omega \;,$$
$$\delta\omega = \tfrac{i}{2}(D^{(+)})^4 \left[\epsilon_A^k\, u_k^-\, \Theta^{A(-)}\, A^{(-2)} \right] \;, \qquad (3.1)$$

which are analogous to the D=4, N=4 transformations [2].

The real coordinates in D=6, N=2 superspace are x^{AB}, Θ_k^A, $\tilde{\Theta}_A^{k'}$, where k and k' are indices of the SU(2)xSU'(2) automorphism group. The superfield constraints of SYM_6^2 have the following form [3] :

$$\{\nabla_A^k , \nabla_B^l \} = i\,\epsilon^{kl}\, \nabla_{AB} \;,\; \{\tilde{\nabla}_{k'}^A , \tilde{\nabla}_{l'}^B = -i\,\epsilon_{k'l'}\, \nabla^{AB}, \qquad (3.2)$$

$$\{ \nabla^k_A , \tilde{\nabla}^B_{l'} \} = \delta^B_A W^k_{l'} , \tag{3.3}$$

where $W^k_{l'}$ is the on-shell supergauge tensor. The reality condition on $W^k_{l'}$ is a consequence of (3.3).

In accordance with general principles of the formal harmonic calculus [5] we consider the $(S\overset{*}{U}(4)/SL(2,C) \times U(1)$-harmonics which are the components of $SU(4)$ matrices L and $\tilde{L}$ satisfying the relations:

$$\varepsilon^{ABCD} L^{\alpha(+)}_A L^{\beta(+)}_B L^{\dot{\gamma}(-)}_C L^{\dot{\delta}(-)}_D = \varepsilon^{\alpha\beta} \varepsilon^{\dot{\gamma}\dot{\delta}}, \tag{3.4}$$

$$L^{\alpha(+)}_A \tilde{L}^{A(-)}_\beta = \delta^\alpha_\beta , \quad L^{\alpha(+)}_A \tilde{L}^{A(+)}_{\dot{\beta}} = 0 , \tag{3.5}$$

$$L^{\dot{\alpha}(-)}_A \tilde{L}^{A(-)}_\beta = 0 , \quad L^{\dot{\alpha}(-)}_A \tilde{L}^{A(+)}_{\dot{\beta}} = \delta^{\dot{\alpha}}_{\dot{\beta}} \tag{3.6}$$

where $\alpha, \dot{\alpha} \ldots$ and $\pm$ are $SL(2,C) \times U(1)$ indices.

Important harmonic $SL(2,C) \times U(1)$-derivatives are:

$$D^{(+2)\alpha}_{\dot{\beta}} = L^{\alpha(+)}_A \partial / \partial L^{\dot{\beta}(-)}_A - \tilde{L}^{A(+)}_{\dot{\beta}} \partial / \partial \tilde{L}^{A(-)}_\alpha , \tag{3.7}$$

$$D^{(-2)\dot{\alpha}}_\beta = L^{\dot{\alpha}(-)}_A \partial / \partial L^{\beta(+)}_A - \tilde{L}^{A(-)}_\beta \partial / \partial \tilde{L}^{A(+)}_{\dot{\alpha}} . \tag{3.8}$$

At the first step of the SYM^2_6 harmonization we reduce the constraints (3.2,3) to relations between the covariant derivatives in (1,8)-dimensional hyperplane

$$\tilde{L}^{A(+)}_{\dot{\alpha}} \nabla^k_A , \quad L^{\alpha(+)}_A \tilde{\nabla}^A_{k'} , \tag{3.9}$$

$$\tfrac{1}{2} \tilde{L}^{A(+)}_{\dot{\alpha}} \tilde{L}^{B\dot{\alpha}(+)} \nabla_{AB} . \tag{3.10}$$

An analogous (1,8)-system was discussed in [6] for D=4, 10 supergauge theories. Note, that the choice of a pure gauge solution for (1,8) derivatives provides a particular class of SYM^2_6 solutions only. The dimension of the corresponding analytical superspace for this class of solutions is (5,8).

The second step of SYM_6^2 interpretation is the harmonization of $SU(2)xSU'(2)$ with the help of harmonics $u_k^{(\pm,0)}$ and $u_{k'}^{(0,\pm)}$. Now we can construct basic covariant derivatives in the hyperplane of dimension (0,4):

$$\nabla_{\dot\alpha}^{(+,+,0)} = \tilde{L}_{\dot\alpha}^{A(+)} u_k^{(+,0)} \nabla_A^k , \qquad \tilde{\nabla}^{(+,0,+)} = L^{\alpha}{}_A^{(+)} u^{k'(0,+)} \tilde{\nabla}_{k'}^A \tag{3.11}$$

The SYM_6^2-constraints (3.2,3) are equivalent to the following conditions of the integrability of this hyperplane:

$$\{\nabla_{\dot\alpha}^{(+,+,0)} , \nabla_{\dot\beta}^{(+,+,0)}\} = \{\tilde{\nabla}_{\alpha}^{(+,0,+)} \tilde{\nabla}_{\beta}^{(+,0,+)}\} = 0$$

$$\{ \nabla_{\dot\alpha}^{(+,+,0)} , \tilde{\nabla}_{\beta}^{(+,0,+)}\} = 0. \tag{3.12}$$

These equations have pure gauge solutions only. The dimension of the analytical superspace for this interpretation is (6,12).

The alternative harmonization of $SU^*(4)/L(1)xU(1)xU(1)$ can be also used for the interpretation of SYM_6^2-constraints.

4. Six-dimensional supergravity in the harmonic superspace. The four-dimensional N=2 supergravity SG_4^2 in the harmonic approach is discussed in [1] . A natural formulation of SG_4^2 can be obtained with the help of dimensional reduction from 5D-space. We consider the 6-dimensional generalization of supergravity SG_6^1, that corresponds to the interaction of SG_4^2 with the Maxwell superfield after the reduction $x^5,x^6 \longrightarrow 0$.

Let us consider analytical coordinates $\zeta^\Sigma = (\zeta^{AB} , \theta^{A(+)}), \theta^{A(-)}$. One can use the following gauge supergroup of coordinate transformations in SG_6^1 :

$$\delta\zeta^{\Sigma} = \lambda^{\Sigma}(\zeta, u) , \qquad \delta u_i^{\pm} = 0 , \tag{4.1}$$

$$\delta\theta^{A(-)} = \lambda^{A(-)}(\zeta, \theta^{(-)}, u) . \tag{4.2}$$

A covariant harmonic derivative is

$$\Delta^{(+2)} = D^{(+2)} + V^{(+2)\Sigma}(\zeta, u)\, \partial/\partial\zeta^{\Sigma} + \\ + V^{(+)A}(\zeta, \theta^{(-)}, u)\, \partial/\partial\theta^{A(-)} , \tag{4.3}$$

where $V^{(+2)AB}$, $V^{(+3)A}$ are analytical prepotentials, and $V^{(+)A}$ is a compensating superfield. This derivative conserves analyticity.

It is convenient to introduce real coordinates $Z_o^M = (x_o^{AB}, \theta_{oi}^A)$ related to the analytical coordinates ζ, $\theta^{(-)}$, u by Eq.(2.1). Covariant differential operators in the flat basis can be expanded in terms of derivatives $D^{(+2)}$, $D^{(-2)}$, $D_M = (\partial_{AB}^o , D_A^{oi})$, for instance

$$\Delta^{(+2)} = D^{(+2)} + V^{(+2)M} D_M = D^{(+2)} + V^{(+2)} , \tag{4.4}$$

$$\Delta^{(-2)} = D^{(-2)} + A^{(-2)M} D_M = D^{(-2)} + A^{(-2)} . \tag{4.5}$$

The supertetrad components $A^{(-2)M}$ should be written in terms of prepotentials $V^{(+2)M}$. For this we use the fundamental constraint

$$[D^{(+2)}, A^{(-2)}] - [D^{(-2)}, V^{(+2)}] + [V^{(+2)}, A^{(-2)}] = 0 \tag{4.6}$$

Taking into account the relation $\{ D_A^{ok} , D_B^{ol} \} = i\varepsilon^{ki}\, \partial_{AB}^o$ one can obtain the following component equation

$$D^{(+2)}A^{(-2)M} - D^{(-2)}V^{(+2)M} + V^{(+2)}A^{(-2)M} - A^{(-2)}V^{(+2)M} + \\ + C_{pN}^{M} A^{(-2)N} V^{(+2)p} = 0 \tag{4.7}$$

where C^{M}_{pN} are the flat torsion components.

An iterative solution $A^{(-2)}$ of the operator equation (4.6) can be written in a form analogous to the solution (2.8) in SYM^{1}_{6}. The component solution is more complicated, so we write a linearized solution only

$$A^{(-2)M}_{1}(z_o,u) = \int du_1 \frac{V^{(+2)M}(z_o,u_1)}{(u^{+}u^{+}_{1})^{2}} . \tag{4.8}$$

The transformation of the covariant spinor derivative $\Delta^{(+)}_{A} = u^{+}_{i}D^{oi}_{A}$ is of the following form:

$$\delta\, \Delta^{(+)}_{A} = -\, (\, \Delta^{(+)}_{A}\, \lambda^{B(-)})\, \Delta^{(+)}_{B} = \omega^{B}_{A}\, \Delta^{(+)}_{B} . \tag{4.9}$$

A spinor derivative of the compensator can be treated as a connection with the parameter ω^{B}_{A} and derivative $\Delta^{(+2)}$

$$\gamma^{(+2)B}_{A} = \Delta^{(+)}_{A}\, V^{(+)B} , \tag{4.10}$$

The other components of the spinor connection can be calculated by the iterative method using the equations

$$\Delta^{(+2)}\, \gamma^{(+)C}_{AB} - \Delta^{(+)}_{A}\, \gamma^{(+2)C}_{B} + [\gamma^{(+2)}, \gamma^{(+)}_{A}]^{C}_{B} = 0, \tag{4.11}$$

$$\Delta^{(-2)}\, \gamma^{(+2)B}_{A} - \Delta^{(+2)}\, \gamma^{(-2)}_{A} + [\, \gamma^{(-2)}, \gamma^{(+2)}]^{B}_{A} = 0, \tag{4.12}$$

$$\gamma^{(-)C}_{AB} = \Delta^{(-2)}\, \gamma^{(+)C}_{AB} - \Delta^{(+)}_{A}\, \gamma^{(-2)C}_{B} + [\, \gamma^{(-2)}, \gamma^{(+)}_{A}]^{C}_{B} . \tag{4.13}$$

Using these connections in the covariant derivatives one can construct the components of superfield torsion in terms of prepotentials.

Aknowledgements. The author would like to thank cordially A.S.Galperin, E.A.Ivanov, V.I.Ogievetsky, E.S.Sokatchev for stimulating discussions of the harmonic superspaces.

References

1. Galperin,A., Ivanov,E., Kalitzin,S., Ogievetsky,V., Sokatchev,E. (1984). Class. Quant. Grav. 1, 469.

2. Galperin,A., Ivanov,E., Ogievetsky,V., Sokatchev,E. (1985). JINR preprints, E2-85-127, 128, Dubna.

3. Howe,P.S., Stelle,K.S.(1984). Phys. Lett., 137B, 175.

4. Zupnik,B. (1984). Doklady Uzbek. Acad. Scien. 7,28 (Russian).

5. Ivanov,E.,Kalitzin,S., Nguen Ai Viet, Ogievetsky,V. JINR preprint, E2-84-720, Dubna, 1984.

6. Witten,E. (1978). Phys. Lett. 77B, Princeton Univ. Preprint.

NOETHER IDENTITIES IN SUPERGRAVITY MODEL CONSTRUCTION

V.I.Kuvshinov, F.I.Fedorov

Institute of Physics of the BSSR Academy of Sciences, Minsk

We consider Utiyama-Kibble's programme [1,2] of constructing gauge-invariant field interaction theory in N=1 supergravity on the base of universal matrix first-order equations (UME) with quadratic non-linearities [3-9]. Fixing universal form of Lagrangian and equations which differ for various gauge models only by form of constant matrices defining physical properties of the model and field component structure, we get a possibility to formulate in general form the conditions of action invariance under gauge and Lorentz transformations as equations for the matrix components. For a given gauge group a solution of these equations completely determines the gauge field sector of the model (e.g. gluodynamics for $SU_c(3)$), i.e. the tensor structure of field function under corresponding gauge and Lorentz group actions, the form of generators of representations, the form of the main matrices which in turn define Lagrangian and equations [7,8]. In this report such approach is developed for supergravity.

A feature of this approach relevant to SUSY is that by definition [3,4] unified field $\Psi(x) = (\Psi_A(x)$ satisfying UME contains on equal footing bose- and fermi-fields as multiplet components and realise Grassmann algebra elements. Since the order of the equation is 1, Ψ contains both potentials and strengths of the fields. Since we wish to describe fermion-gravity interaction, the tetrad approach is relevant in supergravity, and, since gauge field transformed by adding derivatives of local Lorentz parameters is the Ricci coefficient then so is Palatini rormalism. Thus the 1st order equations together with lowering non-linearity to the quadratic one allow introducing tetrads, Ricci coefficients and Riemann tensor into the Ψ-function. Variation of $\Psi(x)$ under local SUSY transformations is linear on $\Psi(x)$ [5,9]. UME features, in principle, give a possibility to develop a systematic general method for supergravity model constructing on the base of local SUSY invariance conditions. The use of the 1.5 order formalism [10] resolves difficulties connected with auxiliary fields transformations.

Action for UME Lagrangian

$$L = \frac{1}{2}\bar{\Psi}(\alpha^{\mu}\partial_{\mu} + \alpha^{0}) + \frac{g}{3!}\bar{\Psi}\Lambda\Psi\Psi \tag{1}$$

is invariant up to surface terms if corresponding invariance conditions are satisfied under local SUSY transformations with the following variation of the unified field and coordinates

$$\bar{\delta}\Psi(x) = (\xi^{\Omega}N_1^{\Omega} + \xi^{\Omega}N_2^{\mu,\Omega}\partial_{\mu} + \partial_{\mu}\xi^{\Omega}N_3^{\mu,\Omega})\Psi(x) + M(\xi),$$

$$\delta x^{\mu} = \xi^{\mu}, \tag{2}$$

$$N_1 = (O^{\mu} \oplus N_1^{[mn]} \oplus N_1^{\alpha}),\ N_2^{\mu} = (N_2^{\mu,\nu} \oplus O^{\mu,[mn]} \oplus O^{\mu,\alpha}),$$

$$N_3^{\mu} = (N_3^{\mu,\nu} \oplus O^{\mu,[mn]} \oplus O^{\mu,\alpha}) \tag{3}$$

Here $\alpha^{\mu} = (\alpha^{\mu}_{AB})$, $\alpha^{0} = (\alpha^{0}_{AB})$, $\Lambda = (\Lambda_{ABC})$ are constant matrices to be derived, $(\xi^{\Omega}) = (\xi^{\mu}, \xi^{[mn]}, \xi^{\alpha})$ parameters of transformations, $M(\xi)$ the function of non-homogeneous transformations determined from the invariance conditions. In the expressions (2,3) new parameters of local translations are introduced and 4-translation $N_2^{\mu,\nu}\partial_{\nu}$ and 4-rotation $N_1^{[mn]}$ generators commute and have different indices: $\mu, \nu; m, n; \alpha, \beta$. The quantities $\partial_{\mu}\xi^{\nu}N_3^{\mu,\nu}$ include variation of Ψ -components having global indices under local 4-translations. A feature of the local SUSY transformations in supergravity is the equation $N_2^{\mu,\alpha} = 0$ connected with dimension of the coupling included in N_1^{α}.

According to the second Noether theorem the action invariance up to the surface term is

$$\left(\frac{\partial L}{\partial \Psi} - \partial_\mu \frac{\partial L}{\partial(\partial_\mu \Psi)}\right)\bar{\delta}\Psi + \partial_\mu \theta_\mu = \partial_\mu \Phi_\mu, \tag{4}$$

$$\theta_\mu = \frac{\partial L}{\partial(\partial_\mu \Psi)}\bar{\delta}\Psi + L\,\delta x^\mu \tag{5}$$

From (4) using (1-3) and equating to 0 coefficients of ξ^Ω, $\partial_\mu \xi^\Omega$ we obtain the invariance conditions as equations for matrices α^μ, α^0, Λ. Total derivative $\partial_\mu \Phi_\mu$ in (4) is as known SUSY, characteristic of first-order equations and leads to dynamic variables $\theta'_\mu = \theta_\mu - \Phi_\mu$. Since adding total derivative to Lagrangian $L \to L' = L + \partial_\mu f_\mu$ does not change equations but changes the variation of action, we can make the action manifestly invariant making the surface term into zero with the help of $\partial_\mu f_\mu$: $\bar{\delta} L' + \partial_\mu (L' x^\mu) = 0$. Fixing Φ_μ in the form leading to the standard spin-vector current form we get equations for f_μ :

$$\begin{aligned}&\frac{\partial f_\mu}{\partial \Psi}(N_1\Psi + N_2^\nu \partial_\nu \Psi) + \frac{\partial f_\mu}{\partial(\partial_\nu \Psi)}(N_1 \partial_\nu \Psi + N_2^\sigma \partial_\nu \partial_\sigma \Psi) = \\ &= \frac{1}{2}\Psi\alpha^\nu N_2^\mu \partial_\nu \Psi + \frac{1}{2}\Psi\alpha^0 N_2^\mu \Psi + \frac{g}{3!}\Lambda N_2^\mu \Psi\Psi\Psi.\end{aligned} \tag{6}$$

Consider now how we can find matrices and component structure of Ψ from the invariance conditions. From

$$\alpha^\mu \partial_\mu \xi^\Omega N_1^\Omega + \Lambda M(\xi) = 0 \tag{7}$$

we get

$$M(\xi) = \partial_\nu \xi^{[mn]} e^{\nu[mn]} + \partial_\nu \xi^{\alpha} e^{\nu\alpha} \tag{8}$$

$$\Lambda = \sum_{\pi} e^{\mu\Omega} \alpha^{\mu} N_1^{\Omega} \tag{9}$$

Here $(e^{AB})_{CD} = \delta_{AC}\delta_{BD}$ are basic elements of general matrix algebra. Consider at first boson part of Ψ and corresponding matrices. By (9) gauge field components connected with local $\xi^{[mn]}$ are Ricci coefficients $\omega_\nu^{[mn]}$. Using invariance condition

$$\alpha^{\nu} \partial_\nu M(\xi) = 0 \tag{10}$$

and previously known symmetry properties of α^{μ} [4,5] we have

$$\alpha^{\nu}_{(\omega)} = \frac{1}{2}(e^{[mn][\mu\nu]\,,\,\mu[mn]} - e^{\mu[mn]\,,\,[mn][\mu\nu]}) \tag{11}$$

From the symmetry properties of α^{0}, Λ and the universal form (1) we can recover auxiliary fields $\Phi^{[\nu\mu]}_{[mn]}$ and $R_{[\nu\mu]}^{[mn]}$. Introducing also Ψ -components connected with local ξ^{μ} in the form $h_\mu{}^n$, which transform homogeneously according to (2),(9), we can fix the following structure of $\Phi^{[\nu\mu]}_{[mn]}$ and $\alpha^{\nu}_{(h)}$

$$\Phi^{[\nu\mu]}_{[mn]} = q\varepsilon_{mnkl}\varepsilon^{\nu\mu\varkappa\tau} h_{\varkappa}{}^{k} h_{\tau}{}^{l} \tag{12}$$

$$\alpha^{\nu}_{(h)} = \frac{1}{2}(e^{l[\tau\nu]\,,\,l\tau} - e^{l\tau\,,\,l[\tau\nu]}) \tag{13}$$

Just as above, we may introduce also auxiliary fields $\Phi_1^{[\nu\tau]}$ and $C_{[\nu\tau]}^{l}{}^{1}$ and consequently the bose-part of Ψ is determined in the form

$$\Psi_{BOSE}(x) = (h_\mu{}^n, \omega_\nu^{[mn]}, \Phi_1^{[\nu\tau]}, C_{[\nu\tau]}{}^{1}, \Phi_{[mn]}^{[\nu\mu]}, R_{[\nu\mu]}{}^{[mn]}) \tag{14}$$

The fermi-component of the gauge field non-homogeneous with respect to local SUSY transformations is by (8) a spin-vector and

$$\alpha^{\nu}_{(\Psi)} = \frac{1}{2}\left(e^{\alpha[\nu\mu],\mu\alpha} - e^{\mu\alpha,[\nu\mu]\alpha}\right) \tag{15}$$

Then the fermi-part of Ψ is

$$\Psi_{FERMI}(x) = (\Psi^{\alpha}_{\nu}, \Phi^{[\nu\mu]}_{\alpha}, \Phi^{\alpha}_{[\nu\mu]}) \tag{16}$$

where $\Phi^{[\sigma\rho]}_{\beta}$ has the following simplest form $\Phi^{[\sigma\rho]}_{\beta} = \varepsilon^{\nu\mu\sigma\rho}\Psi^{\alpha}_{\nu} h_{\mu}{}^{n}(c\gamma^{5}\gamma^{n}\eta_{n'n})_{\alpha\beta}$. Introducing 4-rotations generators for $\omega_{\mu}{}^{[mn]}$: $N_1{}^{[mn]} = S^{[mn][\mu\nu]}\alpha^{[\mu}_{(h,\omega)}\alpha^{\nu]}_{(h,\omega)}$ (where $S^{[mn][\mu\nu]}$ is a quasidiagonal matrix with constant elements) and SUSY generator for Ψ^{α}_{ν} : $N^{\alpha}_{1} = \frac{1}{2}(\gamma^{n}\gamma^{m} - \gamma^{m}\gamma^{n})C^{\alpha\beta}\eta_{km'}\eta_{mn'}e^{\nu\beta,\nu[mn]}$, we can derive a cubic matrix via (9). The matrix Λ is determined up to a matrix P such that $(e^{\mu[mn]} + e^{\mu\alpha})P = 0$. The matrix P is found from the invariance conditions connecting Λ and N_1, symmetry properties of Λ, α^{0}.

Taking into account symmetry properties and the structure of Ψ we can also obtain the form of α^{0} from the invariance conditions. Finally we have the following expressions for Ψ, α^{μ}, α^{0}, Λ :

$$\Psi(x) = (\Psi_{BOSE}(x), \Psi_{FERMI}(x)), \tag{17}$$

$$\alpha^{\nu} = \alpha^{\nu}_{(\omega)} + \alpha^{\nu}_{(h)} + \alpha^{\nu}_{(\Psi)}, \tag{18}$$

$$\alpha^{0} = -\frac{1}{2}\left(e^{1[\nu\tau],[\nu\tau]1} + e^{[\nu\tau]1,1[\nu\tau]} + e^{[\nu\mu][mk],[mk][\nu\mu]} + e^{[mk][\nu\mu],[\nu\mu][mk]} + e^{\alpha[\mu\nu],[\mu\nu]\alpha} - e^{[\mu\nu]\alpha,\alpha[\mu\nu]}\right. \tag{19}$$

$$\Lambda = \sum_{\pi} \frac{1}{2} \Big(\frac{1}{g} e^{\mu[mn]\,,\,[mk][\mu\nu]\qquad ,\nu[pk]} - \frac{1}{g} e^{\tau[lm]\,,\nu m,\,l[\tau\nu]} +$$

$$+ \frac{1}{8g} e^{\sigma\alpha,\ \beta[\sigma\rho],\ \rho[nm]} C_{\beta\gamma} \sigma^{[nm]} \eta_{nn'} \eta_{mm'} -$$

$$- \varepsilon_{mknl} \varepsilon^{\nu\mu\varkappa\tau} e^{[\nu\mu][mk]\,,\varkappa n\,,\tau l} - \tag{20}$$

$$- \frac{1}{8g} e^{\nu\mu\gamma\rho} (C\gamma^5 \gamma^{n'} \eta_{n'n})_{\alpha\beta}\ e^{\nu\alpha,\,[\nu\tau]\ \ l,\,[\sigma\rho]\beta} +$$

$$\varepsilon_{lpnm} \varepsilon^{\tau\varkappa\nu\mu} e^{[\nu\tau]l,\,\mu[mn]\,,\,\varkappa\rho} \Big).$$

Note that matrices α satisfy the following simple equations

$$(\alpha^{\mu}\alpha^{\nu}\alpha^{\sigma} + \alpha^{\sigma}\alpha^{\nu}\alpha^{\mu})_{AB} = -2(-1)^{(A)(B)} (\delta_{\mu\nu}\alpha^{\sigma} + \delta_{\sigma\nu}\alpha^{\mu})_{AB} \tag{21}$$

The expressions (17)-(20) determine a Lagrangian of N =1 supergravity (automaticaly invariant under local SUSY transformations (2-3) and in terms of $h_{\mu}{}^{n}$, $\Psi_{\nu}{}^{\alpha}$ equal to the well-known Lagrangian up to a total derivative) and equations for all components of the unified field $\Psi(x)$. If we consider g as Newton's constant and $h_{\mu}{}^{n}$ as tetrads we get from the equations for $\omega_{\mu}{}^{[mn]}$, $R_{[\nu\mu]}{}^{[mn]}$, $C_{[\nu\tau]}{}^{l}$ expressions allowing one to consider them as Ricci coefficients, Riemann tensor and torsion tensor respectively and the bose-part of the Lagrangian as Einstein's Lagrangian. From the equation which relates $\omega_{\mu}{}^{[mn]}$, $\Phi_{[mn]}{}^{[\nu\mu]}$, $\Psi_{\mu}{}^{\alpha}$, and $\Phi_{[\nu\mu]}{}^{\alpha}$ we see that the torsion does not vanish. Note that introducing Ricci coefficients in $\Psi(x)$ leads to first-order equations. The other auxiliary components are needed for lowering non-linearities to quadratic ones in the equations (cubic in the Lagrangian). This fact is connected with the larger non-linearity of the tetrad formulation as compared with the metric formulation of General Relativity. It is interesting that matrices α^{μ} in tetrad formulation form nevertheless a much simpler algebra and the minimal polynomial is of less degree than in the metric formulation [3].

References

1. Utiyama,R. (1956). Phys.Rev., 101, 1597.
2. Kibble,T.W. (1961). Journ. of Math. Phys., 2, 212.
3. Fedorov,F.I. (1968). DAN SSSR, 179, 802. (Russian).
4. Kuvshinov,V.I., Fedorov,F.I. (1970). Izvestiya Acad. Sc. BSSR, ser.ph.-m., N 2, 69. (Russian).
5. Babichev,L.F., Kuvshinov,V.I., Fedorov,F.I. (1980). DAN SSSR, 253, 1088. (Russian).
6. Okubo,S., Tosa,Y. (1979). Phys. Rev., D20, 462.
7. Babich,A.A., Kuvshinov,V.I., Fedorov,F.I. (1982). DAN SSSR, 267, 1093. (Russian).
8. Kuvshinov,V.I., Fedorov,F.I. (1984). In: Proc. of VII Int. Sem. Protvino, 1, p.267. (Russian).
9. Grigoryev,A.A., Kuvshinov,V.I., Fedorov,F.I. (1985). DAN SSSR, 281, N 6. (Russian).
10. West,P.C., Chamseddine,A.H. (1977). Nucl. Phys., B129, 39.

SPONTANEOUS COMPACTIFICATION OF EXTRA SPACE DIMENSIONS IN SUPERGRAVITIES WITH $d>4$

D.V.Volkov, D.P.Sorokin and V.I.Tkach

Institute of Physics and Technology, the Ukrainian Academy of Sciences, Kharkov 310108

1. The revival of considerable interest in the Kaluza-Klein (KK) approach [1] for constructing a unified theory is connected with the construction of supergravity (SG) theories in higher dimensional spacetimes ($d>4$). In particular, SGs in d=11,10 have been constructed [2,3], and intensive studies of possible spontaneous compactification (SC) mechanisms of extra space dimensions in these theories have been carried out. It was shown that the d=11 spacetime M^{11} may be compactified into AdS^4xM^7 (M^7 - being the 7-dimensional compact space) as a result of the Freund-Rubin or Englert (FRE) compactification mechanisms [4]. Note, that besides the FRE compactification, d=11 SG solutions were found, which correspond to the compactification of M^{11} into AdS^5xM^6 (M^6 being the symmetric space with a holonomy group containing an abelian normal subgroup: CP^2xS^2, CP^3, S^4xS^2, $S^2xS^2xS^2$, SO(5)/(SO(3)xSO(2) [6].

A non-chiral N=2, d=10 SG was also explored from the KK point of view. As a result, some nontrivial d=10 vacuum solutions of the form AdS^4xM^6 (M^6 being a compact space of (6)) were found in [3,7]. For N=2 a d=10 SG is obtained from a d=11 SG after dimensional reduction of one coordinate and all the d=10 solutions arise as a result of a combined SC mechanism, including the FR field and the abelian gauge field (arising in d=10 as a concequence of the compactification of one coordinate into S^1), it is natural to suggest that all the d=10 vacuum solutions are connected with the corresponding FR solutions of the d=11 SG.

In this report we show that M^7 subspaces of the d=11 SG can be treated as bundle manifolds $P(M^6,S^1)$, where M^6 is a solution of the d=10 SG, and the d=10 "Maxwell" field plays the role of components of this fibre bundle connection. From the physical point of view, this means that SC in the d=11 SG can happen consecutively, i.e., at the first step of the compactification the S^1 curle and the "Maxwell" field arise (as it was in the classical d=5 KK theory); this field ensures the second step, i.e., the compactification of six coordinates with the help of the background FR field. Moreover, in the case of S^7 and SU(3)/U(1) spaces, a three-step compactification can happen: the first step is similar to that considered above, but at the second one the "Maxwell" field provides the compactification of a 2-dimensional subspace into S^2 and the formation of SU(2) gauge fields which, in their turn, lead to the third step, the compactification of the M^4 subspace. Geometrically, it means that M^6 can be the associated total space of the fibre bundle $E(M^4,S^2,SU(2))$, where M^4 is the base, S^2 the fibre, and SU(2) the structure group.

Note that the induced U(1) and SU(2) gauge field take part in the compactification due to the SC mechanism of Einstein-Yang-Mills theories considered in [8].

We emphasize that though vacua of the both theories agree with each other, the effective d=4 theories, arising as a result of d=10 and d=11 compactifications, are different, in general, because owing to the reduction the N=2, d=10 SG contains only a part of the d=11 SG field spectrum. This difference becomes apparent as soon as the SUSYs of the corresponding d=4 theories are studied.

2. Let us consider how one can establish a correspondence between N=2, d=10 and N=1, d=11 vacua.

The supermultiplet of the d=11 SG consists of the gravitational field $e^{(\hat{M})}_{\hat{N}}$, Majorana Rarita-Schwinger field $\psi_{\hat{\mu}}$ and antisymmetric field $A_{\hat{M}\hat{N}\hat{L}}$. After the compactification of one coordinate into S^1 and subsequent dimensional reduction one obtains a non-chiral N=2, d=10 SG [3] comtaining $e_N^{(M)}$, antisymmetric fields A_{MNL}, A_{MN}, "Maxwell' field A_M, scalar ϕ and Majorana fields ψ_μ and λ that can be decomposed into Weyl components.

The fields A_M and ϕ appear in d=10 as a result of the follosing ansatz for the d=11 metric:

$$g_{\hat{M}\hat{N}} = \phi(X)^{-\frac{1}{4}} \begin{pmatrix} g_{MN}(X) + \phi^2 R_o^2 A_M A_N & R_o\phi\ A_M(X) \\ R_o\phi\ A_N(X) & \phi(X) \end{pmatrix} \tag{1}$$

Where $g_{MN}(X)$ is the d=10 metric, R_o is the radius of S^1.

We consider the solutions of d=11 (or 10) SG field equation corresponding to the background, where $M^{11} = AdS^4 x M^7$ ($M^{10} = AdS^4 x M^6$) and the compactification is caused by the FR mechanism [4]. Thus, in d=11 the field strength $F_{\hat{M}\hat{N}\hat{L}\hat{P}} = 4\ \partial_{[\hat{M}} A_{\hat{N}\hat{L}\hat{P}]}$ has the following nonzero background components:

$$\langle F_{\mu\nu\lambda\rho}\rangle = 3m(\det e_\delta^{(\sigma)})\ _{\mu\nu\lambda\rho} \tag{2}$$

As usual, the fermionic background is equated to zero ($\langle\psi_{\hat{\mu}}\rangle=0$). In this case the solutions of the bosonic part of the d=11 SG equations are the vacuum states characterized by M^{11} which is the direct product of 4-dimensional (usually, AdS^4) and 7-dimensional Einstein spaces with the following Ricci tensores:

$$R_{\mu\nu}(x) = 12\ m^2 g_{\mu\nu}(x), \tag{3}$$

$$R_{\hat{m}\hat{n}}(Y) = -6\ m^2 g_{\hat{m}\hat{n}}(\hat{Y}) \tag{4}$$

(The scalar curvature of the sphere is negative in our notation). The classification of all the known FR solutions is in ref. [9].

In the N=2, d=10 SG not only FR background field is needed for the compactification of the M^6 subspace, but the nonzero background values of A_M are required as well. This is connected with the following correlation for the vacuum values of A_{MNL} and A_M, demanded by the field ϕ equation of motion ($<\phi>=1$, $<A_{MN}>=0$) [3]:

$$F^2_{MNLP} + 8R_o^2 F^2_{MN} = 0 \qquad (F_{MN} = \partial_M A_N - \partial_N A_M) \tag{5}$$

As shown in [8], the abelian field A_M can cause a compactification of extra dimensions into coset spaces $G/(HxH_o)$ with the symmetry group HxH_o containing normal abelian subgroup H_o. The field strength F_{MN} background values, contributing to the vacuum SC solution, are constructed from structure constants of the group G with one of the indices corresponding to H_o. In our case of N=2, d=10 SG the equations of motion allow the compactification into $M^{10} = AdS^4 x M^6$, where $M^6 = CP^3$, $CP^2 x S^2$, $S^4 x S^2$, $S^2 x S^2 x S^2$, $\frac{SO(5)}{SO(3)xSO(2)}$, $\frac{Sp(2)}{Sp(1)xU(1)}$, $\frac{SU(3)}{U(1)xU(1)}$ (6)

if F_{MNLP} is of the FR-type and F_{MN} has the following components in the orthonormal basis [3,7]:

$$< F_{(\mu)(\nu)} > = F_{(\mu)(m)} = 0, \quad F_{(m)(n)} = -\frac{\rho}{R_o^2} C_{(m)(n)\circ} \tag{7}$$

where $C_{(m)(n)\circ}$ are the structure constants of the symmetry group of M^6; (m)(n) correspond to the coset space $G/(HxH_o) = M^6$; $\circ$ is the index of H_o; ρ the parameter the values of which can be restricted by the topology of M^6 (recall the monopole solution on S^2).

From the d=11 SG point of view, the choice of A_M field background values as in (7) means that the d=11 metric has the from:

$$g_{\hat{M}\hat{N}} = \begin{matrix} g_{\mu\nu}(x) & 0 \\ 0 & g^{M^7}_{\hat{m}\hat{n}}(\hat{Y}) \end{matrix} ,$$

$$g^{M^7}_{\hat{m}\hat{n}} = \begin{pmatrix} g^{M^6}_{mn}(Y) + R_o^2 A_m(Y) A_n(Y) & R_o A_m(Y) \\ R_o A_n(Y) & 1 \end{pmatrix} \tag{8}$$

(note that $<\varphi>= 1$).

Thus, $M^{11}=AdS^4 x M^7$ and we return to the FR solutions (3), (4). From the geometrical point of view, spaces M^7 characterized by the metric (9) from a class of fibre bundles $P(M^6, S^1)$ (see, e.g. [10]).

Calculating the curvature tensor components of $P(M^6,S^1)$ using curvature tensors of M^6, eqs.(7), (4), one can make sure all these spaces are equivalent to already known FR solutions of the d=11 SG. The results are listed in table:

d=11 SG subspaces $M^7=P(M^6,S^1)$	Subspaces M^6 of the N=2, d=10 SG
$T^7 = T^6 x S^1$	T^6
$K3xT^3 = K3xT^2xS^1$	$K3xT^2$
$S^7 \sim P(CP^3,S^1) \sim P(\frac{Sp(2)}{Sp(1)xU(1)},S^1) \sim$ $\sim P(E(S^4,S^2,SU(2)),S^1) \sim$ $\sim E(S^4,S^3,SU(2))\ P(S^4,SU(2))$	$CP^3 = \frac{SU(4)}{SU(3)xU(1)}$, $\frac{Sp(2)}{Sp(1)xU(1)} \sim E(S^4,S^2,SU(2))$
$\frac{SU(3)}{U(1)} \sim P(E(CP^2,S^2,SU(2)),S^1) \sim$ $\sim E(CP^2,S^3,SU(2))\ P(CP^2,SU(2))$	$\frac{SU(3)}{U(1)xU'(1)} \sim E(CP^2,S^2,SU(2))$
$\frac{SU(3)xU(1)}{U'(1)xU''(1)} \sim P\frac{SU(3)}{U'(1)xU''(1)}, S^1)$	$\frac{SU(3)}{U(1)xU'(1)}$
$\frac{SU(3)xSU(2)xU(1)}{SU'(2)xU'(1)xU''(1)} \sim P(CP^2,S^2,S^1)$	CP^2S^2
$\frac{SU(2)xSU(2)xSU(2)xU(1)}{U'(2)xU''(1)xU'''(1)} \sim P(S^2xS^2xS^2,S^1)$	$S^2xS^2xS^2$
$\frac{SO(5)xSO(2)}{SO(3)xSO(2)} \sim P(\frac{SO(5)}{SO(3)xSO(2)},S^1)$	$\frac{SO(5)}{SO(3)xSO(2)}$
$S^4xS^3 \sim P(S^4xS^2,S^1)$	S^4xS^2

It is interesting that in the case of S^7 and SU(3)/U(1) the associated fibre bundles $E(S^4,S^2,SU(2))$ and $E(CP^2,S^2,SU(2))$ can be taken as the base M^6 of the corresponding $P(M^6,S^1)$.

Thus, we have $P(E(M^4,S^2,SU(2)),S^1) \sim E(M^4,S^3,SU(2)) \sim$ $\sim P(M^4,SU(2))$, where $M^4=S^4$, CP^2. The consideration of S^7 and SU(3)/U(1) as instanton bundles was proposed in [11].

3. Now let us discuss the question of how many SUSYs can be conserved in the effective d=4 theory after the com-

pactification of M^{10} into $AdS^4 x M^6$.

The d=11 SG vacuum solution is supersymmetric if supertransformations of $\Psi_{\hat{\mu}}$ do not violate the condition $<\Psi_{\hat{\mu}}>=0$ (i.e., $\delta\Psi_{\hat{\mu}}=0$ for the background). For the FR solutions the number of SUSYs in d=4 equals the number of Killing spinors η of M^7 that are solutions of the equations [11,12]:

$$D_m\eta - \frac{1}{2}m\eta e_m^{(\hat{n})}\Gamma_{(\hat{n})} = 0, \tag{10a}$$

$$D_{11}\eta - \frac{1}{2}m\eta e_{11}^{(\hat{n})}\Gamma_{(\hat{n})} = 0 \tag{10b}$$

($\Gamma_{(\hat{m})}$ are the 8x8 Dirac matrices. We have decomposed the 11th component in (10) to compare with d=10 SG solutions). N=8,4,3,2,1 SUSY vacuum configurations are known in the d=11 SG [9,11,12].

The spinor fields Ψ_μ and λ of the N=2, d=10 SG are constructed as linear combinations of d=11 Rarita-Schwinger field components Ψ_μ and Ψ_{11} [3]. Thus, the Ψ_μ, λ field transformations, corresponding to the background configurations that arise due to the action of the combined FR mechanisms, are the linear combinations of d=11 supertransformations, in particular, of eqs. (10). The only difference between the d=11 and d=10 supertransformations is the absence of the derivative $\partial_{11}\eta$ in λ field transformations (all functions of the d=11 SG are independent of the 11th component in the sector corresponding to the reduction of the theory to the N=2, d=10 SG). Therefore, if the d=10 background is supersymmetric, then so is the corresponding d=11 background. In general however, the opposite statement fails.

Since all d=11 vacuum cinfigurations conserving N$\geq$1 SUSY ("round" S^7, SU(3)/U(1), $M^{p,q,r}$, ...) in the chosen coordinate system with the metric of M^7 in the form (9) have the solutions of (10) depending on $\hat{x}^{11}$ (the latter acting as a parameter of the U(1) transformations of the Killing spinors), then the equations corresponding to eqs. (10) in d=10 have no solutions transforming non-trivially by the SO(2) subgroup.

Thus, the non-trivial vacuum configurations of N=2, d=10 SG, corresponding to the 11-dimensional ones with N>1, have 2 SUSYs less in the effective d=4 theory. For example, the d=10 vacuum configuration with $M^6=CP^3$ corresponding to the "round" S^7 (N=8) in the d=11 SG, preserves N=6 SUSY in d=4, while the "round" S^7 (N=8) of the d=11 SG, preserves N=8 SUSY in d=4. The d=10 SG ground state with $M^6 = SU(3)/U(1)xU'(1)$ corresponding to the right SU(3)/U(1) space of the d=11 SG with N=3 has only N=1 SUSY and the N=2, d=10 vacuum configurations corresponding to the vacuums of the d=11 SG with N=2 SUSY have no SUSY preserved in the d=4 theory. In the case of N=1 SUSY vacuum configurations of the d=11 SG ("squashed" S^7, SU(3)xU(1)/U(1)xU'(1) and also when $M^7=T^7$ (N=8) or $M^7 = K3xT^3$ (N=4), the spinors η, the solutions of eqs. (10), are not only covariantly constant, but exactly constant on M^7 in the chosen coordinate system, i.e., $\partial_{\hat{\mu}}\eta=0$. Hence the vacuum solutions of the N=2, d=10 SG corresponding to such vacuum configurations of the d=11 SG, preserve the same number of SUSYs as the initial d=11 SG solutions.

References

1. Kaluza, T. (1921). Sitz. Preuss. Acad. Wiss. Berlin Math. Phys. Kl. 966; Klein, O. (1929). Z. Phys. 37. 895
2. Gremmer, E., Julia, B., Scherk, J. (1978). Phys. Lett., 76B, 409; Chamseddine, A.H. (1981). Nucl. Phys. B185, 403; Bergshoeff, B. et al. (1982). Nucl. Phys. B195, 97
3. Campbell, I.C.G., West, P.C. (1984). Nucl. Phys. B243, 112; Giani, F., Pernici, M. (1984). Phys. Rev. D30, 325
4. Freund, P.G.O., Rubin, M.A. (1980). Phys. Lett. 97B,233
5. Englert, F. (1982). Phys. Lett., 119B, 339
6. Sorokin D.P., Tkach, V.I., Volkov, D.V. (1984). Yad. Fiz. 39. 1306; Dolan, D. (1984). Phys. Lett. 140B. 304
7. Watamura, S. (1983). Phys. Let. 188; Sorokin, D.P., Tkach, V.I., Volkov, D.V. (1983). Zh.E.T.Ph. Lett. 38.

397 (Russian); Sorokin, D.P., Tkach, V.I. (1984). Ukr. Fiz. Zh. 28. 1765 (Russian)

8. Tkach, V.I., Volkov, D.V. (1980). Zh. E. T. Ph. Lett. 32. 681 (Russian); (1982) Theor. Math. Phys. 51. 171 (Russian); Sorokin, D.P., Tkach, V.I., Volkov, D.V. (1983). Theor. Math. Phys. 56. 171; (1984) 61 241 (Russian); Randjbar-Daemi, S., Percassi, R. (1982). Phys. Lett. 117B 378
9. Duff, M.J., Wilsson, B.E.W., Pope, C.N. (1984). Nucl. Phys., B233 433; Castellani, L., Romans, L.J., Warner, N.P. (1984). Nucl. Phys., B241 429
10. Percassi, R., Randjbar-Daemi, S. (1983). J. Math. Phys. 24 807
11. Awada, M.A., Duff, M.J., Pope, C.N. (1983). Phys. Rev. Lett. 50 294; Yamagishi, K. (1984). Phys. Lett. 134B 47; Sorokin, D.P., Tkach, V.I., Volkov, D.V. (1984). Zh. ETPh. Lett. 356 (Russian); Rajpoop, S., Robb, T., Taylor, J.G. (1984). Phys. Lett. 146B 318; Page, D.N., Pope, C.N. (1984). Phax. Lett. 147B 55; Sorokin, D.P., Tkach, V.I., Volkov, D.V. (1985). Yad. Fiz.(Soviet Nucl. Phys.). 1371 (Russian)
12. Duff, M.J., Nilsson, B.E.W., Pope, C.N. (1983). Phys. Lett. 129B 39.

CONFORMALLY COVARIANT EQUATIONS IN FIELD THEORY AND GENERAL RELATIVITY

A.A. Leonovich

Institute of Applied Physics, BeSSR Academy of Sciences, Minsk

1. Introduction. In recent years much more attention has been paid to studying antisymmetric tensor fields due to their inherent geometric, topological and group-theoretical properties as well as in connection with their possible physics applications [1]. Antisymmetric tensor fields are widely used in mathematics and physics, e.g. 2nd rank antisymmetric tensor field is a mathematically expressed electromagnetic field. By Maxwell equations both the exterior differential d and codifferential δ of the tensor field vanish in the absence of charges. These facts suggest that if besides Maxwell equations there exist some other relativistic equations formulated in terms of these differential operators, the latter equations might also be of some interest from the point of view of physics.

This paper deals with 1st order equations for a set of antisymmetric tensor fields, Dirac-Kahler equations [2-3], interpreted in the paper [2] as generally covariant formulation of the Dirac equation whereas the main subject of the study is a field which is the direct sum of antisymmetric tensor fields of all possible ranks. The 1st order differential equations studied here have a number of remarkable properties. The generators of their internal symmetry group, unlike those of the usually considered inner symmetry groups, are transformed via tensor representations of the Lorentz group [4] . Antisymmetric tensor fields may become a source of non-linear tensor fields of the Yang-Mills type [5]. The paper [6] describes an invariant quantization scheme and proposes an expression for the commutator function of these fields.

The role of conformal transformations in theoretical physics is well-known. By the conformal invariance principle equations for massless fields must be conformally invariant and the trace of the energy-momentum tensor must vanish [7]. The conformal invariance principle is of great significance since it leads to strong constraints for admissible types of interactions. The ideas pertaining to conformal invariance prove to be also rather important in the quantum field theory on a Riemannian space-time.

2. Conformal momentum of the Dirac-Kahler field. Operators d, δ are determined in the following way

$$(d F_{(p)})_{\alpha_1 \dots \alpha_{p+1}} = (p+1)\, \partial_{[\alpha_1} f_{\alpha_2 \dots \alpha_{p+1}]},$$
$$(\delta F_{(p)})_{\alpha_1 \dots \alpha_{p-1}} = \partial^{\alpha} f_{\alpha \alpha_1 \dots \alpha_{p-1}}, \tag{1}$$

where $F_{(p)}$ is an antisymmetric tensor field of rank (p-vector) p, $f_{\alpha_1 \dots \alpha_p}(X)$ are its covariant components and the metric of the n-dimensional flat space-time is determined by tensor $g_{\mu\nu}$. Wave function is a nonhomogeneous differential form, the sum of homogeneous differential forms

$$F = \sum_{p=0}^{n} \oplus F_{(p)} \tag{2}$$

Dirac-Kahler equation [3] is written as

$$\Lambda F = mF, \quad \Lambda = d + \delta \tag{3}$$

Composition of tensor antisymmetric fields into one field is of informal character. Eq. (3) is a set of 1st order differential equations.

Assuming K is a vector field, $K_\mu(X)$ are its covariant components and substituting ∂_μ for $K_\mu(X)$ in the righthand side of eq. (1) we obtain operators Kd, $K\delta$. Together with d, δ the vector operators Kd, $K\delta$ enable us to obtain operators connected with the space-time symmetries and valid for Dirac-Kahler equation solutions. These operators are

$$K_1 = K_\delta d + dK_\delta,\quad K_2 = K_d\delta + \delta K_d,\quad K_3 = K_d d + dK_d,\quad K_4 = K_\delta\delta + \delta K_\delta \tag{4}$$

Then, using the identity

$$(p+1)\,\partial_{[\mu} f_{\alpha_1\ldots\alpha_p]} = \partial_\mu f_{\alpha_1\ldots\alpha_p} - p\,\partial_{[\alpha_1} f_{|\mu|\alpha_2\ldots\alpha_p]} \tag{5}$$

we can prove that

$$K_1 - K_2 = (n-2p)\mathcal{F}, \tag{6}$$

where the vector field obeys the conformal Killing equations

$$\partial_\mu K_\nu + \partial_\nu K_\mu = 2\mathcal{F} g_{\mu\nu},\quad \mathcal{F} = \frac{1}{n}\partial^\nu K_\nu \tag{7}$$

Now, consider the operator

$$\hat{K} = K_1 + K_2 + K_3 + K_4 + (2-n)\mathcal{F} - 2K^\nu\partial_\nu \tag{8}$$

and determine the commutator $[\Lambda, \hat{K}]$ for the conformal Killing vector. We have

$$[\Lambda\hat{K}] = [\delta, K_1] + [d, K_2] + [\delta, K_3] + [d, K_4] + (2-n)[\delta,\mathcal{F}] + (2-n)[d,\mathcal{F}] - 2[\delta, K^\nu\partial_\nu] - 2[d, K^\nu\partial_\nu] = -4\mathcal{F}\Lambda \tag{9}$$

which proves that the Dirac-Kahler equation is conformal invariant for $m=0$. The proof of (9) is done with the help of the relations (5-7) which yield

$$[\delta, K_1] = (n-2p-2)\mathcal{F}\delta - 2\mathcal{F}\delta,\quad [d, K_2] = -(n-2p+2)\mathcal{F}d - 2\mathcal{F}d,$$
$$[\delta, K_3] - 2[d, K^\nu\partial_\nu] = 2(p-n)\mathcal{F}d - 2\mathcal{F}d,\quad [\delta,\mathcal{F}] = \mathcal{F}\delta, \tag{10}$$
$$[d, K_4] - 2[\delta, K^\nu\partial_\nu] = 2p\mathcal{F}\delta - 2\mathcal{F}\delta,\quad [d,\mathcal{F}] = \mathcal{F}d,$$

where $\mathcal{F}d$, $\mathcal{F}\delta$ are determined by the vector field $\mathcal{F}_\mu = \partial_\mu \mathcal{F}$

A symmetric traceless energy-momentum tensor is written as

$$T_{\mu\nu} = T^B_{\mu\nu} + \partial^\lambda (S_{\mu\lambda\nu} + S_{\nu\lambda\mu}), \qquad (11)$$

where $T^B_{\mu\nu}$ is the Belinfante energy-momentum tensor, $S_{\mu\lambda\nu} = -S_{\mu\nu\lambda}$ is a 3rd rank tensor whose conservation law $\partial^\mu S_{\mu\nu\lambda} = 0$ is determined by the Dirac-Kahler equation internal symmetry group [4]. If the equations of motion (3) hold $\partial^\mu T_{\mu\nu} = 0$, $g^{\mu\nu} T_{\mu\nu} = 0$, $m = 0$

The formulas (9-11) constitute the main result. The peculiarity lies in the fact that antisymmetric tensor fields of all possible ranks are considered as a single geometric object. Both the equations of motion (3) and the operator of conformal momentum (8) "mix up" tensor fields of different ranks.

3. Conformal covariant 2nd order equations for a set of antisymmetric tensor fields. A characteristic feature of the Penrose equation for a scalar field and the Dirac equation for a spinor is their being conformally covariant in a space-time of any dimension n [7]. For the Maxwell equations $n = 4$ is essential. The paper [8] proposes equations for vector and bivector.

The problem of conformally covariant equations for antisymmetric tensor fields of any rank in the Riemannian space-time of an arbitrary dimension is studied because the formulas obtained may also be interesting from a purely geometric point of view [9-10]. Besides, it is necessary for studying conformal anomalies as well as for constructing theories of Kaluza-Klein type.

Let us consider a linear differential equation

$$\Lambda F = (b\delta d + c d\delta + \ell R + h R d\delta) F = 0 \qquad (12)$$

and prove that it is conformal covariant for some dimensionless constants b, c, l, h. Under the conformal mapping of Riemannian spaces

$$'g_{\mu\nu} = e^{2\sigma} g_{\mu\nu}, \quad 'g^{\mu\nu} = e^{-2\sigma} g^{\mu\nu} \tag{13}$$

a p-vector F is transformed as

$$F' = e^{a\sigma} F. \tag{14}$$

Hence

$$\Lambda' F' = e^{(a-2)\sigma} \Lambda F \tag{15}$$

if

$$\begin{aligned} &ba^2 + l(n-1)(n-2) - hp(n-2) = 0, && b(2+2p-a-n) + c(a+n-2p) = 0, \\ &ba - 2l(n-1) + hp = 0, && ba - c(a+n-2p) - h(n-2) = 0 \\ &ba - c(a-2) = 0, && ba^2 + c(4p-2n-a^2) - h(n-2) = 0 \\ &b(2a-2-2p+n) = 0, && \end{aligned} \tag{16}$$

The solution of system (16) is

$$b = a - 2, \quad c = a = \frac{2-n}{2} + p,$$
$$l = a(a-2)\frac{n-2+2p}{2(n-1)(n-2)}, \quad h = a(a-2)\frac{2}{n-2} \tag{17}$$

For $p = 0$ the equation (12) with the coefficients (17) reduces to the Penrose equation for a scalar field, and with p=1, n=4 - to the Maxwell equation for a vector field.

A symmetric traceless energy-momentum tensor is easily found by the algebraic method proposed in [11] and based on the noncommutativity of covariant derivatives.

The conformal momentum operator $Q = K\delta d + dK\delta + aJ$, where $K\delta$ is determined by the Killing conformal vector and acts in the space of solutions of (12).

For eqs. (12, 17) it is possible to determine scalar product in the space of solutions and prove the theorems on integrals of the energy-momentum tensor as for the Dirac, Maxwell, Penrose equations [7]. Nonlinear generalization of eq. (12), provided the conformal symmetry is retained, is carried out by the same lines as in the case of a scalar field. The above approach might be useful for constructing conformal covariant 1st order operator for a set of spin-forms.

The principle peculiarity of the proposed eq. (12) is that only the components with curvature providing for conformal covariance are present.

4. Conformal covariant 4th order equation for a 2nd-rank symmetric tensor in a four-dimensional Riemannian space-time. In [12] a conformally covariant formulation of the theory of 2nd rank symmetric tensor in the Riemannian space-time of arbitrary dimension is given. A conformal and gauge invariant 4th order equation for $A_{\mu\nu}$ in the Minkowski space-time (linearized conformal gravity or linearized Bach tensor) is presently under study. Therefore it is worthwhile to study the problem of existence of conformal covariant equations for $A_{\mu\nu}$ which may be responsible for the general gravitation interaction.

Let us recover tensors $Y_{\mu\alpha\beta\nu}$ and $X_{\mu\alpha\beta\nu}$ with algebraic properties of the Weyl conformal curvature and Riemannian tensors respectively from $A_{\mu\nu}=A_{\nu\mu}$ and $g_{\mu\nu}$.

$$Y_{\mu\alpha\beta\nu} = X_{\mu\alpha\beta\nu} + X_{\mu[\nu} g_{\beta]\alpha} + X_{\alpha[\beta} g_{\nu]\mu} + \frac{1}{3} g_{\mu[\beta} g_{\nu]\alpha}, \quad (18)$$

$$X_{\mu\beta} = g^{\alpha\nu} X_{\mu\alpha\beta\nu}, \quad X = g^{\mu\beta} X_{\mu\beta}, \quad g^{\alpha\nu} Y_{\mu\alpha\beta\nu} \equiv 0, \quad (19)$$

$$X_{\mu\alpha\beta\nu} = \frac{1}{2} \left[(\nabla_\alpha \nabla_\beta + \nabla_\beta \nabla_\alpha - R_{\alpha\beta}) A_{\mu\nu} + (\nabla_\mu \nabla_\nu + \nabla_\nu \nabla_\mu - R_{\mu\nu}) A_{\alpha\beta} \right] - [\mu \leftrightarrow \alpha]$$

Consider the Lagrangian and derive the equation of motion

$$L = Y_{\mu\alpha\beta\nu} Y^{\mu\alpha\beta\nu} \quad (20)$$

$$\Lambda_{\mu\nu} = 4 (\nabla_\alpha \nabla_\beta + \nabla_\beta \nabla_\alpha - R_{\alpha\beta}) Y_{\mu\ \ \ \nu}^{\ \alpha\beta} = 0 \quad (21)$$

Under conformal mapping (13) we get

$$\begin{aligned} A'_{\mu\nu} &= e^{2\sigma} A_{\mu\nu} && \text{— a metric tensor,} \\ Y'^{\cdot\mu}_{\cdot\alpha\beta\nu} &= Y^{\cdot\mu}_{\cdot\alpha\beta\nu} && \text{— a Weyl tensor,} \\ \Lambda'_{\mu\nu} &= e^{-2\sigma} \Lambda_{\mu\nu} && \text{— a Bach tensor.} \end{aligned} \tag{22}$$

Using the formalism [11] for theories with higher order derivatives we find a symmetric energy-momentum tensor

$$T_{\alpha\beta} = \Pi_{\mu\nu(\alpha}\nabla_{\beta)}A^{\mu\nu} + \Pi_{\mu\nu\cdot(\alpha}^{\ \ \ \lambda}\nabla_{\beta)}\nabla_\lambda A^{\mu\nu} + \tag{23}$$
$$+ 2\nabla^\lambda(\Pi_{\mu\nu(\alpha}S_{\beta)\lambda}A^{\mu\nu} + \Pi_{\mu\nu\cdot(\alpha}^{\ \ \ \delta}S_{\beta)\lambda}\nabla_\delta A^{\mu\nu} - g_{\alpha\beta}L +$$
$$+ 2Z^{\cdot\mu\nu}R_{\mu\alpha\nu\beta} - g_{\alpha\beta}\nabla^{\cdot\mu}\nabla^\nu Z_{\mu\nu} + \nabla_\alpha\nabla^\nu Z_{\nu\beta} + \nabla_\beta\nabla^\nu Z_{\nu\alpha} -$$
$$- \Box Z_{\alpha\beta},$$

where $Z^{\alpha\beta} = \dfrac{\partial L}{\partial R_{\alpha\beta}} = -4Y^{\cdot\mu\alpha\beta\nu}A_{\mu\nu}$, $\Pi^{\mu\nu\beta} = -\nabla_\alpha \Pi^{\cdot\mu\nu\beta\alpha}$,

$$\Pi^{\cdot\mu\nu\beta\alpha} = \frac{\partial L}{\partial(\nabla_\alpha\nabla_\beta A_{\mu\nu})} = 4(Y^{\cdot\mu\alpha\beta\nu} + Y^{\nu\alpha\beta\mu}),$$

$$S_{\alpha\beta}A_{\mu\nu} = g_{\mu[\alpha}A_{\beta]\nu} + g_{\nu[\alpha}A_{\beta]\mu}.$$

If (21) are satisfied then $\nabla^\alpha T_{\alpha\beta} = 0$, $g^{\alpha\beta}T_{\alpha\beta} \equiv 0$

The following analogy is to be pointed out. The Maxwell equations are conformal covariant for both the field strengths and vector-potential. For eq. (21) the situation is similar. In the Minkowski space-time eq. (21) reduces to an equation of linearized conformal gravity. The conformal momentum operator is then

$$\hat{K}A_{\alpha\beta} = K^\nu\nabla_\nu A_{\alpha\beta} + A_{\alpha\nu}\nabla_\beta K^\nu + A_{\beta\nu}\nabla_\alpha K^\nu - 2\mathcal{F}A_{\alpha\beta}$$

5. Equations of conformal gravity in space-time of dimension $n = 4K-1$. The Pontryagin characteristic classes for the Riemannian manifolds with the metric $g_{\mu\nu}$ of dimension $n = 4K$ (the connection is

symmetric and agrees with the metric) are [1]

$$P_K = \delta^{j_1 \dots j_{2K}}_{i_1 \dots i_{2K}} \in^{\alpha_1 \dots \alpha_{4K}} R^{i_1}_{\cdot j_1 \alpha_1 \alpha_2} \cdots R^{i_{2K}}_{\cdot j_{2K} \alpha_{4K-1} \alpha_{4K}}, \qquad (24)$$

where the Riemannian tensor obeys both the Jacobi $R_{[\mu\nu\alpha]\beta} \equiv 0$ and Bianchi $\nabla_{[\lambda} R_{\mu\nu]\alpha\beta} \equiv 0$ identities. The Einstein generalized tensor [13] for P_K vanishes identically.

Let us consider the vector

$$Z_\nu = \delta^{j_1 \dots j_{2K}}_{i_1 \dots i_{2K}} \in^{\alpha_1 \dots \alpha_{4K-1}} R^{i_1}_{\cdot j_1 \alpha_1 \alpha_2} \cdots R^{i_{2K}}_{\cdot j_{2K} \alpha_{4K-1} \nu} = X^{\alpha\ j}_{\cdot i \cdot} R^{i}_{\cdot j \alpha \nu} \qquad (25)$$

in a space-time of dimension $n = 4K-1$. Using the formula $\in^{[\alpha_1 \dots \alpha_{4K-1}} \delta^{\nu]} \equiv 0$ prove that

$$Z_\nu \equiv 0$$

Using the expression for the commutator of covariant derivatives

$$(\nabla_K \nabla_\ell - \nabla_\ell \nabla_K) X^{i\ \beta}_{\cdot \alpha \cdot} =$$
$$= -R^{i}_{\cdot qK\ell} X^{q\ \beta}_{\cdot \alpha \cdot} + R^{q}_{\cdot \alpha K\ell} X^{i\ \beta}_{\cdot q \cdot} - R^{\beta}_{\cdot qK\ell} X^{i\ q}_{\cdot \alpha \cdot},$$

we obtain

$$Z_\nu = \nabla^{\cdot\mu} \nabla_\lambda (X^{\lambda}_{\cdot \mu\nu} - X^{\ \lambda}_{\mu \cdot \nu} - X^{\ \lambda}_{\nu \cdot \mu}). \qquad (27)$$

For the first component in (27) the Bianchi identities give

$$\nabla_\lambda X^{\lambda}_{\cdot \mu\nu} \equiv 0$$

Thus, a differential identity is obtained

$$\nabla_\mu G^{\mu\nu} \equiv 0,$$

where

$$G^{\mu\nu} = \nabla^\lambda (X^{\mu \cdot \nu}_{\cdot \lambda} + X^{\nu \cdot \mu}_{\cdot \lambda}) \qquad (29)$$

and $X^{\mu \cdot \nu}_{\cdot \lambda}$ is determined by formula (25). Using the Bianchi and Jacobi identities prove that the trace of symmetric tensor $G^{\mu\nu}$ (the Einstein generalized tensor) vanishes identically

$$g_{\mu\nu} G^{\mu\nu} \equiv 0 \qquad (30)$$

Under conformal mapping of the Riemannian spaces (13)

$$'\nabla_\mu \, 'G^{\mu\nu} = e^{-(n+2)\sigma} \nabla_\mu G^{\mu\nu}, \qquad (31)$$

$$'G^{\mu\nu} = e^{-(n+2)\sigma} G^{\mu\nu}. \qquad (32)$$

The above scheme is useful for all the Pontryagin classes P_K and products of lower dimension classes, i.e. $n=4-P_1$, $n=8-P_2$, P_1^2, $n=12-P_3$, $P_1 P_2$, P_1^3, etc.

It is worthwhile to consider a generalization of the proposed approach for the case of the Riemann-Cartan-Weyl geometry. In line with everything described above it might be interesting to study the Euler-Poincare invariant.

Note that in [11] there is proposed a simple non-variational method of obtaining the Hilbert symmetric energy-momentum tensor and the Einstein generalized tensor for arbitrary Lagrangians based on the study of the Lie derivative of Lagrangian.

Conformally covariant gravitation equations

$$G^{\mu\nu} = 0$$

may evidently be obtained by means of metric variation of the Chern-Simons invariant [14-15]. But since the explicit form of the invariant is complex and the variation procedure cumbersome, this is done only for n=3 [14].

Acknowledgements. The author expresses his deep gratitude to Academician S.P. Novikov for clarification of certain geometrical and topological problems. He is also thankful to Profs. B.M. Barbashov, V.N. Ponomariev, Ja.A. Smorodinsky and N.A. Chernikov for their critical and helpful remarks.

References

1. Pestov, A.B. (1980). In: Abstracts of contributes papers GR 9 Conference. Jena, p. 607.
2. Kahler, E. (1962). Rendiconti di Mat. 21, 425.
3. Pestov, A.B. (1978). Theor. Math. Phys., 34, N 1, 48 (Russian).
4. Pestov, A.B. (1983). JINR P2-83-506.
5. Leonovich, A.A., Pestov, A.B. (1980). JINR P2-80-823, Dubna; (1981) DAN BSSR, 25, N 10, 892 (Russian).
6. Leonovich, A.A. (1983). Theor. Math. Phys., 57, N 2, 265.
7. Chernikov, N.A., Shavokhina, N.S. (1974). Theor. Math. Phys., 18, N 3, 310 (Russian).
8. Barbashov, B.M., Leonovich, A.A. (1983). JINR P2-83-524, Dubna.
9. Penrose, R. (1964). In: Relativity, Groups and Topology. Les Houches Summer School 1963. Gordon and Breach. N.Y., p. 565-584.
10. Branson, T.P. (1983). Lett. Math. Phys., 7, 63.
11. Barbashov, B.M., Leonovich, A.A., Pestov A.B. (1982). JINR P2-82-151, Dubna; (1983) Yad. Fiz. (Sov. Nucl. Phys.), 38, 261 (Russian).
12. Leonovich, A.A., Nesterenko, V.V. (1984). JINR E2-84-11, Dubna.
13. Barbashov, B.M., Leonovich, A.A. (1983). JINR P5-83-398, Dubna (Russian).
14. Deser, S., Jackiw, R., Templeton, S. (1982). Ann. Phys., 140, 372.
15. Wu, Y.S. (1984). Ann. Phys., 156, 194.

STOCHASTIC QUANTIZATION, SUPERSYMMETRY AND RENORMALIZATION

V.Ya.Fainberg[+], I.N.Maliev, V.P.Spiridonov
[+]P.N.Lebedev Physical Institute, Moscow

It is known [1,2] from stochastic quantization of an "equilibrium" potential field theory in dth dimensions that the generating functional of a(d+1)-dimensional ("non-equilibrium") field theory possesses supersymmetry cf. [3]. Moreover if the basic equilibrium theory is renormalizable then so is the non-equilibrium one. The question arises whether supersymmetry and multiplicative renormalizability is conserved in the (d+1)-dimensional theory when radiative corrections are considered (d-dimensional theory is supposed to be multiplicatively renormalizable). The aim of this note is to show that in a non-equilibrium theory in λ^2-order (λ is the coupling constant) the supersymmetry is conserved but there is no multiplicative renormalizability.

For simplicity we consider stochastic quantization of the Euclidean scalar field theory $\varphi(x)$ with action $S[\varphi]$. The Langevin equation in $d+1$ dimensions is of the form:

$$\frac{\partial\varphi(x,\tau)}{\partial\tau} + \frac{\delta S[\varphi]}{\delta\varphi(x,\tau)} = \eta(x,\tau) \qquad (1)$$

where $\eta(\cdot)$ is a Gaussian white noise. Set the following initial value

$$\varphi(x,\tau) \equiv \dot{\varphi}(x,\tau) = 0 \qquad \text{as } \tau \to -\infty. \qquad (2)$$

Correlation functions of the (d+1)-dimensional theory are given by:

$$< \varphi(x_1,\tau_1)\dots\varphi(x_n,\tau_n)> \equiv$$

$$\equiv N \int \mathcal{D}\eta \exp\{-\frac{1}{2}\int_{(x,\tau)} \eta^2(x,\tau)dxd\tau)\}\varphi(x_1,\tau_1)\dots\varphi(x_n\tau_n) \quad (3)$$

where $\varphi(x_i,\tau_i)$ is a solution of equation (1) with initial value (2) and N is the normalization factor. Due to (2) equal time $(\tau_1 = \dots = \tau_n)$ correlators (3) coincide with Euclidean Green functions of the d-dimensional theory [4]. Passing in (3) to the field variables and introducing external source J, we obtain for the generating functional the following expression

$$Z[J]=\tilde{N}\int d\varphi Dc D\bar{c} \exp\{-\frac{1}{2}\int_{(x,\tau)}(\dot{\varphi}+\frac{\delta S}{\delta\varphi})^2+\int_{(x,\tau)}\bar{c}(x_1,\tau)(\partial_\tau \delta(x_1-x_2)+$$

$$+\frac{\delta^2 S}{\delta\varphi(x_1,\tau)\,\delta\varphi(x_2\delta)})c(x_2,\tau)+\int_{(x,\tau)} J(x,\tau)\varphi(x,\delta)\} \quad (4)$$

where c, $\bar{c}$ are Grassmann scalar fields. To analyze the supersymmetry in (4). it is convenient to introduce an auxiliary field B:

$$Z[J]=\tilde{N}\int D\varphi DBDc D\bar{c} \exp\{\int_{(x,\tau)}[B^2-B(\dot{\varphi}+\frac{\delta S}{\delta\varphi})]-\int_{(x,\tau)}\bar{c}(x_1,\tau)(\partial_\tau\delta(x_1-x_2)$$

$$+\frac{\delta^2 S}{\delta\varphi(x_1,\tau)\,\delta\varphi(x_2\tau)})c(x_2,\tau)+\int_{(x,\tau)} J(x,\tau)\varphi(x,\tau)\} \quad (5)$$

where we integrate over B along imagynary axis.

The action W in (5) is invariant with respect to the following supersymmetry transformations:

$$\delta\varphi=\bar{\varepsilon}c,\ \delta\bar{c}=\bar{\varepsilon}B,\ \delta c=0,\ \delta B=0 \quad (6a)$$

or

$$\delta\varphi = \bar{c}\,\varepsilon,\ \delta c = (B - \dot{\varphi})\varepsilon,\ \delta\bar{c} = 0,\ \delta B = \dot{\bar{c}}\,\varepsilon. \qquad (6b)$$

These transformations form a supergroup with generators:

$$Q = c\,\frac{\delta}{\delta\varphi} + B\frac{\delta}{\delta\bar{c}}\,,\quad \bar{Q} = \bar{c}\,\frac{\delta}{\delta\varphi} + (\dot{\varphi} - B)\frac{\delta}{\delta c} + \dot{\bar{c}}\,\frac{\delta}{\delta B} \qquad (7)$$

such that

$$\{\bar{Q}, Q\} = \frac{\partial}{\partial\tau}\,,\quad Q^2 = \bar{Q}^2 = 0. \qquad (8)$$

Adding to (5) the source of the fields $c, \bar{c}, B$ and chainging the integration variables according to the (6), we find the simplest Ward-Fradkin-Takahashi-Slavnov-Taylor identities:

$$\langle B(z')\varphi(z)\rangle + \langle c(z)\bar{c}(z')\rangle = 0 \qquad (9a)$$

$$\langle B(z)B(z')\rangle = 0 \qquad (9b)$$

$$\langle\varphi(z)B(z')\rangle - \langle B(z)\varphi(z')\rangle = \langle\varphi(z)\varphi(z')\rangle \qquad (9c)$$

$$\langle B(z')\dot{\varphi}(z)\rangle + \langle c(z)\bar{c}(z')\rangle = 0,\ z = (x,\tau). \qquad (9d)$$

From the translation invariance one can see that (9d) is a consequence of (9a). From the identities (9) and some more complicated ones it follows that the generating functional for the proper Green functions:

$$\Gamma(\Phi, c, \bar{c}) = \ln z - J\Phi - \eta\bar{c} - \bar{c}\eta \qquad (10)$$

is also supersymmetric. Therefore radiative corrections do not break down the supersymmetry provided the identities (9) are true in all orders of perturbation theory. For example, the two-point contribution to Γ has the following form

$$\Gamma_2=\int\{B(z)K(z,z')B(z')\rangle - B(z)\langle B(z)\varphi(z')\rangle^{-1}\varphi(z')$$

$$-\bar{c}(z)\langle c(z)\bar{c}(z')\rangle^{-1}c(z')\}\,dz\,dz' \qquad (11)$$

where $K(z,z')=\int dz_1 dz_2 \langle B(z)\varphi(z_1)\rangle^{-1}\langle\varphi(z_1)\varphi(z_2)\rangle\langle\varphi(z_2)B(z')\rangle^{-1}$

It is not difficult to verify that $Q\Gamma_2=\bar{Q}\Gamma_2=0$ due to (9a-9d).

The calculations of the one-particle Green functions entering the propagators of (9) and (11) were petformed for $\lambda\varphi^3/3$ and $\lambda\varphi^4/4$ -models in the second-order perturbation theory. Let us introduce the following notation for free propagators:

$\langle B\varphi\rangle_0 =$  ; $\langle c\bar{c}\rangle_0 =$  ; $\langle\varphi\varphi\rangle_0 =$ 

in momentum space

$$\langle B\varphi\rangle_0=-\langle c\bar{c}\rangle_0=(i\omega+k^2)^{-1},\ \langle\varphi\varphi\rangle_0=2(\omega^2+k^4)^{-1}.$$

The 2nd-order corrections to propagators are the following ones:

$\langle B(z)\varphi(z')\rangle_{(2)} = 9$ 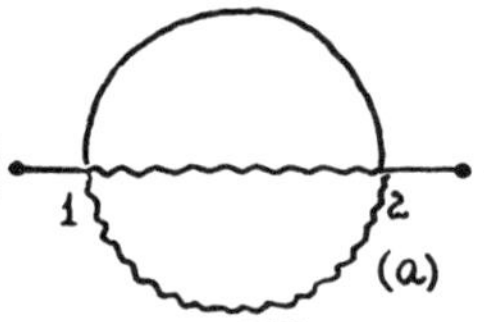 $+\,9$ 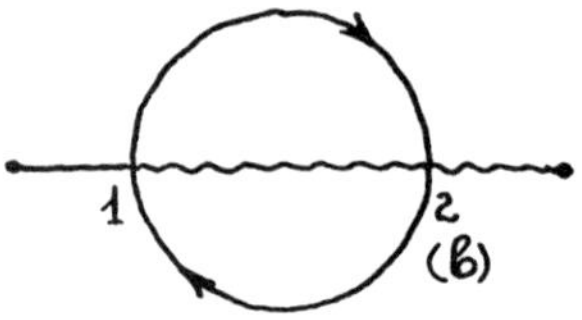

$\langle c(z)\bar{c}(z')\rangle_{(2)} = 9$ 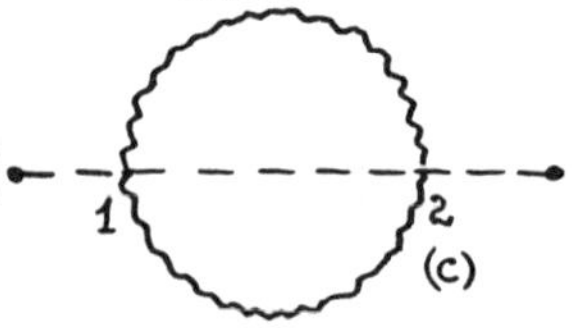 $+\ (1\longleftrightarrow 2)$

$\langle B(z)B(z')\rangle_{(2)} =$ 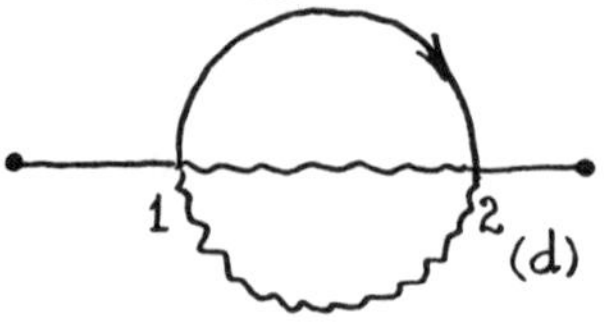 $+\ (1\longleftrightarrow 2)$

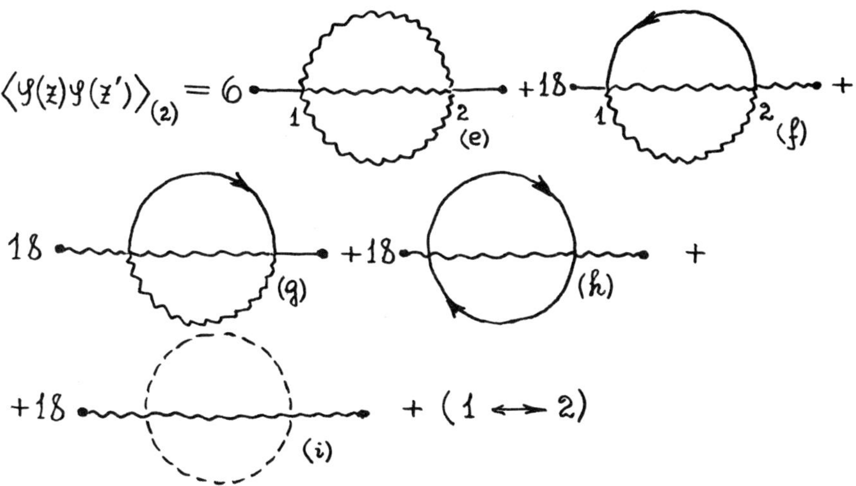

The diagrams (a) and (c) differ in sign, and the diagram (b) gives no contribution. Hence, (9a) holds. Diagrams (d) vanish also. Therefore (9b) holds. The identity (9c) is easily verified by calculating the diagramms (a) and (e-g). We wish to emphasize that diagramms (h) and (i) equal zero separately.

A divergent contribution to Γ_2 can be represented as $\Gamma_2 = QF$. From dimensional considerations and $\bar{Q}\Gamma_2 = 0$, we obtain:

$$F=\int dz\,\bar{c}\,(\alpha\varphi^3+\beta(\dot{\varphi}-B)+\gamma\,\Box\,\varphi).$$

The direct calculation of the divergent constants β and γ in the λ^2-order (we use the dimensional regularization) gives:

$$\beta \underset{\varepsilon\to 0}{=} \left(-\frac{3}{2}+\frac{9}{2}\ln\frac{4}{3}\right)\frac{1}{\varepsilon}\left(\frac{\lambda}{16\pi^2}\right)^2;\quad \gamma \underset{\varepsilon\to 0}{=} \left(-\frac{9}{2}\ln\frac{4}{3}\right)\frac{1}{\varepsilon}\left(\frac{\lambda}{16\pi^2}\right)^2.$$

One can see that $\beta \neq \gamma$. This means that the theory is not multiplicatively renormalizable in the non-equilibrium phase. It can be shown that in the equilibrium limit wave-function renormalization constant is

$$Z_\varphi^{-1} = 1+\beta+\gamma = 1-\frac{3}{2\varepsilon}\left(\frac{\lambda}{16\pi^2}\right)^2$$

and coincides with Z_φ^{-1} in the corresponding 4-dimensional $\lambda\varphi^4/4$-theory. The same situation takes place in the $\lambda\varphi^3/3$ -theory in the λ^2 -approximation.

We should emphasize that in the analysis of the supersymmetry of non-equilibrium theory we have never used the fact that initial equilibrium theory is potential. Therefore the non-equilibrium theory arising from stochastic quantization always has supersymmetry. But the corresponding action can be expressed in superfield form only for the potential models [1,2]. To conclude, note that vanisning of the diagram (i) containing closed fermion loop (formed of c-fields) is not accidential. It can be regorously proved that the Jacobian of the change $\eta \to \varphi$ in (3) with our initial values is 1 for the scalar theories. It means, that c-fields give no contribution to real scattering processes of the φ-particles among themselves neither in d+1 nor in d-dimensional theories. This statement is of general nature and reflects non-relativistic nature of the c-particle propagation. In some gauges an analogous situation appears from stochastic quantization of the non-abelian gauge fiels. These questions and the calculation of Γ_2 will be considered in detail elsewhere.

References

1. Feigel'man. M.V. Tavelick A.M. (1983). Phys. Lett. 95A, 469.

2. Egoryan E.Sh. (1984). Erevan Phys. Inst. Preprint 751 (66).-84 (Russian).

3. Parisi, G., Sourlas, N., (1979). Phys. Rev. Lett., 43, 744.

4. Gadgiev, S.A., Fainberg, V.Ya. (1984). Short reports on physics. Lebedev Phys. Inst., 6, 36 (Russian).

ON QUANTIZATION OF SUPERSYMMETRIC THEORIES

A.Smilga
ITEP, B.Cheremushkinskaya 25, Moscow 117259

It is well known that reconstruction of a quantum Hamiltonian by its classical counterpart involves ambiguities in sufficiently complicated theories. The simplest example where such an ambiguity appears is a free motion on a Riemannian manifold. In this case the classical Hamiltonian is given by the expression

$$H^{cl} = \tfrac{1}{2}\, g^{ij} p_i p_j \tag{1}$$

where the metrics $g^{ij}(q)$ is a nontrivial coordinate function. Trying to quantize this Hamiltonian we immediately run into the problem: which order of the noncommuting quantum operators $\hat{p}_i = -i\partial/\partial q^i$ and $g^{ij}(q)$ should be chosen. A number of different quantum operators: $\tfrac{1}{2}\, g^{ij}\hat{p}_i\hat{p}_j$, $\tfrac{1}{2}\,\hat{p}_i g^{ij}\hat{p}_j$, etc. have the same classical limit. Auxiliary arguments are necessary to make a choice between these possibilities. A popular choice is

$$\hat{H}^{qu} = \frac{1}{2g^{\frac{1}{4}}}\, \hat{p}_i \sqrt{g}\, g^{ij}\hat{p}_j g^{-\frac{1}{4}} \tag{2}$$

so that the operator $g^{-\frac{1}{4}}\hat{H}^{qu}g^{\frac{1}{4}}$ coincides with the covariant Laplacian acting on the wave functions $\psi(q)$ normalized by the condition

$$\int |\psi(q)|^2\, d^n q\, \sqrt{g(q)} = 1 \tag{3}$$

(hereafter we assume the manifold to be compact). It should be noted, however, that the covariance requirement does not preclude adding to the r.h.s. of eq. (2) the term $\sim t^2 R$ up to a factor. Different ways of reasoning result in different values of this factor [1-3] . Note also that the operator ordering ambiguity in a Hamiltonian is in one-to-one correspondence with the ambiguity in the path integral definition of the evolution operator 1,4,5 .

In supersymmetric theories the situation is completely different. We should require the classical supersymmetric algebra

$$i\,\{Q,\bar{Q}\}_{P.B.} = 2H;\ \{Q,Q\}_{P.B.} = \{\bar{Q},\bar{Q}\}_{P.B.} = 0,$$
$$\{Q,H\}_{P.B.} = \{\bar{Q},H\}_{P.B.} = 0 \tag{4}$$

(the symbol $\{\ \}_{P.B.}$ stands for the Poisson bracket) to be preserved on the quantum level. This requirement dictates a definite way of ordering in the operators $\hat{Q}^{qu}$, $\hat{\bar{Q}}^{qu}$.

The latter statement is not new (see e.g. the discussion of particular supersymmetric models in [6]). Our observation is that a general recipe of quantization is applicable in any studied example and seemingly holds for any supersymmetric theory.

We employ the operator ordering procedure suggested by Weyl more than half a century ago [7] elaborated by Moyal [8] and generalized to the fermion degrees of freedom case by Berezin and Marinov [9]. The Weyl procedure is in many respects "the most natural". It is discussed in details in the excellent review [5] and here we confine ourselves to a brief recitation of the most important definitions and properties.

Let a function $A(p,q;\ \bar{\psi},\psi)$ be defined on the phase space of a dynamical system involving both bosonic (p,q) and fermionic $(\bar{\psi},\psi)$ degrees of freedom. The correspondence between the quantum operator $\hat{A}(\hat{p},q;\hat{\bar{\psi}},\psi)$ ($\hat{p}_i \equiv -i\partial/\partial q_i$; $\hat{\bar{\psi}}^a \equiv \partial/\partial\psi^a$) and the classical function $A(p,q;\bar{\psi},\psi)$ is established by the following rule:

$$\hat{A}(\hat{p},q;\hat{\bar{\psi}},\psi) = \int \prod_i d\alpha_i d\beta_i \prod d\varepsilon^a d\bar{\varepsilon}^a h(\alpha,\beta; \varepsilon,\bar{\varepsilon}) \exp\{\sqrt{-1}\left[\sum_i(\alpha_i\hat{p}_i+\beta_i q_i) + \sum_a(\varepsilon^a\hat{\bar{\psi}}^a + \psi^a\bar{\varepsilon}^a)\right]\} \tag{5a}$$

provided

$$A(p,q;\bar{\psi},\psi) = \int \prod_i d\alpha_i d\beta_i \prod_a d\varepsilon^a d\bar{\varepsilon}^a \; h(\alpha,\beta; \varepsilon,\bar{\varepsilon}) \exp\{\sqrt{-1}\left[\sum_i(\alpha_i p_i+\beta_i q_i) + \sum_a(\varepsilon^a\bar{\psi}^a+\psi^a\bar{\varepsilon}^a)\right]\} \tag{5b}$$

Then $\hat{A}$ is called the Weyl ordered operator corresponding to the classical phase space function A and, in turn, A is the Weyl symbol of $\hat{A}$. The definition looks somewhat awkward but its physical sense is rather clear: the canonical coordinates and momenta should enter the operator $\hat{A}$ symmetrically in every case where ordering controversion appears. E.g. the function pq corresponds to the operator $(\hat{p}\hat{q}+\hat{q}\hat{p})/2$, etc.

The Weyl symbol of the product of two operators does not coincide generally with the product of their Weyl symbols (e.g. $\hat{p}\hat{q}$ has the Weyl symbol $pq-i\hbar/2 \neq pq$). We write down the general expression for the Weyl symbol of the (super)commutator of two operators which we will need later.

$$[\hat{A},\hat{B}]_{\pm} \overset{W}{=} >i\hbar\,\{A,B\}_{M.B.} \equiv 2\sin h\left[\frac{\hbar}{2}\sum_a\left(\frac{\partial^2}{\partial\psi_a^{(2)}\partial\bar{\psi}_a^{(1)}} - \frac{\partial^2}{\partial\psi_a^{(1)}\partial\bar{\psi}_a^{(2)}}\right) + \frac{i\hbar}{2}\sum_i\left(\frac{\partial^2}{\partial q_i^{(1)}\partial p_i^{(2)}} - \right.\right.$$

$$- \frac{\partial^2}{\partial q_i^{(2)} \partial p_i^{(1)}})\Big] A(p^{(1)}, q^{(1)}; \bar{\psi}^{(1)}, \psi^{(1)}) B(p^{(2)},$$

$$q^{(2)}; \bar{\psi}^{(2)}, \psi^{(2)})|_{(1)=(2)} \qquad (6)$$

Here $\{A,B\}_{M.B.}$ defines the Moyal bracket of two operators originally introduces in ref.[8] for bosonic systems. In the lowest order in h the Moyal bracket coincides with the standard Poisson bracket. If A and B are "linear enoung" the contribution of the higher terms is zero. Generally, they should be accounted for.

So, we return to the question of quantization of supersymmetric theories. Our prescription consists in choosing Weyl-ordered supercharges as a quantum counterpart to the classical supercharges expressed via canonical coordinates and momenta. We may ask: is the classical supersymmetry algebra (4) preserved as an algebra of (super)commutators of quantum operators defined in such a way? It is so indeed in all the considered examples and thus the supersymmetry algebra is preserved on the quantum level. We cannot, however, present any general proof for this fact. Moreover, it may be not true in some cases. E.g. in refs. [10-11] an argumentat is given for the N=2 supersymmetric two-dimensional σ- model to involve a quantum supersymmetric anomaly and central charges to be spontaneously generated. (Of course, this does not take place in the quantum mechanics of the N=2 SUSY σ - model, the anomaly being specific of the field theory).

Consider now the particular examples. Let us dwell upon (0+1)-dimensional N=1 σ - model. The classical supercharges and the Hamiltonian have the following form in this model:

$$\begin{cases} Q^{cl} = \psi^a l^i_a (p_i + i\, \omega_{ibc}\, \bar\psi^b\, \psi^c) \\ \bar Q^{cl} = \bar\psi^a l^i_a (p_i + i\, \omega_{ibc}\, \bar\psi^b\, \psi^c) \end{cases} \qquad (7)$$

$$H^{cl} = \tfrac{1}{2}\, g^{ij} (p_i + i\, \omega_{iab}\, \bar\psi^a\, \psi^b)(p_j + i\, \omega_{jcd}\, \bar\psi^c\, \psi^d) -$$

$$- \tfrac{1}{2}\, R_{abcd}\, \bar\psi^a\, \psi^b\, \bar\psi^c\, \psi^d \qquad (8)$$

where $\omega_{iqb} = -e_{aj}(\, \partial_i\, e^j_b + \Gamma^j_{ik}\, e^k_b\,)$ are spin connections. The quantum operators corresponding to eqs. (7,8) were built earlier basing on the differential form interpretation of wave functions [12-14] . We build them following our own prescription. One can be easily convinced that the Weyl-ordered operators corresponding to eq. (7) have the form

$$\hat Q^{qu} = g^{\frac14}\, \psi_a e^i_a\, (\hat p_i + i\, \omega_{ibc}\, \hat{\bar\psi}^b\, \psi^c)\, g^{-\frac14}$$

$$\hat{\bar Q}^{qu} = g^{\frac14}\, \bar\psi^a\, e^i_a\, (p_i + i\, \omega_{ibc}\, \hat{\bar\psi}^b\, \psi^c)\, g^{-\frac14} \qquad (9)$$

The "sandwich" $\sim g^{\frac14} \ldots g^{-\frac14}$ have the same origin as in eq. (2) due to the fact that the quantum operators thus defined act on the wave functions $\psi\,(q,\psi\,)$ normalized by the condition

$$\int \psi(q,\psi\,)\, \psi^+(q,\, \bar\psi) \exp\{ - \sum_a \bar\psi^a\, \psi^a\} \cdot$$

$$\cdot \prod_i dq^i \prod_a d\, \bar\psi^a\, d\, \psi^a = 1 \qquad (10)$$

A conventional definition includes the factor $\sqrt{g(q)}$ in the integration measure (cf. eq. (3)). Under this choice the "sandwiches"isappear and quantum supercharges coincide with the known expressions. The quantum Hamiltonian can be found by a straightforward calculation of the anticommutator of supercharges (9). The result is

$$\hat{H}^{qu} = g^{\frac{1}{4}}\hat{H}^{cov} g^{-\frac{1}{4}} = \tfrac{1}{2} g^{-\frac{1}{4}} \hat{\Pi}_i \sqrt{g}\, g^{ij} \hat{\Pi}_j g^{-\frac{1}{4}} -$$

$$- \tfrac{1}{2} R_{abcd} \bar{\psi}^a \psi^b \bar{\psi}^c \psi^d \qquad (11)$$

where $\hat{\Pi}_i = \hat{p}_i + i\,\omega_{iab} \hat{\bar{\psi}}^a \psi^b$. The meaning of $\hat{H}^{cov}$ is that of a Laplacian acting on differential forms i.e. components of $\psi^{cov}(q,\psi)$.

Now we will obtain this result in another way calculating first the Weyl symbol of the quantum Hamiltonian as the Moyal bracket (6) of the classical supercharges (7). The first term of the hyperbolic sine expansion is just the Poisson bracket of classical supercharges, i.e. the (doubled) classical Hamiltonian. The case under consideration is nontrivial enough for the nonzero higher terms in t. In particular, with the term $\sim t^3$ preserved the answer is

$$\{\hat{Q}^{qu}, \hat{\bar{Q}}^{qu}\}_+ = 2t\, \hat{H}^{qu} \overset{W}{=\!\!>} 2t \Big[H^{cl} +$$

$$+ \frac{t^2}{2} g^{ij} (\Gamma^l_{ik} \Gamma^k_{jl} + \omega_{iab}\, \omega_{jab}) \qquad (12)$$

The appendix $\sim t^2$ to H^{cl} is noncovariant and looks rather weird. Nonetheless, it is necessary. To make sure it is so let us find the Weyl-ordered operator corresponding to H^{cl}. We see that besides $g^{\frac{1}{4}}\hat{H}^{cov}g^{-\frac{1}{4}}$ a noncovariant appendix appears which cancels exactly the second term in the r.h.s. of eq.(12).

Note that the Weyl-ordered operator corresponding to the classical Hamiltonian of the standard σ-model (1) is also noncovariant and equals

$$\hat{H}^W = \frac{1}{2} g^{-\frac{1}{4}} \hat{p}_i \sqrt{g}\, g^{ij} \hat{p}_j g^{-\frac{1}{4}} + \frac{t^2}{8}\left[R - g^{ij}\, \Gamma^k_{il}\, \Gamma^l_{jk}\right] \tag{13}$$

The noncovariance of $\hat{H}^W$ is due to the noncovariance of the path integral for the evolution operator which corresponds to the Weyl prescription of operator ordering. It is given by [5]

$$G(q_{(iu)}, q_{(f)}; \Delta t) = \lim_{N\to\infty} \int \prod_{n=1}^{N-1} \prod_i \frac{dq^i_{(n)}}{\sqrt{2\pi\, t\varepsilon}} \cdot$$

$$\cdot \prod_{k=o}^{N-1} \sqrt{g\left(\frac{q^i_{(k)} + q^i_{(k+1)}}{2}\right)} \exp\left\{\frac{i}{2\varepsilon\, t}\, g_{ij}\left(\frac{q^e_{(k)} + q^e_{(k+1)}}{2}\right) \cdot\right.$$

$$\left.\cdot\ (q^i_{(k+1)} - q^i_{(k)})\ (q^i_{(k+1)} - q^j_{(k)})\right\} \tag{14}$$

where $\varepsilon = \Delta t/N$; $q^i_{(o)} \equiv q^i_{(in)}$, $q^i_{(N)} \equiv q^i_{(f)}$. In refs. [2-3] a covariant expression for $G(q_{(in)}, q_{(f)}; \Delta t)$ was postulated which resulted in the covariant expression for the corresponding quantum Hamiltonian.

Consider now the N=2 SUSY σ-model. For simplicity we confine ourselves to the case of a two-dimensional manifold. Reducing the two-dimensional Lagrangian of ref. [15] we get

$$L = h\dot{\varphi}\dot{\bar{\varphi}} + \frac{ih}{2}\left(\bar{\psi}\dot{\psi} - \dot{\bar{\psi}}\psi\right) + \frac{i\bar{\Psi}\Psi}{2}\left(\dot{\varphi}\,\partial ht - \right.$$
$$\left. -\dot{\bar{\varphi}}\,\bar{\partial} h\right) - \frac{R}{4}\,\psi^2\ \bar{\psi}^2 \tag{15}$$

where $h(\varphi, \bar{\varphi})$ is an arbitrary function of complex coordinates φ and $\bar{\varphi}$, $R = (-\partial\bar{\partial} h + h^{-1}\, \partial h\, \bar{\partial} h)/h^2$ is the curvatute of the manifold, ψ_α is the two-component spinor (or, rather, is a spinor in the (1+1)-dimensional theory) with the properties

$$\bar\psi^{\alpha} = (\psi_{\alpha})^{+}, \psi^{\alpha} = \varepsilon^{\alpha\beta}\psi_{\beta}, \psi\chi \equiv \psi_{\alpha}\chi^{\alpha}, \bar\psi\bar\chi \equiv \bar\psi^{\alpha}\bar\chi_{\alpha},$$

$$\bar\psi\chi \equiv \bar\psi^{\alpha}\chi_{\alpha}$$

The supercharges and the Hamiltonian can be found in a standard way

$$\begin{cases} Q_{\alpha} = \sqrt{2}\left[\pi - \frac{i\bar\psi\psi}{2h}\partial h\right]\frac{\psi_{\alpha}}{\sqrt{h}} \\ \bar Q^{\beta} = \frac{\sqrt{2}\bar\psi^{\beta}}{\sqrt{h}}\left[\bar\pi + \frac{i\bar\psi\psi}{2h}\bar\partial h\right] \end{cases} \tag{16}$$

$$H^{cl} = h^{-1}\left[\pi - \frac{i\bar\psi\psi}{2h}\partial h\right]\left[\bar\pi + \frac{i\bar\psi\psi}{2h}\bar\partial h\right] - \frac{\psi^{2}\bar\psi^{2}}{4}R \tag{17}$$

where $\pi(\bar\pi)$ are bosonic canonical momenta and $\Psi = \sqrt{h}\psi$, $\bar\Psi = \sqrt{h}\bar\psi$ are canonically conjugated variables. The expressions for the Weyl-ordered quantum supercharges coincide with eq. (16), the classical momenta $\pi(\bar\pi)$ and $\bar\psi^{\alpha}$ being substitutted by the corresponding differential operators and the order of symbols in eq. (16) being preserved. Calculating directly the Moyal brackets (6) of the supercharges we can see that the classical extended supersymmetry algebra is preserved on the quantum level. However, by the full analogy with N=1 case the Moyal bracket $\{Q_{\alpha}, \bar Q^{\beta}\}_{M.B.}$ acquires a contribution from the higher in $\hbar$ terms and is given by

$$i\{Q_{\alpha}, \bar Q^{\beta}\}_{M.B.} = 2\delta_{\alpha}^{\beta}\left[H^{cl} + \frac{h^{2}}{8\hbar^{3}}\partial h\bar\partial h\right] \tag{18}$$

(Recall, h is metrics and $\hbar$ is the Planck constant). The quantum Hamiltonian with the Weyl symbol given by the square bracket in eq. (18) has the form (cf. ref. [6b])

$$\hat{H}^{qu} = \left[\hat{\pi} - \frac{i\,\hat{\bar{\psi}}\,\psi}{2h}\,\partial h \right] h^{-1} \left[\hat{\bar{\pi}} + \frac{i\hat{\bar{\psi}}\,\psi}{2h}\,\bar{\partial} h \right] - \frac{R}{4}\,\hat{\bar{\psi}}^2\,\psi^2 \tag{19}$$

The suggested procedure can be applied not only to σ - models. The most interest realm of its application is perhaps gauge theories. The Yang-Mills Hamiltonian $H = = \frac{1}{2}\,\mathrm{Sp} \int (E_i^2+B_i^2)\,d^3x$ is simple only from the first sight. In fact, resolving of constraints $D_iE_i = 0$ results in an essentially nonlinear Hamiltonian, where coordinates and momenta are sophistically interlaced rather similar to the σ- model Hamiltonian (see, e.g. [16]). The same ordering problems as in σ- model appear in quantization of this Hamiltonian. Just as in σ - model there is no golden rule to quantize the Yang-Mills theory. The most natural way to do it is to follow the Dirac prescription which consists in quantizing the unconstrained Hamiltonian and subsequent imposing the operator restrictions $D_i\hat{E}_i\,\phi(A_i) = 0$. Some other possibilities may exist.

For the supersymmetric Yang-Mills theory our procedure can be followed to obtain the SUSY-preserving quantum Hamiltonian. This Hamiltonian appears to coincide with the Hamiltonian which follows from the Dirac prescription. A detailed analysis of this interesting question will be presented elsewhere.

I am indebted to M.A.Olshanetsky, M.A,Shifman, Yu.A.Simonov and A.I.Vainstein for the useful discussions.

References

1. Marinov,M.S. (1980). Phys. Reps. 60, 1.
2. DeWitt,B.S. (1959). Rev. Mod. Phys. 29, 377.
3. Cheng,K.S. (1972). J. Math. Phys. 13, 1723.
4. Feynman,R. (1948). Rev. Mod. Phys. 20, 367.
5. Berezin,F.A. (1980). Usp. Fiz. Nauk 132,497.
6. Nakamura,A.,Palumbo,F. (1984). Phys.Lett. 14, 96.

Macfarlane,A.J., Popat,P.C. (1984). J. Phys. A17,2955.

7. Weyl,H. (1931). Gruppentheorie und Quantenmechanik, Leipzig,Hirzel.

8. Moyal,I.E. (1949). Proc. Cambr. Phil. Soc. 45, 99.

9. Berezin,F.A. , Marinov,M.S. (1977). Ann. Phys.104,336.

10. Witten,E. (1979). Nucl. Phys. B149, 285.

11. Krivoshchekov,V.K., Medvedev,P.B. (1984). Preprint ITEP-6

12. Witten,E. (1982). J. of Diff. Geom. 17, 661,

13. Witten,E. (1982). Nucl. Phys. B202, 253.

14. Claudson,M., Halpern,M.B. (1985). Nucl. Phys. B250, 689.

15. Zumino,B. (1984). Phys. Lett. 87B, 203.

16. Simonov,Yu.A. (1984). Preprint ITEP-108.

Asatryan,H.M., Savvidy,G.K. (1983). Phys. Lett. 99A, 290.

INTEGRALS OF MOTION IN SUPERSYMMETRIC QUANTUM MECHANICS

R.G.Agayeva

Institute of Physics, the Azerbaijan SSR Academy of Sciences, Baku

In modern practice supersymmetry is extensively used in various domains of physics: nuclear physics, elementary particle physics, condensed matter physics, statistical physics, and gravity [1]. To a large extent this has been provided by the development of the supersymmetric quantum mechanics [2].

It is known that a system possesses supersymmetry if there are operators generating bose-fermi transformations and integrals of motion. Therefore constructing integrals of motion is very important.

The principal object of this paper is calculating integrals of motion for supersymmetric generalizations of non-stationary 1-dimensional harmonic oscillator and of motion of a charged particle in a non-stationary quantising magnetic field. Some motion integrals for these systems are well known in quantum mechanics [3]. In [4] the motion integrals are briefly considered for non-stationary quadratic fermi-bose systems. To construct the motion integrals for the non-stationary supersymmetric systems we, according to [3], demand the commutativity of the motion integrals with the operator $(i\hbar \frac{\partial}{\partial t} - \hat{\mathcal{H}}_S)$, where $\hat{\mathcal{H}}_S$ is the Hamiltonian of the corresponding supersymmetric problem.

The 1-dimensional supersymmetric Hamiltonian of a non-stationary harmonic oscillator is of the form [2]:

$$\hat{\mathcal{H}}_S = \hat{\mathcal{H}} - \frac{\hbar\omega(t)}{2}[\hat{\psi}^+, \hat{\psi}^-] \tag{1}$$

where

$$\hat{\mathcal{H}} = \frac{\hat{p}^2}{2m} + \frac{m\omega^2(t)\hat{x}^2}{2} \tag{2}$$

$\hat{x}$ is a Bose operator, $\hat{p}$ its conjugate, $[\hat{p},\hat{x}] = -i\hbar$, and $\psi^{\pm}$ are two Fermi operators such that $\{\hat{\psi}^+, \hat{\psi}^-\} = 1$, $\hat{\psi}^{-2} = \hat{\psi}^{+2} = 0$. These Fermi operators are usually chosen in the following matrix representation:

$$\hat{\psi}^+ = \begin{pmatrix} 0 & 0 \\ 1 & 0 \end{pmatrix}, \quad \hat{\psi}^- = \begin{pmatrix} 0 & 1 \\ 0 & 0 \end{pmatrix}. \tag{3}$$

In this representation we have

$$\hat{\mathcal{H}}_S = \begin{pmatrix} \hat{\mathcal{H}} + \frac{\hbar\omega(t)}{2} & 0 \\ 0 & \hat{\mathcal{H}} - \frac{\hbar\omega(t)}{2} \end{pmatrix}. \tag{4}$$

Write a solution of the Schrödinger equation

$$\left(i\hbar\frac{\partial}{\partial t} - \hat{\mathcal{H}}_S\right)\varphi = 0 \tag{5}$$

as vector

$$\varphi = \begin{pmatrix} \varphi^f \\ \varphi^b \end{pmatrix}. \tag{6}$$

Henceforth the superscript "f" corresponds to the Fermi and "b" to the Bose state.

Let us introduce the supercharges

$$\hat{Q}^{\pm} = \hat{A}^{\pm} \exp\left(\pm i \int^t \omega(t)\, dt\right) \hat{\psi}^{\pm} \tag{7}$$

where the operator

$$\hat{A}^{-} = \frac{i}{\sqrt{2\hbar}}\left[\varepsilon(t)\frac{\hat{p}}{m} - \dot{\varepsilon}(t)x\sqrt{m}\right] \tag{8}$$

and its adjoint, $\hat{A}^{+}$, are integrals of motion for the corresponding non-supersymmetric problem, i.e.

$$\left[i\hbar\frac{\partial}{\partial t} - \hat{\mathcal{H}}, \hat{A}^{\pm}\right] = 0 , \tag{9}$$

and the function $\varepsilon(t)$ is a solution of the classical harmonic oscillator equation $\ddot{\varepsilon} + \omega^2(t)\varepsilon = 0$ [3].

It is easy to verify that

$$\left[i\hbar\frac{\partial}{\partial t} - \hat{\mathcal{H}}_S, \hat{Q}^{\pm}\right] = 0 \tag{10}$$

hence these supercharges are integrals of motion. Besides,

$$\hat{Q}^{-2} = \hat{Q}^{+2} = 0. \tag{11}$$

From the non-Hermitian operators $\hat{Q}^{-}$ and $\hat{Q}^{+}$ we form the Hermitian operator

$$\hat{K} = \{\hat{Q}^{-}, \hat{Q}^{+}\} = \begin{pmatrix} \hat{A}^{-}\hat{A}^{+} & 0 \\ 0 & \hat{A}^{+}\hat{A}^{-} \end{pmatrix} \tag{12}$$

which is an invariant of our supersymmetric problem. Hence, the solutions ψ of the Schrödinger equation (5) are the eigenfunctions of the operator $\hat{K}$, and

$$[\hat{K}, \hat{Q}^{\pm}] = 0 . \tag{13}$$

From the expression (12) it is clear that the eigenfunctions of $\hat{A}^{-}\hat{A}^{+}$ and $\hat{A}^{+}\hat{A}^{-}$ are degenerate because [3]

$$[\hat{A}^-, \hat{A}^+] = 1 \tag{14}$$

$$\hat{A}^+ \hat{A}^- |n,t\rangle = n\,|n,t\rangle \qquad (n=0,1,2\ \dots) \tag{15}$$

where $|n,t\rangle$ is a solution of the non-stationary Schrödinger equation for the harmonic oscillator with the non-supersymmetric Hamiltonian $\hat{\mathcal{H}}$. Then we get the following expressions for the eigenfunctions of $\hat{K}$:

$$\varphi \equiv \varphi_n = \begin{pmatrix} \varphi_n^f \\ \varphi_n^b \end{pmatrix}, \tag{16}$$

$$\varphi_n^f = \exp\left(-\frac{i}{2}\int^t \omega(t)\,dt\right)\cdot |n-1,t\rangle,$$

$$\varphi_n^b = \exp\left(\frac{i}{2}\int^t \omega(t)\,dt\right)\cdot |n,t\rangle$$

and

$$\hat{K}\varphi_n = n\varphi_n. \tag{17}$$

The supersymmetry takes place if the vacuum state φ_0 is $\hat{Q}^\pm$-invariant:

$$\hat{Q}^\pm \varphi_0 = 0. \tag{18}$$

We determine the state φ_0 starting from (18):

$$\varphi_0 = \begin{pmatrix} 0 \\ \varphi_0^b \end{pmatrix}, \quad \varphi_0^b = \exp\left(\frac{i}{2}\int^t \omega(t)\,dt\right)\cdot |0,t\rangle \tag{19}$$

where $|0,t\rangle$ is the vacuum state of the system described by the Hamiltonian $\hat{\mathcal{H}}$. This state is given by the expression [3]:

$$|0,t\rangle = \sqrt[4]{\frac{m}{\pi\hbar\varepsilon^2}}\exp\left(\frac{i\dot{\varepsilon}x^2 m}{2\hbar\varepsilon}\right). \tag{20}$$

The function φ_0 obeys the Schrödinger equation (5), the (18) and the normalization condition.

Thus, it is the normalised vacuum state consisting of a single boson state. The other states are paired and consist of one Bose and one Fermi states. This situation is analogous to the stationary case [2].

It is not difficult to make sure that in the presence supersymmetry the motion integrals $\hat{Q}^{\pm}$ transform Fermi and Bose states of the same pair into each other:

$$\begin{aligned} \hat{Q}^{-}\varphi_n &= \hat{Q}^{-}\varphi_n^{b} = \sqrt{n}\,\varphi_n^{f}, \\ \hat{Q}^{+}\varphi_n &= \hat{Q}^{+}\varphi_n^{f} = \sqrt{n}\,\varphi_n^{b}. \end{aligned} \tag{21}$$

Hence, two independent motion integrals $\hat{Q}^{-}$ and $\hat{Q}^{+}$ of the one-dimensional system under discussion are actually supercharges.

Now, consider the motion of a charged particle in a non-stationary magnetic field described by the Hamiltonian

$$\hat{\mathcal{H}}_H = \frac{1}{2m}\left[\left(\hat{p}_x + m\omega(t)\hat{y}\right)^2 + \left(\hat{p}_y - m\omega(t)\hat{x}\right)^2\right]. \tag{22}$$

Here we choose the z-axis along the magnetic field $\vec{H}$. The motion along the z -axis can be neglected.

Taking into account the spin-dependent term $\mu\sigma_z\vec{H} = -\frac{\hbar\omega(t)}{2}\left[\hat{\psi}^{+}, \hat{\psi}^{-}\right]$ we get a supersymmetric generalisation of our problem [4]:

$$\hat{\mathcal{H}}_{H_S} = \hat{\mathcal{H}}_H - \frac{\hbar\omega(t)}{2}\left[\hat{\psi}^+, \hat{\psi}^-\right]. \tag{23}$$

It is straightforward that operators

$$\hat{Q}^{\pm}_H = \hat{A}^{\pm}_H \exp\left(\pm i\int^t \omega(t)\,dt\right)\hat{\psi}^{\pm}$$

and (24)

$$\hat{\Theta}^{\pm}_H = \hat{B}^{\pm}_H \exp\left(\pm i\int^t \omega(t)\,dt\right)\hat{\psi}^{\pm}$$

commute with the operator $\left(i\hbar\frac{\partial}{\partial t} - \hat{\mathcal{H}}_{H_S}\right)$, i.e. they are invariants. In the expression (24)

$$\hat{A}^-_H = \frac{1}{2\sqrt{\hbar m}}\exp\left(i\int^t \omega(t)dt\right)\left[\varepsilon(t)(\hat{p}_x + i\hat{p}_y) - im\dot{\varepsilon}(t)(\hat{y} - i\hat{x})\right],$$

(25)

$$\hat{B}^-_H = \frac{1}{2\sqrt{\hbar m}}\exp\left(i\int^t \omega(t)dt\right)\left[\varepsilon(t)(\hat{p}_y + i\hat{p}_x) - im\dot{\varepsilon}(t)(\hat{x} - i\hat{y})\right]$$

$\hat{A}^{\pm}_H$, $\hat{B}^{\pm}_H$ are the motion integrals for the corresponding non-supersymmetric problem [3] and satisfy Bose Commutation relations.

The operators $\hat{Q}^{\pm}_H$ are supercharges. Indeed, let us construct the Hermitian operator

$$\hat{K}_H = \left\{\hat{Q}^+_H, \hat{Q}^-_H\right\} \tag{26}$$

which is an integral of motion. It is easy to verify that the operator $\hat{K}_H$ obeys the following relations:

$$\left[\hat{K}_H, \hat{Q}^{\pm}_H\right] = \left[\hat{K}_H, \hat{\Theta}^{\pm}_H\right] = 0$$

$$\hat{Q}^{\pm 2}_H = \hat{\Theta}^{\pm 2}_H = 0. \tag{27}$$

Clearly, the solutions φ_H of the non-stationary Schrödinger equation with $\hat{\mathcal{H}}_{H_S}$ are also the eigenfunctions of the operator $\hat{K}_H$. The solution of the eigenvalue problem for $\hat{K}_H$ is

$$\varphi_H = \begin{pmatrix} \varphi_H^f = \exp\left(-\frac{i}{2}\int^t \omega(t)dt\right) \cdot |n_1 - 1, n_2, t\rangle \\ \\ \varphi_H^b = \exp\left(\frac{i}{2}\int^t \omega(t)dt\right) \cdot |n_1, n_2, t\rangle \end{pmatrix}. \quad (28)$$

Here the states $|n_1, n_2, t\rangle$ are the solutions of the non-stationary Schrödinger equation with $\hat{\mathcal{H}}_H$ and of the equations:

$$\begin{aligned} \hat{A}_H^- \hat{A}_H^+ |n_1, n_2, t\rangle &= n_1 |n_1, n_2, t\rangle, \\ \hat{B}_H^- \hat{B}_H^+ |n_1, n_2, t\rangle &= n_2 |n_1, n_2, t\rangle. \end{aligned} \quad (29)$$

The supersymmetry is not broken if

$$\hat{Q}_H^\pm \varphi_{0H} = 0. \quad (30)$$

It is easy to show that there are normalized vacuum states φ_{0H}:

$$\varphi_{0H} = \begin{pmatrix} 0 \\ \varphi_{0H}^b \end{pmatrix}, \; \varphi_{0H}^b = \exp\left(\frac{i}{2}\int^t \omega(t)\,dt\right) \cdot |0, n_2, t\rangle \quad (31)$$

where (see [3]) $|0, n_2, t\rangle$ are the vacuum states of the corresponding non-supersymmetric problem

$$|0,n_2,t\rangle=\left(\frac{m}{\pi\hbar\varepsilon^2 n_2!}\right)^{1/2}\left(\frac{\varepsilon^*}{\varepsilon}\right)^{n_2/2}\left[\sqrt{\frac{m}{\hbar|\varepsilon|^2}}(x+iy)\exp\left(i\int^t\omega(t)dt\right)\right]^{n_2}\cdot$$

$$\cdot\exp\left[\frac{i\dot{\varepsilon}m}{2\hbar\varepsilon}(x^2+y^2)\right]. \qquad (32)$$

We finally get

$$\hat{Q}_H^-\varphi_H=\hat{Q}_H^-\varphi_H^b=\sqrt{n_1}\,\varphi_H^f,\quad \hat{Q}_H^+\varphi_H=\hat{Q}_H^+\varphi_H^f=\sqrt{n_2}\,\varphi_H^b \qquad (33)$$

i.e. all the states of the system constitute the set of Bose-Fermi pairs except the vacuum states of single bosons.

Thus, for the 2-dimensional supersymmetric problem four independent invariants are constructed. Two of them are supercharges. These supercharges generate the supersymmetric transformations which map Fermi and Bose states of the same pair into themselves. Similar results are discussed in the review [6].

Acknowledgments. I wish to express sincere thanks to V.I.Man'ko for the discussions and attention to this study.

References

1. Supersymmetry in Physics. (1985). Physica, 15D, 1-295.

2. Cooper, F., Freedman, B. (1983). Ann. Phys., 146, 262-288.

3. Malkin, I.A., Man'ko, V.I. (1979). Dynamic Symmetries and Coherent States of Quantum Systems. Nauka, Moscow (Russian).

4. Dodonov, V.V., Man'ko, V.I. (1983). Integrals of Motion and Dynamics of General Non-Stationary Quadratic Fermi-Bose Systems. - Trudy PhIAN, Nauka, Moscow, 153, 145-193 (Russian).

5. Grombrugghe, M., Rittenberg, V. (1983). Ann. Phys., 151, 99-126.

6. Gendenstein, L.E., Krive, I.V. (1985). Russian Phys. Surveys, 146, 4, 553-736 (Russian).

Part II

QUANTUM FIELD THEORY

CONFORMAL INVARIANCE IN TWO-DIMENSIONAL FIELD THEORY

E.S. Fradkin, M.Ya. Palchik[+]

Lebedev Physical Institute of the USSR Academy of Sciences, Moscow

[+]Institute of Automation and Electrometry of Siberian Branch of the USSR Academy of Sciences, Novosibirsk

Abstract. Papers [1] and [2] formulate the plan of constructing conformal solution for quantum field theories in space of any dimension. Important points of the programme are using the Ward identities for conserved current and energy-momentum tensor, and also resolving uncertainties $0 \times \infty$ in the operator products $Z_1 \varphi(x)\chi(x)$, $Z_1 Z_2^{-1} \varphi(x)\chi(x)$, $Z_1 \varphi^2(x)$, etc.

Here we show that to correctly resolve those uncertainties a regulation should be introduced into exact equations. The theory $L_{int} = \lambda\varphi^4$ in two-dimensional space is considered and its solution found. Closed equations for the Green functions in this theory are obtained after the regularization is removed. These equations agree with those of Ref. [3], which determine correlation functions of the Ising model. A generalization of the corresponding procedure to the case of $\mathcal{D}$ - dimensional space is discussed.

1. Introduction. Papers [1,2,4] show that, with conserved current j_μ or energy-momentum tensor $T_{\mu\nu}$ taken into account, an infinite set of tensor fields appears in the operator algebra, along with the fundamental field φ ,

$$P_s = P^{d+s}_{\mu_1\ldots\mu_s}(x) \tag{1.1}$$

with scale dimensions

$$d_s = d + s \tag{1.2}$$

These fields appear in operator expansions of the products

$$j_\mu(x)\varphi(x+\varepsilon), \quad T_{\mu\nu}(x)\varphi(x+\varepsilon) \tag{1.3}$$

encountered in the Ward identity for the Green functions

$$\langle 0|P^{\ell}_{\mu_1\ldots\mu_s}(x_1)\,\varphi(x_2)\,j_\mu(x_3)|0\rangle,$$

$$\langle 0|P^{\ell}_{\mu_1\ldots\mu_s}(x_1)\,\varphi(x_2)\,T_{\mu\nu}(x_3)|0\rangle$$

(For $T_{\mu\nu}$ the calculations were published only partially. The result presented here can be obtained, as in [1, 4] for the case of j_μ, if we use formulas (4.9) and (4.10) of [5], assuming $A = \mathcal{D}(\mathcal{D}-2)$, $B = \mathcal{D}$, $C = 1$. For $\mathcal{D} \neq 2$ there is, probably, another set of tensor fields whose dimensions differ from $d + s$ by an integer.)
Only if $\ell = d + s$ in the right-hand side of the Ward identities do quasi-local (Schwinger) terms of admissible structure appear, see [1, 4] and also Sect.6.

An important feature of the conformal theory [2,1] is the fact that the fields P_s make also a contribution to the operator product of fundamental fields entering the Lagrangian, e.g.

$$L_{int} = \lambda\varphi^2\chi,$$
$$\varphi(x)\chi(x+\varepsilon) = \{\varphi(x)\} + \sum_s \{P_s(x)\} \tag{1.4}$$

where $\{\varphi(x)\} = \int dy\, Q(x, x+\varepsilon; y)\,\varphi(y)$ and $\{P_s(x)\}$

denotes the contribution of the field and of all its derivatives (details see in [1]). In the second term the fields (1.1) contribute to the Green function (as $x_{12} \to 0$)

$$G^{T}_{\mu\nu}(x_1, x_2, x_3, x_4) = \langle 0 | \varphi(x_1) \chi(x_2) \varphi(x_3) T_{\mu\nu}(x_4) | 0 \rangle \quad (1.5)$$

i.e. appear in the operator expansion of each of the products $\varphi(x) T_{\mu\nu}(x+\varepsilon)$ and $\varphi(x)\chi(x+\varepsilon)$.

Expansion (1.4) in the conformal theory allows one [1] to determine the renormalized operator products of the type $Z_1 \ \varphi(x)\chi(x)$ entering dynamic field equations. The contribution of the fields P_s is significant. Dynamic equations for the Green functions of the theory can be written as relations among matrix elements of some of the fields P_s . In simple cases [1, 2, 6] closed equations for the Green functions can be written in the form: $R(x) = 0$ or

$$\langle 0 | R(x_1) \varphi(x_2) \ldots \varphi(x_n) \chi(y_1) \ldots \chi(y_n) | 0 \rangle = 0 \quad (1.6)$$

where $R(x)$ is one of the fields P_s or a certain combination of them. This equation appears in the Thirring model where the role of $R(x)$ is playcd by a spinor field $\psi_1(x)$ of dimension $d_1 = d + 1$, arising in the expansion

$$\gamma_\mu j_\mu(x) \psi(x+\varepsilon) = a \frac{\hat{\varepsilon}}{\varepsilon^2} \psi(x) + b \hat{\partial} \psi(x) + \psi_1(x) + \ldots$$

where a, b are constants, and equation of type (1.6) appears when this expansion is substituted into the right-hand side of the field equation: $\hat{\partial} \psi(x) =$

$= \lambda \gamma_\mu j_\mu(x)\, \psi(x)$. As a result we have: $\psi_1(x) = 0$, or $\langle 0|\psi_1(x_1)\psi(x_2)\dots\bar{\psi}(x_{2n})|0\rangle = 0$.

It is convenient to rewrite it in the form specific for the conformal theory

$$\langle 0|\psi_1(x_1)\psi(x_2)\dots\bar{\psi}(x_{2n})|0\rangle = \operatorname*{res}_{\ell = d+1}$$ 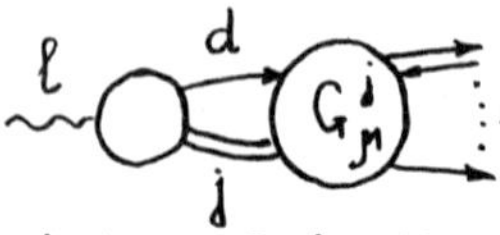 $= 0$ (1.7)

where $G^j_\mu = \langle 0|j_\mu \psi \dots \bar{\psi}|0\rangle$. The integral in the right-hand side can be easily calculated with the help of the Ward identities reducing (1.7) to a closed differential equation for the Green function $\langle 0|\psi\dots\bar{\psi}|0\rangle$ (details see in [1,6]).

A generalization of this method to more realistic models in D-dimensional space is discussed in [1,2]. As it is mentioned in [1], for $D > 3$ it looks like the equations

$$P^{d+s}_{\mu_1\dots\mu_s}(x) = 0 \tag{1.8}$$

hold for all s . In fact, the positivity gives [1]: $d_s \geqslant d+s$. On the other hand, the dimension of the fundamental field φ is bounded by the constraint [1]: $d < D/2$. It is consistent with the previous one only for $D > 3$. The theory, however, can be redefined so as to avoid the restriction $d < D/2$. (See [7] where the Thirring model is discussed in this connection; in this model d depends on the coupling constant and may take any values.) But then it follows from the positivity that either $d > D-2$ or (1.8) holds. The contribution of the fields P_s to operator products of the fundamental fields and their role in solving dynamic equations is discussed in [2]. In that paper the authors use two inequivalent representations of the conformal group of the Euclidean D-dimensional space $(D \neq 2)$, one of which can be related with a "longitudinal" part $T^{\ell}_{\mu\nu}$ of the Euclidean energy-momentum tensor, and the other one with the transverse part $T^{tz}_{\mu\nu}$ (details see in Sect.6). Further it was assumed that the total $T^{tz}_{\mu\nu} + T^{\ell}_{\mu\nu}$ is transformed as the direct sum of those representations. In a sense this assumption is complementary to the definition of the theory. Here another way to complete the definition is discussed (see Sect. 2).

It has been shown recently [3] that correlation functions at a critical point and critical indices of the two-dimensional Ising model are determined by closed equations of type (1.6). Such equations were elaborated in [3] as a consequence of infinite parameter conformal symmetry of the two-dimensional theory. Later in [8] and [9] similar equations were obtained for other statistical models. In the simplest case of the Ising model closed equations can be represented in the form, see Sect.4, $P_{\mu\nu}^{d+2}(x) = 0$. The aim of this paper is to show that this equation is a result of the dynamic equations if we appropriately define the product of quantum fields at coincident points, and it can be obtained within the frames of the approach developed in [1,2] to construct conformal solutions. We confine ourselves to calculations for the theory in 2-dimensional space. The next step consists in generalizing these results for a D-dimensional case. Some preliminary remarks on this generalization are given in Sec.6.

2. Regularization and definition of operator products of renormalized fields. Let us consider the interaction $L_{int} = \lambda' \varphi^4$ (2.1) of a massless field in the D-dimensional Euclidean space. Introducing an intermediate field $\chi = \sqrt{\lambda'}\varphi^2$, rewrite (2.1) in the form

$$L_{int} = \frac{1}{2}\lambda\varphi^2\chi - \chi^2. \tag{2.2}$$

Interactions (2.1) and (2.2) are equivalent, if $\lambda^2 = -8\lambda'^2$. If ultraviolet divergences are available in Feynman graphs, introduction of some regularization to the initial

Lagrangian is necessary for complete determining the theory. In our case the regularization with the help of higher derivatives is convenient:

$$L_{reg} = -\frac{1}{2}\left[(\partial_\mu \varphi)^2 + \alpha(\Box \varphi)^2\right] + L_{int} \tag{2.3}$$

where α is the regularization parameter. It will be tended to zero at the end of calculations. Passing to the renormalized fields

$$\varphi(x) = z_2^{1/2}\varphi_R(x),\ \chi(x) = z_3^{1/2}\chi_R(x),\quad \lambda = z_1 z_2^{-1} z_3^{-1/2}\lambda_R$$

we obtain: (2.4)

$$L_{reg} = -z_2\left\{\frac{1}{2}\left[(\partial_\mu \varphi_R)^2 + \alpha(\Box\varphi_R)^2 - \delta m^2 \varphi_R^2\right]\right\} + \frac{1}{2} z_1 \lambda_R \varphi_R^2 \chi_R - z_3 \chi_R^2$$

where δm^2 is the mass counterterm. Below only renormalized fields are considered and the index R is omitted.

Now we discuss the role of regularization in the conformal theory. The fields φ and χ in the perturbation theory have scale dimensions $D/2-1$ and $D/2$, respectively. When $D>3$ there are ultraviolet divergences that are compensated by the renormalizations of the fields and the coupling constant. At D = 3 there is only mass divergence, while at D = 2 there is no divergence at all. At $D < 4$, however, infrared divergences appear (since $m_R = 0$), and converging graphs are singular at $p^2 = 0$. It is usually supposed that conformal regime in these theories is caused by infrared behaviour and corresponds to a zero of the Gell-Mann-Low function. The fields take on anomalous dimensions that weaken infrared singularities and at the same time "change for the worse" (since $m_R = 0$) the ultraviolet behaviour. Renormalization constants tend to zero or infinity (depending on space dimension) as in the case of ultraviolet divergences. As a result, operator products in the field equations, such as

$$z_1\varphi(x)\chi(x),\quad z_1 z_2^{-1}\varphi(x)\chi(x),\quad z_1\varphi^2(x)\quad \text{etc.} \tag{2.5}$$

contain uncertainties $0\times\infty$. This situation arises as a non-perturbative effect.

Correct resolving of the uncertainties in (2.5) requires introducing a regularization independent of ultraviolet divergences in the perturbation theory. Operator products (2.5) in the conformal theory should be defined as limits of the products of the fields with separated arguments, renormalization constants functions of the argument separation. For example,

$$z_1\varphi(x)\chi(x) = \lim_{\varepsilon\to 0}\int d\Omega_\varepsilon\, z_1(\varepsilon)\,\varphi(x)\chi(x+\varepsilon), \tag{2.6}$$

where $\int d\Omega_\varepsilon$ means averaging over the directions of the vector ε_μ . The renormalization constants appear to be power functions (details see in [1])

$$z_1(\varepsilon)\sim(\varepsilon^2)^{d+\frac{\Delta}{2}-\frac{D}{2}},\quad z_2(\varepsilon)\sim(\varepsilon^2)^{d-D/2+1},\quad z_3(\varepsilon)\sim(\varepsilon^2)^{\Delta-\frac{D}{2}},\quad \delta m^2(\varepsilon)\sim\frac{1}{\varepsilon^2} \tag{2.7}$$

where d and Δ are the scale dimensions of the fields φ and χ , respectively. It is shown below that the parameter ε_μ is expressed through the regularization parameter α and tends to zero together with it: $\varepsilon^2\sim\alpha$. This also can be seen from dimension considerations. Therefore for $\alpha\neq 0$ we can substitute

$$z_1\varphi(x)\chi(x)\to\int d\Omega_\varepsilon\, z_1(\varepsilon)\,\varphi(x)\chi(x+\varepsilon) \tag{2.8}$$

where $\varepsilon^2\sim\alpha$, and instead of the operator field equation

$$z_2\left[-\delta m^2\varphi(x)+\Box\varphi(x)\right]=\lambda z_1\varphi(x)\chi(x) \tag{2.9}$$

consider the regularized equation

$$z_2(\varepsilon)\left[-\delta m^2(\varepsilon)\,\varphi(x)+\Box\varphi(x)+\alpha(\varepsilon)\Box^2\varphi(x)\right]=$$

$$=\lambda z_1(\varepsilon)\int d\Omega_\varepsilon \varphi(x)\chi(x+\varepsilon). \tag{2.10}$$

It is more convenient to consider vector ε_μ as an independent parameter, and consider the regularization parameter α as a function of ε_μ:

$$\alpha(\varepsilon)=a_2\varepsilon^2 \tag{2.11}$$

where a_2 is a constant. In this case removing regularization can be achieved by passing to the limit $\varepsilon^2 \to 0$. This is assumed everywhere below.

Introducing regularization allows to formulate more consistently statements of [1] according to which some summands in the left-hand side of (2.9) should be considered as the first terms of expansion of the operator product $\int d\Omega_\varepsilon \varphi(x)\chi(x+\varepsilon)$ in the powers of ε_μ. In fact, substituting (2.7) into the regularized equation (2.10) we can rewrite (2.10) in the form:

$$\lambda\int d\Omega_\varepsilon \varphi(x)\chi(x+\varepsilon)\sim(\varepsilon^2)^{-\frac{\Delta}{2}+1}\left\{\frac{a_0}{\varepsilon^2}\varphi(x)+\Box\varphi(x)+a_2\varepsilon^{2l}\Box^{l}\varphi(x)\right\}. \tag{2.12}$$

The role of regularization, however, is not exhausted by this fact. Eq. (2.9) appears to be very sensitive to the term $\alpha\Box\varphi$: if from the very beginning we put $\alpha=0$ in (2.10) then it is impossible to obtain an unambiguous solution (details see in [1]), while introducing this term with subsequent passing to the limit $\alpha\to 0$ singles out a unique solution at least in case D = 2 considered below.

Thus the regularization should be considered as a part of the definition of operator products (2.5), that makes the field equations more meaningful (see the end of Sect.3) and does not contradict conformal invariance.

3. Contribution of $P^{d+2}_{\mu\nu}$. Let us consider the Green function

$$G^T_{\mu\nu}(x_1x_2x_3x_4)=\langle 0|\varphi(x_1)\chi(x_2)\varphi(x_3)T_{\mu\nu}(x_4)|0\rangle. \tag{3.1}$$

Its "longitudinal" part can be determined from the Ward identity (which in general is defined by the formulas from Sect.6) and the contribution of the fields P_s can be calculated. At D = 2 it can be unambiguously determined by the Ward identity

$$\overset{x_4}{\partial_\nu} G^{T}_{\mu\nu}(x_1x_2x_3x_4) = -\Big\{\delta(x_{14})\overset{x_1}{\partial_\mu} + \delta(x_{24})\overset{x_2}{\partial_\mu} + \delta(x_{34})\overset{x_3}{\partial_\mu} - $$

$$-\overset{x_4}{\partial_\mu}\Big[\frac{d}{D}\,\delta(x_{14}) + \frac{d}{D}\,\delta(x_{24}) + \frac{\Delta}{D}\,\delta(x_{34})\Big]\Big\} G_3(x_1x_2x_3) \qquad (3.2)$$

where

$$G_3(x_1x_2x_3) = \langle 0|\varphi(x_1)\varphi(x_2)\chi(x_3)|0\rangle = g\,(x_{12}^2)^{-d+\frac{\Delta}{2}}(x_{13}^2x_{23}^2)^{-\frac{\Delta}{2}}. \qquad (3.3)$$

We are interested in the contribution of the symmetrical traceless tensor $P^{d+2}_{\mu\nu}$ since it contributes in the same order in ε_μ as $\varepsilon^2\Box^2\varphi$ in (2.12) does. In two-dimensional space only the field $P^{d+2}_{\mu\nu}$ contributes in this order. For D ≠ 2, as stated above, other fields can also exist. Then the scheme shown below can be easily generalized.

Let us consider the operator product $\varphi(x)\chi(x+\varepsilon)$ with regard to the field $P^{d+2}_{\mu\nu}(x)$:

$$\varphi(x)\chi(x+\varepsilon) \sim (\varepsilon^2)^{-\frac{\Delta}{2}+1}\Big\{\alpha_0\frac{1}{\varepsilon^2}\varphi(x) + \alpha_1\frac{\varepsilon_\mu}{\varepsilon^2}\partial_\mu\varphi(x) +$$

$$+\Big[\alpha_2'\Box\varphi(x) + \alpha_2''\frac{\varepsilon_\mu\varepsilon_\nu}{\varepsilon^2}\partial_\mu\partial_\nu\varphi(x) + \beta_0\frac{\varepsilon_\mu\varepsilon_\nu}{\varepsilon^2}P^{d+2}_{\mu\nu}(x) + \dots\Big] +$$

$$+\Big[\alpha_3'\varepsilon_\mu\partial_\mu\Box\varphi(x) + \beta_1'\varepsilon_\mu\partial_\nu P^{d+2}_{\mu\nu}(x) + \dots\Big] + \qquad (3.4)$$

$$+\varepsilon^2\Big[\alpha_4'\Box^2\varphi(x) + \beta_2'\partial_\mu\partial_\nu P^{d+2}_{\mu\nu}(x) + \alpha_4''\frac{\varepsilon_\mu\varepsilon_\nu}{\varepsilon^2}\partial_\mu\partial_\nu\Box\varphi(x) +$$

$$+\beta_2''\frac{\varepsilon_\mu\varepsilon_\nu}{\varepsilon^2}\partial_\mu\partial_\rho P^{d+2}_{\nu\rho}(x) + \dots\Big]\Big\} + \dots$$

where constant coefficients are determined from conformal invariance. Averaging this expression in directions of the vector ε_μ we have

$$\int d\Omega_\varepsilon \varphi(x)\chi(x+\varepsilon) \sim (\varepsilon^2)^{-\frac{\Delta}{2}+1}\Big\{\alpha_0 \frac{1}{\varepsilon^2}\varphi(x) + \alpha_2 \Box \varphi(x) + $$
$$+\varepsilon^2\big[\alpha_4 \Box^2 \varphi(x) + \beta_2 \partial_\mu \partial_\nu P^{d+2}_{\mu\nu}(x)\big] + \ldots\Big\}. \tag{3.5}$$

In (3.4) we did not take into account the terms of the type

$$(\varepsilon^2)^{-\frac{\Delta}{2}+1}\Big\{\frac{\varepsilon_\mu\varepsilon_\nu\varepsilon_\rho}{(\varepsilon^2)^2} P^{d+3}_{\mu\nu\rho}(x) + \frac{\varepsilon_\mu\varepsilon_\nu}{\varepsilon^2}\partial_\rho P^{d+3}_{\mu\nu\rho}(x) + \ldots$$
$$\ldots + \frac{\varepsilon_\mu\varepsilon_\nu\varepsilon_\rho\varepsilon_\tau}{(\varepsilon^2)^2} P^{d+4}_{\mu\nu\rho\tau}(x)\Big\}. \tag{3.6}$$

All of them disappear on averaging in the directions of ε_μ because of tracelessness of the tensors, and do not contribute to (3.5).

Expansion (3.5) obtained from the solution of the Ward identity should coincide with expansion (2.12), that has been obtained from the regularized field equation. Equating (2.12) to (3.5) we find: $\delta m^2(\delta) = \frac{\alpha_0}{\varepsilon^2}$, $\alpha_2 z_2(\varepsilon) = (\varepsilon^2)^{d-D/2+1}$ and also

$$(\alpha_4 - a_2)\Box^2 \varphi(x) + \beta_2 \partial_\mu \partial_\nu P^{d+2}_{\mu\nu}(x) = 0 .$$

This is an operator equation. Verifying it for the conformal 3-point functions $\langle 0|\varphi\varphi\chi|0\rangle$ and $\langle 0|P_{\mu\nu}\varphi\chi|0\rangle$ we can show that it holds only if $\alpha_4 = a_2$ and $\partial_\mu \partial_\nu P^{d+2}_{\mu\nu}(x) = 0$. This implies the operator equation

$$P^{d+2}_{\mu\nu}(x) = 0. \tag{3.7}$$

It must be true for any matrix elements

$$\langle 0|P^{d+2}_{\mu\nu}(x_1)\varphi(x_2)\ldots\varphi(x_n)|0\rangle = 0 \tag{3.8}$$

and is a source of closed equations for the Green functions $\langle 0|\varphi(x_1)\dots\varphi(x_n)|0\rangle$.

As it was said at the end of the previous Section, introducing regularization is significant for our method. If we put from the very beginning $\alpha(\varepsilon)=0$ in (2.10) then there is no reason to take into account the terms $\sim\varepsilon^2$ in the brackets in (2.12) and (3.5), hence equation (3.7) is absent. (It would be possible to demand cancellation of the contributions $\sim\varepsilon^2$ at $\alpha=0$. This would lead to equation $\alpha_4\Box^2\varphi+\beta_2\partial_\mu\partial_\nu P^{d+2}_{\mu\nu}=0$. But as we have mentioned this equation is inconsistent with conformal invariance for 3-point Green functions.) Thus the possibility to derive some information from the existence of $P^{d+2}_{\mu\nu}$ is lost. Partially this is caused by averaging in directions of vector ε_μ defining the operator product (2.6). The point is that the field $P^{d+2}_{\mu\nu}$ appears even in zero order in ε_μ in brackets in (3.5) and makes the same contribution as the term $\Box\varphi$. This contribution, however, disappears after averaging in the ε_μ-directions. Thus, introducing the term $\alpha\Box^2\varphi$ can be considered as a way to compensate the information loss which is inevitably caused by averaging in directions of ε_μ. The necessity of averaging is caused by desire to retain Lorentz invariance (in Euclidean space the invariance with respect to rotations). It is possible to refuse averaging and Lorentz invariance of the Lagrangian (introducing in this case the counterterms $\varphi\partial_\mu\varphi$ and $\partial_\mu\varphi\partial_\nu\varphi$) if we preserve the conformal invariance of solution. Then we would also come to equation (2.7) without introducing $\alpha\Box^2\varphi$. And what is more, in this case introducing $\alpha\Box^2\varphi$ causes contradiction, since the terms of (3.6) should vanish, $P^{d+3}_{\mu\nu\rho}(x)=P^{d+4}_{\mu\nu\rho\tau}(x)=0$, which contradicts (3.7).

4. Equations in two-dimensional space. A feature of two-dimensional theories is the fact that the Ward identities unambiguously determine the Green functions comprising the conserved quantities j_μ and $T_{\mu\nu}$ through the Green functions of fundamental fields. In case of the energy-momentum tensor we have

$$T_{\mu\nu}(x) = \partial_\mu T_\nu(x) + \partial_\nu T_\mu(x) - \delta_{\mu\nu}\partial_\rho T_\rho(x) \tag{4.1}$$

where $T_\mu(x) = \frac{1}{\Box}\partial_\nu T_{\mu\nu}(x)$. These relations are true for any traceless symmetric tensor in two-dimensional space. Substituting (3.2) at D = 2 into (4.1) and using the identity $\frac{1}{\Box}\delta(x) = -\frac{1}{4\pi}\ln x^2$ we obtain for the Green function (3.1):

$$\begin{aligned} G^T_{\mu\nu}(x_1x_2x_3x_4) &= \frac{1}{2\pi}\Big\{d\frac{1}{x^2_{14}}g_{\mu\nu}(x_{14}) + d\frac{1}{x^2_{24}}g_{\mu\nu}(x_{24}) + \\ &+ \Delta\frac{1}{x^2_{34}}g_{\mu\nu}(x_{34}) + \sum_{i=1}^{3}\frac{1}{x^2_{i3}}\Big[(x_{i3})_\mu\partial^{x_i}_\nu + (x_{i3})_\nu\partial^{x_i}_\mu - \\ &- \delta_{\mu\nu}(x_{i3})_\lambda\partial^{x_i}_\lambda\Big]\Big\}G_3(x_1x_2x_3), \end{aligned} \tag{4.2}$$

where $g_{\mu\nu}(x) = \delta_{\mu\nu} - 2\frac{x_\mu x_\nu}{x^2}$. Besides, we have

$$\begin{aligned} G^T_{\mu\nu}(x_1x_2x_3) &= \langle 0|\varphi(x_1)\varphi(x_2)T_{\mu\nu}(x_3)|0\rangle = \\ &= \frac{1}{2\pi}\Big\{d\Big[\frac{1}{x^2_{13}}g_{\mu\nu}(x_{13}) + \frac{1}{x^2_{23}}g_{\mu\nu}(x_{23})\Big] + \\ &+ \sum_{i=1}^{2}\frac{1}{x^2_{i3}}\Big[(x_{i3})_\mu\partial^{x_i}_\nu + (x_{i3})_\nu\partial^{x_i}_\mu - \delta_{\mu\nu}(x_{i3})_\lambda\partial^{x_i}_\lambda\Big]\Big\}G(x_{12}) \end{aligned} \tag{4.3}$$

where

$$G(x_{12}) = (x^2_{12})^{-d}. \tag{4.4}$$

Let us consider the regularized equation for the Green function $\langle 0|\varphi\varphi T_{\mu\nu}|0\rangle$:

$$z_2(\varepsilon)\{-\delta m^2(\varepsilon) + \Box_{x_1} + \alpha(\varepsilon)\Box^2_{x_1}\}\langle 0|\varphi(x_1)\varphi(x_2)T_{\mu\nu}(x_3)|0\rangle =$$

$$=(\dots)+\lambda z_1(\varepsilon)\int d\Omega_\varepsilon \langle 0|\varphi(x_1)\chi(x_1+\varepsilon)\varphi(x_2)T_{\mu\nu}(x_3)|0\rangle \quad (4.5)$$

where (...) denotes quasi-local terms that at D = 2 are of no interest and will not be considered. According to Sect.3 this equation is equivalent to an equation of the type (3.7), which in our case is of the form

$$\langle 0|P^{d+2}_{\sigma\tau}(x_1)\varphi(x_2)T_{\mu\nu}(x_3)|0\rangle = 0 . \quad (4.6)$$

This Green function is determined according to (3.5) by the equation:

$$\partial^{x_1}_\sigma \partial^{x_1}_\tau \langle 0|P^{d+2}_{\sigma\tau}(x_1)\varphi(x_2)T_{\mu\nu}(x_3)|0\rangle \sim$$

$$\sim\lambda \lim_{\varepsilon\to 0}\Big\{z_1(\varepsilon)z_2^{-1}(\varepsilon)\frac{1}{\varepsilon^2}\langle 0|\varphi(x_1)\chi(x_1+\varepsilon)\varphi(x_2)T_{\mu\nu}(x_3)|0\rangle$$

$$-\Big[\frac{a_0}{(\varepsilon^2)^2}+\frac{1}{\varepsilon^2}\Box_{x_1}+a_2\Box^2_{x_1}\Big]\langle 0|\varphi(x_1)\varphi(x_2)T_{\mu\nu}(x_3)|0\rangle\Big\}. \quad (4.7)$$

Substituting expressions (4.2) and (3.3) into (4.5) or, which is the same, into (4.6) we obtain after simple calculations with the help of (4.7)

$$d=\frac{3}{8}\Delta-\frac{1}{4}. \quad (4.8)$$

In the calculations it is convenient to use the variables

$$x^{\pm}=x_1\pm ix_2,\ \partial_\pm=\frac{1}{2}(\partial_1\mp i\partial_2),\ T_\pm=T_{11}\pm iT_{12},\ P_\pm=P_{11}\pm iP_{12},$$

$$\frac{1}{x^2}[g_{11}(x)\pm ig_{12}(x)]=-\frac{1}{(x^\pm)^2},$$

$$\frac{1}{x^2}(x_\mu\partial_\nu+x_\nu\partial_\mu-\delta_{\mu\nu}x_\lambda\partial_\lambda)\to\frac{1}{2}\frac{1}{x^\pm}\partial_\pm . \quad (4.9)$$

Now let us consider eq. (3.8). Similarly to (3.13) we have (details see in [1])

$$\langle 0|P^{d+2}_{\mu\nu}(x_1)\varphi(x_2)\varphi(x_3)\varphi(x_4)|0\rangle = \underset{\ell=d+2}{\mathrm{res}} \; \overset{x_1}{\sim\!\sim}\bigcirc\!=\!\left(G^T_{\mu\nu}\right)\!\begin{smallmatrix} x_2 \\ x_3 \\ x_4 \end{smallmatrix} \tag{4.10}$$

where

$$G^T_{\mu\nu} = \langle 0|T_{\mu\nu}(x_1)\varphi(x_2)\varphi(x_3)\varphi(x_4)\varphi(x_5)|0\rangle \tag{4.11}$$

$$\overset{\ell}{\sim\!\sim}(C)\!\!< \; = \langle 0|P^{\ell}_{\mu\nu}(x_1)\tilde{\varphi}^{d}(x_2)\tilde{T}_{\sigma\tau}(x_3)|0\rangle \sim \langle 0|P^{\ell}_{\mu\nu}\varphi^{2-d}T_{\sigma\tau}|0\rangle \tag{4.12}$$

is the conformal-invariant 3-point function. The point in the line and the tilde $(\tilde{\varphi}, \tilde{T}_{\sigma\tau})$ indicate amputation of the corresponding leg. Integration is ment over internal lines. An expression for the function (4.12) is given in [5]. It comprises three different structures. But in two-dimensional space only two of them are independent and the pole in (4.9) occurs in one of them only. In variables (4.5) we have

$$\overset{\ell}{\sim\!\sim}(C)\!=\!\left(G^T_{\mu\nu}\right)\!= \; = \int dx_6^+ dx_6^- dx_7^+ dx_7^- \langle 0|P^{\ell}_{\pm}(x_1)\varphi^{2-d}(x_6)\tilde{T}_{\mp}(x_7)|0\rangle \tag{4.13}$$

$$\langle 0|T_{\pm}(x_7)\varphi(x_6)\varphi(x_3)\varphi(x_4)\varphi(x_5)|0\rangle$$

where

$$\langle 0|P^{\ell}_{\pm}(x_1)\varphi^{2-d}(x_6)\tilde{T}_{\mp}(x_7)|0\rangle \sim \left[\left(\partial^{x_6}_{\mp}\right)^3 \frac{1}{(x^2_{16})^{\frac{\ell-d}{2}}}\right] \frac{(x^{\mp}_{67})^{\frac{\ell+d}{2}+1}}{(x^{\mp}_{17})^{\frac{\ell+d}{2}-1}} \partial^{x_7}_{\pm}\left(\frac{x^{\pm}_{67}}{x^{\pm}_{17}}\right)^{\frac{\ell+d}{2}-2}. \tag{4.14}$$

The component $\langle 0|P^{\ell}_{\pm}\varphi^{2-d}\tilde{T}_{\pm}|0\rangle$ makes no contribution to

(4.13) since it is the same for both independent structures, and such combination of them can be chosen that the pole contributions cancel.

To calculate integral (4.13) we use the Ward identity that in variables (4.9) is of the form

$$\partial_{\pm}^{x_7}\langle 0|T_{\pm}(x_7)\varphi(x_1)\varphi(x_2)\varphi(x_3)\varphi(x_4)|0\rangle = \tag{4.15}$$

$$= -\left\{\sum_{i=1}^{4}\delta(x_{i7})\partial_{\mp}^{x_i} - \frac{d}{2}\partial_{\mp}^{x_7}\sum_{i=1}^{4}\delta(x_{i7})\right\}\langle 0|\varphi(x_1)\varphi(x_2)\varphi(x_3)\varphi(x_4)|0\rangle .$$

Let us substitute (4.14) into (4.13) and integrate by parts over x_7^+. As a result the integral over x_7^+ and x_7^- is evaluated thanks to δ-functions in the Ward identity (4.15). Now we take the residue of the remained expression at $\ell = d+2$. The pole occurs owing to the factor $(x_{16}^2)^{\frac{\ell-d}{2}}$ in (4.14); we have [10]

$$\underset{\ell=d+2}{\mathrm{res}}\ \frac{1}{(x_{16}^2)^{\frac{\ell-d}{2}}} = -2\pi\delta(x_{16}).$$

The remained integral over x_6 is evaluated thanks to $\delta(x_{16})$ and we find

$$\langle 0|P_{\pm}^{d+2}(x_1)\varphi(x_2)\varphi(x_3)\varphi(x_4)|0\rangle \sim d(d+1)(d+2)$$

$$\left\{\frac{3}{2(d+1)}(\partial_{\pm}^{x_1})^2 - \sum_{i=2}^{4}\frac{1}{x_{1i}^{\pm}}\partial_{\pm}^{x_i} - \frac{d}{2}\sum_{i=2}^{4}\frac{1}{(x_{1i}^{\pm})^2}\right\}\langle d\varphi(x_1)\varphi(x_2)\varphi(x_3)\varphi(x_4)|0\rangle .$$

Assuming $P_{\pm}^{d+2} = 0$ we obtain the equation of paper [3]

$$\left\{\frac{3}{2(d+1)}(\partial_{\pm}^{x_1})^2 - \sum_{i=2}^{4}\frac{1}{x_{1i}^{\pm}}\partial_{\pm}^{x_i} - \frac{d}{2}\sum_{i=2}^{4}\frac{1}{(x_{1i}^{\pm})^2}\right\}G_4(x_1x_2x_3x_4) = 0 \tag{4.16}$$

Similarly, a closed equation for the Green function

$$\tilde{G}_4(x_1x_2x_3x_4) = \langle 0|\varphi(x_1)\chi(x_2)\varphi(x_3)\chi(x_4)|0\rangle$$

is obtained from the condition:

$$\langle 0|P^{d+2}_{\mu\nu}(x_1)\chi(x_2)\varphi(x_3)\chi(x_4)|0\rangle = 0$$

and is of the form

$$\Big\{\frac{3}{2(d+1)}(\partial^{x_1}_{\pm})^2 - \sum_{i=2}^{4}\frac{1}{x^{\pm}_{1i}}\partial^{x_i}_{\pm} - \frac{d}{2}\frac{1}{(x^{\pm}_{13})^2} - \tag{4.17}$$
$$-\frac{\Delta}{2}\Big[\frac{1}{(x^{\pm}_{12})^2}+\frac{1}{(x^{\pm}_{14})^2}\Big]\Big\}\tilde{G}_4(x_1x_2x_3x_4) = 0$$

Its most general solution is of the form (see also [3,9])

$$\tilde{G}_4(x_1x_2x_3x_4) = \frac{1}{(x^2_{13})^d(x^2_{24})^\Delta}\left[\xi_+\xi_-(1-\xi_+)(1-\xi_-)\right]^{-\mu}\times$$

$$\times\{a_{11}G^+_1G^-_1 + a_{12}G^+_1G^-_2 + a_{21}G^+_2G^-_1 + a_{22}G^+_2G^-_2\} \tag{4.18}$$

where

$$G^{\pm}_1 = F(\alpha,\beta;\gamma;\xi_{\pm})$$

$$G^{\pm}_2 = [\xi_{\pm}(1-\xi_{\pm})]^{1-\gamma}F(1-\alpha,1-\beta;2-\gamma;\xi_{\pm})$$

$$\xi_{\pm} = \frac{x^{\pm}_{12}x^{\pm}_{34}}{x^{\pm}_{13}x^{\pm}_{24}}$$

$F(\alpha,\beta;\gamma;\xi)$ is the hypergeometric function. If d and Δ are arbitrary the parameters are determined by the equations

$$\mu^2 + \frac{1}{3}(1-2d)\mu - \frac{1}{3}\Delta(d+1) = 0 \tag{4.19}$$

$$\gamma = \frac{2}{3}(d+1) - 2\mu,\ \alpha\beta = 2\mu\left(2\mu - \frac{4}{3}d - \frac{1}{3}\right),\ \alpha+\beta+1 = 2\gamma.$$

Let us introduce instead of Δ a new free parameter h:

$$\Delta = \frac{1}{3(d+1)}\left[h^2 - \left(d-\frac{1}{2}\right)^2\right].$$

From (4.19), (4.20) we have

$$\mu = \frac{1}{3}\left[\left(d-\frac{1}{2}\right)+h\right] \tag{4.20}$$

$$\alpha = -\frac{2}{3}\left(d-\frac{1}{2}+h\right),\ \beta = +\frac{2}{3}(d+1-h),\ \gamma = 1-\frac{2}{3}h. \tag{4.21}$$

If here we put $\Delta = d$ then a general solution of equation (4.16) is obtained. Assuming $\Delta = \frac{2}{3}(4d+1)$, see (4.8), we obtain a solution of (4.17). In the first case: $h_1 = (2d+\frac{1}{2}),\ \mu_1 = d$ implying

$$G_4(x_1x_2x_2x_4) = \frac{1}{(x_{13}^2 x_{24}^2)^d}\left[\xi_+\xi_-(1-\xi_+)(1-\xi_-)\right]^{-d}\Big\{\left|F(\alpha_1,\beta_1;\gamma_1;\xi_+)\right|^2 -$$

$$-\left|\xi_+\xi_-(1-\xi_+)(1-\xi_-)\right|^{1-\gamma_1}\frac{\Gamma^2(\gamma_1)}{\Gamma^2(2-\gamma_1)}\frac{\Gamma(1-\alpha_1)\Gamma(1-\beta_1)}{\Gamma(\alpha_1)\Gamma(\beta_1)}\left|F(1-\alpha_1,1-\beta_1;2-\gamma_1;\xi_+)\right|^2\Big\}$$

where $\alpha_1 = -2d,\ \beta_1 = -\frac{1}{3}(2d-1),\ \gamma_1 = -\frac{2}{3}(2d-1).$ (4.22)

In the second case we find: $h_2 = 3(d+\frac{1}{2}),\ \mu_2 = \frac{\Delta}{2} = \frac{1}{3}(4d+1),$

$$\tilde{G}_4(x_1x_2x_3x_4) = \frac{\Gamma(\alpha_2)\Gamma(\beta_2)}{\Gamma^2(\gamma_2)}\frac{\Gamma^2(\gamma_2-\alpha_2)}{\Gamma^2(\beta_2-\alpha_2)}\frac{\Gamma(1-\alpha_2)\Gamma(\beta_2)}{\{\Gamma(\alpha_2)\Gamma(1-\alpha_2)-\Gamma(\beta_2)\Gamma(1-\beta_2)\}}$$

$$-\frac{1}{(x_{13}^2)^d(x_{24}^2)^\Delta}\left[\xi_+\xi_-(1-\xi_+)(1-\xi_-)\right]^{-\frac{\Delta}{2}}\Big\{\left|F(\alpha_2,\beta_2;\gamma_2;\xi_+)\right|^2 - \tag{4.23}$$

$$-\left[\xi_+\xi_-(1-\xi_+)(1-\xi_-)\right]^{1-\gamma_2}\frac{\Gamma^2(\gamma_2)}{\Gamma^2(2-\gamma_2)}\frac{\Gamma(1-\alpha_2)\Gamma(1-\beta_2)}{\Gamma(\alpha_2)\Gamma(\beta_2)}\left|F(1-\alpha_2,1-\beta_2;2-\gamma_2;\xi_+)\right|^2\Big\}$$

where $\alpha_2 = -\Delta,\ \beta_2 = -\frac{1}{2}\Delta,\ \gamma_2 = -2d.$ (4.24)

These expressions satisfy locality and asymptotic conditions

$$G_4(x_1 x_2 x_3 x_4)\big|_{x_{12}\to 0} \simeq G(x_{12})G(x_{34}) \tag{4.25}$$

$$\tilde{G}_4(x_1 x_2 x_3 x_4)\big|_{x_{13}\to 0} \simeq G(x_{13})D(x_{24}) \tag{4.26}$$

where $D(x_{24}) = \langle 0|\chi(x_2)\chi(x_4)|0\rangle$.

5. <u>Equation for intermediate field</u>. One of the conditions of conformal bootstrap for interaction (2.2) is of the form (see [1, 11, 12] and references therein)

$$\text{---}\overset{\chi}{\text{---}}(G_3)\genfrac{}{}{0pt}{}{\varphi}{\varphi} = \underset{\ell=d}{\operatorname{res}} \overset{\ell}{\sim\!\sim}(C)\text{===}(G_4) \tag{5.1}$$

where G_3 and G_4 are Green functions (3.3) and (3.17), and the function C is obtained from (3.3) by substituting $\Delta\to\ell$. The functions G_3 and G_4 are known, therefore (5.1) relates d, Δ and the "coupling" g. The condition relating d and Δ in (5.1) appears from the requirement that for $\ell=\Delta$ the right-hand side has a pole.

Eq. (5.1) means that the operator expansion of the product $\varphi(x_1)\varphi(x_2)$ comprises the field $\chi(x)$ with dimension Δ. In practice it is convenient to write it in different form [1]. According to (2.4) we have

$$\chi(x) = \frac{1}{4}\lambda z_1 z_3^{-1}\varphi^2(x). \tag{5.2}$$

Let us determine operator products in the right-hand side as in Sect.1, separating the arguments:

$$z_1 z_3^{-1}\varphi^2(x) = \lim_{\varepsilon\to 0}\int d\Omega_\varepsilon\, z_1(\varepsilon) z_3^{-1}(\varepsilon)\left[\varphi(x)\varphi(x+\varepsilon) - \frac{1}{(\varepsilon^2)^d} I\right]. \tag{5.3}$$

In the right-hand side a subtraction term $\frac{1}{(\varepsilon^2)^d} I$ is introduced, where I is the unit operator. This is necessary for $\langle 0|\varphi^2(x)|0\rangle = 0$. The point is that the disconnected part of the Green function $\langle 0|\varphi(x_1)\varphi(x_1 +$

$+\varepsilon)\varphi(x_2)\varphi(x_3)|0\rangle$ contains a term $\langle 0|\varphi(x_1)\varphi(x_1+\varepsilon)|0\rangle$ $\langle d\varphi(x_2)\varphi(x_3)|0\rangle$. Therefore operator expansion of identical fields $\varphi(x_1)\varphi(x_2)$ always begins with the unit operator (which is absent in $\varphi(x_1)\chi(x_2)$). In the perturbation theory subtraction of the unit operator occurs in passing to the normal product in (5.2):

$\varphi(x)\varphi(x)\rightarrow :\varphi(x)\varphi(x):$

For the Green functions G_3 and G_4 we have from (5.2), (4.3) and (2.6a):

$$\langle 0|\chi(x_1)\varphi(x_2)\varphi(x_3)|0\rangle \sim \lim_{\varepsilon\to 0}\int d\Omega_\varepsilon (\varepsilon^2)^{d-\frac{\Delta}{2}} \tag{5.4}$$

$$\left\{\langle 0|\varphi(x_1)\varphi(x_1+\varepsilon)\varphi(x_2)\varphi(x_3)|0\rangle - \frac{1}{(\varepsilon^2)^d}\langle 0|\varphi(x_2)\varphi(x_3)|0\rangle\right\}.$$

Let us substitute here expression (4.18) at $d=\Delta$. The first term makes the contribution $(\varepsilon^2)^{-d}\langle 0|\varphi(x_3)\varphi(x_4)|0\rangle$ and is compensated by subtraction in (5.4). The second and third terms contribute $\sim(\varepsilon^{\pm})^{\frac{d+1}{3}}(\varepsilon^{\mp})^{-d}$. Multiplying by $z_1 z_3^{-1}\sim(\varepsilon^2)^{d-\frac{\Delta}{2}}=(\varepsilon_+\varepsilon_-)^{d-\frac{\Delta}{2}}$ in (5.3) they remain singular $\sim(\varepsilon^-)^{-\frac{\Delta}{2}}, (\varepsilon^+)^{-\Delta/2}$ as $\varepsilon_\pm\to 0$. So it is necessary to put $\alpha_{12}=\alpha_{21}=0$ in (4.18). Finally, the fourth term in (4.18) contributes $\sim(\varepsilon^2)^{\frac{d+1}{3}}(x_{13}^2 x_{24}^2)^{-\frac{4d+1}{3}}$. For the limit (5.4) to be finite set $d-\frac{\Delta}{2}=-\frac{d+1}{3}$. This again gives condition (4.8): $d=\frac{3}{8}\Delta-1/4$ and the limit (5.4) coincides with the three-point function (3.3)(up to a multiple).

From (5.3) the "coupling constant" g entering (3.3) can also be calculated. For this we consider the renormalized equation for the propagator $\langle 0|\chi(x_1)\chi(x_2)|0\rangle$ and find the proportionality coefficients in the formulas $z_1\sim(\varepsilon^2)^{d+\frac{\Delta}{2}-\frac{D}{2}}$ and $z_3\sim(\varepsilon^2)^{\Delta-D/2}$, see [1]. It can be shown that in the normalization adopted here $\frac{1}{4}\lambda z_1(\varepsilon)z_3^{-1}(\varepsilon)g=(\varepsilon^2)^{d-\frac{\Delta}{2}}$. Substituting this into (4.4) we get

$$g^2=-\frac{\Gamma^2(\gamma_1)\Gamma(1-\alpha_1)\Gamma(1-\beta_1)}{\Gamma(\alpha_1)\Gamma(\beta_1)\Gamma^2(2-\gamma_1)}\,. \qquad (5.5)$$

This constant can be calculated also from the equation $\varphi(x)=\lambda z_1 z_2^{-1}(\delta m^2)^{-1}\varphi(x)\chi(x)$. In this case expression (4.23) for the Green function $\tilde{G}_4$ and the equation $\frac{1}{4}\lambda z_1(\varepsilon)z_2^{-1}(\varepsilon)[\delta m^2(\varepsilon)]^{-1}=(\varepsilon^2)^{d-\frac{\Delta}{2}}$ are used. As a result

$$g^2=\frac{\Gamma(\alpha_2)\Gamma(\beta_2)\Gamma^2(\gamma_2-\alpha_2)\Gamma(1-\alpha_2)\Gamma(1-\beta_2)}{\Gamma^2(\gamma_2)\Gamma^2(\beta_2-\alpha_2)\{\Gamma(\alpha_2)\Gamma(1-\alpha_2)-\Gamma(\beta_2)\Gamma(1-\beta_2)\}}\,.$$

It can be easily verified substituting (4.22) and (4.24) that this expression coincides with (5.5) if d and Δ are related by (4.8).

Emphasize that in each of these two versions of the calculation the use of the asymptotic condition is significant: (4.25) in the former case and (4.26) in the latter.

Concluding discuss the problem of regularization required for resolving uncertainty in the terms $z_1\varphi^2(x)$, $z_1 z_3^{-1}\varphi^2(x)$. It is obvious that addition of the term $\sim\tilde{\alpha}(\partial_\mu\chi)^2$, where $\tilde{\alpha}\to 0$, to Lagrangian (2.4) will give no new constraints. This indicates that the theory is not defined well enough. In the perturbation theory an analogous situation occurs when introducing a given regularization does not eliminate all divergences. To completely define the theory in this case an additional regularization is required consisting e.g. in adding a term with a higher power of derivative. As it was explained in Sect.2 for the conformal theory it is impossible to indicate beforehand which terms are required, since the power of divergence of integrals in exact equations changes as compared with the perturbation theory. Therefore, we assume that in exact equations as many terms should be added to the initial Lagrangian as it is required to single out an unambiguous solution. In the given case the two terms are needed: $\alpha_0(\partial_\mu \chi)^2+\alpha_1(\Box\chi)^2$. Acting as in Sects. 2,3 we can show that the field $\partial_\mu\partial_\nu \tilde{P}^{\Delta+2}_{\mu\nu}$ enters the operator expansion of the product $\varphi(x)\varphi(x+\varepsilon)$ in the same order in ε as the term $\Box^2\chi$. This can be seen by expansion of the Green function

$$\langle 0|\varphi(x_1)\varphi(x_1+\varepsilon)\chi(x_2)T_{\mu\nu}(x_3)|0\rangle$$

in a series in powers of ε_μ. Let us calculate from this expansion the Green function $\langle 0|\tilde{P}^{\Delta+2}_{\tau\lambda}(x_1)\chi(x_2)T_{\mu\nu}(x_3)|0\rangle$ and equate it to zero. This gives a second condition on the scale dimensions, see also [9a]

$$d=\frac{1}{4}\,\frac{\Delta(2-\Delta)}{\Delta+1}.$$

Solving this equation together with (3.8) we obtain values of dimensions in the Ising model : $d=\frac{1}{8}$, $\Delta=1$. For higher Green functions additional equations are obtained

$$\langle 0|\tilde{P}^{\Delta+2}_{\mu\nu}(x_1)\varphi(x_2)\ldots\varphi(x_n)\chi(y_1)\ldots\chi(y_m)|0\rangle=0$$

which coincide, as can be easily seen, with the equations in [3, 9a].

6. Conclusions. As it was mentioned in Introduction there are two unequivalent representations of the conformal group of D-dimensional Euclidean space, corresponding to the energy-momentum tensor. One of them acts on the space of "longitudinal" tensor functions $T^{\ell}_{\mu\nu}$. The special conformal transformations in this representation are nonlocal. The other representations act on transverse functions.

These representations are constructed with the help of the following projections [2,5]

$$P^{\ell}_{\mu\nu,\tau\lambda}=\partial_\mu P_{\nu,\tau\lambda}+\partial_\nu P_{\mu,\tau\lambda}-\frac{2}{D}\delta_{\mu\nu}\partial_\rho P_{\rho,\tau\lambda} \qquad (6.1)$$

where

$$P_{\mu,\tau\lambda} = -\frac{D-2}{2(D-1)}\frac{\partial_\mu\partial_\tau\partial_\lambda}{\Box^2} + \frac{1}{2}\left(\delta_{\mu\lambda}\frac{\partial_\tau}{\Box} + \delta_{\mu\tau}\frac{\partial_\lambda}{\Box}\right) - \frac{1}{2(D-1)}\delta_{\tau\lambda}\frac{\partial_\mu}{\Box}$$

$$P^{\ell}_{\mu\nu,\rho\sigma}P^{\ell}_{\rho\sigma,\tau\lambda} = P^{\ell}_{\mu\nu,\tau\lambda}$$

$$P^{\ell}_{\mu\nu,\tau\lambda} = P^{\ell}_{\nu\mu,\tau\lambda} = P^{\ell}_{\tau\lambda,\mu\nu}\,,\quad P^{\ell}_{\mu\mu,\tau\lambda} = 0\,.$$

It can be verified that

$$P^{tr}_{\mu\nu,\tau\lambda} = I_{\mu\nu,\tau\lambda} - P^{\ell}_{\mu\nu,\tau\lambda} \tag{6.2}$$

where $I_{\mu\nu,\tau\lambda} = \frac{1}{2}(\delta_{\mu\tau}\delta_{\nu\lambda} + \delta_{\mu\lambda}\delta_{\nu\tau} - \frac{2}{D}\delta_{\mu\nu}\delta_{\tau\lambda})$

$$\partial_\mu P^{tr}_{\mu\nu,\tau\lambda} = 0\,,\quad P^{tr}_{\mu\nu,\rho\sigma}P^{tr}_{\rho\sigma,\tau\lambda} = P^{tr}_{\mu\nu,\tau\lambda}\,.$$

Any Euclidean symmetric and traceless tensor of the 2-nd rank can be represented in the form

$$T_{\mu\nu} = T^{tr}_{\mu\nu} + T^{\ell}_{\mu\nu} \tag{6.3}$$

The "longitudinal" part is of the form

$$T^{\ell}_{\mu\nu}(x) = \partial_\mu T_\nu(x) + \partial_\nu T_\mu(x) - \frac{2}{D}\delta_{\mu\nu}\partial_\lambda T_\lambda(x) \tag{6.4}$$

where

$$T_\mu(x) = P_{\mu,\tau\lambda}T_{\tau\lambda}(x)\,. \tag{6.5}$$

Accordingly, two Green functions can be considered

$$G^{tr}_{\mu\nu} = \langle 0|T^{tr}_{\mu\nu}\varphi\varphi\chi|0\rangle\,,\quad G^{\ell}_{\mu\nu} = \langle 0|T^{\ell}_{\mu\nu}\varphi\varphi\chi|0\rangle\,. \tag{6.6}$$

Each of them is invariant [2,5] under the corresponding representation of the conformal group; $\partial_\mu G^{tr}_{\mu\nu} = 0$, and $G^{\ell}_{\mu\nu}$ is unambiguously defined by the Ward identities. Projections (6.1) and (6.2) applied to equations of the type (4.5) yield independent equations for the "longitudinal" and transverse parts.

All said above is true for any space dimension except D = 2. In two-dimensional space only the first of the two irreducible representations remains, the second is absent. It can be verified that

$$\left.\begin{array}{ll} P^{\ell}_{\mu\nu,\sigma\tau} = I_{\mu\nu,\sigma\tau}, & T^{\ell}_{\mu\nu}(x) = T_{\mu\nu}(x) \\ P^{tr}_{\mu\nu,\sigma\tau} = 0, & T^{tr}_{\mu\nu}(x) = 0 \end{array}\right\} \quad \text{for D = 2} \tag{6.7}$$

Since $P^{\ell} = I$, transformations in this representation become local and coincide with usual ones, but act on the space of "longitudinal" tensors.

In general case for any D with "longitudinal" function $G^{\ell}_{\mu\nu}$ the set of the fields P_s can be correlated

$$\langle 0|P^{d+s}_{\mu_1\cdots\mu_s}(x_1)\varphi(x_2)T^{\ell}_{\mu\nu}(x_3)|0\rangle = \underset{\ell=d+s}{\mathrm{res}} \quad [\text{diagram: } \ell,s \sim (C) \text{---} (G^{\ell}_{\mu\nu})]$$

Calculating from this the Green function $\langle 0|P^{d+2}_{\mu_1\mu_2}\varphi T^{\ell}_{\mu\nu}|0\rangle$ and equating it to zero we obtain conditions on dimensions d and Δ, that at D = 2 coincide with those considered above. From the conditions

$$\langle 0|P^{d+2}_{\mu_1\mu_2}(x_1)\varphi(x_2)\ldots\varphi(x_n)|0\rangle = \underset{\ell=d+2}{\mathrm{res}} \quad [\text{diagram: } \ell \sim (C) \overset{T^{\ell}}{=\!=} (G^{\ell}_{\mu\nu})] \tag{6.8}$$

we obtain closed differential equations for the Green functions $\langle 0|\varphi(x_1)\ldots\varphi(x_n)|0\rangle$. (In fact, however, equations can become more complicated than it is said here. In the equations the contribution of quasi-local terms, omitted from equations of the type (4.5), should be taken into account. At D = 2 these terms are insignificant thanks to (6.7). But for D ≠ 2 after action of a nonlocal projection (6.1) they can make a finite contribution.)

In conclusion we point to a certain analogy between the conformal theory in the D-dimensional space discussed above and the structure of the two-dimensional model of papers [3,8,9] defined by the infinite parametric conformal symmetry. This analogy is due to the fact that primary and secondary fields (in terms of [3]) exist also in D-dimensional space. Fundamental fields φ and χ are primary since the energy-momentum tensor commutator with either of them is expressed through the same field and its derivative. Fields P_s discussed above are secondary with respect to φ . In fact, let us consider the 3-point Green function

$$G_{\mu\nu,\mu_1\ldots\mu_s}(x_1x_2x_3)=\langle 0|T_{\mu\nu}(x_3)\varphi(x_2)P_{\mu_1\ldots\mu_s}(x)|0\rangle .$$

Using the general conformal-invariant expression introduced in [5] and assuming in it $\ell=d+s$ and $A=D(D-2)g_T$, $B=Dg_T$, $C=g_T$

we obtain

$$G_{\mu\nu,\mu_1\ldots\mu_s}(x_1x_2x_3)=g_T \lim_{\ell\to d+s}\frac{1}{(x_{13}^2)^{\frac{\ell-d-s+D-2}{2}}}\frac{1}{(x_{23}^2)^{\frac{d-\ell+s+D-2}{2}}}$$

$$\frac{1}{(x_{23}^2)^{\frac{\ell+d-s-D+2}{2}}}\Big\{D(D-2)\lambda^{x_1}_{\mu_1\ldots\mu_s}(x_3x_2)\lambda^{x_3}_{\nu_1\nu_2}(x_1x_2)+ \tag{6.9}$$

$$+D\frac{1}{x_{13}^2}\sum_k \lambda^{x_1}_{\mu_1\ldots\hat{\mu}_k\ldots\mu_s}(x_3x_2)\Big[g_{\mu_k\nu_1}(x_{13})\lambda^{x_3}_{\nu_2}(x_1x_2)+g_{\mu_k x_2}(x_{13})\times$$

$$\times\lambda^{x_3}_{\nu_1}(x_1x_2)+\frac{2}{D}\delta_{\nu_1\nu_2}\frac{x_{12}^2}{x_{23}^2}\lambda^{x_1}_{\mu_k}(x_2x_3)\Big]+\frac{1}{(x_{13}^2)^2}\sum_{k\neq 2}\lambda^{x_1}_{\mu_1\ldots\hat{\mu}_k\ldots\hat{\mu}_2\ldots\mu_s}(x_3x_2)$$

$$\Big[g_{\mu_k\nu_1}(x_{13})g_{\mu_2\nu_2}(x_{13})-\frac{1}{D}\delta_{\nu_1\nu_2}\delta_{\mu_k\mu_2}\Big]\Big\}$$

where g_T is some constant,

$$\lambda^{x_1}_{\mu}(x_2x_3)=\frac{(x_{21})_\mu}{x^2_{12}}-\frac{(x_{31})_\mu}{x^2_{13}},\ \lambda^{x_1}_{\mu_1\ldots\mu_s}(x_2x_3)=\lambda^{x_1}_{\mu_1}(x_2x_3)\cdots\lambda^{x_1}_{\mu_s}(x_2x_3)$$

are traces.

(This expression contains uncertainty $0\times\infty$ resulting from $(x^2)^{-\frac{D}{2}+\varepsilon}\Big|_{\varepsilon\to 0}\sim\frac{1}{\varepsilon}\delta(x)$ which can be resolved in the way similar to that used in analogous Green functions with the current, see [2, 5]). We can show that for $s\geqslant 2$ Green functions (6.9) satisfy the Ward identities which in symbolic form are

$$\partial^{x_3}_{\nu}G_{\mu\nu,\mu_1\ldots\mu_s}(x_1x_2x_3)=g_T\sum_{k=2}^{s}sym\Big\{\alpha_k(\partial^{x_3})^k\delta(x_{13})(\partial^{x_2})^{s-k}\frac{1}{(x^2_{12})^d}\Big\} \quad (6.10)$$

where $(\partial^{x_3})^k=\partial^{x_3}_{\nu_1}\cdots\partial^{x_3}_{\nu_k}$, α_k are some constants determined by the conformal symmetry. The symbol $sym\{\ldots\}$ denotes symmetrization and subtraction of traces over all the indices $\mu_1\ldots\mu_s$. Formulas (4.9), (4.10) from [5] and the relation [10] $\underset{\alpha=0}{res}\,(x^2)^{-\frac{D}{2}+\alpha}=\frac{\pi^{D/2}}{\Gamma(D/2)}\,\delta(x)$ have been used. The Ward identity for the current analogous to (6.9) and the details of calculations see in [1, 4]. From (6.9) it follows that an equal-time commutator $[P^{d+s}_{\mu_1\ldots\mu_s}(x_1),\ \varphi(x_2)]$ may be presented as the sum of terms which contain the field $\varphi(x)$ and its derivatives (up to $(s-1)$-th order) and all the fields $P_{s'}$ with $2\leqslant s'\leqslant s-1$ and their derivatives. Only the field $\varphi(x)$ and its derivatives contribute to the right-hand side of (6.9) since $\langle 0|\varphi(x_1)P_{s'}(x_2)|0\rangle=0$ for $s'\neq 0$. Fields $P_{s'}$ enter similar identities for the Green functions $\langle 0|T_{\mu\nu}P_{s'}P_s|0\rangle$, where $s'<s$. We can show that for $s=1$ the term $\delta(x_{23})$ appears in the right-hand side of Ward's identity (it might be possible if the equal-time commutator $[T_{\mu 0}(x_1),\varphi(x_2)]$ is expressed in terms of the field P^{d+1}_{μ}). Because of

this the field P_μ^{d+1} is not the secondary field, it does not contribute to the operator expansion of the product $T_{\mu\nu}(x)\varphi(x+\varepsilon)$.

As in [3] the primary field φ and the infinite set of the secondary fields P_s, $s \geq 2$ (and their derivatives) can be united into a conformal family. A similar family is associated with χ . It is natural to suppose now that the whole conformal family $\{\varphi\}$ corresponds to the representation of some infinite-dimensional algebra similar to the Virasoro algebra. Then eq. (6.8) may be interpreted (as in [5]) as the consequence of the existence of null-vector in this representation space (see [13]). As for $D=2$ there exists an infinite set of such models. Each model is defined by the requirement for one of the P_s -fields to generate the null-vector.

$$\langle 0|P^{d+s}_{\mu_1\cdots\mu_s}(x_1)\varphi(x_2)\cdots\varphi(x_n)|0\rangle = \operatorname{res}_{\ell=d+s} \; \overset{\ell,s}{\sim}(C)\overset{T^\ell}{=}(G)\equiv \qquad (6.11)$$

The main thing now is to find an infinite-parameter group of D-dimensional space which is an analogue of the group of complex-analytical transformations of the two-dimensional space that served a basis of the approach used in [3,8,9]. One of the possible variants is to use some generalization of complex numbers for D-dimensional case.

It is essential that equations (6.8) and (6.11) do not contain the whole tensor $T_{\mu\nu}$ as for D = 2, but its longitudinal part only. If there exists an infinite-dimensional algebra with the properties described above it should have a representation corresponding to the longitudinal part $T^{\ell}_{\mu\nu}$. It is not difficult to see that any infinite-dimensional algebra which is a subalgebra in the algebra of the general covariance group has such a representation. This results from the following statements:

1) The infinite-dimensional algebra of the general covariance group is the closure of finite-dimensional algebras of the conformal and affine groups [14].

2) The conformal algebra has an irreducible representation in the space of "longitudinal" tensors $T^{\ell}_{\mu\nu}$ (it was constructed in [5] and mentioned above).

3) A representation of the affine algebra in this space also exists. This representation can be obtained in the following way. Consider now the tensors with upper indices. It is not difficult to verify that the condition $\partial_\nu T^{\mu\nu} = 0$ is invariant with respect to affine transformations. Thus in the subspace $H^{tr} \subset H$ which consists of transverse functions $(T^{\mu\nu})^{tr}$ the affine group acts. The "longitudinal" tensors $T^{\ell}_{\mu\nu}$ do not form an invariant subspace. Therefore the corresponding representation acts in the space of equivalence classes H/H^{tr}. Using the technique of [5] this representation may be realized in the subspace of H consisting of "longitudinal" tensors $(T^{\mu\nu})^{\ell}$. In this realization the affine generators are expressed via the projections (6.1) (the details see in [5]). In the case of tensors with superscripts the situation is opposite: the space H^{ℓ} consisting of the "longitudinal" tensors $T^{\ell}_{\mu\nu} = = \partial_\mu T_\nu + \partial_\nu T_\mu - \frac{2}{D} \delta_{\mu\nu} \partial_\lambda T_\lambda$ is invariant with respect to affine transformations. An irreducible representation acts in the quotient space $\tilde{H}/H^{\ell}$ where $\tilde{H}$ is the whole space of tensors with superscripts. This representation can be realized in the space of transverse tensors $T^{tr}_{\mu\nu}$ with the help of the transverse projection (6.2).

References

1. Fradkin, E.S., Palchik, M.Ya. (1978). Phys. Reports, C44, 249.

2. Palchik, M.Ya., Fradkin, E.S. (1983). In: Group Theoretical Methods in Physics. 1, Nauka, Moscow pp. (Russian).

3. Belavin, A.A., Polyakov, A.M., Zamolodchikov, A.B. (1984). Nucl. Phys. B241, 333.

4. Fradkin, E.S., Palchik, M.Ya. (1976). Nuovo Cim. 34A, 438.

5. Fradkin, E.S., Palchik, M.Ya. (1984). Class Quantum, Grav.1, 131.

6. Palchik, M.Ya. (1980). Short Comm. on Phys.4. 19, Lebedev Phys. Inst., Moscow (Russian).

7. Palchik, M.Ya., Prati, M.C., Zaikin, V.N. (1982). Nuovo Cimento 72A, 87.

8. Friedan, D., Qiu, Z., Shenker, S. (1984). Phys. Rev. Lett. 52. 1575.

9. Dotsenko, V.S. (1984). Nucl. Phys. B235, 54, Dotsenko, V.S., Fateev, V.A. (1984). Nucl. Phys. B240, 312.

10. Gelfand, I.M., Shilov, G.E. (1975). Generalized Functions, 1. State Publishing House of Physico-Mathematical Literature, Moscow.

11. Dobrev, V.K., Mack, G., Petkova, V.B., Petrova, S.G., Todorov, I.T. (1977). L. in Phys. 63, Springer, Berlin et al.

12. Todorov, I.T., Mintchev, M.C., Petkova, V.B. (1978). Conformal Invariance in Quantum Field Theory, SNS, Piza.

13. Feigin, B.L., Fuks, D.B. (1982). Functz. Analiz. 16, 47.

14. Ogievetsky, V.I. (1973). Lett. Nuovo Cim. 8, 988.

ARRANGEMENTS OF REAL HYPERPLANES AND ZAMOLODCHIKOV EQUATIONS

Yu.I.Manin, V.V.Schechtman[+]

V.Steklov Mathematical Institute, Vavilova 42, Moscow,
[+]Institute of problem of microelectronic technology,Chernogolovka, 142432

This paper is devoted to the study of n-symplex equations - multydimensional generalization of the known triangle or Yang-Baxter equations. Solutions of triangle equations play the key role in computing the statistical sum of some two-dimensional models of statistical mechanics via the method of commuting transfer matrices [1]. Equations of n-symplices arise at the attempt to generalize this method to higher dimensions. They were first introduced (for $n = 3$) in remarkable papers by A.B.Zamolodchikov [2,3] . In these works Zamolodchikov exhibited a non-trivial solution of tetrahedron equations, cf.[6]. Recently R.Baxter [4] has used this solution to construct an exactly solvable 3-dimensional model. General n-symplex equations were considered in [5].

In [2] it is shown that, like Yang-Baxter equations, the tetrahedron equation may be interpreted as the factorizability condition in a scattering theory of straight "strings" on a plane. Here we evolve this approach using a convenient language of configurational spaces. §1 is pure geometry. Its main result is Theorem 1.5 ("On a dial") which enables us to express a general n-symplex equation in an explicit form and to find a "hidden" $\mathbf{Z}/2n$-symmetry of the structure of these equations.

Coordinatewise the n-symplex equation is a system of functional equations whose number grows twice exponentially with n and the number of unknowns exponentially.

As far as we know the only solution (for $n > 2$) known so far is the above-mentioned Zamolodchikov solution. In § 2 of this paper we construct a large linear manifold of solutions for an arbitrary n, see (2.5). Its colour space is the space of the regular representation of an arbitrary abelian group. The theorem on a dial provides with a short proof that it really is a solution. (A generalization of this solution is presented in Appendix. There is also shown the place for these solution in statistical picture).

As is known, the triangle equation may be read as the factorizability condition of a matrix-valued function on the symmetric group Σ_m, $m \geq 3$, [7]. In § 3 which formally does not depend on the preceding sections we construct partially ordered sets $B(m,k)$ related to the (k+1)-symplex equation as Σ_m is related to the triangle equation. The main result of the section is a purely combinatorial theorem 3.3 which states analogues for $B(m,k)$ of the classical properties of Σm with respect to the weak Brua ordering, cf. [8,9, ch. 1].

Acknowledgements are due to O.V.Ogievetsky who helped us to realize how to prove Theorem 3.3 and A.V.Zelevinsky, who drew ourattention to very interesting papers [8] and [10], for the valuable criticism.

1. Theorem on a dial

1.1 Let H be a real k-dimensional affine space, $\mathcal{H}$ a set of affine hyperplanes in H. Such a set will be called an arrangement. For $J = \{i_1,\dots,i_p\} \subset I$ set $H_J = H_{i_1\dots i_p} = H_{i_1} \cap \dots \cap H_{i_p}$. In what follows we will assume that H_i are in general position, i.e. $\operatorname{codim}_H H_J = |J|$ (the number of elements of J) for all $J \subset I$ unless otherwise stated.

Set $F(\mathcal{H}) = \pi_0(H - \bigcup_{i\in I} H_i)$; elements of $F(\mathcal{H})$ will be called (0-) edges (zero is the codimension). Let $F^b(\mathcal{H}) \subset F(\mathcal{H})$ be the subset of bounded edges, and $F^u(\mathcal{H}) = F(\mathcal{H}) - F^b(\mathcal{H})$. A family $f = (f_i)_{i\in I}$ of affine functions on H such that $f_i = 0$ is an equation of H_i is called an equation of $\mathcal{H}$. Fix such an equation. Then each $a \in F(\mathcal{H})$ is determined by a system of inequalities $\{\varepsilon_i f_i > 0\}_{i\in I}$, where $\varepsilon = (\varepsilon_i) \in \{\pm 1\}^I$. We call ε the (f-) number of a. The edge $\tau(a)$ with the f-number $-\varepsilon = (-\varepsilon_i)_{i\in I}$ is opposite to a. For an arbitrary $\varepsilon \in \{\pm 1\}^I$ the system of inequalities $\{\varepsilon_i f_i > 0\}$ either determines the edge $F(\varepsilon) \in F(\mathcal{H})$ or has an empty set of solutions. In the latter case we will say that ε is the number of an empty edge or just an empty edge.

It is easy to prove the following

1.2. Lemma. An edge $a \in F(\mathcal{H})$ is bounded if and only if $\tau(a)$ is empty.

1.3. In H fix a point O, the origin, and denote $Gr(H;I)$ the space of sets $\theta = (\theta_i)_{i\in I}$ of hypersurfaces through O in general position. For $\theta \in Gr(H;I)$ denote by D_θ the affine space of arrangements $\mathcal{H} = (H_i)$ such that H_i is parallel to θ_i for all i. Any subset $J \subset I$ defines the affine subspace $D(J) = \{(H_i) \mid \bigcap_{i\in J} H_i \neq \varphi\}$ in D_θ of codimension $|J| - k$; $D_\theta = D(\varphi)$; $D(I) = \bigcap_j D(J)$. Set $D_\theta^{(i)} = D_\theta - \bigcup_{|J|=k+i} D(J)$ for $i \geqslant 0$.

1.4. Let $|I| = n = k+2$, $\theta \in Gr(H;I)$. In this case in D_θ there are n hypersurfaces $D_i = D(I-\{i\})$, $i \in I$ whose intersection is the subspace $D(I)$ of dimension 2 and that divide D_θ into $2n$ sectors. Fix $x_o \in D_\theta^{(1)}$ and consider a loop $\varphi(t)$, $0 \leqslant t \leqslant 1$ in $D_\theta^{(2)}$ with the beginning and the end at x_o encircling $D(I)$ once and piercing twice each D_i.

The data uniquely identify $I \cong \{1,2,\dots,n\}$ if the order of piercing is the following: $D_n, D_{n-1}, \dots, D_1, D_n, D_{n-1}, \dots, D_1$. Accordingly, let us renumber sectors $a_j \in \pi_0 D_\theta^{(1)}$ so that φ pass them in the following: $a_1, a_2, \dots, a_{2n}, a_1$.

Consider the arrangement $x_o = (H_1,\dots,H_n)$. Clearly

H_n does not intersect the symplex Δ_n bounded by $H_1,\dots,H_{n-1}$. Therefore we may choose an equation $f = (f_1,\dots,f_n)$ for x_o so as the f-number of Δ_n were $(1,1,\dots,1,-1)$. Fix such an f. Then each $x = (H_1',\dots,H_n') \in D_\theta$ is determined by equations $f' = (f_1+t_1,\dots,f_n+t_n)$, $t_i \in \mathbb{R}$; therefore it is possible to speak about numbers of edges $F(x)$.

1.5. Theorem. Let $\varepsilon_i = 1$ for $1 \leq i \leq n$, $\varepsilon_i = -1$ for $n+1 \leq i \leq 2n$; $\delta^i = (\varepsilon_{i+1},\dots,\varepsilon_{i+n})$, $1 \leq i \leq 2n$ (indices of ε_j are reduced mod 2n). Then for $x \in a_i$ numbers of $F^b(x)$ are $\delta^i, \delta^{i+1},\dots, \delta^{i+n-2}$; where numbers δ^j, $j \in [1, 2n] - [i,i+n-2]$ determine empty edges and the rest numbers determine unbounded edges.

Proof. First consider an arrangement x_o. Let Δ_i, $1 \leq i \leq n$, be a symplex bounded by planes H_j, $j \neq i$.

Statement. The plane H_i is situated so that on one side of it there are i-1 vertices of Δ_i and on the other one there are n-i-1 of them. H_i divides Δ_i into two parts with numbers δ^{n-i}, δ^{n-i+1} (edges δ^0 and δ^n are empty).

Proof. In H_i introduce a ccordinate system $(t_1,\dots,t_k)$, where $f_i = t_i$ for $i=1,\dots,k$, $f_{k+1} = 1-\sum t_i$, and $f_{k+2} = \sum a_i t_i + B$. By hypothesis, H_i moving parallelly to itself torwords Δ_i will encounter vertices of Δ_n in the order $\alpha_{n-1}, \alpha_{n-2},\dots, \alpha_1$, where $\alpha_j = H_{1\dots\hat{j}\dots n-1}$. It follows that $b < 0 < a_k < a_{k-1} < \dots < a_1$. Now the rest is quite straight-forward.

Further, it is not difficult to see that faces $\delta^1,\dots, \delta^{n-1}$ exhaust $F^b(x_o)$. This implies the statement of the theorem for the first sector.

Now let us move from the first sector to the second one piercing D_n. The symplex Δ_n will be turned inside out; the numbers of bounded faces are $\delta^2,\dots,\delta^n$ and after a cyclic renumbering of planes we return to the initial position.

2. Equation of (n-1)-symplices

2.1. The elementary scattering operator. Let $I = \{1,2,\dots,k+1\}$, $\theta \in Gr(H,I)$. Then $\pi_0 \mathcal{D}_\theta^{(1)}$ consists of two elements a a and $\tau(a)$. Let $x_0 = (H_1,\dots,H_{k+1}) \in a$ and $|F(x_0)| = 2^{k+1}$. Clearly F^b consists of one element, the symplex $\Delta(x_0)$.

Consider a path $f(t)$ in D_θ joining x_0 with a point $x_1 \in \tau(a)$. It corresponds to the movement of planes H_i so that $\Delta(x_0)$ is contracted to a point and then turned inside out turning into $\Delta(x_1)$.

Let us identify by continuity sets $F(x)$ for x belonging to the same component $D_\theta^{(1)}$ and, separately, all $F^u(x)$, $x \in D_\theta$. We get three sets: $F(a)$, $F(\tau(a))$ and $F^u = F^u(a) = F(a) \cap F(\tau(a))$.

Fix a vector space V. If C is a set then $V^{\otimes C}$ is the tensor product of "C copies" of V, i.e. the quotient of the free abelian group spanned by Hom(C,V) modulo the usual multilinearity relations.

A matrix of an elementary scattering at θ is a pair of linear operators $S(\theta,a): V^{\otimes F(a)} \rightleftarrows V^{\otimes F(\tau(a))} : S(\theta,\tau(a))$ identical on $V^{\otimes F^u}$, i.e. $S(\theta,a)(v) = \otimes v_i'$ where $v_i' = v_i$ for $i \in F^u$ for all $v = \otimes v_i \in V^{\otimes F(a)}$ and the same holds for $S(\theta,\tau(a))$. We say that S satisfies the unitarity condition if $S(\theta,a)S(\theta,\tau(a)) = Id = S(\theta,\tau(a))S(\theta,a)$.

2.2. Equation of (n-1)-symplices. Return to the situation of (1.4) and assume that a matrix of an elementary scattering is defined on $Gr(H;\{1,\dots,k+1\})$. Then to each path $f(t)$ in $D_\theta^{(2)}$ such that $f(i) = x_i$, $i = 0,1$, we assign the operator $S(f): V^{\otimes F(x_0)} \to V^{\otimes F(x_1)}$. First suppose that x_0 and x_1 belong to the neighbour sectors and f pierces $\mathcal{D}_i$ once. Then f corresponds to the movement of hyper-planes H_i that corresponds to the turning the symplex Δ_i bounded by $H_1,\dots,\hat{H}_i,\dots,H_n$ inside out.. The symplex Δ_i is situated on one side of H_i . This defines the embedding $\chi : \pi_0(H - \bigcup_{j\neq i} H_j \to \pi_0 (H - \bigcup H_j)$. Identify $I - \{i\}$ with $\{1,2,\ldots,k+1\}$ with the help of the ordering inherited from l I. Let $\Psi_i : Gr(H,I) \to Gr(H,I-\{i\})$ be an obvious projection. Set $S(f) = S(\Psi_i(\theta),x_o) \otimes Id(V^{\otimes F(x_o)-Im\chi})$. For an arbitrary $f = \prod f_i$ where f_i is the passage to the neighbour sector set $S(f) = \prod S(f_i)$. If S is unitary then $S(f)$ depends only on the homotopic class of f (in $D_\theta^{(2)}$).

Now let f_1 and f_2 be two paths in $D_\theta^{(2)}$ from a_1 to a_{n+1} such that f_1 pierces D_i in the order $D_n, D_{n-1},\ldots,D_1$ and f_2 in the order $D_1,D_2,\ldots,D_n$. (Therefore φ from 1.4 is $f_2^{-1} f_1$). We will say that S satisfies (n-1)-symplex equation if

$$S(f_1)=S(f_2).$$

If S is unitary then (1) is equivalent to

$$S(\varphi) = Id.$$

Remark. a) Conventionally (1) is called "the (n-1)-symplex equation in static limit with colours on O-faces" A obvious modification of the above definitions gives the complete (n-1)-symplex equation with colours on faces of any codimension.

b) Note that the identity scattering operator (the colour of the turned inside out symplex does not vary) is not a solution of (1) for $n > 3$. It is also interesting to consider a "twisted" equation for which the identity operator is a solution. We do not deal with this here.

2.3. Constant solutions. Return to the situation of (2.1). Let G be an abelian group, V the vector space with the basis $\{e(g)\}$, $g \in G$; let $h \in G$, $2h = 0$. Define

$$S^h = S^h(\theta,a) : V^{\otimes F(a)} \to V^{\otimes F(\tau(a))}$$

setting

$$S^h(\otimes e(g_i)) = \otimes e(g'_i)_{i \in F(\Gamma(a))},$$

where $g'_i = g_i$ for $i \in F^u$ and

$$g'_{\Delta(\tau(a))} = \Big(\sum_{i \in F(a)} \varepsilon_i g_i\Big) + h$$

where $\varepsilon_i = -1$ if in the arrangement $a = (H_1, \dots, H_n)$ components i and $\Delta(a)$ are situated on different sides of H_1 and $\varepsilon_i = 1$ otherwise.

2.4. Proposition. The matrix S^h satisfies the unitarity condition and the $(n-1)$-symplex equation (1).

Proof. The unitarity condition follows immediately from the definition. Let us verify (1) for the case $h = 0$ (the case $h \neq 0$ is quite similar).

Making use of the numbering of domains from the theorem on dial (1.5) we see that the verification is reduced to the following. There are 2 systems of n linear equations over G with 2^n unknowns x_ε, $\varepsilon = (\varepsilon_1, \dots, \varepsilon_n) \in \in \{\pm 1\}^n$.

The first system ("the left-hand side of (1)"): the i-th equation is of the form $\sum \varepsilon_1 x_\varepsilon = 0$ for $1 \leqslant \leqslant i \leqslant n$; $\sum \varepsilon_2 x_2 = 0$ for $i = n$ the sum runs ε such that $\varepsilon_{n+1-i} = 0$. The second system ("the right-hand side of (1)"): the i-th equation $\sum \varepsilon_1 x_\varepsilon = 0$ for $1 < i \leqslant n$, $\sum \varepsilon_2 x_\varepsilon = 0$ for $i = 1$ the sum runs ε such that $\varepsilon_i = 1$.

The problem is as follows. Given $x_\varepsilon \in G$ for all ε except $\varepsilon = \varepsilon^i$, $\varepsilon^i_j = -1$ for $j < i$ and $\varepsilon^i_j = 1$ for $j \geqslant i$, $1 \leqslant i \leqslant n+1$ ("initial colouring"), we may define $x_{\varepsilon^2}, x_{\varepsilon^3}, \dots, x_{\varepsilon^{n+1}}$ from the 1st system and $x_{\varepsilon^1}, x_{\varepsilon^2}, \dots, x_{\varepsilon^n}$ from the 2nd one (e.g. by subtracting from the i-th equation the (i-1)-th one). The values $x_{\varepsilon^2}, \dots, x_{\varepsilon^n}$ determined from the 1st and 2nd systems ("final colourings") should coincide. This is subject to a straightforward verification.

2.5. Theorem. Set $G' = \{h \in G \mid 2h = 0\}$. Suppose that for any $h \in G'$ we are given an arbitrary function φ^h of angles and velocities of the $(k+1)$-th string. Then $S = \sum \varphi^h S^h$ where S^h are solutions constructed in (2.3) is a solution of (1).

Remark. S does not satisfy in general the unitarity condition.

Proof. Set $A = \{\pm 1\}^n - \{\delta^n, \delta^{n+1}, \dots, \delta^{2n}\}$. $B = \{\delta^{n+1}, \delta^{n+2}, \dots, \delta^{2n-1}\}$, where δ^i are the same as in (1.5). Clearly, for a fixed initial colouring $\{x_\varepsilon\}_{\varepsilon \in A}$ non-zero elements of matrices $S(f_1)$, $S(f_2)$ might occur only for finite colours of the form $\{x'_\varepsilon\}_{\varepsilon \in B}$, where $x'_\varepsilon = x_\varepsilon + h_\varepsilon$, $h_\varepsilon \in G'$, $\{x_\varepsilon\}_{\varepsilon \in B}$ is the finite colouring for the solution S^o. Therefore, there are $|G'|^{|B|}$ equations for φ^h one for each set $\{h_\varepsilon\}_{\varepsilon \in B}$, $h_\varepsilon \in G'$. Suppose such a set is given. Let us write the corresponding equation. Let φ^h_j, $j = 1,2,\dots, n$ be a function of arguments corresponding to the strings with numbers $1,2,\dots,\hat{j},\dots,n$.

Set ${}^h\varphi = \varphi_n^{h_0} \varphi_{n-1}^{h_1} \dots \varphi_1^{h_{n-1}}$, where $h_o = h$ and $h_j = h_{j-1} + h_{\delta^{n+j}}$.

Similarly set $h_{\bar{\varphi}} = \varphi_1^{h_0} \dots \varphi_n^{h_{n-1}}$, where $h_o = h$, $h_j = h_j - 1 + h_{\delta^{2n-j}}$. Then the equation corresponding to the finite colouring $\{x_\varepsilon + h_\varepsilon\}_{\varepsilon \in B}$ is of the form

$$\sum_{h \in G'} {}^h\varphi = \sum_{h \in G'} {}^h\bar{\varphi}'.$$

Clearly this is identically satisfied.

Therefore only the unitarity condition is a nontrivial constrain onto functions φ^h, $h \in G$.

3. Higher Bruhat orderings

3,1. Let W be a finite Coxeter group with the system of generators S [11]. In W introduce a partial ordering setting $w \leqslant w'$ if there exist $\tau_1, \ldots, \tau_m \in S$ such that $w' = w\tau_1 \cdot \ldots \cdot \tau_m$ and $l(w') = l(w)+m$ where $l(w)$ is the length of w. This ordering is called a weak Bruhat ordering, [8]. It is well known that W has unique minimal and maximal elements with respect to this ordering and is a ranged partially ordered set (poset) where rank is a function l, i.e. for $w \leqslant w'$ the length of any non-condensable chain connecting w with w' equals $l(w') - l(w)$.

In this section for the case W of type A_{n-1} we introduce the system of ranged posets $B(n,k)$, $k = 1,\ldots,n$, such that $B(n,1) = W$ with a weak Bruhat ordering and $B(n,k+1)$ is a slightly modified set of maximal chains in $B(n,k)$. The set $B(n,k)$ is closely connected with the geometry of general arrangements of n affine hyperplanes in and the structure of equations of $(k+1)$-symplices and also with the combinatorical structure of the convex envelope of a generic orbit of W in $\mathbb{R}^n$, [10].

3.2. Let n,k be integers $n \geqslant k \geqslant 1$. Let $C(n,k)$ be the set of subsets of $\{1,2,\ldots,n\}$, containing k elements. Elements of $C(n,k)$ will be denoted $(i_1 i_2 \ldots i_k)$, $1 \leqslant i_1 < i_2 < \ldots < i_k \leqslant n$. For $C = (i_1 \ldots i_k) \in C(n,k)$, $1 \leqslant j \leqslant k$ set $\hat{c_j} = (i_1 \ldots \hat{i_j} \ldots i_k) \in C(n,k-a)$.

3.2.1. <u>Definition</u>. $A(n,k)$ is the set of complete orderings of $C(n,k)$ such that for any $d \in C(n,k+1)$ either $\hat{d_1} < \hat{d_2} < \ldots < \hat{d}_{k+1}$ or $\hat{d_1} > \hat{d_2} > \ldots > \hat{d}_{k+1}$.

Elements $a \in A(n,k)$ will be written in the form of a chain $a = C_1 C_2 \ldots C_N$, $C_i \in C(n,k)$, $C_1 < C_2 < \ldots < C_N$,

$N = \binom{k}{n}$.

We set $\mathrm{Inv}(a) = \{d \in C(n,k+1) \mid \hat{d_1} < \ldots < \hat{d}_{k+1}\}$, $\mathrm{inv}(a) = |\mathrm{Inv}(a)|$.

Elements of $\mathrm{Inv}(a)$ will be called inversions of a.

3.2.2. <u>Example</u>. If a_{min} - is the lexicographic ordering of $C(n,k)$ then $\mathrm{Inv}(a_{min}) = \Phi$. If a_{max} is the ordering inverse to the lexicographic one, then $\mathrm{Inv}(a_{max}) = C(n,k+1)$.

We say that $a \sim a'$ for $a, a' \in A(n,k)$ if a' is obtained from $a = C_1 \ldots C_N$ by transposition of two neighbouring C_j and C_{j+1}, and $|C_j \cup C_{j+1}| \geqslant k+2$.

3.2.3. Definition. $B(n,k)$ is the quotient of $A(n,k)$ modulo the equivalence relation generated by $\sim$. Let $\pi : A(n,k) \to B(n,k)$ be the projection.

Clearly, Inv factors through π.

Let $a = C_1 \ldots C_N \in A(n,k)$, $d \in C(n,k+1) - \mathrm{Inv}(a)$ be such that $\hat{d_i} = C_{j+i}$ for some j and all i: $1 \leqslant i \leqslant k+1$. Set $a' = p_d(a) = C_1' \ldots C_N'$, where $C_i' = C_i$ for $i \in [j+1, j+k+1]$ and $C_{j+i}' = C_{j+k+2-i}$ for $1 \leqslant i \leqslant k+1$

Clearly, $\mathrm{Inv}(a') = \mathrm{Inv}(a) \cup \{d\}$. We will say that a' is obtained from a by an elementary modification with the help of d. For $b, b' \in B(n,k)$ we will write $b' = p_d(b)$ if there exist $a, a' \in A(n,k)$ such that $\pi(a) = b$, $\pi(a') = b'$, $a' = p_d(a)$.

Now introduce a partial ordering on $B(n,k)$.

3.2.4. Definition. Say that $b \leqslant b'$ if there exist $d_1, \ldots, d_m \in C(n,k+1)$ such that $b' = p_{d_m} p_{d_{m-1}} \ldots p_{d_1}(b)$.

3.2.5. Example. $B(n,1) = A(n,1)$ is the symmetric group with the weak Bruhat ordering.

3.3. Theorem. a) $B(n,k)$ is the ranged poset with respect to the above ordering where rank is inv. There are the unique minimal $b_{min} = \pi(a_{min})$ and maximal $b_{max} = \pi(a_{max})$ elements in $B(n,k)$ (cf. Example 1).

b) The correspondence $(b_{min} < p_{d_1}(b_{min}) < \ldots p_{d_M} \ldots$ $\ldots p_{d_1}(b_{min}) = b_{max}) \mapsto (d_1 \ldots d_M)$ establishes a bijection (maximal chains of $B(n,k) \overset{\sim}{\to} A(n,k+1)$.

c) Inv : $B(n,k)$ the set of subsets of $C(n,k+1)$ is an embedding.

3.3.1. Example. $B(n,n) = A(n,n) = \{a = (12\ldots n)\}$; $B(n,n-1) = A(n,n-1) = \{\hat{a_1} \ldots \hat{a_n}, \hat{a_n} \ldots \hat{a_1}\}$, $B(n,n-2)$ consists of $2n$ elements and is just 2 non-intersecting chains of length n joining b_{min} and b_{max} (see Lemma 3.3.3 below).

The last circumstance is closely connected with the fact that the $(n-1)$-symplex equation contains the product of n-operators in its left and right-hand sides.

<u>Sketch of the proof of Theorem (3.3)</u>. Proof consists of a chain of lemmas whose verification we mostly leave to the reader. In what follows we will say that C_1, $C_2 \in C(n,k)$ commute if $|C_1 \cup C_2| \geq k+2$, cf. 3.2.2 below.

3.2.2. <u>Lemma</u>. Any $b \in B(n,k)$ has a representative $a = (C_1 \ldots C_N) \in A(n,k)$ with all the C_i - containing 1 going one after the other.

This is the key fact on which all the subsequent inductive arguments are based. From it we easily deduce.

3.3.3. <u>Lemma</u>. $B(n,n-2)$ consists of $2n$ elements. Its structure is as follows: (cf. 3.3.1):

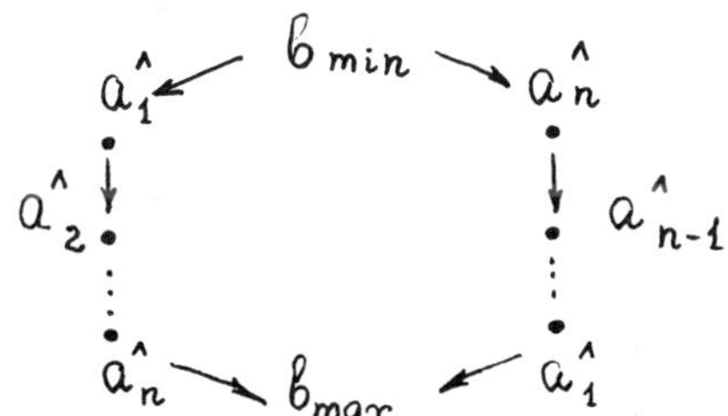

3.3.4. <u>Lemma</u>. a) The following properties of $b \in B(n,k)$ are equivalent:

1) $b = \pi(a_{min})$ (resp. $b = \pi(a_{max})$) (cf. 3.2.2).

2) $Inv(b) = \varphi$ (resp. $Inv(b) = C(n,k+1)$).

3) b is the minimal (resp. maximal) element of $B(n,k)$.

b) Let $B_{min} = \pi(a_{min}) < p_{d_1}(b_{min}) < \ldots < p_{d_N} \ldots$

$\ldots p_{d_1}(b_{min}) = \pi(a_{max})$

be the maximal chain. Then $(d_1 \ldots d_N) \in A(n,k+1)$.

<u>Comment</u>. b) is a corollary of Lemma 3.3.3.

Another key technical result is

3.3.5. <u>Lemma</u>. Let $a \in A(n,k)$, $d \in C(n,k+1) - Inv(a)$.

Supposeing for any $C \in C(n,k)$ such that $C \in \{\hat{d_1}, \ldots$ $\ldots, \hat{d_{k+1}}\}$ and $\hat{d_{k+1}}$ is situated in a to the left of C and $\hat{d_1}$ to the right we have: either C commutes with all $\hat{d_i}$ to the left of it or with all $\hat{d_i}$ to the right of it. Then there exists $a' \sim a$ with all the $\hat{d_i}$ going one after another.

Proof is done by induction in n+k with the help of Lemma 3.3.2. From here we deduce a) and b) of Theorem 3.3 We deduce c) by induction and with Lemma 3.3.2.

3.4. Relation with arrangement of hyperplanes. Let $\mathbb{R}^2$ be a fixed basis in $\{e_1, e_2\}$. Let $\mathcal{B}(n,2)$ be the space whose points are srts of n affine lines in $\mathbb{R}^2$ in general position and non-parallel to e_2 and $\mathcal{A}(n,2)$ $\subset \mathcal{B}(n,2)$ the subspace of sets (l_i) such that for any quadruple of pairwise different l_1, l_2, l_3, l_4 the line $l_1 \cap l_2 \quad l_3 \cap l_4$ is not parallel to e_2. Let $(l_i) \in$ $\mathcal{A}(n,2)$; $t_o \in \mathbb{R}$ be such that the abcissa of all points $l_i \cap l_j$ $(i \neq j)$ is $> t_o$. Let us enumerate the lines in accordance with the growth of the ordinate of the points of their intersection with the line $t_o e_1 + \mathbb{R}e_2$. Setting $(ij) < (km)$ if the abscissa $(l_i \cap l_j$ is less then the abscissa of $(l_k \cap l_m)$ we get an element of A(n,2). This gives the maps $\varphi(n,2): \pi_0 A(n,2) \rightarrow A(n,2)$. The maps $\psi(n,2): \pi_0 \; \mathcal{B}(n,2) \rightarrow B(n,2)$ are defined similarly.

3.4.1. Conjecture. $\varphi(n,2)$ and $\psi(n,2)$ are bijections.

Similarly A(n,k) and B(n,k) are interpreted as arrangements of n hyperplanes in $\mathbb{R}^k$.

3.4.2. Remark. It would be interesting to study linear representations of posets B(n,k). The constant solutions of (k+1)-symplex equations constructed in § 2 are examples of such representations.

Appendix. On certain higher dimensional models behaving as one-dimensional ones (V.V.Schechtman)

In this Appendix which is independent of the main text, we present some solutions of n-symplex equations which generalize the solutions obtained in §3 , and solve the corresponding statistical models. A though formally n-dimensional, these models resemble one-dimensional systems more than n-dimensional ones.

Fix integers n,k such that $0 \leqslant k \leqslant n$. Let G be an abelian group, and f an arbitrary number function on G. If k=0 or n, suppose that 2g=0 for any $g \in G$. Consider a cubic lattice L in $\mathbb{R}^n$. We assume that "spins" of our statistical model are situated on k-dimensional faces of L and take their values in G. For a given configuration σ of spins and a vertex x of L set

$$W_\sigma(x) = f(\Sigma\sigma_F) \qquad (1)$$

where $\sigma_F \in G$ is the spin corresponding to a face F, the sum (with values in G) runs all the k-dimensional faces F, containing x. Set

$$W(\sigma) = \prod_x W_\sigma(x),$$

the product over all the vertices x, and

$$Z(f) = \sum_\sigma W(\sigma),$$

the sum over all the configurations σ with cyclic boundary values. Denote by T(f) the layer-to-layer transfer matrix of this model (layers are (n-1)-dimensional, cf. [5]).

Theorem A1. T(f) and T(f') commute for every two functions f,f' .

Proof follows from the main result of [5] and

Theorem A2. The S-matrix defined by (1) satisfies the n-symplex equation.

Exercise. Note that for k=0 this theorem reduces to Theorem 2,5.

<u>Theorem A3</u>. Suppose that G is finite and all the values of f are real and positive. Let N be the number of k-dimensional faces of L. Then

$$Z(f) \sim \Big(\sum_{g \in G} f(g)\Big)^N \qquad \text{as } N \to \infty,$$

i.e. free energy per face is $f = \ln(\sum f(g))$.

<u>Proof</u>. This follows from the fact that the vector with all components equal to 1 is the eigenvector of T(f) with the greatest eigenvalue.

<u>Examples</u>. Suppose that $G = \{\pm 1\}$. Then if n=1 our model is equivalent to the one-dimensional Ising model. If n=2 , k=0 or 2, then our model reduces to the degenerate case of eight vertex model,corresponding to disordered phase, cf. [1, 10.11.6].

Theorems A1-A3 show that our models are "effectively one-dimensional". A model equivalent to ours for $G = \{\pm 1\}$ is considered in [12] , where an analogue of Theorem A3 is announced. In [12] there is also considered a certain n-dimensional model whose statistical sum is simply related to the statistical sum of the two-dimensional Ising model.This model might be called "effectively two-dimensional", By analogy with Theorems A1-A2 we may suggest that the corresponding n-symplex equations have solutions in elliptic functions.

References

1. Baxter,R. (1982). Exactly solved models in Statistical Mechanics. Acad. Press, London et al.

2. Zamolodchikov, A.B. ZhETF (1980). 79, 2(8), 641-664 (Russian).

3. Zamolodchikov,A.B. (1981). Comm. Math. Phys. 79, 489-505.

4. Baxter,R. (1984). Phys. Rev. Lett. 23, 19, 1795-1798.

5. Bazhanov,V.V., Stroganov,Yu.G. (1982). Theor. Math. Phys. 52, 1, 105-113 (Russian).

6. Baxter,R. (1983). Comm. Math. Phys. 88, 185-205.

7. Cherednik,I.V. (1984). Theor. Math. Phys. 61, 1,35-44 (Russian).

8. Stanley,R.P. (1984). On the number of neduced decompositions of elements of Coxeter groups. Preprint.

9. Knuth,D. (1973). The art of computer programming. 3. Sorting and searching. Addison-Wesley, Reading.

10. Berezin,F.A., Gelfand,I.M. (1956). Trudy MMO (Moscow Math. Soc. Transactions),5, 312-351.

11. Bourbaki,N. (1968). Groupes et algebres de Lie, Ch. IV-VI. Hermann, Paris.

12. Nencka-Ficek,H. (1983). LN in Phys., 180, 543.

BACKGROUND FIELD METHOD IN COVARIANT SECOND-QUANTIZED STRING THEORY

A.A. Tseytlin
P.N. Lebedev Physical Institute, Moscow 117924

1. Introduction. Superstring models have become very popular as possible candidates for a unified theory incorporating all known particles and interactions in a way consistent with quantum mechanics [1,2]. However, a covariant formulation of these theories is presently poorly understood. Such a formulation is necessary for developing a string theory analogue to the background field method and hence for determining the true ground state. There are two covariant formulations of string theories which are being developed. The "first-quantized" one is based on the amplitudes given by sums over 2-surfaces. The generating functional Γ of all possible off-shell scattering amplitudes was introduced in [3,4] under the name of "effective action". Γ is a functional of the infinite number of local background fields corresponding to string modes. Problems of the "first-quantized" approach (like an ad hoc prescription for weights in summation over topologies and a lack of a direct proof of unitarity) can be resolved starting with more fundamental "second-quantized" approach based on a covariant action for string functionals [5-8]. Here we suggest a new approach to covariant functional string field theory starting from the condition that the resulting covariant perturbation theory should have a representation in terms of sums over regular two-dimensional surfaces, defined according to [9]. Then we define an effective action for this functional theory and establish

its connection with the effective action proposed in [3,4]. In this paper we only consider the case of closed oriented Bose strings.

We start with a particle-theory analogue and derive a path integral representation for the effective action in the ϕ^3-field theory (sect.2). Then we define the propagator and the 3-vertex of the functional string field theory (sect.3). In sect.4 we introduce the effective string field action. In Sect.5 we discuss non-trivial equivalence between the effective action and source functional approaches in string theory.

2. Path integral representation of ϕ^3-field theory.

Starting with the covariant particle action

$$I_0 = \int_C^1 d\tau \,(e^{-1}\dot{x}^\mu \dot{x}^\mu + m^2 e), \tag{1}$$

where e is the square root of a one-dimensional metric, we define the propagator

$$G(x_1, x_2) = \int_{\partial C = (x_1, x_2)} DC \;\exp(-I_0). \tag{2}$$

Here $DC = [de(\tau)\, dx^\mu(\tau)]$. Inverting G we get the kinetic operator in the free-field action

$$S_0 = \frac{1}{2}\int d^D x_1\, d^D x_2\, \phi(x_1)\,\Delta(x_1,x_2)\,\phi(x_2) \tag{3}$$

It is easy to prove using the "proper-time" gauge $e = const$ that $\Delta = (-\Box + m^2)\delta(x_1 - x_2)$. The interaction term in the field action in the particle theory case cannot be fixed only on the basis of path integral considerations. We choose the simplest local ϕ^3-interaction since it corresponds to the unique cubic interaction in the closed string field theory (and since only in this case it is possible to give a closed path integral expression for the quantum effective action)

$$S = \int d^D x \left\{ \frac{1}{2}\phi(-\Box + m^2)\phi + \frac{1}{3!} g\,\phi^3 \right\} \tag{4}$$

Introducing a background field $\bar{\phi} = g^{-1}\tilde{\phi}$ we define the effective action as the sum of all irreducible diagrams on the background,

$$\exp(-\Gamma[\bar{\phi}]) = \int \mathcal{D}\phi \, \exp\left(-S[\bar{\phi}+\phi] + \frac{\delta\Gamma[\bar{\phi}]}{\delta\bar{\phi}}\phi\right) \qquad (5)$$

In the ϕ^3-theory $\bar{\phi}$ enters the diagrams only through the propagator $\tilde{G} = (\Delta + \tilde{\phi})^{-1}$, which has the following integral representation

$$\tilde{G}(x_1, x_2) = \int_{\partial C = (x_1, x_2)} \mathcal{D}C \, \exp(-I_c - I_{int}), \quad I_{int} = \int_c^1 d\tau \, e \, \tilde{\phi}(x(\tau)) \qquad (6)$$

An analogous representation exists also for log det$(\Delta + \tilde{\phi})$ and hence for the full effective action (see also [4])

$$\Gamma[\tilde{\phi}] = g^{-2} S[\tilde{\phi}] - \frac{1}{2} \operatorname{tr} \log \tilde{G} - \\ - \sum_{\gamma_n} g^n d_n^{-1} \int d^D x_1 \ldots d^D x_n \, \tilde{G}(x_1, x_2) \ldots \tilde{G}(x_{n-1}, x_n) \qquad (7)$$

Here the sum runs all irreducible graphs. The number of vertices in the graph γ_n is $n = 2L - 2 = 0, 2, \ldots$, where L is the number of loops, d_n is the order of the symmetry group of γ_n. Instead of summing over paths for each individual propagator and then integrating over the boundary points we can directly path integrate over classes of graphs of a given topology. It is plausible that factors d_n cancel out and hence "different topologies" are summed with "equal weight"

$$\Gamma = g^{-2} S + \Gamma_q, \quad \Gamma_q = -\sum_{\gamma_n} g^n \int \mathcal{D}C \, \exp(-I_c - I_{int}) \qquad (8)$$

We can also consider the generating functional for all the related Green functions

$$\exp(-W[J]) = \int \mathcal{D}\phi \, \exp(-S[\phi] + \phi \cdot J) \qquad (9)$$

Then

$$W = W_c + W_q, \quad W_c = S[\bar{\phi}(J)] - \bar{\phi}(J) \cdot J, \\ W_q = \tilde{W}_q[\bar{\phi}(J)], \quad \frac{\delta S[\bar{\phi}]}{\delta\bar{\phi}} = J, \qquad (10)$$

where $\tilde{W}_q$ is the "background functional" [10]

$$\exp(-\tilde{W}_q[\bar{\phi}]) = \int \mathcal{D}\phi \, \exp\left(-S[\bar{\phi}+\phi] + S[\bar{\phi}] + \frac{\delta S}{\delta\bar{\phi}}\phi\right) \qquad (11)$$

given by the sum of all connected vacuum diagrams computed on the background $\bar{\varphi}$ and also has a path integral representation analogous to (8). Though each term in the expansion of $\bar{\varphi}(J)$ in powers of J (each tree diagram) has a path integral representation, it seems that a compact path integral expression for $\bar{\varphi}(J)$ analogous to (8) and hence for $W_c[J]$ and $W_\varphi[J]$ does not exist.

3. Covariant functional string field action. We determine the "kinetic operator" as well as the interaction term in the functional string action starting with the condition that the resulting perturbation theory represents in terms of sums over regular 2-surfaces with the weight given by the exponent of the covariant string action [11,9]

$$I_o = \frac{1}{4\pi\alpha'} \int d^2 z \sqrt{g}\, g^{ab} \partial_a x^M \partial_b x^M . \tag{12}$$

We use the functional measure $\mathcal{D}M = \left[d g_{ab}(z) dx^M(z) \right]$ introduced in [9]. The basic string field is the functional $\Phi[C] = \Phi\left[x^M(\sigma), e(\sigma)\right]$ of a parametrized loop $(0 \leqslant \sigma \leqslant 1)$ and of one-dimensional metric. The propagator $\langle \Phi[C_1] \Phi[C_2] \rangle$ is defined by analogy with (2) as a sum over cylindrical 2-surfaces

$$G\left[C_1, C_2\right] = \int_{\partial M = (C_1, C_2)} \mathcal{D}M \, \exp(-I_o). \tag{13}$$

Inverting G in the loop space we get the kinetic operator in the free part of the string field action

$$S_o = \frac{1}{2} \int \mathcal{D}C_1 \mathcal{D}C_2 \, \Phi[C_1] \, \Delta\left[C_1, C_2\right] \, \Phi\left[C_2\right]. \tag{14}$$

Computing (13) it is useful to select the "cylindrical" gauge: $ds^2 = g_{ab}\, dz^a dz^b = e^2(\tau,\sigma) d\sigma^2 + d\tau^2$, $0 \leqslant \tau \leqslant a$, $0 \leqslant \sigma \leqslant 1$, so that $C_1 = \{x^M(\sigma,0), e(\sigma,0)\}$, $C_2 = \{x^M(\sigma,a), e(\sigma,a)\}$. Then the integral over g_{ab} reduces to $\int [de] \int da\, f(a)$, where f is a function (a plays the role of the moduli parameter of a cylinder). If $D = 26$ the integral over

e is trivial while the integral over a produces poles of the propagator for specific choices of C_1 and C_2 (e.g. point-like, cf. [12]). It appears likely that for $D = 26$ Δ is a local operator, $\Delta[C_1,C_2]=\Delta_0[C_1]\delta[C_1-C_2]$. Consider a complete set of orthonormal functions on the circle

$$\int_0^1 d\sigma\, e(\sigma)\, f_n(\sigma)\, f_m(\sigma) = \delta_{nm}, \quad x^\mu(\sigma) = \sum_n x_n^\mu\, f_n(\sigma) \tag{15}$$

Then the covariant functional derivative is given by (cf. [6])

$$\frac{\hat{\delta}}{\delta x^\mu(\sigma)} = \sum_n f_n(\sigma)\frac{\partial}{\partial x_n^\mu}, \quad \left[\frac{\hat{\delta}}{\delta x^\mu(\sigma)},\, x^\nu(\sigma')\right] = \delta_\mu^\nu\, \delta(\sigma,\sigma') \tag{16}$$

where $\delta(\sigma,\sigma') = e^{-1}\delta(\sigma-\sigma')$. The obvious covariant analogue of the kinetic operators Δ_0 in [5,6] is

$$\Delta_0[C] = \int_0^1 d\sigma\, e(\sigma)\left[-\left(\frac{\hat{\delta}}{\delta x^\mu(\sigma)}\right)^2 + \left(\frac{x'^\mu}{2\pi\alpha'}\right)^2 e^{-2}\right] \tag{17}$$

where $x' = dx/d\sigma$. This expression follows from substituting $\dot{x}^\mu \to -i\,\hat{\delta}/\delta x^\mu$ in the string action (12) computed in the "cylindrical" gauge: $I_c = \frac{1}{2}\int d\tau' \int d\sigma\, e\left[\dot{x}^2 + \left(\frac{x'}{2\pi\alpha'}\right)^2 e^{-2}\right]$, $\tau' = 2\pi\alpha'\tau$. Though at present I cannot claim that (17) is the exact consequence of (13) it may turn out that the action (14) with Δ_0 given by (17) is the correct covariant free action describing physical as well as auxiliary degrees of freedom and equivalent to the actions discussed in [7,8]. Our basic field $\Phi[x^\mu(\sigma), e(\sigma)]$ corresponds to an infinite number of functionals of $x^\mu(\sigma)$. This is to be compared with the observation of Siegel and Zweibach [8] that the whole system of functionals present in their covariant action may be represented as a single functional of $x^\mu(\sigma)$ and of one additional bosonic variable, depending on σ.

Now let us determine the interaction terms in the string field action. It is well known that all regular closed orientable two-dimensional manifolds (which by our assumption (14) represent perturbation theory vacuum diagrams) have topology of a sphere with handles and hence can be "glued" out of 3-tube "vertices". Slicing a 2-manifold by a system of parallel planes it is easy to see that the only "interaction" of contours which can occur in a slice and is stable under small variations of a slicing plane is that of two contours touching at one point and forming a single contour. "Interactions" in which contours touch at several distinct points or in which more than two contours touch at one point all have "zero measure" (cf. [5,6]). Thus the fundamental closed string interaction is uniquely determined to be the Φ^3-one

$$S_{int} = \frac{g}{3!} \int \mathcal{D}C_1 \mathcal{D}C_2 \mathcal{D}C_3 \, K[C_1, C_2, C_3] \, \Phi[C_1] \, \Phi[C_2] \, \Phi[C_3] \qquad (18)$$

where the kernel K is non-zero only if one contour has a self-crossing and the other two coincide with its simple parts. We define K in terms of sum over infinitesimal 3-tube surfaces

$$K[C_1, C_2, C_3] = \lim_{\varepsilon \to 0} \int_{\partial M_\varepsilon = (C_1, C_2, C_3)} \mathcal{D}M \, \exp(-I_0) \qquad (19)$$

where we assume that $K=0$ if C_1, C_2, C_3 cannot form the boundary of an infinitesimal surface (an analogous representation of the interaction kernel in the light-cone gauge is given in [5]). K can also be represented in the following reparametrization invariant form probably following from (19) (cf. [6])

$$K = \int_0^1 d\lambda \, e_1(\lambda) \int_0^1 d\lambda' \, e_1(\lambda') \, \delta\big[C_1(\sigma_1) - \Theta(\sigma_1 - \lambda)\,\Theta(\lambda' - \sigma_1)\, C_2(\sigma_2) - \Theta(\lambda - \sigma_1)\,\Theta(\sigma_1 - \lambda')\, C_3(\sigma_3)\big],$$

where $C_i = \{x_i^M(\sigma_i), e_i(\sigma_i)\}$. As follows from (13), (19) the whole Φ^3-perturbation theory can be expressed in terms of sums over regular surfaces (just as the perturbation theory for the particle ϕ^3-theory can be represented in terms of sums over paths). For example, the correlators $\langle \Phi[C_1] \ldots \Phi[C_N] \rangle$ are given by sums over arbitrary "N-tube" surfaces with the topology (in the tree approximation) of a sphere with N discs cut out.

4. Effective string field action. To define the quantum effective action Γ_q in the theory (14), (18) note first that unlike the particle theory case the notion of an irreducible vacuum graph is meaningless in the string theory (any 2-surface can be cut into two pieces by first deforming it and then cutting the connecting tube). Hence Γ_q is to be given by the sum of all connected vacuum diagrams on a background

$$\exp(-\Gamma_q[\bar{\Phi}]) = \int \mathcal{D}\Phi \exp\Big(-S[\bar{\Phi}+\Phi] + S[\bar{\Phi}] + \frac{\delta S[\bar{\Phi}]}{\delta \bar{\Phi}}\Phi\Big), \tag{20}$$

Again unlike the particle theory case all graphs with a given number of loops are now topologically equivalent and thus there is essentially one diagram in each loop order. (The propagator (13) and the kernel (19) satisfy certain identities which make it possible to prove various relations between diagrams, e.g. ⊖ = O--O).

Expanding S at $\bar{\Phi}$ we see that as in the particle Φ^3-theory the background enters the perturbation theory only through the propagator which can be represented as sum over surfaces (cf. (6), (13))

$$\tilde{G}[C_1,C_2] = (\Delta + K\cdot\tilde{\bar{\Phi}})^{-1} = \int_{\partial M=(C_1,C_2)} \mathcal{D}M \exp(-I_0 - I_{int}). \tag{21}$$

Here $\hat{\Phi}$ is a functional of $g_{ab}(z)$ and $x^M(z)$ related to $\tilde{\bar{\Phi}}, I_{int}=\hat{\Phi}$. To justify (22) one may expand $\tilde{G}$ in powers of $\tilde{\bar{\Phi}}$ and to use the "cylindrical" gauge $ds^2 = e^2(\sigma,\tau)d\sigma^2 + d\tau^2$. In this parametrization $\hat{\Phi}[x^M(z), g_{ab}(z)] = \int d\tau \tilde{\bar{\Phi}}_\tau[x^M(\sigma,\tau), e(\sigma,\tau)] + \dots$ (cf. (5)), where $\tilde{\bar{\Phi}}_\tau$ is the functional $\tilde{\Phi}$ on the one-parameter family of loops $C(\tau) = \{x^M(\sigma,\tau), e(\sigma,\tau)\}$, $C(0)=C_1$, $C(a)=C_2$ (exact form of the relation between $\hat{\Phi}$ and $\tilde{\bar{\Phi}}$ depends on the form of Δ). In a general coordinate system $\hat{\Phi}$ can be taken as an arbitrary covariant local functional of $x^{\mu}(z)$ and $g_{ab}(z)$ (non-trivially related to $\tilde{\Phi}$)

$$
\begin{aligned}
I_{int} = \hat{\Phi} = \int d^2z \sqrt{g} \Big\{ & \varphi(x) + \Big[\partial_a x^\mu \partial_a x^\nu H_{\mu\nu}(x) + \\
& + \frac{i}{\sqrt{g}} \varepsilon^{ab} \partial_a x^\mu \partial_b x^\nu A_{\mu\nu}(x) + R\,\phi(x) \Big] + \\
& + \Big[\partial_a x^\mu \partial_a x^\nu \partial_b x^\lambda \partial_b x^\rho B_{\mu\nu\lambda\rho}(x) + D^2 x^\mu \partial_a x^\nu \partial_a x^\lambda B_{\mu\nu\lambda}(x) \\
& + D^2 x^\mu D^2 x^\nu B_{\mu\nu}(x) + R\, \partial_a x^\mu \partial_a x^\nu C_{\mu\nu}(x) + \\
& + R D^2 x^\mu C_\mu(x) + R^2 C(x) + (\text{term with } \varepsilon^{ab}) \Big] + \dots \Big\}
\end{aligned}
\tag{22}
$$

Here R is the scalar curvature of $g_{ab}(z)$ and all possible covariant operators built of ∂x^μ and ∂g_{ab} are multiplied by arbitrary local functions of $x^\mu(z)$ (we included only the terms which cannot be reduced to each other by integrating by parts). The terms in (23) are grouped according to the dimension of the fields, i.e. the number of derivatives in the operator (we take x^μ to be dimensionless).

The path integral with the action $I_0 + I_{int}$ was interpreted in [3,4] as an "effective action" for the infinite system of fields corresponding to the string "vibration" modes

$$\hat{\Gamma}_c [\varphi, H_{\mu\nu}, A_{\mu\nu}, \phi] = - g^{-2} \int_{\partial M_2 = 0} DM \exp(-I_c - I_{int}) \tag{23}$$

(here we consider $\hat{\Gamma}$ in the tree approximation, so that the integration goes over surfaces with the topology of a sphere, $\chi = 2$). The reason for introducing (24) was the observation that I_{int} can be represented as

$$I_{int} = -\sum \int d^D x \; V_i(x)\, \varphi_i(x) \tag{24}$$

where φ_i stand for the system of background fields in (23) and V_i are the standard vertex operators in coordinate representation [9, 13, 4] (e.g. for the tachyon $V \sim \int d^2z \sqrt{g}\, \delta^D(x(z) - x_0)$). Hence $A_n \sim \delta_n \hat{\Gamma} / \delta\varphi \ldots \delta\varphi$ with the fields restricted on shell give the standard on-shell scattering amplitudes ($\sim \langle V \ldots V \rangle$) . Since the dual theory diagrams are certainly "irreducible" from the ordinary field theory point of view, we are tempted to interpret $\hat{\Gamma}$ (extended off-shell) as a sort of effective action (see, however, sect.5).

The fields φ, $G_{\mu\nu} = \tilde{c}_{\mu\nu}$, $A_{\mu\nu}$ and ϕ were identified in [3,4] as the background values of a "tachyon", the D-dimensional metric, the antisymmetric tensor [14] and the "dilaton" [15] respectively. The other fields in (23) correspond to higher level (dimension) massive string modes (to the physical degrees of freedom as well as to auxiliary modes needed in a covariant formulation [7,8]). The fields multiplying the operators depending on the 2-curvature R are necessarily auxiliary. In fact, they are absent in the light-cone gauge in which one represents the 2-surface as a flat domain of the complex plane (with necessary identifications of points), i.e. takes $g_{ab} = \delta_{ab}$. For the dilaton this observation is in agreement with [8,7], where ϕ appeared as a component of an auxiliary functional field (in the light-cone gauge the dilaton comes from $H_{\mu\nu}$). It is interesting to note also that in (23) there are enough candidates for auxiliary fields in a covariant massive spin-4 (and spin-2) theory (the minimal set of spin-4 auxiliary fields is $C_{\mu\nu}$, C_μ and C [16]).

Given the expression for the propagator on a background (22) it is straightforward to write down the sum over surfaces representation for the quantum effective action (21) (cf. (7), (8))

$$\Gamma_q[\bar{\Phi}] = -\sum_{\chi} g^{-\chi} \int_{\partial M_\chi = 0} DM \exp(-I_0 - I_{int}) \tag{25}$$

Here the sum runs the non-positive values of the Euler

number of 2-surfaces with the topology of sphere with L handles ($\chi = 2 - 2L = -n$, $n = 0, 2, \ldots$ is the number of Φ^3-vertices in a "graph", L is the number of "loops"). It is plausible that if one carefully eliminates multiple countings of virtual surfaces of a given topology, numerical weight factors like d_n in (7) cancel out so that different topologies are summed simply with "equal weight" (up to a factor of $g^{-\chi}$). The weight $g^{-\chi}$ is conjectured in [3,4] to be correct because $\chi \ln g$ can be represented as a local addition to the string action (12), (23) ($\chi = \frac{1}{4\pi}\int d^2z \sqrt{g} R$) so that the coupling constant ($\ln g$) could be absorbed in the vacuum value of the dilaton (cf. [2,3,4]).

Let us now discuss the meaning of functional derivatives of Γ_q with respect to the background field $\bar{\Phi}$. The functional derivative of the propagator (22)

$$\delta \tilde{G}[C_1,C_2] / \delta \tilde{\Phi}[C_3] = -\int DC_1' DC_2' \tilde{G}[C_1,C_1'] K[C_1',C_2',C_3] G[C_2',C_2] \quad (26)$$

corresponds geometrically to cutting a disc bounded by C_3 out of a cylinder connecting C_1 and C_2. The "vertex" that does this operation in the sum over surfaces (22) is $\hat{V} = \delta \hat{\Phi}[M] / \delta \tilde{\Phi}[C_3]$ (its particle-theory analogue $\int d\tau\, e\, \delta^D(x(\tau) - x_3)$ places a point x_3 on a path connecting x_1 and x_2, cf. (6)). Unlike the particle-theory where it is not possible to generate "trees" attached to vacuum loops by inserting external "points" on the quantum lines here we can always deform a surface representing L-loop diagram and having N discs cut out in order to make it equivalent to a surface representing an arbitrary L-loop connected diagram with N external strings. This suggests that $\delta^n \Gamma_q / \delta \tilde{\Phi} \ldots \delta \tilde{\Phi}$ represent <u>loop corrections to connected Green functions in the theory (14), (18)</u>.

The propagator (22) differentiated $N-2$ times with respect to $\tilde{\Phi}$ corresponds to an arbitrary connected tree diagram with N external strings ("tubes"). Introduce a functional $\Gamma_c[\tilde{\Phi}]$ such that

$$\tilde{G}[C_1, C_2] = -g^2 \frac{\delta^2 \Gamma_0}{\delta \tilde{\Phi}[C_1] \delta \tilde{\Phi}[C_2]} . \qquad (27)$$

Since each $\delta/\delta\tilde{\Phi}$ is equivalent to inserting the "vertex operator" $\hat{V}$ which cuts a disc out of the 2-surface, (28) is solved by Γ_0 given by an integral over closed surfaces with the topology of a sphere

$$\Gamma_0 = -g^{-2} \int_{\partial M_2 = 0} \mathcal{D}M \; exp(-I_0 - I_{int}). \qquad (28)$$

Comparing (24) with (29) we see that $\Gamma_0 = \hat{\Gamma}_c$. Meanwhile Γ_0 is a generating functional for connected tree Green functions.

It is remarkable that the tree "effective action" has the same functional integral representation as the quantum one (26). This result is peculiar to string theory. Though it is possible to express the particle theory tree diagrams as sums over paths, no path integral representation analogous to (29) exists for the corresponding generating functional $W_c(J)$ (10) (the effective action Γ (5) coincides in the tree approximation with the classical action and hence its derivatives do not reproduce all the connected tree diagrams).

5. Effective action or source functional? The above discussion of effective action may seem very confusing from the standard field theory point of view. It may seem that in order to define a generating functional for string-theory connected Green functions we should introduce a source $J[C]$ and consider the functional of the source $W[J]$ satisfying the relations analogous to (9)-(11). In particular, $W = W_c + W_q$,

$$W_c = S[\bar{\Phi}(J)] - \bar{\Phi}(J)\cdot J, \qquad \frac{\delta S[\bar{\Phi}]}{\delta \bar{\Phi}} = J, \qquad (29)$$

$$W_q[J] = \Gamma_q[\bar{\Phi}(J)], \qquad (30)$$

where Γ_q is the "effective action" (21) of the previous section. Then

$$\frac{\delta W_c}{\delta J} = -\bar{\Phi}(J), \quad \frac{\delta^2 W_c}{\delta J[C_1]\delta J[C_2]} = -\frac{\delta\bar{\Phi}}{\delta J} = -\bar{G}[C_1,C_2], \qquad (31)$$

$$\bar{G}^{-1}[C_1,C_2] = \frac{\delta^2 S[\bar{\Phi}]}{\delta\bar{\Phi}[C_1]\delta\bar{\Phi}[C_2]} = \Delta + K\cdot\tilde{\bar{\Phi}}(J), \quad \tilde{\bar{\Phi}} = g\,\bar{\Phi}, \qquad (32)$$

i.e. $\bar{G}$ is the propagator $\tilde{G}$ (22) with $\bar{\Phi}$ replaced by $\bar{\Phi}(J)$. Higher derivatives of W_c give connected (tree) Green functions on the background $\bar{\Phi}(J)$

$$\bar{G}_n[C_1,\ldots,C_n] = -\frac{\delta^n W_c}{\delta J[C_1]\ldots\delta J[C_n]}. \qquad (33)$$

In particular (cf. (27))

$$\bar{G}_3 = \frac{\delta\bar{G}}{\delta\bar{\Phi}}\bar{G} = g\,K\cdot\bar{G}\cdot\bar{G}\cdot\bar{G},$$

$$\bar{G}_4 = \frac{\delta^2\bar{G}}{\delta\bar{\Phi}\,\delta\bar{\Phi}}\cdot\bar{G}\cdot\bar{G} + \frac{\delta\bar{G}}{\delta\bar{\Phi}}\cdot\frac{\delta\bar{G}}{\delta\bar{\Phi}}\cdot\bar{G} \qquad (34)$$

Due to (22), (27) $\bar{G}_n$ have sum over surfaces representations

$$\bar{G}_n[C_1,\ldots,C_n] = \int_{\partial M_2=(C_1,\ldots,C_n)} DM\ \exp(-I_c - \bar{I}_{int}) \qquad (35)$$

where $\bar{I}_{int}$ is a functional of g_{ab} and x^{μ} related to $\bar{\Phi}(J)$ in the same way as I_{int} in (22) is related to $\bar{\Phi}$. An analogous representation for $W_q[J]$ holds (cf. (31), (26)) and it presumably exists also for

$$W_0[J] = -g^{-2}\int_{\partial M_2=0} DM\ \exp(-I_c - \bar{I}_{int}) \qquad (36)$$

In fact, (37) is consistent with (34) and (36) if the insertion of the "vertex operator" $\delta\bar{I}_{int}[M]/\delta J[C]$ is equivalent to cutting a disc out of the 2-surface. It would be desirable to derive (37) directly from the definition (30) of W_c. For this it may be useful to suppose that $J[C]$ has a path integral representation $J[C] = \int_{\partial M=C} DM\, e^{-I_0}\,\mathcal{J}[M]$ and to solve the classical equation $\delta S/\delta\bar{\Phi} = J$ perturbatively in J in order to find an analogous representation for $\bar{\Phi}(J) = \int_{\partial M = C} \mathcal{D}M\, e^{-I_0} F[M]$. The final step is to use $\delta W_c / \delta J = -\bar{\Phi}(J)$ in order to identify F with $e^{-I_{int}}$ in (37).

Using $W_c[J]$ we can compute the corresponding on-shell scattering amplitudes and generating functional for (tree) S-matrix, $\hat{\Gamma}_0[\Phi_{in}] = W[J]\big|_{J \to \Phi_{in}\Delta}$ (Δ is the free kinetic operator (14)). $\hat{\Gamma}_0$ is nothing but the functional (24) with the arguments restricted on shell. It was suggested in [17, 15, 18] that to study the field theory limit of a dual model one may try to identify the (appropriate limit of) dual-theory S-matrix with the field theory S-matrix. Thus it appears that having found $\hat{\Gamma}_0[\varphi_{in}]$ (φ denote the "component" fields in (24)) we still should make a Legendre transformation to eliminate all reducible tree diagrams and find the corresponding field theory action $S[\varphi]$ (cf. [19])

$$\hat{\Gamma}_0[\varphi_{in}] = S[\varphi(\varphi_{in})] - \varphi_{in}\Delta\varphi, \qquad (37)$$
$$\varphi = \varphi_{in} - \Delta^{-1}\frac{\delta S_{int}[\varphi_{in}]}{\delta\varphi_{in}} + \ldots , \qquad \Delta\varphi = \frac{\delta S_0}{\delta\varphi}$$

This recipe obviously contradicts the suggestion of [3,4] to interpret $\hat{\Gamma}_0$ or Γ_0 (29) (and Γ_q (26)) as an action of an effective field theory, approximating string dynamics.

Now, is it possible to reconcile the approach of [3,4] with that of [17, 15, 18]? A positive answer is based on the following statement: there exists a non-trivial equivalence between the source and background field functionals in string theory. This equivalence is a manifestation of duality or the two-dimensional geometrical nature of string theory perturbation diagrams. Its existence and hence "triviality" of the Legendre transformation in string theory is prompt already by the indistinguishability of "reducible" and "irreducible" string diagrams noted above. The connected Green functions are generated, according to sect. 4, by $\Gamma_0[\tilde{\Phi}]$ (29) and $\Gamma_q[\tilde{\Phi}]$ (26), while according to the present section they are generated by $W_c[J]$ (30) and $W_q[J]$ (31). To prove that the objects obtained by differentiating Γ and W with respect to $\tilde{\Phi}$ and J (and putting $\tilde{\Phi}=0$ and $J=0$) are in fact the same, one should use various identities between products of G and K which follow from their geometrical definitions (13), (19) (and express the property of duality). The equivalence between Γ and W must be understood in the sense that they have the sum over surfaces representations which are of the same form (cf. (29) and (37)) the only difference being in the different arguments and hence in the different local fields in I_{int} (23) and in $\bar{I}_{int}=\int d^2z\sqrt{g}\,\bar{\varphi}(x)+\dots$. The precise connection between $\bar{\Phi}$ and J (or between $\varphi, H_{\mu\nu}\dots$ and $\bar{\varphi}, \bar{H}_{\mu\nu}\dots$) is not important since being related by a redefinition of their arguments Γ and W lead to the same "mass-shell", $\frac{\delta\Gamma}{\delta\Phi}=0$ or $\frac{\delta W}{\delta J}=0$. This equivalence stated in terms of $\hat{\Gamma}_0[\varphi]$ in (24) and $S[\varphi]$ (38) implies that $\hat{S}$ has the same sum over surfaces representation as $\hat{\Gamma}_0$ (24) but with different system of local fields. The theories with the actions $\hat{\Gamma}_0$ and $\hat{S}$ are equivalent on-shell and have equivalent S-matrices (as two theories related by field redefinitions). Thus it does not matter whether we call the arguments of Γ "sources" or "background fields" and to study the string theory ground state problem we may start directly with the equation $\frac{\delta\Gamma}{\delta\tilde{\Phi}}=0$ or $\delta\hat{\Gamma}/\delta\varphi=0$ which is exactly the ground-state equation of [3,4].

Concluding note that it would be desirable to understand a connection between the tree effective action (24) or (29) and the classical string field action $g^{-2}S[\tilde{\Phi}]$ (14), (18). In fact, the low-energy field theory actions derived by "integrating out" massive string modes in $S[\tilde{\Phi}]$

(using $\frac{\delta S}{\delta \tilde{\Phi}} = 0$) and by direct computing $\Gamma_c [\tilde{\Phi}]$ should agree. It also remains to prove that $\delta \Gamma_c / \delta \tilde{\Phi} = 0$ implies that the "on-shell" local fields in (23) are such that the string action $I_c + I_{int}$ is that of a conformally invariant two-dimensional field theory.

Though there remain many principal as well as technical problems it seems that covariant sums over surfaces provide a viable approach for constructing a second-quantized string theory and serve as a basis for developing the background string field method.

We acknowledge useful conversations with S.M. Apenko, E.S. Fradkin, R.E. Kallosh, A.M. Semikhatov, I.V. Tyutin, M.A. Vasiliev and B.L. Voronov.

Remark. After this paper had been written we received [20] containing an analogous suggestion to use Polyakov's path integral to define the propagator of the second-quantized string theory.

References

1. Schwarz, J.H. (1982). Phys. Rep. 89, 223; Green, M.B., Schwarz, J.H. (1984). Phys. Lett. 149B, 117; (1985). 151 B, 21.

2. Candelas, P., Horowitz, G.T., Strominger, A., Witten, E. (1985). Nucl. Phys. B258, 46.

3. Fradkin, E.S., Tseytlin, A.A. (1985). ZhETFPisma (Russian), 41, 169; (1985). Phys. Lett. 158B, 316; (1985). Lebedev Inst. preprint N 150; (1985). Nucl. Phys. B261, 1.

4. Tseytlin, A.A. (1985). Lebedev Inst. preprint N 153; (1986). Yad. Fiz. 43 (Russian, to appear).

5. Kaku, M., Kikkawa, K. (1974). Phys. Rev. D10, 1110, 1823.

6. Marshall, C., Ramond, P. (1975). Nucl. Phys. B85, 375.

7. Banks, T., Peskin, M.E. (1985), preprint SLAC-PUB-3740; Kaku, M. (1985). City College preprint; Friedan, D. (1985). Chicago preprint EFI-85-27; Neveu, A., West, P.C. (1985). CERN preprint TH. 4200.

8. Siegel, W. (1984). Phys. Lett. 149B, 157, 162; (1985). 151B, 391, 396; Siegel, W., Zwiebach, B. (1985). Berkeley preprint UCB-PTH 85/30.

9. Polyakov, A.M. (1981). Phys. Lett. 103 B, 207; Friedan, D. (1984). In: Recent Advances in field theory and statistical mechanics North-Holland; Alvarez, O., (1983). Nucl. Phys. B216, 125; Fradkin, E.S. Tseytlin, A.A. (1982). Ann. of Physics 143, 413.

10. Witt, B.S.De. (1965). Dynamical Theory of Groups and Fields, Gordon and Breach.

11. Brink, L., Vecchia, P. di, Howe, P.S. (1976). Phys. Lett. 65B, 471.

12. Polyakov, A.M., unpublished; Goldshmidt, Y., Tan, C.I. (1982). Phys. Lett. 114B, 257.

13. Aoyama, H., Dhar, A., Namazie, M. (1985). Preprint SLAC-PUB-3623; Weinberg, S. (1985). Phys. Lett. 156B, 309.

14. Ogievetsky, V.I., Polubarinov, I.V. (1966). Yad. Fiz. (Nucl. Phys.), 4, 202 (Russian); Kalb, M., Ramond, P. (1974). Phys. Rev. D9, 2273.

15. Scherk, J., Schwarz, J.H. (1974). Nucl. Phys. B81, 118; Ademollo, M., d'Adda, A.D., Auria, R.D., Gliozzi, F., Napolitano, E., Sciuto, S., Vecchia, P. di. (1975). Nucl. Phys. B94, 221.

16. Singh, L.P.S., Hagen, C.R. (1974). Phys. Rev. D9, 898, 910

17. Scherk, J. (1971). Nucl. Phys. B31, 222; Nakanishi, N. (1972). Progr. Theor. Phys. 48, 355; Neveu, A., Scherk, J. (1972). Nucl. Phys. B36, 155.

18. Yoneya, T. (1974). Progr. Theor. Phys. 51, 1907.

19. Kikkawa, K., Yamasaki, M. (1985). Osaka Univ. preprint. OU-HET-83.

20. Cohen, A., Moore, G., Nelson, P., Polchinski, J. (1985) Harvard preprint HUTP-85/A058.

STRING MODEL OF HADRONS AND THEIR INTERACTIONS

V.I.Borodulin, G.P.Pron'ko, A.V.Razumov, L.D.Soloviev, O.L.Zorin
Institute for High Energy Physics, Serpukhov, 142284

Quantum chromodynamics is presently the most popular theory of strong interactions. On the other hand, the problem of confinement and description of hadrons whose constituent quarks are spaced far from each other awaits for its solution.

It has been shown in [1] that the relativistic string in one-mode (rectilinear) approximation can be consecutively quantized in four-dimensional space-time whereupon it describes the states belonging to the leading Regge trajectory. From the physics viewpoint such an approximation is fairly reasonable to be used to describe the low-energy phenomenology of mesons, i.e. their spectra, decays, etc.

Similar models of extended particles were offered earlier [2,3] . However, the fact that when obtaining our model we proceed from the relativistic string allows to conserve its geometric interpretation and to introduce particle interactions in a natural way. Using a relative simplicity of the model one can take into account quark masses. This brings about changes in the Regge trajectory. Account of the spin of quarks splits the Regge trajectory into four trajectories: the one,corresponding to the single state, contains particles with an abnormal parity and the remaining three correspond to the triplet state and contain particles with a normal parity, To describe the flavour of quarks, we introduce singlet and octet string fields in which case the mixing angle is a free parameter. The model constructed in such a way is used to describe meson decays. Two coupling constants corresponding to F and D couplings are introduced as interaction parameters for all decays. These constants are determined from two decays and are further used to calculate the decay widths.

Let us considere the action of the string

$$S = -(2\pi\alpha')^{-1} \int d\tau \int d\sigma \{(\dot{y}y')^2 - \dot{y}^2 y'^2\}^{1/2}. \qquad (1)$$

Here $y_\mu(\sigma,\tau)$ is the position of the string point in four-dimensional space-time, σ characterizes its extent $(y(\sigma,\tau) - y(\sigma',\tau)$ is a time-like vector), and τ characterizes the time evolution $(y(\sigma,\tau) - y(\sigma,\tau')$ is a time-like vector). The prime and the dot mean the derivatives over σ and τ, respectively.

For the straight-line string

$$y_\mu(\sigma,\tau) = X_{\mu(\tau)} + q_\mu(\tau) f(\sigma,\tau). \qquad (2)$$

Inserte (2) into (1). The integrand takes the form

$$f'(\sigma,\tau)\left[-c(f(\sigma,\tau)-f_1(\tau))(f(\sigma,\tau)-f_2(\tau))\right]^{1/2},$$

where $f_{1,2}(\tau) = -\eta \pm (\xi+\eta^2)^{1/2}$ correspond to the string ends and

$$\xi = a/c, \quad \eta = b/c, \quad a = (\dot{X}q)^2 - q^2\dot{X}^2,$$
$$b = (\dot{X}\dot{q})q^2 - (\dot{X}q)(q\dot{q}), \quad c = \dot{q}^2q^2 - (q\dot{q})^2.$$

The integration over σ in (1) is reduced to that over f. Integrating from f_1 to f_2 we obtain the action and the Lagrangian of the straight-line string:

$$S = \int d\tau \; L \qquad (3)$$
$$L = -(4\alpha')^{-1} c^{1/2}(\xi+\eta^2) \qquad (4)$$

It may be shown that the string equations of motion and boundary conditions following from (1) are fulfilled, provided so are the equations of motion following from (3) and (4).

Let us find the momenta P_μ and r_μ canonically conjugated with the coordinates X_μ and q_μ. Lagrangian (4) is invariant under reparametrizations with arbitrary τ-dependent functions, therefore the canonical variables satisfy the constraints

$$\phi_1 = Pq \approx 0 \qquad (5)$$

$$\phi_2 = rq \approx 0 \qquad (6)$$

$$\phi_3 = \alpha' P^2 - \left[q^2(r^2-(Pr)^2/P^2)\right]^{1/2} \approx 0 \qquad (7)$$

Here the wavy equality sign means that the constraints will be considered to equal zero on calculating all the Poisson brackets.

The Poisson brackets of these constraints are:

$$\{\phi_1, \phi_2\} = \phi_1, \quad \{\phi_1, \phi_3\} = \{\phi_2, \phi_3\} = 0 \qquad (8)$$

The canonical Hamiltonian of the system is zero, therefore (5)-(7) are first class constraints. The evolution of the variables is determined by the generalized Hamiltonian which is a linear combination of the constraints

$$H = \sum_{1\leq i\leq 3} V_i\phi_i, \quad g = \{g, H\} \qquad (9)$$

The conserved Lorentz group generators found from (3) by the Noether theorem are of the form

$$M_{\mu\nu} = X_\mu P_\nu - X_\nu P_\mu + q_\mu r_\nu - q_\nu r_\mu, \qquad (10)$$

and the translation generator coincides with P_μ. The spin of the system is determined by the pseudovector W_μ ($\epsilon_{0123} = -1$)

$$W_\mu = \frac{1}{2}\epsilon_{\mu\nu\rho\sigma}P^\nu M^{\rho\sigma}/(P^2)^{1/2} \qquad (11)$$

As may easily be verified, if constraints (5)-(7) hold

$$\phi_3 \approx \alpha' P^2 - (-W^2)^{1/2}.$$

Constraints (5)-(7) are the generators of a gauge group.

In our further investigation we can act in two way. The first one is to fix the gauge amtiguity by introducing gauge-constraints. If conditions

$$X_1 = Pr \approx 0, \quad X_2 = r^2-q^2 \approx 0 \qquad (12)$$

are imposed, then constraints (5)-(7) and (12) coincide with conditions (12) and (13) of [1]. In accordance with the general theory for constrained systems, the Poisson brackets should be replaced in this case by the Dirac ones, determined with regard for (5), (6), and (12), whereupon constraints (5), (6) and (12) should be considered rigorously and X_μ, P_μ, q_μ, r_μ should be replaced by independent variables. For this introduce the tetrad $e^{(\alpha)}(P)$, $\alpha = 0,1,2,3$, associated with the total momentum as $e^{0}_{\mu} = P_\mu/(P^2)^{1/2}$, $e^{(\alpha)} e^{(\beta)} = g^{\alpha\beta}$, $g_{\alpha\beta} e^{(\alpha)}_{\mu} e^{(\beta)}_{\nu} = g_{\mu\nu}$ and consider the variables P_μ, Z_μ, n_i, Li, $i = 1,2,3$, which will be referred to as independent ones,

$$n_i = e^{(i)} q/(-q^2)^{1/2}, \quad L_i = e^{(i)} W,$$
$$Z_\mu = X_\mu + \frac{1}{2} \epsilon_{ijk} L_i e^{(j)} \partial e^{(k)}/\partial P^\mu, \quad P_\mu, \tag{13}$$

where W_μ is the pseudovector from (11), associated with the spin. According to the construction technique, $\vec{n}^2=1$, $\vec{L}\vec{n}=0$. For the new variables, the non-zero Dirac brackets are of the form

$$\{Z_\mu, P_\nu\} = g_{\mu\nu}, \quad \{L_i, L_j\} = -\epsilon_{ijk} L_k, \quad \{L_i n_j\} = -\epsilon_{ijk} n_k \tag{14}$$

Relations (13) can be inverted e.g. as follows:

$$q_\mu = -e^{(i)}_{\mu} n_i (\vec{L}^2)^{1/4}, \quad r_\mu = -e^{(i)}_{\mu} (\vec{L} \times \vec{n})_i (\vec{L}^2)^{-1/4} \tag{15}$$

In terms of the independent variables, the remaining constraint (7) and Lorentz group generators (10) have the form

$$\alpha' P^2 - (\vec{L}^2)^{1/2} \approx 0, \tag{16}$$

$$M_{\mu\nu} = Z_\mu P_\nu - Z_\nu P_\mu - \frac{1}{2} \epsilon_{ijk} L_i \Big[e^{(j)} (\partial e^{(k)}/\partial P^\mu) P_\nu - e^{(j)} (\partial e^{(k)}/\partial P^\nu) P_\mu - 2 e^{(i)}_{\mu} e^{(j)}_{\nu} \Big]. \tag{17}$$

To quantize the system, one should replace brackets (14) by commutation relations for the operators and impose conditions (16) on the wave functions. The construction of the operators satisfying (14) presents no difficulty. The solution of (16) yields the mass-spin relation

$$\alpha' M^2 = \big[L(L+1)\big]^{1/2}, \quad L = 0,1,\ldots . \tag{18}$$

Note, however, that under an appropriate choice of variables in the classical theory and with a proper ordering of the operators in quantum equation (16), one can get the linear Regge trajectory $\alpha' M^2 = L$ instead of (18). To be consistent, we shall treat in our further consideration the quantum theory corresponding to (18).

The construction of the quantities q_μ and r_μ in the quantum theory is ambiguous, in view of the necessity to fix the order of the operators in (15). It is important that with the ordering of the operators in accordance with (17), in the quantum theory the algebra of the Poincare group generators is fulfilled and the theory is relativistic invariant.

Thus, for the straight-line string one succeeds in choosing independent variables (13), among which the spin plays an important role, enabling to construct the quantum theory in four-dimensional space. Just here lies the difference between our quantization procedure and that of an arbitrary string in the transverse gauge, when the quantum theory proves to be relativistic invariant in 26-dimensional space only [4].

The other way is to consider the canonical variables S, P, q, r as ranging the values in the whole phase space and evolving in accordance with (9), while the physical subspace is defined by constraints (5)-(7).

In the quantum theory, the canonical variables become the operators satisfying ordinary commutation relations. They are defined in the whole state space. The physical subspace is singled out by the conditions

$$\phi_i \quad \Phi_{ph} = 0, \tag{19}$$

where ϕ_i are constraints (5)-(7) with ordered, eg symmetrized, operators instead of the classical variables. Such an ordering does not violate the commutation relations of constraints (8), and thus conditions (19) are consistent.

One can construct a general solution of (19). Then the physical state space, from the viewpoint of the Poincare group representations, turns out to be a direct sum of the subspaces corresponding to the particles of arbitrary spins and masses, given by formula (18). Thus, one obtains the physical state space coinciding with the state space, evolving with the first quantization procedure used. The advantage of the second method is its manifest relativistic invariance.

Our further step is the second quantization, i.e. transition to the quantized string field, whose elementary excitations, unlike those of a local field, are extended rather than point-like objects.

In terms of the independent variables, the field equation is

$$(\alpha'\Box - (\vec{L}^2)^{1/2})\ \phi(Z, \vec{n}) = 0 \tag{20}$$

One can easily write the free field action leading to (20) and perform the canonical quantization [1]. As a result, the quantized field $\phi(Z, \vec{n})$ is obtained whose excitations constitute an infinite set of particles belonging to the Regge trajectory (18). The expansion of this field in creation-annihilation operators is of the form

$$\phi(Z,\vec{n}) = \sum_{L,m} \left(\frac{2L+1}{4\pi}\right)^{1/2} \int d^3\tilde{P}_L \Big[a^{+}{}_L^m(\vec{P})\ Y_L^m(\vec{n}) e^{i(\omega_L(\vec{P})Z_0 - \vec{P}\vec{Z})} + a_L^m(\vec{P})\ Y^{*}{}_L^m(\vec{n}) e^{-i(\omega_L(\vec{P})Z_0 - \vec{P}\vec{Z})}\Big], \tag{21}$$

Here $\omega_L^2(\vec{P}) = \vec{P} + M_L^2$, $M_L^2 = \frac{1}{\alpha'}\left[L(L+1)\right]^{1/2}$, $d^3\tilde{P}_L = \dfrac{d^3P}{(2\pi)^3 2\omega_L(\vec{P})}$,

and $Y_L^m(\vec{n})$ are spheric harmonics. The operators of particle creation and annihilation with the momentum $\vec{P}$, mass M_L, spin L and spin projection m satisfy the commutation relations:

$$\left[a_L^m(\vec{P}), a_{L'}^{+m'}(\vec{P}')\right] = (2\pi)^3 2\omega_L(\vec{P})\,\delta(\vec{P}-\vec{P}')\,\delta_{LL'}\delta_{mm'} \tag{22}$$

In the covariant formulation, the string field depends on two 4-vectors, X_μ and q_μ, and satisfies

$$(\alpha'\Box - (-W^2)^{1/2})\,\phi(X,q) = 0$$

and

$$q^\mu \frac{\partial}{\partial X^\mu}\ \phi(X,q) = 0, \quad q^\mu \frac{\partial}{\partial q^\mu}\ \phi(X,q) = 0 \tag{23}$$

Conditions (22) mean that $\phi(X,q)$ is independent of both the q_μ-direction of X_μ and q^2. Therefore one has to use some other measure, e.g. the one from [1], instead of d^4Xd^4q:

$$d^4XD(q) = 2d^4Xd^4q\delta(q^2+1)\delta(\frac{1}{i}\frac{\partial}{\partial X^\mu}\frac{1}{i}\frac{\partial}{\partial q_\mu})\sqrt{\Box} \tag{24}$$

The connection between the two ways, used to describe the string field, is fairly simple:

$$\Phi(P,q) = (-P^2q^2)^{1/2}\delta(Pq)\varphi(P, \frac{e^{(i)}(P)q}{(-q^2)^{1/2}}). \tag{25}$$

Here $\Phi(P,q)$ and $\varphi(P,n_i)$ are the momentum representations of the covariant field and the field in independent variables, respectively.

To introduce the straight-line string interaction, one can make use of the geometric picture of their decay and fusion. In this case one should take into account, that at the instant of the decay the decaying string orientation coincides with those of the decay products and the initial string length equals the sum of the product lengths. The latter condition is, however, inessential because by (22) the field is q^2-independent. So, the simplest interaction vertex reduces to $\phi^3(X,q)$ up to the formfactor. The contribution from interaction to the action may be written as

$$A_I = g\int d^4X\phi^2(X,q)D(q)\phi(X,q). \tag{26}$$

Defining the scattering matrix as a formal Wick T-exponent, we can construct the perturbation theory and derive the Feynmann rules for the processes whose particles belong to the Regge trajectory. Yet, our model is far from being able to describe the real world, because it does not take into account quarks on the string ends.

To estimate the effect of the masses of the string ends, we use the fact that the Lagrange equations for a string with point-like masses m_1 and m_2 on its ends have the solution presentable as a rectilinear rotating rod [5]. In this case the Regge trajectory for small (large) spins is of the form

$$L = \frac{4\pi m_1 m_2}{3(m_1+m_2)^2}\alpha'(M^2-(m_1+m_2)^2),$$

$$(L = \alpha' M^2 (1 - \frac{\pi^{1/2}}{3} ((\frac{m_1}{M})^{3/2} + (\frac{m_2}{M})^{3/2}) + 0((\frac{m_i}{M})^{5/2})).$$

So, the effect of the masses is essential only of the beginning of the trajectory, which can be considered, up to a sufficient accuracy, as linear with the intercept changed.

To describe the real mesons, we should also introduce the quark flavours and spins within the frames of the formalism available.

Considering only the mesons, containing just u, d, s quarks, it suffices to introduced the nonet of string fields $\Phi^\alpha(X,q)$ or the matrix field

$$\Phi_i{}^j(X,q) = \sum_{1 \leqslant d \leqslant q} \Phi^\alpha(X,q)(\Lambda^\alpha)_i{}^j \tag{27}$$

Here $(\Lambda^\alpha)_i{}^j$ are the linear combinations of $[1]$ and eight λ-matrices, which correspond to the states with the definite values of isospin I, I_3 and hypercharge Y. (For states with I=0 the matrices Λ^α are constructed with regard for phenomenological singlet-octet mixing).

Since quarks have a half-interger spin, they may be either in a singlet (S=0) or in triplet (S=1) state. In the latter case, the spin of the meson is bound with the orbital momentum of the quarks by one of the three equalities:

$$J = L-1,\ L,\ L+1,\ (\text{if} L=0,\ \text{then } J=1).$$

In a singlet state, J=L. The mesons with equal spins J, but different S and L have unequal masses, i.e. there are four different Regge trajectories. The mesons with an abnormal parity belong to the trajectory S=0, which begins with the known nonet of pseudoscalar mesons, (π, K, η, η'), and is described by the pseudoscalar string field $\bar{\phi}(X,q)$.

In the case of S=1 we have three trajectories with the same slope, out of which two trajectories, $J=L\pm 1$, have a normal parity and the third one, J=L, has an abnormal one. The mesons, belonging to these trajectories, will be described by the vector string field $\Psi_\mu(X,q)$, which satisfies both subsidiary conditions (23) and the transversity condition

$$\frac{\partial}{\partial X_\mu} \psi_\mu(X,q) = 0.$$

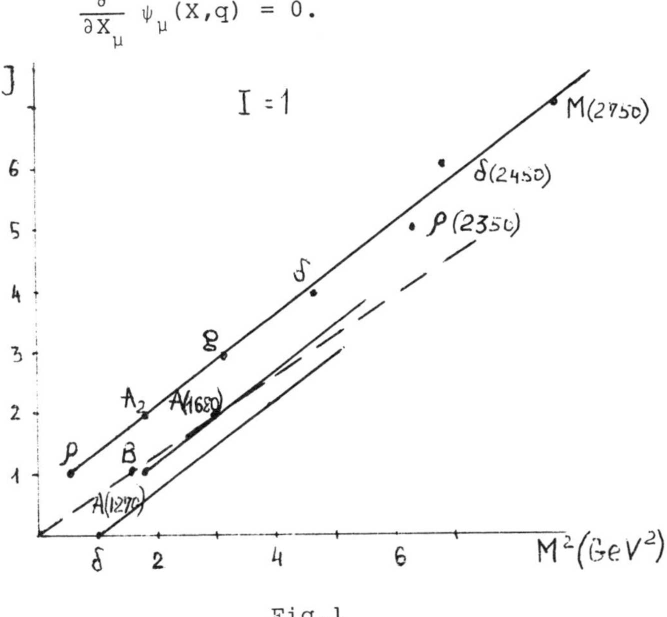

Fig.1

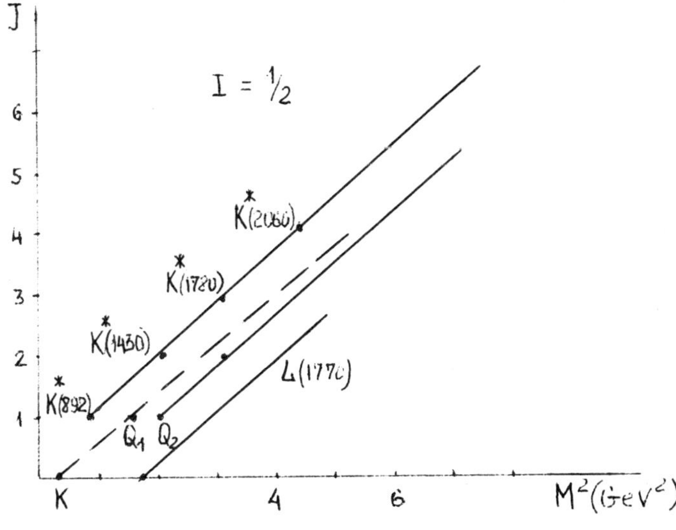

Fig.2

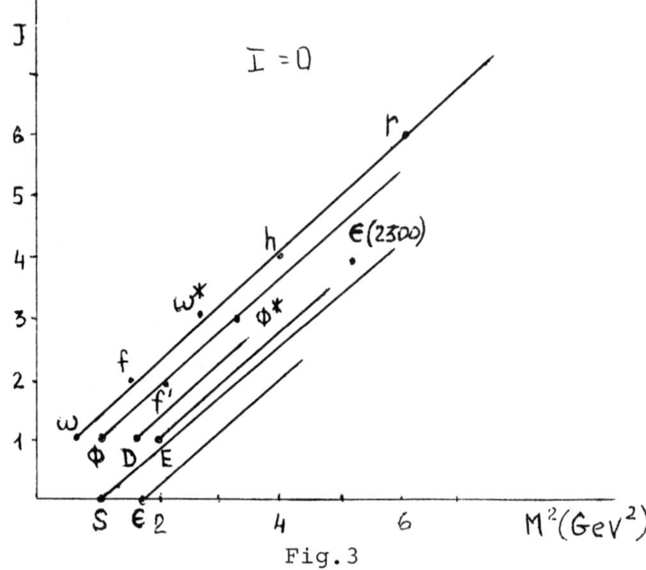

Fig.3

In the momentum representation, the field $\psi_\mu(X,q)$ can be expressed through independent components $\vec{\psi}(Z,\vec{n})$ (cf. with (25)) as follows:

$$\psi_\mu(P,q) = \vec{e}_\mu(P)\vec{\psi}(P,\frac{e^{(i)}(P)q}{\sqrt{-q^2}})\ \delta(Pq)\ \sqrt{-q^2P^2}, \qquad (28)$$

which, in their turn, are expanded in the local field $\psi^m_{L,J}(P)$, describing the mesons with the orbital moment L, spin J and the third projection of spin m:

$$\vec{\psi}(P,n^i) = \sum_{J,m} (\frac{2J+1}{4})^{1/2} \Big[(\frac{J}{2J+1})^{1/2} \psi^m_{J-1,J}(P)(\frac{1}{J}\frac{\partial}{\partial\vec{n}} + \vec{n}) +$$

$$+ (J(J+1))^{-1/2}\ \psi^m_{J,J}(P)\Big[\vec{n}\times\frac{\partial}{\partial\vec{n}}\Big] + (\frac{J+1}{2J+1})^{1/2}\ \psi^m_{J+1,J}(P)$$

$$(\frac{1}{J+1}\frac{\partial}{\partial\vec{n}} - \vec{n})\Big]\ Y^m_J(\vec{n}). \qquad (29)$$

To describe the interaction of strings, corresponding to real mesons let us make use of the fields introduced. Consider the vertex, containing two pseudoscalar and one vector string field. Proceeding from (26) we put it down in the form

$$A_I = \int d^4X D(q)\,Sp(\psi_\mu(X,q)\,(g_D q^\mu + g_F r^\mu(P_1,P_2))\bar{\phi}(X_1,q)\bar{\phi}(X_2,q)) \qquad (30)$$

$$X_1 = X_2 = X.$$

Here the trace is taken over the SU(3) indices, the measure of integration, D(q), is the same as in (24), and the operator

$$r^{\mu}(P_1,P_2) = (P_1-P_2)^{\mu}\,\frac{1}{2}\left[\frac{(P_1+P_2)^2}{(P_1P_2)^2-P_1^2P_2^2}\right]^{1/2},$$

where $P_{i\mu} = \frac{1}{i}\frac{\partial}{\partial X^{i\mu}}$ acts to the right.

The first term is (30) is symmetric with respect to the fields $\overline{\phi}(X,q)$, and describes the interaction with D-coupling, while the second one is antisymmetric with respect to them describing the interaction with F-coupling. Since the integration measure is an even function of q_{μ}, we obtain straight away the selection rules for the vertex with respect to the orbital moment L, i.e. with respect to the polynomial power q_{μ} in the expansion of the fields $\overline{\phi}(X,q)$ and $\psi_{\mu}(X,q)$ over the traceless multitensors. These rules are as follows: if the sum of the orbital moments of quarks in mesons is an odd number, $(L_1+L_2+L_3=2n+1)$, the decay is described by the vertex with D-coupling and if it is even, $L_1+L_2+L_3=2n$, the decay is described by the one with F-coupling. Besides, since the product $q_{\mu}\psi^{\mu}(X,q)$ contains no states with an abnormal parity, decays of such mesons are determined only by the vertex with F-coupling. The widths of the admissible decays are given by the expressions

$$\Gamma^{(D)}_{(L_3\neq J_3)} = V_{ph}g_D^2\,\frac{(2J_1+1)(2J_2+1)(J_3+L_3+1)}{2(2J_i+1)}\,I^D(J_1J_2J_3)\sum_{\{\alpha_f\}}\frac{(d^{\alpha_1\alpha_2\alpha_3})^2}{1+\eta}$$

$$\Gamma^{(F)}_{(L_3\neq J_3)} = V_{ph}g_F^2\,\frac{2(2J_1+1)(2J_2+1)}{(2J_i+1)(J_3+L_3+1)}\,I_1^F(J_1J_2J_3)\sum_{\{\alpha_f\}}(f^{\alpha_1\alpha_2\alpha_3})^2, \tag{31}$$

$$\Gamma^{(F)}_{(L_3=J_3)} = V_{ph}g_F^2\,\frac{(2J_1+1)(2J_2+1)(2J_3+1)}{(2J_i+1)\,J_3(J_3+1)}\,I_2^F(J_1J_2J_3)\sum_{\{\alpha_f\}}(f^{\alpha_1\alpha_2\alpha_3})^2.$$

Here V_{ph} is the phase space, η equals 0 or 1 (in the latter case there are two identical particles in the final state); J, L, α are the quantum numbers of the mesons:

J is the spin, L the orbital moment of quarks, and α the projection of the unitary spin. For the mesons with the quark spin S=0, the orbital moment coincides with the total spin L=J, therefore the quantum numbers of a pair of such mesons participating in the decay are $J_{1,2}$ and $\alpha_{1,2}$. To set the meson state with S=1, three quantum numbers, L_3, J_3, and α_3, are required. The spin of the decaying meson is denoted by J_i, the unitary spin projections of the mesons in the final state are α_f's and $f^{\alpha_1\alpha_2\alpha_3}$ and $d^{\alpha_1\alpha_2\alpha_3}$ are the structure constants of SU(3). The factors $I(J_1,J_2,J_3)$ are the integrals of the product of the Legendre polynomials:

$$I^D(J_1,J_2;J_3) = \frac{1}{\pi}\int_0^\pi d\theta P_{J_1}(\cos\theta)P_{J_2}(\cos\theta)P_{J_3}(\cos\theta),$$

$$I_1^F(J_1,J_2;J_3) = \frac{1}{\pi}\int_0^\pi d\theta P_{J_1}(\cos\theta)P_{J_2}(\cos\theta)P'_{J_3}(\cos\theta), \qquad (32)$$

$$I_2^F(J_1,J_2;J_3)=\frac{1}{\pi}\int_0^\pi d\theta P_{J_1}(\cos\theta)P_{J_2}(\cos\theta)\Big[J_3(J_3+1)P_{J_3}(\cos\theta)- - \cos\theta P'_{J_3}(\cos\theta)\Big].$$

The calculational results are presented in Tables.

The expressions for the widths of the decays with isoscalar mesons have the corresponding phenomenological angles θ of singlet-octet mixing. $\theta_0\approx 35.3°$ is the ideal mixing angle ($Tg\theta_0=1/\sqrt{2}$), $\theta_1\approx 19.5°$ ($tg\theta_1=1/2\sqrt{2}$). The expressions for the widths of the decays h (2030) and ϵ (2300), marked with a star, have been obtained under the assumption on h (2030) being a more singlet state, and ϵ (2300) being a more octet one with the mixing angle $\theta\approx 30°$. The estimates for the decay widths have been obbained for the following values of the coupling constants: $g_D^2\approx 44$ GeV^2, $g_F^2\approx 24$ GeV^2.

J^{PC}	Mode	Γ/Vph	Γ^{th}(MeV)	Γ^{exp}(MeV)	
1^-	$K^*(892)\to K\pi$	$0.25\ g_F^2$	85	51.3	
2^+	$K^*(1430)\to K\pi$	$0.08\ g_D^2$	40	45 ± 5	
3^-	$K^*(1780)\to K\pi$	$0.08\ g_F^2$	20	27 ± 11	I=1/2
4^+	$K^*(2060)\to K\pi$	$0.05\ g_D^2$	18	15 ± 5	S=1
0^+	$K(1350)\to K\pi$	$0.75\ g_D^2$	413	~400	
2^-	$L(1770)\to K\pi$	0	0	0	
1^{--}	$\rho\to\pi\pi$	$0.33\ g_F^2$	189	154	
3^{--}	$g\to\pi\pi$	$0.11\ g_F^2$	30	48 ± 7	I=1
5^{--}	$\rho(2350)\to\pi\pi$	$0.064\ g_F^2$	24	?	S=1
7^{--}	$M(2750)\to\pi\pi$	$0.045\ g_F^2$	15	?	
1^{+-}	$B(1235)\to\omega\pi$	$0.5g_D^2\cos^2(\theta-\theta_o)$	202	<150	
1^{+-}	$B(1235)\to\phi\pi$	$0.5g_D^2\sin^2(\theta-\theta_o)$	~0	0	
2^{-+}	$A(1680)\to f\pi$	$0.69g_D^2\cos^2(\theta-\theta_o)$	138	133 ± 27	I=1
2^{-+}	$A(1680)\to f'\pi$	$0.69g_D^2\sin^2(\theta-\theta_o)$	0.6	0	S=0
2^{-+}	$A(1680)\to\rho\pi$	$0.5\ g_F^2$	110	85 ± 32	
2^{-+}	$A(1680)\to K^*\bar{K}+\bar{K}^*K$	$0.25g_F^2$	38	10 ± 5	
2^{++}	$f\to\pi\pi$	$0.15g_D^2\cos^2(\theta-\theta_o)$	101	150 ± 15	
2^{++}	$f'\to\pi\pi$	$0.15g_D^2\sin^2(\theta-\theta_o)$	2	poss.seen	
4^{++}	$h(2030)\to\pi\pi$	$0.09g_D^2\cos^2(\theta-\theta_o)$*	40	37 ± 9	I=0
4^{++}	$\epsilon(2300)\to\pi\pi$	$0.09g_D^2\sin^2(\theta-\theta_o)$*	0.5	?	S=1
6^{++}	$r(2510)\to\pi\pi$	$0.07g_D^2\cos^2(\theta-\theta_o)$	24	?	
1^{--}	$\phi\to K^L K^S$	$0.25g_F^2\cos^2\theta$	16	1.5	
1^{--}	$\phi\to K^+K^-$	$0.25g_D^2\cos^2\theta$	19	2.1	
2^{++}	$f\to K\ \bar{K}$	$0.15g_D^2\cos^2(\theta+\theta_1)$	32	5 ± 1	I=0
2^{++}	$f'\to K\ \bar{K}$	$0.15g_D^2\sin^2(\theta+\theta_1)$	33	< 70	S=1

3^{--}	$\omega(1670)\to K\bar{K}$	$0.16g_F^2\sin^2\theta$	8	~ 0	
3^{--}	$\phi(1850)\to K\bar{K}$	$0.16g_F^2\cos^2\theta$	28	seen	
4^{++}	$h(2030)\to K\bar{K}$	$0.09g_D^2\cos^2(\theta+\theta_1)$*	16	$1.5^{+0.9}_{-0.5}$	
4^{++}	$\epsilon(2300)\to K\bar{K}$	$0.09g_D^2\sin^2(\theta+\theta_1)$*	17	?	I=0
6^{++}	$r(2510)\to K\bar{K}$	$0.07g_D^2\cos^2(\theta+\theta_1)$	9	?	S=1

References.

1. Borodulin, V.I., Pron'ko, G.P., Razumov, A.V., Soloviev, L.D., Zorin, O.L. (1984). Preprint IHEP 84-202, Serpukhov; id. (1984). Proc. of XXII Intern. Conf. on High Energy Physics, Leipzig, v.1, pp. 129-130.
2. Mukunda, N., Dam van, H., Biedenharn, L.C. (1980). Phys. Rev., 22, 1938-1951.
3. Aldinger, R.R., Bohm, A. et al. (1984). Phys. Rev., 29, 2828-2837.
4. Rebbi, C. (1974). Phys. Rep., 12C, 1-73.
5. Barbashov, B.M. (1984). In: Proc. of VI Intern. Workshop on High Energy Physics and Quantum Field Theory (Protvino, July 1983), Serpukhov, 2, pp. 3-14.

SYMMETRY BREAKING FOR ROTATING CLOSED RELATIVISTIC STRING AND REGGE TRAJECTORY'S SINGULAR POINTS

V.V.Vladimirsky, I.L.Ilyina, E.A.Kiklevitch, E.S.Nicolaevsky

Institute of Theoretical and Experimental Physics, Moscow

1°. A closed string is used in the dual theory as a pomeron model for high-energy hadron scattering. In the topological expansion approach [1,2] the vacuum singularity exchange between hadrons is described by a cylindrical topology diagram. The longitudinal cutting of a cylindrical diagram and consideration of decay of the two arising strings with quarks at the ends is a useful approach for description of multiparticle production. There is a closed string without quarks in the transversal section of the cylindrical diagram. In the annihilation channel this string may be considered as a glueball, i.e. a multigluon state, singlet in colour. However, the string with Namnbu Lagrangian neither in classical nor in quantum formulation gives rise to solutions that might be interpreted as the cylindrical topology diagram continuation to the gluon sector. The complication introduced for the string Lagrangian [4] enables us to obtain missing rotation-type solutions for the closed string.

Writing the Lagrangian [4] there was used a concept of string with finite but small transversal size of the order of the confinement radius r_c. Such a consideration is based on the assumption that the string length highly exceeds r_c. The solutions found here violate this condition. So the Lagrangian is properties were extrapolated beyond the direct application region of speculations used write to the Lagrangian. So far as a form of the Lagrangian is quite allowed from the general physical point of view, such an extrapolation does not lead to contradiction.

Considering the relativistic string as a limit case of a for-stretched bag (gluon tube) one has additional degrees of freedom describing breathing movements of the bag's wall. In the simplest case [4] it suffices to introduce a single extra variable φ expressed through the transversal section area A of the gluon tube: $\varphi = \log(A/A_o)$, where A_o is the area value in the equilibrium. The variation of φ along the tube and in time is described with a nonlinear wave equation of the Sinh-Gordon type. The deviations of the area A from its mean value can propagate along the string as phonons. This excitations do not carry the gluon field quantum numbers although their origin is the gluon field flow along the string. Their properties must be like those of gluonium which is a gluon field state, singlet in colour. The purest kind of gluophonen excitation is realized on a closed string which does not carry quarks and may be considered as a single multigluon formation. In the report a particular case of propagation of gluophonon excitations is considered, namely rotations: the string disposes along a planar closed line and rotates with a constant angular velocity. The problem is solved in a classical (nonquantum) approach. This consideration has allowed us to obtain the connection between the angular momentum and the energy of the closed string. The obtained Regge trajectories have small slopes and rather small nonlinearities. The maximum angular momentum for a given energy may be expected to result just from pure rotation.

2°. The action for the string [4] is:

$$S = \int d\zeta^o d\zeta^1 \, LG \qquad (1)$$

where ζ^{o}, ζ^{1} are coordinates on the surface described by the string in the Minkovsky space $x^{\mu} = x^{\mu}(\zeta)$. The metric on the surface

$$g_{ab} = \frac{\partial x_{\mu}}{\partial \zeta^{a}} \frac{\partial x^{\mu}}{\partial \zeta^{b}} \tag{2}$$

has the signature (+ -), the signature in the Minkovsky space being (+ - - -). The geometric factor

$$G = \sqrt{-\det g_{ab}} = \sqrt{-\left(\frac{\partial x}{\partial \zeta^{o}}\right)^{2}\left(\frac{\partial x}{\partial \zeta^{1}}\right)^{2} + \left(\frac{\partial x}{\partial \zeta^{o}} \frac{\partial x}{\partial \zeta^{1}}\right)^{2}} \tag{3}$$

expresses the surface area element through increments of coordinates. Under such a definition the density of the Lagrangian is a scalar in a curvilinear space

$$L = \frac{K}{2} \frac{\partial \varphi}{\partial \zeta^{a}} \frac{\partial \varphi}{\partial \zeta^{b}} g^{ab} - \mu^{2} \cosh(\varphi) \tag{4}$$

and the action is invariant under changes of coordinates. A constant μ^{2} is connected with the slope of Regge trajectories for ordinary hadrons by the relation $2\pi\mu^{2}\alpha' = 1$, it has the dimension of mass square. In the following calculations we suppose $\mu^{2} = (2\pi)^{-1}$, the leading Regge trajectory slope being 1 (with $\alpha' = 0.87\ \mathrm{GeV}^{-2}$; energy is measured in units 1,07 GeV). A dimensionless constant K defines for a fixed μ^{2} the gluophonon excitation mass $m = \mu/\sqrt{K}$. A comparison with the bag theory allows us to connect K with the strong interaction constant: $K = C_{2}\alpha_{s}/2$. Using $\alpha_{s} = 0.225$ and the Casimir operator $C_{2} = 4/3$ we have $K = 0.15$. Just this value was taken for numerical calculations. This gives the gluophonon mass 1.o3 in units $(\alpha')^{-\frac{1}{2}}$, i.e. somewhat higher than 1 GeV. Solutions with other values of the constant K are obtained from the found solutions through the coordinate and time scaling: with $x \to \lambda x$, $\to$, $\mu^{2} \to \mu^{2}$ we have $K \to \Lambda^{2}K$. Considering the string as a gluon tube we have for the transversal size of the tube $r_{c} = 4\sqrt{\pi K \alpha'}$ (2.75 for the chosen string parameters).

To analyse the closed string it is convenient to choose as coordinates on the surface the time $\zeta^0 = t$ and the azimuth with respect to the rotation centre $\zeta^1 = \theta$, and take the radius $r = r(t, \theta)$ as a variable describing the motion. Coordinates of the string's points are $x^1 = r\cos(\theta)$, $x^2 = r\sin(\theta)$, $x^3 = 0$. Equations of motion obtained for $r(t,\theta)$ and $\varphi(t,\theta)$ are compatible with the assumption of pure rotation $\partial/\partial t = -\nu\partial/\partial\theta$ (ν is a constant angular velocity). In this case r and φ depend on a single variable $\theta - \nu t$ which we denote by θ as before. The density of the Lagrangian is

$$L = \frac{KW}{2G^2} - U, \quad \text{where} \quad U = \mu^2 \cosh(\varphi), \quad W = N'^2,$$
$$G = \sqrt{r^2 - Nr'^2}, \quad N = \nu^2 r^2 - 1 \qquad (5)$$

Introducing the momenta

$$\pi = \frac{\partial LG}{\partial \varphi'} = \frac{KN}{G}\varphi', \quad p = \frac{\partial LG}{\partial r'} = \frac{FN}{G} r', \quad \text{where}$$
$$F = \frac{KW}{2G^2} + U \qquad (6)$$

one easily obtains an invariant quantity - the Hamiltonian which in this case does not coincide with the energy:

$$H = \pi + pr' - LG = r^2 F/G . \qquad (7)$$

All the introduced functions are expressed via the momenta:

$$F = \frac{\pi^2}{2KN} + U, \quad G = \frac{rF}{\sqrt{F^2 + p^2/N}}, \quad H = r\sqrt{F^2 + p^2/N} \qquad (8)$$

The Hamilton equations

$$r' = \frac{pG}{FN}, \quad p' = -\frac{H}{r} + \frac{r}{H}\frac{N+1}{N^2}\left(\frac{F\pi^2}{K} + p^2\right), \quad \varphi' = \frac{\pi G}{KN},$$
$$\pi' = -G\mu^2 \sinh(\varphi) \qquad (9)$$

were numerically solved with the help of Runge-Kutta

method. The Hamiltonian conservation was used as an accuracy test.

The closed atring motions take place in the region $\nu r > 1$ $(N > 0)$. Peripheral velocities of points with constant values of the variables $r, p, , \pi$ exceed the speed of in the absolute value light. It is not a paradox because they are phase velocities of a finite mass excitation. In the physical region of the motion we have $G^2 > 0$ $(r^2 > Nr'^2)$ or $|v| < 1$, where v is the component of the velocity of the string points in the normal direction. Besides pure geometrical quantities, the angular momentum and the string energy values were calculated:

$$J = \int r^2 T^{01} G d\theta = \int d\theta \; \nu r^2 \left| \frac{K\varphi'^2}{2G^3} (G^2+r^2) + \frac{r'^2 U}{G} \right| \tag{10}$$

$$E = \int T^{00} G d\theta = \int d\theta \left| \frac{K\varphi'^2}{2G^3} (G^2+\nu^2 r^4) + \frac{r'^2+r^2}{G} U \right| \tag{11}$$

Here T^{ab} are the tension tensor components connected with the Hamiltonian by relation $H = G(T^{00} - T^{01}/\nu)$. Under the scaling $x \to \Lambda x$ the momentum J acquires the factor Λ^2 and E aquires the factor Λ. The Regge trajectories in the diagram $J(E^2)$ (Fig. 1) are extended, the trajectory slope at corresponding points conserves. Note that

$$dE/dJ = \nu \,, \quad dJ/d(E^2) = (2E\nu)^{-1} \tag{12}$$

3°. The closed string must be a periodic solution of Eqs. (9); besides, since in a momental photograph of the string the independent variable is interpreted as the azimuth, then for the geometrical closure of the string the period T must equal $2\pi/n$. In all there is a 5-parameter family of solutions of Eqs. (9): the parameters are $r_o, p_o, \varphi_o, \pi_o$ and ν^2; conditions for the given solutions to be periodic are:

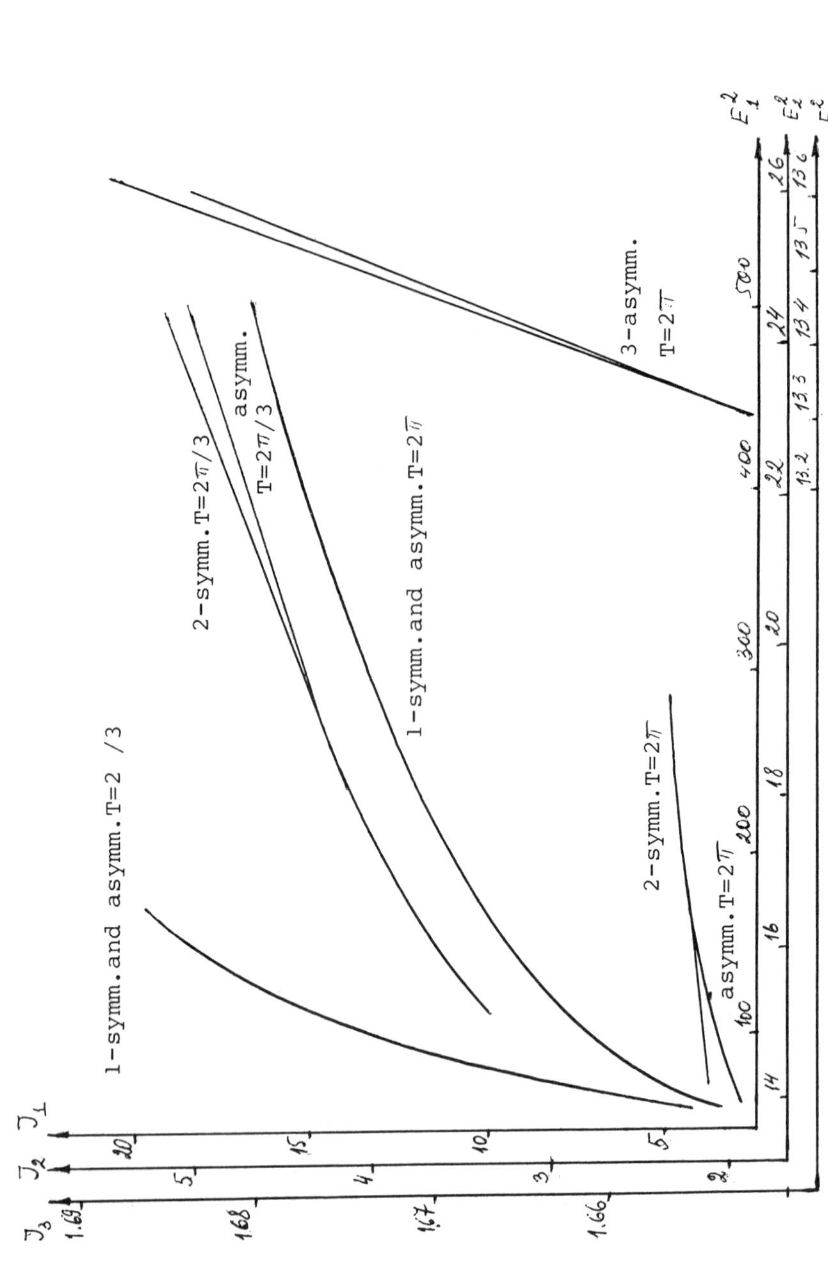

Fig. 1. Gluon Regge Trajectories for Different Periods with Singularoties Singled out

$$r(\theta = T) = r_o\ ,\ p(\theta = T) = p_o,\ \ \varphi(\theta = T) = \varphi_o,$$

$$\pi = (\theta = T) = \pi_o \tag{13a}$$

and we set

$$T = 2\pi/n \tag{13b}$$

Relations (13a) contain 3 conditions since there is a free parameter T; in all, relations (13a-b) contain 4 conditions. Since the Hamiltonian is conserved, we see that relations (13a) contain 2 conditions; therefore relations (13a-b) contain 3 conditions. Calculation of the number of parameters and conditions gives 3-parametrical families of periodic solutions and 2-parametrical families of solutions with period $T = 2\pi/n$. In real calculations such families are actually found. Arnold beleives [5] that such a behavoir is typical for 1-parametrical families of autonomous Hamiltonians with 2 degrees of freedom.

There is a helpful symmetry map preserving Eqs. (9) and some of their solutions:

$$r \to r,\ p \to p\ ,\ \ \to \ \ \ ,\ \pi \to \pi\ ,\ \theta \to T/2+\theta\ . \tag{14}$$

The solutions invariant under (14) will be called symmetrical. In symmetrical T-periodic solutions variables r and p manage to perform at least 2 identical oscillations during the period of an oscillation of variables φ and π. Asymmetrical T-periodic solutions come in pairs: if one applies (14) to such a solution then one again obtains an asymmetrical T-periodic solution.

Since the Hamiltonian is autonomous then there are 1-parametrical families of solutions differing from each other only in variation of the independent variable θ. Such families are usually called orbits. In other words, an orbit is a set of points in the phase space corresponding to the given solution. Closed orbits are called cycles. We have a 4-parameter family of orbits, 2-parameter families of cycles and 1-parameter families of cycles with period $2\pi/n$. A cycle corresponding to symmetrical solutions will be called symmetrical solutions will be called asymmetrical.

The initial conditions were as a rule of the 2 following types:

1) $p_o = 0,\ p_o' > 0$ (r_o is a local minimum), $\varphi_o = 0$

$$p_o' > 0 \quad (\text{i.e. } \pi_o > 0), \tag{15a}$$

2) $p_o = 0,\ p_o' < 0$ (r_o is a local maximum), $\pi_o = 0$,

$$\pi_o' < 0 \quad (\text{i.e. } \varphi_o > 0). \tag{15b}$$

Cycles satisfying conditions (15a or b) , i.e. such that at some point of the cycle $p = 0$ and simultaneously $= 0$ or $\pi = 0$ will be called synphased. It is curious that in spite of numerous attempts we did not succeed in obtaining any asynphased cycle. As cycle parameters we took quantities ν^2, r_o and π_o in case (15a) or φ_o in case (15b). In practice it turned out that deviations provocated by variations of different parameters are mutually compensated: if one changes the value of a parameter then an orbit closed before is not closed any more, but one can close it again changing the value of any other parameter or, more over, to choose the two remaining parameters' values so as to obtain again a cycle with period $2\pi/n$. So we obtained 2-parameter families of cycles and 1-parameter families of cycles with periods $2\pi/n$.

The stability of the found cycles was investigated: eigenvalues and eigenvectors of their Jacoby matrix were found, i.e. of the matrix of the derivatives of $r, p,$ and π at T with respect to the initial values r_o, p_o, φ_o and π_o. As it is known, 2 eigenvalues necessarily equal 1 6 ; the product of 2 remaining eigenvalues also equals 1, they may be either complex conjugate and equal to 1 in modulus (the stability case) or one of them may be 1 and the other 1 (the instability case). In the first case orbits near our cycles realize quasiperiodic motion on 2-dimensional tori: every nonclosed orbit is dense on an invariant torus,ergodicity without mixing takes place; fixing the value of the Hamiltonian we have a 1-parameter family of tori imbedded in each other, and our cycle lies inside of each of them. In the second case eigenvectors correspond to directions of extension and contraction on the 3-dimensional surface of a constant value of the Hamiltonian; almost every orbit near unstable cycle is found dense in a 3-dimensional subset of the surface; ergodicity and mixing take place on this subset, there is an exponential sensibility to the initial values. In this sense one may say that stochasticity takes place (there are stochastic regions among families of invariant tori).

Intermediate cases between stability and instability, when the second pair of eigenvalues equals 1 or -1, are interesting for the following reasons: at eigenvalues -1 [7] the period doubling is possible, i.e. breaking of the symmetry prociding with the cycle of period T, dissappearance of the cycle of period T and appearance of the cycle of period 2T, so far transitions like that were not noticed in our system. At eigenvalues 1 [8] appearance or dissappearance of cycles is possible. Indeed, consider the Poincaré map for variable p: for initial values φ_o and π_o fix the values φ and π at the point of return of the orbit to the plane p=0 at $\theta \simeq T$. The map preserves the area and is Hamiltonian. A symmetrical cycle corresponds to a fixed point of such a map, an asymmetrical one to two fixed points. If there is a 1-parameter family of cycles then variation of the parameter does not affect the equations

$$\varphi_{return} - \varphi_o = 0, \quad \pi_{rcturn} - \pi_o = 0 \qquad (16)$$

If the Jacoby matrix does not have additional eigenvalues 1 then the matrix of derivatives of the left hand sides of Eqs. (16) does not have zero eigenvalues and by the implicit function theorem the continious (even analytical) 1-parameter family of cycles extends in a vicinity of a given point to further values of the parameter, i.e. appearence or dissappearance of cycles is not possible. But at points where all the 4 eigenvalues equal 1 the implicit function theorem is inapplicable, appearance and dissappearance of cycles are possible. Such cases really take place in our system: here occure transitions from one fixed point (symmetrical cycle) to three fixed points (symmetrical and asymmetrical cycles). By [8] this corresponds to the winding number $\frac{1}{2}$ (the ratio of periods of oscillators r, p, and φ, π equals 1:2) and the Hamiltonian of the map must in these cases have singularities corresponding to "simple Arnold's singularities" 9 . Selection was made of the cases when a transition from one fixed point to three fixed points is possible and it was made clear which cases do not contradict the symmetry of our Hamiltonian. Now it is under investigation which selected cases are realized and what singularities the total Hamiltonian (8) has in these cases.

4°. Up to now three 1-parameter families of cycles with period 2π and two families of cycles with period $2\pi/3$ have been calculated. Also cycles with periods quite near to π have been found.

A family of symmetric cycles with period 2π is calculated with the initial values (15a), $r_{min} = r_o$, a quantity r_o may be taken as a parameter. The value $\theta = \pi/2$ corresponds to the initial values (15b); then $r = r_{max}$. The family begins on a singular surface $\nu^2 r^2_{min} = 1$ near $r_o \approx 0.62$. It cannot begin at a stability-to-instability transition point as the winfing number is $\frac{1}{2}$ and so a symmetrical cycle cannot appear. From the first calculated point $r_o \approx 0.629$ to the point

$r_o \approx 0.778$ the cycles of the family are stable. At point $r_o \approx 0.6828$ eigenvalues equal -1 but there is neither transition to instability nor period doubling. The transition point $r_o \approx 0.778$ (eigenvalues 1) is investigated. Up to now additional asymmetrical family near this point has not been found but we are not sure that this family does not exist.

From $r_o \approx 0.778$ to $r_o \approx 1.180$ the cycles of the family are unstable. The maximal eigenvalue 7.975 is found at $r_o \approx 0.947$. A stochastic region where unstable cycles are situated is small and squeezed between the families of invariant tori. Separatrices of unstable cycles are tangent to the limiting tori. In the stochastic region the motion is constructed like this: until some time an orbit winds round the torus then approaches an unstable cycle along the separatrix and moves away from it in the direction of the unstable separatrix. Since in a large interval of "time" the orbit returns near the initial point, an intermittence takes place.

At the transition point $r_o \approx 1.180$ all the eigenvalues equal 1. At this point symmetry (14) breaks: a family of asymmetrical cycles appears (it will be described below). From $r_o \approx 1.180$ to $r_o \approx 2.885$ (the latest up-to-now calculated point) the cycles of the symmetrical family are stable. At $r_o \approx 1.516$ eigenvalues equal 1 but there is neither transition to instability nor period doubling. The energy and the angular momentum of the family monotonously increase with the parameter growth and the Regge trajectory slope slowly decreases.

A family of asymmetrical cycles with period 2π having a common point with the considered family of symmetrical cycles at r_o^c 1.180 satisfies $r(\theta = \pi) < r_o^c < r_o$. The values $r(\theta = \pi)$ monotonously decrease until $r_o \approx 2.43$ and reach 0.9007. At the beginning of the growth of r_o from r_o^c the quantity $r(\theta = \pi)$ is the second maximum and, as before, variables r and p manage to perform two oscillations during the period of an oscillation of variables and π . The amplitude and the period of the second oscillation monotonously decrease however and at $r_o \approx 1.754$ the second oscillation dissappears so that with the further growth of r_o the winding number becomes 1.

From the initial point r_o^c to $r_o \approx 2.65$ cycles of the family are stable. Below $r_o \approx 1.92$ the real part of an eigenvalue $Re\,\Lambda$ decreases and reaches -0.244, then begins to increase and at $r_o \approx 2.65$ equals 1. Perhaps if one considered this 2-parameter family of cycles with any periods, cycles with $\Lambda = -1$ would be found in it. At $r_o \approx 2.65$ eigenvalues equal 1, yet at this point appearance or dissappearance of cycles is impossible, as the winding number is 1, and only one symmetrical solution could appear or dissappear, this is impossible because of asymmetry of a cycle corresponding to this point. At higher r_o cycles of the family are unstable. The family ends at $r_o \approx 2.77$ on the singular surface $\nu^2 r_{min}^2 = 1$. The values of energy and angular momentum for the family monotonously decrease with growth of r_o up to the transition point $r_o \approx 2.65$ but monotonously increase for higher r_o . By (12) tangents to the diagram $J(E)$ or the Regge trajectory $J(E^2)$ coincide for these two intervals; the second interval lies below. At the transition point the curves $J(E)$ and $J(E^2)$ have return points.

If one considers the Regge trajectories for the symmetrical and asymmetrical families together, then the transitions point r_o^c is a singular point of the return point type for the joint trajectory, the asymmetrical trajectory being steeper and lying somewhat below, see Fig. 1.

There is another family of cycles with period 2π . On cycles of this family r and p manage to perform 4 oscillations during the period of an oscillation of and π. The cycles are symmetrical, the even maxima of r coincide, the odd maxima coincide too, even and odd maxima differ now in a few percent. The family is calculated with initial values (15b), r_{max} is reached at $\theta = \pi/2$ and $3\pi/2$. We may take φ_o as a parameter, with growth of φ_o from 0.085 to 0.76 (the latest obtained point) $\nu^2 r^2_{min} - 1$, the energy E, the angular momentum J, and the difference of the maxima monotonously increase. Up to now all the cycles of the family are unstable, eigenvalues increase as $_o$ grows and reach approximatly 4.

Let us consider two more families of cycles with period $2\pi/3$. Both the families are calculated with initial values (15b), r_o may be taken as a parameter. The first family is symmetrical, it contains cycles from $r_o \approx 1.6617$ ($\nu^2 r^2_{min} = 0.033$) to $r_o \approx 3.465$ (the calculation is under way now). The second family is asymmetrical, has a common point with the symmetrical one, the transition point $r_o^c \approx 1.863$; this family exists only for $r_o > r_o^c$. Cycles of the symmetrical family are stable for $r_o < r_o^c$, then unstable up to $r_o \approx 2.39$ and stable at higher r_o; near this transition point additional asymmetrical cycles have not been found yet. At $r_o \approx 2.97$ eigenvalues equal -1 but there is neither a transition to instability nor period doubling. When r_o grows the values of E and J monotonously increase. Cycles of the asymmetrical family are stable up to $r_o \approx 2.228$, at this point eigenvalues equal -1 and a transition from stability to instability takes place, there is no period doubling. Further, up to $r_o \approx 2.83$ (the last calculated point, $\nu^2 r^2_{min} - 1 \approx 0.03$) the cycles are unstable. As r_o grows E and J increase but slower than for the symmetrical family, The Regge trajectory $J(E^2)$ is higher than for the symmetrical family, at r_o^c the joint trajectory has a singular point of the return point type (Fig. 1).

5° In the Chew and Low diagram (Fig. 1) the calculated families of cycles form Regge trajectories lying considerably lower than the leading trajectory obtained from the treatment of the scattering differential cross-sections. The slope of the symmetrical trajectory with period 2π decreases with the energy growth from 0.15 to 0.025 and stays all the time smaller than the vacuum trajectory slope which is 0.25 by the scattering data. Regge trajectories of other families Lie higher than the trajectory of this family. Perhaps in the scope of the considered model the vacuum singularity for positive values of the energy square is not transformed into a gluon trajectory but is rather a place of accumulation of the closed string trajectories.

Unlike the other theoretical approaches to the determination of the gluonia spectrum [10-13] the closed string theory is not limited in the region of small values of the angular momentum but, on the contrary, becomes more reliable with higher J. As the angular momentum grows the mean radius $\bar{r}$ of the curve, along which the string moves , increases ; for the symmetrical families the region $\bar{r} > r_c$ is already reached, so the difficulty with the extrapolation of the gluon tube Lagrangian mentioned in Introduction is partially disposed of. Calculation of the trajectories for large radii is not finished yet, but there are reasons to suppose that the picture of a great number of the gluon trajectories would remain.

References

1. Veneziano,G. (1976). Nucl. Phys., B117, 519.

2. Kaydalov,A.B. (1983). In: The 10th school of ITEP physicists. 2, Energoatomizdat, Moscow, p. 3 (Russian).

3. Ter-Martirosyan,K.A. (1983). In: The 10th school of ITEP physicists. 2. Energoatomizdat, Moscow. p. 64 (Russian).

4. Vladimirsky,V.V. (1984). Sov. J. Nucl. Phys., 39, 310

5. Arnold,V.I. (1978). Mathematical Methods of Classical Mechanics, Springer N.Y. et al.

6. Nemytsky,V.V., Stepanov,V.V. (1947). Katchestvennaya teoriya differentsialnykh uravneniy (Qualitative Theory of Differential Equations) Gostekhizdat, Moscow (Russian).

7. Feigenbaum,M.J. (1979). J. Stat. Phys. 19, 25; (1979), 21, 669.

8. Zisook,A.B. (1984). Comm. Math. Phys., 96, 361.

9. Arnold,V.I. (1972). Funct. Anal. Appl. 6, 254 (Russian).

10. Donoghue,J., Johnson,K., Li,B.A. (1981). Phys. Lett. 99B, 416.

11. Carlson,C.E et al. (1983). Phys. Rev. D, 27, 2169.

12. Izgur,N ., Paton,J. (1983). Phys. Lett., 124B, 247.

13. Shifman,M., Vainshtein,A., Zakharov,V. (1979). Nucl+ Phys. B147, 448.

THE INTERACTION HIERARCHY STABILITY IN GUT MODELS

M.V.Burova, O.K.Kalashnikov
Lebedev Physical Institute, Moscow

1. <u>Introduction</u>. At present the most developed class of the theories unifing the electroweak and strong interactions are the Grand Unified Theories (GUT) based on the gauge Yang-Mill's fields and the spontaneous symmetry breaking mechanism[1].The unification electroweak and strong interactions proposes that GUT has two substantially different mass scales: 10^{14} GeV (the mass scale of the Grand Unification and 10^{2} GeV (the electroweak mass scale). The existance of these two mass scales leads to the "hierarchy problem". There are two aspects in this problem. One of them is related to fine tuning of some parameters of the theory (scalar multiplet masses). The second aspect is that the radiative corrections could be considerable and have the same order of heavy particle masses. In this case hierarchy fixed in tree approximation is broken. But in reality this is not so. In GUT the interaction's hierarchy proves to be stable (both Higgs bosons and vector's bosons), and the radiative mass corrections of the light particles because of their interaction with heavy particles vanish if one takes into consideration new extremum equations. This result is model-independent within a wide class of GUT's.

2. <u>The Grand Unification with the single gauge coupling constant,</u> The existence of a unified interaction of high energies (which after breaking the gauge symmetry gives rise to all the known forces) implies that GUT is to be constructed on the base of a single gauge group.

The examples of such GUT schemes realized within the SU(N) and exceptional E-group models are well known[2,3]. The spontaneous symmetry breaking provides the different mass scales for GUT fixed by a choice of a free parameters at the tree level. The "fine tuning" of these parameters is performed as usual with the radiative corrections taken into account. Besides, it is necessary to be sure however that the radiative corrections of mass spectrum do not mix the ealier established hierarchy.

2.1. The mass hierarchy stability for SU(N) - model of GUT. The stability of scalar bosons mass hierarchy is proved for the model-independent SU(N) scalar self-coupling potential. The scalar sector of our GUT-model may be restricted to two multiplets N and Φ, the fundamental and adjoint representations. According to [4] the one-loop effective potential is

$$V = V_o + \sum \frac{\mathfrak{M}_i^4}{64\pi^2}\left[\ln\frac{\mathfrak{M}_i^2}{\mu^2} - \frac{1}{2}\right], \tag{2.1}$$

where μ^2 is the renormalization point and $\mathfrak{M}_i^2$ are the squared masses of scalar bosons in tree approximation. The tree potential V_o can be expressed as

$$V_o = -\frac{1}{2}M^2\Phi^2 - m^2(N^+N) + \frac{1}{8}f\Phi^4 + \frac{1}{8}k(\Phi D\Phi)^2 + \\ + \frac{1}{2}\lambda(N^+N)^2 + \frac{1}{2}\delta(\Phi D\Phi)(N\frac{\tau}{2}N) + \frac{1}{2}h\Phi^2(N^+N) \tag{2.2}$$

where $\tau/2$ and D are the standard generators and structure constants of SU(N). The proof of the mass hierarchy stability is based on the assumption that

$$\mathfrak{M}_i^4 = -\frac{1}{2}M_i^2\Phi^2 - m_i^2(N^+N) + \frac{1}{8}f_i\Phi^4 + \frac{1}{8}k_i(\Phi D\Phi)^2 + \\ + \frac{1}{2}\lambda_i(N^+N)^2 + \frac{1}{2}\delta_i(\Phi D\Phi)(N^+\frac{\tau}{2}N) + \frac{1}{2}h_i\Phi^2(N^+N) + O_i^2 \tag{2.3}$$

(a linear combination of SU(N) invariants), where M_i, m_i, f_i, k_i, λ_i, δ_i, h_i are some coefficients depending on the vacuum expectation value of fields $\langle\Phi\rangle$, $\langle N\rangle$, and Q_i^2 is small parameter. It is subject to direct evaluation of scalar masses for the SU(2)-model. A similar assumption for the SU(5)-model see in [5]. For definetness the tree-level mass hierarchy is fixed so that $\frac{\langle N\rangle}{\langle\Phi\rangle} \ll 1$ and $Q_i^2 \sim \langle N\rangle^6 / \langle\Phi\rangle^2$.

Let us prove that a scalar bosons mass hierarchy fixed in the tree level is nor broken by radiation corrections. In one-loop approximation after differentiating the effective potential (2.1) with respect to fields Φ, N with (2.2), (2.3), we find that the mass matrix e-e-ment with the tree level order of magnitude $\langle N\rangle^2$ is

$$\mathfrak{M}^2_{N_i+N_k} = (-m^2 + \frac{1}{2} h\Phi^2 + (N^+N))\,\delta_{ik} + \frac{1}{2}\delta(\Phi D\Phi)\,\frac{\tau_{ik}}{2} + \lambda N_i N_k^+ + \sum_j \frac{1}{64\pi^2}\Big[((-m_j^2 + \frac{1}{2}h_j\Phi^2 + \frac{1}{2}\lambda_j(N^+N))\cdot \delta_{ik} + \frac{1}{2}\delta_i(\Phi D\Phi)\,\frac{\tau_{ik}}{2} + \lambda_j N_i N_k^+)\ln\frac{\mathfrak{M}_j^2}{\mu^2} + \text{const}\Big], \tag{2.4}$$

This element and a similar one determine the order of magnitude of the lowest mass. Before we use the new extremum conditions, (2.4) contains heavy masses and the tree hierarchy is broken. However the fixed relation between $\langle\Phi\rangle$ and $\langle N\rangle$ at the tree level must be also satisfied at the one-loop approximation and therefore the fine tuning of the parameters m,M is necessary. This tuning is realised with the aid of the new extremum conditions

$$\frac{\partial V_o}{\partial \Phi} + \sum_i \frac{1}{64\pi^2} \frac{\partial \mathfrak{M}_j^4}{\partial \Phi} \ln \frac{\mathfrak{M}_j^2}{\mu^2} = 0$$
$$\frac{\partial V_o}{\partial N} + \sum_j \frac{1}{64\pi^2} \frac{\partial \mathfrak{M}_j^4}{\partial N} \ln \frac{\mathfrak{M}_j^2}{\mu^2} = 0 \qquad (2.5)$$

together with (2.4). Now we find that

$$\mathfrak{M}^2_{N_i^+ N_k} \Big|_{\substack{i=k \\ N_i=\langle N\rangle}} = (\lambda + \sum_j \frac{\lambda_j}{64\pi^2} \ln \frac{\mathfrak{M}_j^2}{\mu^2} + \frac{O(N^2)}{\langle N\rangle^2})\langle N\rangle^2 \sim \langle N\rangle^2 \qquad (2.6)$$

and we see that the radiation corrections do not change the order of the mass matrix element. There is the same cancelations mechanism for other matrix elements. Thus the mass hierarchy stability is proved and this proof is model-independent.

2.2. <u>The effective potential for SU(2)-model</u>. The proofof the mass hierarchy stability in one-loop approximation is based on the assumption that the one-loop effective potential may by expressed as a linear combination of some SU(N)-invariants of scalar fields. We have tested this basic assumption for SU(2) and SU(5)-models and the simpliest case is preseuted here. The SU(2)-model under consideration keeps all the basic features of a realistic models of GUT. However, this SU(2)-model simplifies the calculations considerably. We define the one-loop effective potential with the help of group-invariants and show that the mass matrix found by the differentiation of this potential is the same as the at obtained by the Green-function formalism. The Lagrangian of the SU(2)-model is as follows:

$$L = -\frac{1}{4}(\partial_\mu V_\nu^a - \partial_\nu V_\mu^a + g\varepsilon^{abc} V_\mu^b V_\nu^c)^2 - \Big|(\partial_\mu \delta^{ab} -$$
$$- ig\Big[\frac{\tau^s}{2}\Big]_a^b V_\mu^s) N_b\Big|^2 - \frac{1}{2}\Big[(\partial_\mu \delta^{ac} + g\varepsilon^{abc} V_\mu^b)\Phi^c\Big]^2 -$$

$$-\frac{M^2}{2}\Phi^2 - m^2(N^+N) - \frac{f}{8}\Phi^4 - \frac{\lambda}{2}(N^+N)^2 - \frac{h}{2}(N^+N)\ ^2 - kN^+\frac{\tau}{2}\Phi N, \tag{2.7}$$

where V^a_μ is a vector field triplet, Φ^a and N^a scalar fields. The spontaneous symmetry breaking is determined by the nonzero vacuum expectation values of the scalar fields:

$$\langle\Phi\rangle^c\Big|_{c=3} = Z \ ; \qquad \langle N\rangle^b\Big|_{b=2} = \varkappa \ . \tag{2.8}$$

It is possible to choose them so as $\varkappa^2/Z^2 \ll 1$. From Φ and N we can make three invariants Φ^2, N^+N, $N^+\frac{\tau}{2}\Phi N$. The tree scalar fields masses of the 4-th power are expressed in terms of these invariants and determine Eq. (2.1). The structure of a one-loop effective potential is similar to that of the potential in the tree approximation:

$$V = -\tilde{M}^2/2\,\Phi^2 - \tilde{m}^2N^+N + \frac{1}{8}F\Phi^4 + \frac{1}{2}\Lambda(N^+N)^2 +$$

$$+\frac{1}{2}H(N^+N)\Phi^2 + K(N^+\frac{\tau}{2}\Phi N), \tag{2.9}$$

but the coefficient functions look more complex and can be expressed through the tree masses:

$$\tilde{M}^2 = M^2 + \frac{3M^2f}{32\pi^2}\left[\ln\frac{\theta_+^2}{\mu^2} - \frac{1}{2}\right] + \frac{m^2}{32\pi^2}\left[h + \frac{k}{\langle\Phi\rangle}\right]\left[\ln\frac{\Omega^2}{\mu^2} - \frac{1}{2}\right];$$

$$\tilde{m}^2 = m^2 + (2M^2h + 4m^2\hat{h}^2/f)\,\frac{1}{64\pi^2}\left[\ln\frac{\theta_+^2}{\mu^2} - \frac{1}{2}\right] +$$

$$+\frac{m^2}{64\pi^2}\left[h + 6\lambda - \frac{4\hat{h}^2}{f}\right]\left[\ln\frac{\theta_-^2}{\mu^2} - \frac{1}{2}\right] ;$$

$$F = f + \frac{f^2}{18\pi^2}\left[\ln\frac{\theta_+^2}{\mu^2} - \frac{1}{2}\right] + \frac{h^2}{32\pi^2}\left[\ln\frac{\theta_-^2}{\mu^2} - \frac{1}{2}\right] +$$

$$+\frac{1}{16\pi^2}\left[h + k/\langle\Phi\rangle\right]\left[\ln\frac{\Omega^2}{\mu^2} - \frac{1}{2}\right] + \frac{3g^4}{4\pi^2}\left[\ln\frac{\mu_\perp^2}{\mu^2} - \frac{1}{2}\right];$$

$$K = k + \frac{\hat{h}^2 k}{64\pi^2 f}\left[\ln\frac{\theta_+^2}{\mu^2} - \frac{1}{2}\right] - \frac{\hat{h}^2 k}{64\pi^2 f}\left[\ln\frac{\theta_-^2}{\mu^2} - \frac{1}{2}\right];$$

$$\Lambda = \lambda + \frac{1}{32\pi^2}\left[h^2 + \frac{12\lambda\hat{h}^2}{f} - \frac{4\hat{h}^4}{f^2}\right]\left[\ln\frac{\theta_+^2}{\mu^2} - \frac{1}{2}\right] +$$
$$+ \frac{1}{32\pi^2}\left[9\lambda^2 - \frac{12\lambda\hat{h}^2}{f} + \frac{4\hat{h}^4}{f^2}\right]\left[\ln\frac{\theta_-^2}{\mu^2} - \frac{1}{2}\right] +$$
$$+ \frac{1}{16\pi^2}\left[\lambda^2 + \frac{hk}{\langle\Phi\rangle} - \frac{k^2}{\langle\Phi^2\rangle}\right]\left[\ln\frac{\Omega^2}{\mu^2} - \frac{1}{2}\right] +$$
$$+ \frac{3g^4}{64\pi^2}\left[\frac{1}{2}\ln\frac{\mu_\parallel^2}{\mu^2} + \ln\frac{\mu_\perp^2}{\mu^2} - \frac{3}{4}\right];$$

$$H = h + \frac{1}{32\pi^2}\left[3hf + 4\hat{h}^2 + \frac{2h\hat{h}^2}{f}\right]\left[\ln\frac{\theta_+^2}{\mu^2} - \frac{1}{2}\right] +$$
$$+ \frac{1}{32\pi^2}\left[3\lambda(h - k/\langle\Phi\rangle) - 2h\hat{h}^2/f\right]\left[\ln\frac{\theta_-^2}{\mu^2} - \frac{1}{2}\right] +$$
$$+ \frac{1}{16\pi^2}\left[h\lambda + \frac{hk}{2\langle\Phi\rangle} + k^2/\langle\Phi\rangle^2\right]\left[\ln\Omega^2/\mu^2 - \frac{1}{2}\right] +$$
$$+ \frac{3g^2}{16\pi^2}\left[\ln\mu_\perp^2/\mu^2 - \frac{1}{2}\right], \tag{2.10}$$

where $h = (h - \frac{k}{2\langle\Phi\rangle})$.

Here the scalar field masses θ_+^2 , θ_-^2 , Ω^2 are found in the tree approximation

$$\theta_+^2 \approx f\,\Phi^2 \;;\quad \theta_-^2 \approx (2\lambda - \frac{2h^2}{f})(N^+N)\;;\; \Omega \approx \frac{k}{2\langle\Phi\rangle}(2\Phi^2 - N^+N) \tag{2.11}$$

and $\mu_\parallel^2$, $\mu_\perp^2$ are the vector fields masses in the same approximation

$$\mu_\parallel^2 = \frac{g^2}{2}(N^+N) \;;\quad \mu_\perp^2 \approx g^2(\Phi^2 + \frac{1}{2}N^+N) \tag{2.12}$$

The verification of (2.9), (2.10) can be made by comparing the corresponding mass matrix elements. It is most important to study the mass matrix element $\mathfrak{m}^2_{N_2^+N_2}$ because its order of magnitude is $\varkappa^2$. Other matrix elements are similarly analised. The square mass matrix element is calculated at first by direct differentiation of the effective potential (2.9), (2.10):

$$\mathcal{M}^2_{N_2^+N_2} = -m^2 + \frac{h}{2} z^2 + 3\lambda \varkappa^2 - \frac{kz}{2} + \frac{3g^4}{32\pi^2}(z^2+\varkappa^2)\ln\frac{\mu_\perp^2}{\mu^2} + \frac{3g^4\varkappa^2}{64\pi^2}\ln\frac{\mu_\parallel^2}{\mu^2} + \Big[\frac{hfz^2}{32\pi^2} + \frac{\hat{h}^2 z^2}{16\pi^2} + \frac{h(h+k/z)\varkappa^2}{32\pi^2} + \frac{11\lambda\hat{h}^2\varkappa^2}{16\pi^2 f} - \frac{\hat{h}^4\varkappa^2}{4\pi^2 f}\Big]\Big[\ln\frac{\theta_+^2}{\mu^2}\Big] + \Big[\frac{15\lambda^2\varkappa^2}{32\pi^2} + \frac{\hat{h}^4\varkappa^2}{4\pi^2 f^2} - \frac{11\hat{h}^2\lambda\varkappa^2}{16\pi^2 f}\Big]\ln\frac{\theta_-^2}{\mu^2} + \Big[\frac{\lambda^2\varkappa^2}{16\pi^2} - \frac{3k^2\varkappa^2}{256\pi^2 z^2} + \frac{k^2}{32\pi^2} - \frac{k\lambda z}{16\pi^2} + \frac{3hk\varkappa^2}{128\pi^2}\Big]\ln\frac{\Omega^2}{\mu^2} + O(\varkappa^4/z^2) \tag{2.13}$$

and then the same element is determined with the help of the Green-function technique:

$$\mathcal{M}^2_{N_2^+N_2} = -m^2 + \frac{hz^2}{2} + 3\lambda\varkappa^2 - \frac{kz}{2} - \Pi_{N_2^+N_2} \tag{2.14}$$

Here for simplisity we do not take into account the cube selfinteraction terms and give the final result of our calculations only. The polarisation operator which is fixed by (2.14) may be represented by a sum of nine topologically different diagrams as follows

$$-\Pi_{N_2^+N_2} = \Big[\tfrac{1}{2}\ (\text{diagram}) \sim \frac{hfz^2}{32\pi^2}\ln\frac{\theta_+^2}{\mu^2}\Big] + \Big[(\text{diagram}) \sim \frac{\lambda h^2\varkappa^2}{8\pi^2 f}\ln\frac{\theta_+^2}{\mu^2} + \Big(\frac{\lambda^2\varkappa^2}{8\pi^2} - \frac{\lambda h^2\varkappa^2}{8\pi^2 f}\Big)\ln\frac{\theta_-^2}{\mu^2}\Big] + \Big[(\text{diagram})\ \frac{3g^4}{32\pi^2}\Big(z^2 + \frac{\varkappa^2}{2}\Big)\ln\frac{\mu_\perp^2}{\mu^2} + \frac{3g^4\varkappa^2}{128\ \ 2}\ln\frac{\mu_\parallel^2}{\mu^2}\Big] +$$

$$+\left[\frac{1}{2}\ \sim \frac{3g^4\varkappa^2}{64\pi^2}\ln\frac{\mu_\perp^2}{\mu^2}+\frac{3g^4\varkappa^2}{128\pi^2}\ln\frac{\mu_\parallel^2}{\mu^2}\right]+$$

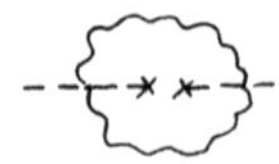

$$+\left[\ \sim\left(\frac{h^2z^2}{16\pi^2}+\frac{h^2\lambda\varkappa^2}{16\pi^2 f}-\frac{3h^4\varkappa^2}{16\pi^2f^2}\right)\ln\frac{\theta_+^2}{\mu^2}+\right.$$

$$\left.+\left(\frac{3h^4\varkappa^2}{16\pi^2f^2}-\frac{h^2\lambda\varkappa^2}{16\ {}^2 f}\right)\ln\frac{\theta_-^2}{\mu^2}\right]+\left[\ \sim -\right.$$

$$\left.-\frac{h^4\varkappa^2}{16\pi^2f^2}\ln\frac{\theta_+^2}{\theta_-^2}\right]+\left[4\ \sim\frac{h^2\lambda\varkappa^2}{2\pi^2 f}\ln\frac{\theta_+^2}{\mu^2}+\right.$$

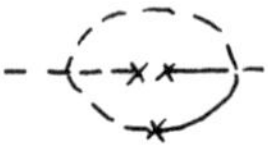

$$\left.+\frac{h^4\varkappa^2}{16\pi^2f^2}\ln\frac{\theta_-^2}{\mu^2}\right]+\left[\frac{1}{2}\ \sim \frac{h^2\lambda\varkappa^2}{2\pi^2 f}\ln\frac{\theta_+^2}{\mu^2}\right]+$$

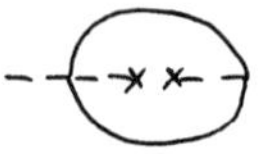

$$+\left[\frac{3}{2}\ \sim\ +2\ -\ +\ \sim\right.$$

$$\left.\frac{11\lambda^2\varkappa^2}{32\pi^2}\ln\frac{\theta_-^2}{\mu^2}\right]. \qquad (2.15)$$

These diagrams have been calculated using the Green-function and vertices in the free approximation. Direct summing of all the terms in (2.15) provides the coincidence of (2.14) with (2.15). This fact proves the equivalence of these matrices and so our assumption is right.

3. The mass hierarchy in the GUT models with a few gauge constants. There is another kind of GUT-models with a few gauge coupling constants. The gauge group defining these models is the direct product of a number of simple groups which determine the low energy phenomenology. The most known example of such a models is the SU(2) x x SU(2) x SU(4)-model first proposed in [6]. For this kind of GUT the mass hierarchy problem was not studied so far. Unfortunately we do not study the SU(2) x SU(2)x x SU(4)-model since obtaining evidence of the mass hierarchy stability for this model proved to be very cunbersome. In what follows we carry out all the calculations for the SU(2) x SU(2) - fragment which is responsible

for the electroweak interactions in the full SU(2) x x SU(2) x SU(4)-model. The Lagrangian for the SU(2) x x SU(2) model is

$$L_g = -\frac{1}{4}\left(\partial_\mu W^a_{1,\nu} - \partial_\nu W^a_{1,\mu} + g_1 \epsilon^{abc} W^b_{1\mu} W^c_{1\nu}\right)^2 -$$
$$- \frac{1}{4}\left(\partial_\mu W^\alpha_{2\nu} - \partial_\nu W^\alpha_{2\mu} + g_2 \epsilon^{\alpha\beta\gamma} W^\beta_{2\mu} W^\gamma_{2\nu}\right)^2 -$$
$$- \Big|\Big(\delta^{ab}\delta^{\alpha\beta}\partial_\mu - ig_1\left[\frac{\tau^t}{2}\right]^b_a \delta^{\alpha\beta} W^t_{1,\mu} +$$
$$+ ig_2\left[\frac{\tau^\sigma}{2}\right]^\alpha_\beta \delta^{ab} W^\sigma_{2,\mu}\Big) Z_b^\beta\Big|^2 - \Big|\Big(\delta^{\alpha\beta}\partial_\mu -$$
$$- ig_2\left[\frac{\tau^\sigma}{2}\right]^\alpha_\beta W^\sigma_{2,\mu}\Big) M^\beta\Big|^2, \tag{3.1}$$

where $W^a_{1\mu}$, $W^\alpha_{2\mu}$ are triplet gauge fields, Z^α_a, M^α scalar fields. The tree scalar fields potential is defined by the standard structures:

$$V_o = -m_1^2(Z^+Z) - m_2^2(M^+M) + \frac{1}{2}\lambda_z^2 (Z^+Z)^2 + \frac{1}{2}\delta_z^2 (Z^{+a}_\alpha Z^\alpha_b)(Z^{+b}_\beta Z^\beta_a) \cdot$$
$$+ \frac{1}{2}\lambda_M^2 (M^+M)^2 + \lambda^2_{ZM}(Z^+Z)(M^+M) + \delta^2_{ZM}(Z^{+a}_\alpha Z^\beta_a)(M^{+\alpha}M_\beta) \tag{3.2}$$

where seven effective constants are embeded. The spontaneous symmetry breaking is fixed by the nonzero vacuum expectation values of Z,M

$$\langle Z\rangle^\alpha_a\Big|_{\substack{\alpha=1\\ \beta=1}} = Z \;;\; \langle Z\rangle^\alpha_a\Big|_{\substack{\alpha=2\\ a=2}} = \varkappa \;;\; \langle M\rangle_\beta\Big|_{\beta=2} = \nu, \tag{3.3}$$

where $\nu^2/z^2 \sim \nu^2/\varkappa^2 \ll 1$ and the minimality condition yilds

$$\delta^2_z (z^2 - \varkappa^2) = \delta^2_{ZM}\nu^2 \tag{3.4}$$

The one-loop potential determines the refined mass hierarchy of the scalar fields. Passing to (2.1) we find:

$$V_{eff} = V_o + \frac{R^4}{32\pi^2}\left[\ln\frac{R^2}{\mu^2} - \frac{1}{2}\right] + \frac{Q_+^4}{64\pi^2}\left[\ln\frac{Q_+^2}{\mu^2} - \frac{1}{2}\right] +$$
$$+ \frac{Q_-^4}{64\pi^2}\left[\ln\frac{Q_-^2}{\mu^2} - \frac{1}{2}\right] + \frac{Q_3^4}{64\pi^2}\left[\ln\frac{Q_3^2}{\mu^2} - \frac{1}{2}\right] +$$
$$+ \frac{g\,\mu_\perp^4}{64\pi^2}\left[\ln\frac{\mu_\perp^2}{\mu^2} - \frac{1}{2}\right] + \frac{g\,\mu_\parallel^4}{64\pi^2}\left[\ln\frac{\mu_\parallel^2}{\mu^2} - \frac{1}{2}\right], \tag{3.5}$$

where R, Q_+, Q_-, Q_3 are tree scalar fields masses:

$$R^2 \approx \delta_z^2(z^2+\varkappa^2) + \delta_{ZM}^4\,\nu^2/\delta_z^2 \quad ;$$
$$Q_+^2 \approx (\delta_{ZM}^2 + \lambda_{ZM}^2)(z^2+\varkappa^2) + 2\lambda_z^2\, z\varkappa \; ;$$
$$Q_-^2 \approx (\delta_{ZM}^2 + \lambda_{ZM}^2)(z^2+\varkappa^2) - 2\lambda_z^2\, z\varkappa ;$$
$$Q_3^2 \approx \nu^2 2[\lambda_M^2 - (\lambda_z^2+\delta_z^2)(\lambda_{ZM}^4 + \;_{ZM}^2)^2((\lambda_z^2+\delta_z^2)^2 - 4\lambda_z^2)^{-1} \tag{3.6}$$

and $\mu_\perp^2$, $\mu_\parallel^2$ are vector fields masses:

$$\mu_\parallel^2 = g_1^2 g_2^2 \nu^2/2(g_1^2+g_2^2) \quad ; \quad \mu_\perp^2 = \frac{1}{2}(g_1^2+g_2^2)(z^2+\varkappa^2) +$$
$$+ \frac{g_2^4\nu^2}{2(g_1^2+g_2^2)} \tag{3.7}$$

One of the new equations of minimum is of the form:

$$-m_2^2 + \delta_{ZM}^2\varkappa^2 + \lambda_{ZM}^2(z^2+\varkappa^2) + \frac{1}{32\pi^2}\Big[\,Q_+^2(2\lambda_{ZM}^2 + \delta_{ZM}^2) +$$
$$+ 2(\lambda_{ZM}^2 z + (\lambda_{ZM}^2 + \delta_{ZM}^2)\varkappa)^2\Big]\ln\frac{Q_+^2}{\mu^2} + \frac{1}{32\pi^2}\Big[\,Q_-^2(2\lambda_{ZM}^2 +$$
$$+ \delta_{ZM}^2) + 2\delta_{ZM}^4\varkappa^2\Big]\ln\frac{Q_-^2}{\mu^2} + \frac{1}{16\pi^2}\Big[\,2\lambda_{ZM}^2 + \lambda_M^2 \;+$$

$$+\delta^2_{ZM}\Big] R^2 \ln \frac{R^2}{\mu^2} + \frac{g}{64\pi^2}\frac{g_2^4}{g_1^2+g_2^2}\,\mu_\perp^2 \ln \frac{\mu_\perp^2}{\mu^2} + O(\nu^2)=0 \tag{3.8}$$

and it is necessary in order to cansel the terms of order z^2 (or $\varkappa^2$) in the "small" (of order ν^2) mass matrix elements. Such mass matrix element is

$$\mathfrak{M}^2 = -m_2^2 + \lambda^2_{ZM}(z^2+\varkappa^2) + \delta^2_{ZM}\varkappa^2 - \Pi + O(\nu^2) \tag{3.9}$$

It determines the mass of the light scalar particles. The polarisation operator in (3.9) is

$$-\Pi = \Big[\, 2$$

$$\sim \frac{\lambda^4_{ZM}\, z^2}{16\pi^2} \ln \frac{Q_+^2 Q_-^2}{\mu^4}\Big] + \Big[\,\frac{1}{2}$$
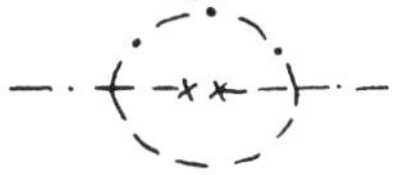
$$\sim \frac{(\lambda^2_{ZM}+\delta^2_{ZM})^2\varkappa^2}{16\pi^2} \ln \frac{Q_+^2 Q_-^2}{\mu^2}\Big] + \Big[\, 8$$
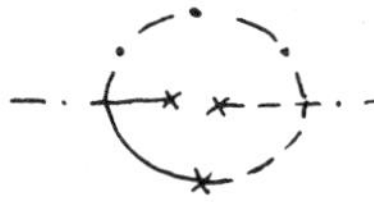
$$\sim \frac{\lambda^2_{ZM}(\lambda^2_{ZM}+\delta^2_{ZM})\, z\varkappa}{8\pi^2} \ln \frac{Q_+^2}{Q_-^2}\Big] + \Big[-\frac{1}{2}$$

$$\sim \frac{\lambda^2_{ZM}}{32\pi^2}\Big(Q_+^2 \ln \frac{Q_+^2}{\mu^2} + Q_-^2 \ln \frac{Q_-^2}{\mu^2}\Big)\Big] + \Big[-\frac{1}{2}$$

$$\sim \frac{(\lambda^2_{ZM}+\delta^2_{ZM})}{32\pi^2} \cdot \Big(Q_+^2 \ln \frac{Q_+^2}{\mu^2} + Q_-^2 \ln \frac{Q_-^2}{\mu^2}\Big) + \Big[$$

$$\sim \frac{g}{64\pi^2}\frac{g_2^4\,\mu_\perp^2}{(g_1^2+g_2^2)} \ln \frac{\mu_\perp^2}{\mu^2}\Big] + \Big[$$

$$\sim \frac{R^2}{16\pi^2}(2\lambda^2_{ZM} + \delta^2_{ZM} + \lambda^2_M) \ln \frac{R^2}{\mu^2}\Big], \tag{3.10}$$

where the new extremum conditions (3.8) are accounted. The terms of order z^2 and $\varkappa^2$ in (3.10) cancel and therefore the resulting matrix element is of order ν^2. Thus we see, that the tree-mass hierarchy is not spoilt.

This result is also true for higher orders in perturbation theory when the corresponding new extremum conditions are used.. Working with Green function the specifics of more complicated gauge group does not show, but it remains ins unclear how to express the effective potential of the model in terms of simple invariants of scalar fields, as in GUT's based on SU(N).

4. Coclusion. The model-independent proof of the mass hierarchy stability for SU(N)-models of GUT is the main result of this work. We also found that this stability takes place for GUT's with the more complex gauge group. Our results show that the "pragmatic" aspect of the mass hierarchy problem is completely solved although the higher-order extremum equations have to be used each time for fine tuning of spontaneous symmetry breaking parameters. But this concerns the "aesthetic" aspect of hierarchy problem and does not affect the progress of the theory.

The developed method of calculating the one-loop effective potential for SU(N)-models of GUT has an independent interest and can be used for many applications e.g. the one-loop potential of the realistic SU(5)-model of GUT [7] built here by analogy with (2.9) has also the simple forme

$$
\begin{aligned}
V_{eff} = & - \left[\tfrac{1}{2}\widetilde{M}^2 \Phi^2 + \widetilde{m}_1^2 (M^+M) + \widetilde{m}_2^2 (N^+N) \right] + \tfrac{1}{8} \widetilde{\lambda}^2_{\Phi} \Phi^4 + \\
& + \tfrac{1}{8} \widetilde{\delta}^2_{\Phi} (\Phi D \Phi)^2 + \left[\tfrac{1}{2} \widetilde{\lambda}^2_{\Phi M} \Phi^2 (M^+M) + \tfrac{1}{2} \widetilde{\delta}^2_{\Phi M} (\Phi D \Phi) \cdot \right. \\
& \left. \cdot (M + \tfrac{\tau}{2} M) + \tfrac{1}{2} \widetilde{\lambda}^2_{M} (M^+M)^2 + (M \leftrightarrow N_{terms}) \right] + \\
& + \widetilde{\lambda}^2_{MN} (M^+M9 (N^+N) + \widetilde{\delta}^2_{MN} (M^+N)(N^+M) \qquad (4.1)
\end{aligned}
$$

although it contains larger number of invariants and more complicated coefficient functions. Calculating these functions is a difficult prob lem, however, it is important that the mass hierarchy stability is proved and it follows that we can propose some founded predictions for the high energy physics. The important consequence of GUT-models is the existence of light (of order 10^2 GeV) scalar bosons and their experimental search is an urgent problem. The control experiment is connected with the search for light charged scalar bosons which are not predicted by the standard SU(5)-theory.

Reference

1. Weinberg,S. (1978). Phys. Lett. 82B, 387.
2. Kalashnikov,O.K., Klimov,V.V. (1978). Phys. Lett. 80B, 75.
3. Burova,M.V., Kalashnikov,O.K. (1984). Sov. Nucl. Phys. 39, 771 (Russian).
4. Coleman,S., Weinberg,E. (1973). Phys. Rev. D7, 1888.
5. Georgalas,B.C. et al. (1983). Nucl. Phys. 212B, 203.
6. Pati,J.C., Salam,A. (1974). Phys. Rev. D10, 275.
7. Fradkin,E.S., Kalashnikov,O.K. (1976). Phys. Lett. 64B, 177; Kalashnikov,O.K. (1980). Prepr. Lebedev Phys. Inst. N 166, Moscow.

SPECTRAL ASYMMETRY AND THE QCD CHIRAL LAGRANGIAN

A.Andrianov, V.Andrianov, V.Novozhilov, Yu.Novozhilov
Department of Theoretical Physics, Leningrad State University

1. Introduction. Spontaneous breaking of chiral symmetry (SBCS) is believed to provide the main source of mass generation in the case of light pseudoscalar mesons (Π - mesons). However, quantitative understanding SBCS as a low energy non-perturbative phenomena is still missing in QCD [1] although recent studies [2] clearly indicate the way it happens. Therefore, it is important to develop approaches, in which masses and low-energy vertices of Π - mesons can be directly expressed in terms of QCD basic low-energy parameters, namely, mass scale Λ and the quark condensate $<\bar{\psi}\,\psi>$ as well as in terms of current quark masses m_q and the number of colours N_c. Evidently, the assumption that there is an almost self-sustaining low-energy region in QCD is crucial for such an approach.

Since early 60th, the description of low-energy processes proceeded in a phenomenological way through bosonization of fermion current matrix elements, i.e. their representation by means of the lightest boson fields. In the chiral model for Π - mesons [3] the whole set of vertices has been written down from symmetry properties while their coefficients have been taken from experiments. At present, QCD is the theory of strong interactions, and a new program for derivation of chiral dynamics from QCD is to be developed. In the QCD chiral dynamics all the numerical coefficients at vertices should appear naturally. Moreover, in the course of its derivation underlying assumptions should be made apparent. Such a program has been recently launched in [4,5] where the case of massless quarks and mesons has been discussed starting directly from the basic QCD functional.

The purpose of this paper is to derive the QCD chiral Lagrangian for massive pseudoscalar mesons in presence of the condensate $<\bar{\psi}\psi>$. We apply the same bosonization method as in the massless case [4,5]. It is necessary now to consider a system not only in external vector V_μ and pseudovector A_μ fields, but also in external scalar S and pseudoscalar P fields while the chiral anomaly [6] is to be calculated with an asymmetrical spectrum of low energy quarks. We identify the quark colorsinglet chiral phase with a chiral field;the flavour group is SU(3). As a result, the QCD chiral Lagrangian is derived which depends on the (chiral breaking) mass scale Λ, asymmetry of quark spectrum M current quark masses m_q and the number of colours N_c as parameters. As a by-product which can have an independent application we hace obtained a gauge invariant expression for the non-topological part of the Adler-Bardeen-Gross-Jackiw anomaly [6] in external fields V_μ, A_μ, S, P with an asymmetrical quark spectrum as well as the relevant anomalous action.

2. Spectral asymmetry and chiral anomaly. We leave aside an interrelation between quark flavour and colour phases. Therefore [5] it is sufficient to consider only the Lagrangian for quarks interacting with external colour-singlet fields V_μ, A_μ, S, P:

$$L = \bar{\psi}i\left[\gamma^\mu(\partial_\mu + V_\mu + \gamma_5 A_\mu) - S + i\gamma_5 P\right]\psi \equiv \bar{\psi}\not{D}\psi \quad (1)$$

where fields $V_\mu = V_\mu^a T_a \ldots$ take values in the flavour SU(3) and T_a are antihermitian generators. The Lagrangian (1) is invariant under gauge transformations as well as under local chiral ones $g_5 = \exp i\lambda(x)\gamma_5$ if external fields are transformed accordingly. Under an infinitesimal g_5 :

$$\delta\psi = i\lambda\,\gamma_5\psi\ , \quad \delta V_\mu = [A_\mu,\lambda]\ , \quad \delta A_\mu = [D^V_\mu,\lambda]\ ,$$

$$\delta S = \{P,\lambda\}\ , \quad \delta P = \{S,\lambda\}, \tag{2}$$

where $D^V_\mu = \partial_\mu + V_\mu$.

The generating functional for Green functions of quark currents is given by

$$Z(V,A,S,P) = \int D\bar{\psi}\, D\psi\, \exp i \int d^4 x \bar{\psi}\, \not{D}\psi \tag{3}$$

Our task is to define in (3) an integration over low-energy quarks in such a manner that it is (i) gauge invariant, and (ii) leads to a quark condensate $\langle\bar{\psi}\psi\rangle \neq 0$. For this we go over to the Euclidean space and define gauge invariant eigenvalues k of the complete Dirac operator $\not{D}$ (including external fields): $\not{D}\,\psi_{k\alpha} = k\psi_{k\alpha}$, where α refers to polarization and colour degrees of freedom. In the massless case [4,5] a gauge invariant definition of the low-energy region employs one mass parameter Λ and is symmetric : $|k| < \Lambda$. Now, it is necessary to introduce an additional parameter, the asymmetry of the quark Dirac operator M and to define a low-energy region by

$$-\Lambda + M \leq k \leq \Lambda + M\ , \quad 0 \leq M \leq \Lambda \tag{4}$$

Both parameters Λ and M are invariant under flavour SU(3)-gauge transformations as well as under local chiral ones. Λ does not depend on quark flavours we suppose that M is the same for all flavours too. It means that we use an approximation where the condensates $\langle\bar{\psi}\psi\rangle$ do not depend on quark flavour within SU(3). Empirical values of $\langle\bar{\psi}\psi\rangle$ support this approximation. The general case when M is a matrix and $\langle\bar{\psi}_s\psi_s\rangle \neq \langle\bar{\psi}_u\psi_u\rangle =$

$=\langle \bar{\psi}_d \psi_d \rangle$ will be treated separately. In the presence of external fields the condensate $\langle \bar{\psi}\psi \rangle$ is equal to VEV of quark density averaged over the volume, In the Euclidean space we get for massless quarks, $m_q = 0$,

$$C_E(\Lambda, M) \equiv \frac{1}{\text{Volume}} \int d^4x \langle \bar{\psi}(x)\, \psi(x) \rangle =$$

$$= - \int d^4x 2N_c \frac{1}{k} \left[\Theta(\Lambda + M - k) - \Theta(\Lambda - M - k) \right] = - \frac{N_c}{2\pi^2} (\Lambda^2 M + \frac{M^3}{3}), \quad (5)$$

where the eigenvalue basis of $\not{D}$ has been utilized. In the Minkowski space, $M \to -iM$:

$$C_M = - \frac{N_c}{2\pi^2} (\Lambda^2 M - \frac{M^3}{3}). \quad (5')$$

Thus, we consider chiral non-invariant quark fluctuations in (3) arising with the condensate (5) in the low-energy region (4). Both the low-energy region (4) and the condensate (5) are determined by chiral invariant parameters Λ and M. The total (regularized) functional $Z = Z^L \tilde{Z}$, where in Z^L an integration over fermions is restricted by the region (4) while $\tilde{Z}$ denotes the rest. Our aim to study non-perturbative chiral non invariant fluctuations means that we are interested not in the entire functional Z^L but only in its chiral non-invariant part $Z_{ch} = Z^L Z^{-1}_{inv}$, where Z_{inv} a chral invariant fermionic functional in the region (4).

The functional Z_{ch} is closely related to the Adler-Bardeen-Gross-Jackiw anomaly [6] in both topologically non-trivial and non-topological [7] parts. To calculate the latter we use the finite-mode regularization scheme [7] generalized to the case of extended set of fields V_μ, A_μ, S, P and quark spectral asymmetry $M \neq 0$. There are no $1/\Lambda^2$ and higher corrections in this scheme, so that calculations are exact and result in a finite-term expression. In the Euclidean space the anomaly $A^a(x)$ is given by

$$A^a(x) = - \frac{\delta \ln Z^L}{\delta \lambda^a(x)} = -2\mathrm{tr}(\gamma_5 T^a \langle x| \Theta(\Lambda^2 - (\not{D}-M)^2)|x\rangle), \tag{6}$$

where $S(x)$ is replaced by $S(x) + iM$. We get $A^a(x) = A^a(V,A) + A^{a'}(x)$, where the term $A^a(V,A)$ which does not depend on fields S,P and parameters Λ, M has been already calculated [7]. The new term in the Euclidean space $A^{a'}$ includes also Λ^2 - contribution,

$$\begin{aligned} A^{a'}(x) = &- \frac{N_c}{4\pi^2} \mathrm{tr}\,(T^a\{ 2\Lambda^2([D_\mu,A_\mu] + \{S,P\} + 2iMP) + \\ &+ (\{[D_\mu,A_\mu], (P^2-S^2)\} + 2\{A^2,\{S,P\}\} + \\ &+ [[D_\mu,S], [A_\mu,S]] + 2[\{A_\mu,S\},[A_\mu,P]\} - \\ &- \{S, [A_\mu, [A_\mu,P]]\} - [A_\mu, \{S, [A_\mu,P]\}] - \\ &- [A_\mu, \{S, [A_\mu,P]\}] - [A_\mu, [A_\mu, \{S,P\}]] - \\ &-[[D_\mu,P], [A_\mu,P]] + \tfrac{1}{3} [D_\mu, [D_\mu, \{S,P\}]] + \\ &+ \tfrac{1}{3} [D_\mu, \{S, [D_\mu,P]\}] + \tfrac{1}{3} \{S, [D_\mu, [D_\mu,P]]\} - \\ &- \tfrac{2}{3} [D_\mu,\{S, \{A_\mu,S\}\}] + \tfrac{4}{3} \{S, \{S,[D_\mu,A_\mu]\}\} - \\ &- \tfrac{2}{3} \{S, [D_\mu, \{A_\mu,S\}]\} + \{\{S,P\}, (P^2-3S^2)\} + \\ &+ \tfrac{4}{3} \{S, \{S, \{S,P\}\}\} + \tfrac{4}{3} iM^3P + 2M^2([D_\mu A_\mu] + \\ &+ \{P,S\}) + 2iM(- \{[D_\mu,A_\mu], S\} - 2\{[D_\mu,S],A_\mu\} + \\ &+ \{A_\mu,\{A_\mu,P\}\} + [D_\mu, [D_\mu,P]]) + 4iM(P^3 - \\ &- \{P,S^2\} + SPS))\}), \end{aligned} \tag{7}$$

herein $D_\mu = \partial_\mu + [V_\mu,*]$, $\{S,P\} = SP+PS$, $[S,P]= SP-PS$.

This anomaly appears from an (anomalous) action $\tilde{W} = W_{-} + W_{+}$ by using (2). The term W_{-} relates to the Wess-Zumini-Witten action 9 while W_{+} stands for the non-topological part of $A^{a}(x)$ and is represented by a local functional of external fields. The anomalous action W_{+}' which corresponds to (7) reads in the Minkowski space

$$W_{+}' = \frac{N_c}{4\pi^2} \text{tr} \{ (\Lambda^2 - M^2)(A_\mu^2 + S^2) + 2(\Lambda^2 M - M^3/3)S - 2M(-iP[D_\mu, A_\mu] + 2SA_\mu^2 + S(P^2+S^2)) + A_\mu P A_\mu P - \frac{1}{2}(D_\mu P)^2 - i\{A_\mu, P\}[D_\mu, S] - A_\mu^2 S^2 - \frac{1}{3}[D_\mu, S]^2 + \frac{1}{6}P^4 - \frac{1}{6}\{S,P\}^2 \} \quad (8)$$

The entire action has such a form [7]

$$W_{+}(V,A,S,P) = W_{+}' - \frac{N_c}{12\pi^2}\left[\frac{1}{4}(F^{V}_{\mu\nu})^2 + A_\mu F^{V}_{\mu\nu} A_\nu - \frac{1}{2}([D_\mu, A_\mu])^2 - \frac{1}{2}A_\mu A_\nu A_\mu A_\nu + \frac{3}{2}(A_\mu^2)^2\right], \quad (9)$$

where $F^{V}_{\mu\nu} = \partial_\mu V_\nu - \partial_\nu V_\mu + [V_\mu V_\nu]$.

The expressions (8),(9) are defined up to chiral invariant terms.

3. <u>Effective QCD - action for chiral field</u>. Let us introduce a quark chiral phase $\tilde{U}$ and a chiral field $U = \tilde{U}^2$ and then derive an effective action W_{eff} for U. The chiral field U describes chiral breaking effects and, at the same time, can be treated also as a chiral transformation in its own right, so that an integral over U of a non-invariant quantity f(U) becomes a chiral invariant. Let's take a chiral non-invariant fermionic functional $Z^{L}(V^{u}, A^{u}, S^{u}, P^{u})$ which contains chirally rotated external fields, integrate over U with the Haar measure DU, then the result is a chiral invariant functional $Z_{inv}(V,A,S,P)$

$$Z_{inv}^{-1}(V,A,S,P) = \int DU Z_L^{-1}(V^U, A^U, S^U, P^U), \qquad (10)$$

The effective action $W_{eff}(U)$ for the chiral field U is defined by

$$Z_{ch} = Z_L(V,A,S,P)\, Z_{inv}^{-1}(V,A,S,P) =$$

$$= \int DU \exp\{iW_{eff}(U,V,A,S,P)\} \qquad (11)$$

$$W_{eff} = \tilde{W}(V,A,S,P) - \tilde{W}(V^U, A^U, S^U, P^U) \qquad (12)$$

where $\tilde{W} = W_- + W_+$ is an anomalous action with the Wess-Zumino -Witten term W_- and the P-preserving term W_+ given by (8), (9). Having in mind applications considered below, we omit W_- and present the resulting Lagrangian $L_+(U)$, which corresponds to W_+

$$L_+(U) = N_c \frac{\Lambda^2 - M^2}{16\pi^2} \operatorname{tr}\{\bar{D}_\mu U(\bar{D}^\mu U)^+\} + L^{(4)}(U) -$$
$$- \frac{N_c}{8\pi^2 B_o} \operatorname{tr}\{(\Lambda^2 M - M^3/3)\chi^+ U + \frac{\Lambda^2 - M^2}{8B_o} U\chi^+ U\chi^+ +$$
$$+ \frac{M}{2}(\bar{D}_\mu \chi)^+ \bar{D}^\mu U - \frac{M}{4B_o^2} U\chi^+ \chi\chi^+ + \text{h.c.}\} + O(m_q^4), \qquad (13)$$

where we have dropped those terms $O(m_q^4)$, which do not depend on M and Λ.

$L^{(4)}$ is the term with four derivatives calculated earlier [4] :

$$L^{(4)} = \frac{N_c}{96\pi^2} \operatorname{tr}\Big(\frac{1}{2}\bar{D}_\mu U(\bar{D}_\nu U)^+ \bar{D}_\mu U(\bar{D}^\nu U)^+ - (\bar{D}_\mu U(\bar{D}^\mu U)^+)^2 +$$
$$+ \bar{D}_\mu^2 U(\bar{D}_\nu^2 U)^+ + 2F_{\mu\nu}^L \bar{D}^\mu U(\bar{D}^\nu U)^+ + 2F_{\mu\nu}^L (\bar{D}^\mu U)^+ \bar{D}^\nu U +$$
$$+ F_{\mu\nu}^L U F_R^{\mu\nu} U^+\Big). \qquad (14)$$

Here $\bar{D}_\mu = D_\mu^+ + \{A_\mu, *\}$, $\chi = 2B_o(S+iP)$ and $B_o =$

$= -\langle\bar{\psi}\psi\rangle / F_o^2$, F_o is the "bare" π- decay constant. The quark mass matrix m_q is hidden in S. Note that $\Lambda^2 - M^2$ in the kinetic term in (13) is proportional to F_o^2. The main terms in (13) and (14) have the same characteristic structure as in the phenomenological chiral Lagrangian [3].

To make clear the physical context of $L_{eff} = L_+(1) - L_+(U)$ we expand the chiral field $U = -\exp(i\Pi/F_o)$ in powers of a meson field $\Pi(x)$ and retain the terms in $L_{eff} \cong -L_+(U)$ which are bilinear and linear in Π.

Bilinear part of L_+ is given by

$$L_+(\Pi^2) = \mathrm{tr}\Big(\frac{N_c(\Lambda^2-M^2)}{16\pi^2F_o^2}(\partial_\mu\Pi)^2 + \frac{N_c}{32\pi^2F^2}(\partial_\mu\{m_q,\Pi\})^2 -$$

$$- \frac{N_c}{32\pi^2F_o^2}\partial_\mu\Pi\partial_\mu\{\Pi, m_q^2\} + \frac{N_c}{96\pi^2F_o^2}(\partial_\mu^2\Pi)^2 +$$

$$+ \frac{N_c}{8\pi^2F_o^2}\Big(\Lambda^2M - \frac{M^3}{3}\Big)\Pi\{m_q,\Pi\} + \frac{N_c(\Lambda^2-M^2)}{16\pi^2F_o^2}\{m_q,\Pi^2\} +$$

$$+ \frac{N_cM}{8\pi^2F_o^2}\Pi\{m_q^3,\Pi\} - \frac{N_c}{16\pi^2F_o^2}\{m_q^3,\Pi\}\{m_q,\Pi\} +$$

$$+ \frac{N_c}{96\pi^2F_o^2}\{m_q,\{m_q,\Pi\}\}^2\Big). \qquad (15)$$

where an external field $S(x)$ is replaced by the quark mass matrix in the limit $m_u = m_d = \hat{m}$, $m_q = \frac{1}{2}(2\hat{m} + m_s)\cdot 1 + \frac{1}{\sqrt{3}}(\hat{m} - m_s)\lambda_8$. Remark that (15) contains a tachyonic term originating in (14). In the momentum space the Lagrangian (15) exhibits the following structure

$$L_+(\Pi^2) \to \frac{1}{2}\Pi^*(p)\,(Ap^4+Bp^2-C)\,\Pi(p) \qquad (16)$$

where A,B and C are diagonal matrices in the physical basis of SU(3). Meson masses can be found from the Eq. $Ap^4+Bp^2-C=0$, So that $m_\Pi^2 = (-B_\Pi + \sqrt{B_\Pi^2 + 4A_\Pi C_\Pi})/2A_\Pi$ and the propogator factorizes near the pole $(Ap^4+Bp^2-C)_\Pi^{-1} \approx$ $\approx (p^2-m_\pi^2)^{-1} - (B^2+4AC)_\pi^{-\frac{1}{2}}$.

Coefficients A,B and C are given by a rather complicated expressions, but they can be simplified in the realistic case when the current mass of u- and d-quarks is much smaller then the strange quark $\hat{m} \ll m_s$. Choosing $\mu_\rho^2 = 6(\Lambda^2-M^2)$ as a $(\text{mass})^2$ unit and putting $F_0^2 = N_c(\Lambda^2-M^2)/4\pi^2 = N_c\mu_\rho^2/24\pi^2$ we have $A = \mu_\rho^{-2}$ and obtain for π- meson

$$B \approx 1 , \quad C = 2\hat{m}B_0 ,$$

for K - meson

$$B = 1+6\hat{m}m_s, \quad C = (\hat{m}+m_s)B_0-m_s^2- 12Mm_s^3 + 5m_s^4$$

and for η - meson

$$B = 1+4m_s^2 , \quad C = \frac{2}{3}(\hat{m}+2m_s)B_0 - \frac{8}{3}m_s^2-16Mm_s^3+ \frac{16}{3} m_s^4 \qquad (17)$$

here $B_0 = N_c/2\pi^2F_0^2(\Lambda^2M-M^3/3) \Rightarrow (8M^3+2M)$ and all mass parameters $\hat{m}$, m_s, m_Π , M, B_0 are considered in μ_ρ - units: $(m \to m/\mu_\rho , \ldots)$. Respectively $m_\Pi^2 = \mu_\rho^2 \frac{1}{2}(-B+ + \sqrt{B^2+4C})_\Pi$.

Linear terms in L^+ describe different interactions of the meson field Π with the axial vector field divergence

$$L^+_{eff}(\Pi) = i \frac{N_c}{4\pi^2F_0} \operatorname{tr} \partial_\mu A^\mu \{(\Lambda^2-M^2)\Pi + M\{m_q,\Pi\}- - \frac{1}{2}\{m_q^2, \Pi\} - \frac{1}{6}\partial_\nu^2 \Pi\} = \sum_\Pi \partial_\mu A_\Pi^\mu d_\Pi \Pi(x). \qquad (18)$$

After integration over Π in the generating functional this term leads to a pole in accordanve with PCAC:

$$\frac{1}{2}\, \partial_\mu A^{*\mu}_{\Pi} \frac{F_\pi^2}{p^2-m_\Pi^2} \partial_\nu \quad A^\nu_\Pi$$

hence the decay constant F_Π is

$$F_\Pi^2 = d_\Pi^2 \quad / \sqrt{B^2+4AC}\Big|_{\Pi},\ -\partial_\nu^2 = m_\Pi^2 \tag{19}$$

where d_Π is defined in (18) and in the approximation $\hat{m} << m_s$ can be expressed by

$$\begin{aligned} d_\pi &= F_o(1+12M\hat{m} + m_\pi^2) \\ d_k &= F_o(1+6Mm_s - 3m_s^2+m_k^2) \\ d_\eta &= F_o(1+ 8Mm_s-4m_s^2+m_\eta^2) \end{aligned} \tag{20}$$

in unit of μ_ρ^2.

Thus, starting from the QCD - functional we have arrived at an effective chiral theory with the Lagrangian $L_{eff}(U)$ depending on parameters Λ, M, $\hat{m}$ and m_s. Λ and M characterize the low-energy region as a domain, where chiral breaking processes dominate. The chiral field U is a quark degree of freedom, for which the Lagrangian L_{eff} with the kinetic term has been obtained from the initial Lagrangian (1) by the bosonization procedure. The parameters are fixed by experimental values of m_π, F_π, m_k, F_k : m_π = 135 MeV, F_π = 93 MeV, m_k = 498 MeV, $F_k : F_\pi$ = 1,23 [3] . Then the computation by means of Eqs. (17), (19), (20) (taking into account small deviations from them due to chiral logarithms [3,8]) leads to the following result for characteristic parameters: M = 740 MeV, Λ = 814 MeV, B_o = 6,8 GeV, $<\bar{\psi}\psi> = -(375\ \text{MeV})^3$ and for current quark masses $\hat{m}$ = = 1,4 MeV, m_s = 50 MeV. We find that the appropriate quark masses appear to be three times less than common ones. However their values are remarkably close to the quark masses which has been recently employed in lattice calculations of meson spectra [11] .

The corresponding mass and decay constant of η - meson takes the value $m_\eta = 540$ (exp. 549) MeV, $F_\eta : F_\pi = 1{,}31$ ($1{,}3 \pm 0{,}05$ from [3]) in a reasonable agreement with experimental fitting.

The part $L^{(4)}$ of the effective Lagrangian (14) which does not depend on quark masses has been already analyzed in [4] . The dimensionless coupling constants in $L^{(4)}$ are involved in description of $\pi\pi$ - scattering amplitude (especially, in the D-wave), radius of π - meson and the electromagnetic π - meson decays [3,4] . The comparison of predictions and experimental fitting of above constants has shown the reliable agreement.

The more detailed examination of physical content of the effective Lagrangian (13), (14) is postponed to the next paper.

4. Discussion. As it follows unambiguiously from data comparision, presented in Sect. 3 and in [4] , the QCD-chiral action W_{eff} is well established to describe $\tilde{\pi}$ - meson physics.

The success of W_{eff} confirms i) the existence of a low-energy region in QCD, ii) the principal role played there by chiral non-invariant processes described by the quark chiral phase.

The present approach can be extended to include: (i) a flavour dependence of the asymmetry parameter M, (ii) effects of joint bosonization of quark flavour and colour chiral phases started already in [5] and (iii) the bosonization of vector degrees of freedom.

In this paper we studied the mesonic sector of W_{eff} only. But the σ - model Lagrangians possess also a Skyrmion sector with a baryon [9] . It is naturally to ask whether there is a Skyrmion in the QCD chiral theory with W_{eff}. The answer is no [10] in the massless case $m_q = 0$, $M = 0$. It is highly probable that this situation also holds in the case $m_q \neq 0$ and $M \neq 0$. However, even if the identity "baryon = Skyrmion" or existence of a Skyrmion sector is established, it cannot be used against the QCD chiral theory with W_{eff}. In fact, the action W_{eff} pretends to give a description of chiral non-invariant dynamics at low-energies but not to represent a complete low-energy dynamics; we recall the relation $Z^L = Z_{inv} \int DU \exp iW_{eff}(U)$, where Z^L is related to the low-energy region and Z_{inv} is its chiral invariant part. A complete low-energy dynamics is governed by the action

$$W_L = W_{eff} + W_{inv}, \qquad W_{inv} = -i \, \mathrm{Ln} \, Z_{inv} \tag{21}$$

It was believed sometimes that chiral invariant processes are negligible at low-energies. It is apparent now that they should not be thrown away from the beginning. There are states and quantities which can be described in terms of $W_{eff}(U)$ as well as in terms of W_{inv}. As, a consequence, the general picture of low-energy phenomena should be a kind of superposition of descriptions provided by W_{eff} and W_{inv}.

As far as baryon states are concerned they should be considered as a superposition of a Skyrmion an a three-quark bound state. Recently, superpositions of such kind were proposed [12] in the phenomenological analyses of static properties of low-lying baryons.

One may conjecture that in the complete low-energy dynamics with the action (21) there will be no tachyon pole and the Skyrmion problem will be clarified. The vacuum baryon number current for the complete low-energy dynamics has already been derived [5].

In conclusion, we stress that the QCD chiral theory with W_{eff} gives us an opportunity of an extensive low-energy verification of QCD in a non-perturbative regime.

References

1. Peskin,M. Lectures given at the Les Houches Summer Scholl. (1982). Preprint SLAC - PUB - 3021;

   Fomin,P.I.,Gusynin, V.P., Miranski, V.A., Sitenko,Yu. (1984). Rivista del Nuovo Cimento, 6, 1;

   Miranski, V,A., Fomin, P.I. (1985). Particles and Nuclei, 16, 469.

2. Casalbuoni,R., Curtis, S.De, Dominici,D., Gatto, R. University of Geneve preprint UGVA-DPT 1984/09-439;

   Arbuzov,B.A. (1985). IHEP preprint 85-55, Serpukhov;

   Castorina,P., Pi, S.Y. (1984). MIT Preprint; id. (1985). Phys. Rev. D31, 411.

3. Gasser,J., Leutwyler,H. (1984). Ann. Phys.(N.Y) 158,142 (1985). Nucl. Phys. B250, 465.

4. Andrianov,A.A. (1985). Phys. Lett. B 157, 425.

5. Andrianov,A.A., Novozhilov,Yu.V. (1985). Phys. Lett. 153B, 422.

6. Treiman,S.B., Jackiw,R., Gross,D.J. (1972). Lectures on Current Algebra and its Applications. Princeton University Press.

7. Andrianov,A.A., Bonora,L. (1984). Nucl. Phys. B233, 232.

8. Langacker,P., Pagels,H. (1973). Phys. Rev. D8,4595; (1974). D10, 2904.;

   Kazakov,D.I., Pervushin,V.N., Pushkin,S.V. (1978). J. Phys. A11, 2093.

9. Witten,E. (1983). Nucl. Phys. B223, 422-433.

10. Andrianov,V.A., Novozhilov,V.Yu. (to be published in ZhETPh).

11. Barkai,D., Moriarty,K.J., Rebbi,C. (1985). Phys. Lett. 156B, 385.

12. Nair,V.P., Rodgers,V.G.J. (1984). Princeton preprint.

TOWARDS NON-PERTURBATIVE CALCULATING HADRONIC FIELD CORRELATORS

G.S.Iroshnikov

Moscow PhysicalTechnical Institute

1. Introduction. In QCD calculation of hadron scattering amplitudes it is necessary to average values of products of colourless operators describing a compound hadron. Here we make use of $1/N_c$-expansion (N_c is the colour number) and semiclassical approximation. It is known 1 that in the leading in $1/N_c$ order by planar diagrams with no internal quark loops. Such a diagram posesses topology of a dual (string) models. Moreover, as was indicated in [2] actually, the only contribution to the path integral is due to only one field configuration (master field) as $N_c \to \infty$. This field may be an extremal of effective action for QCD hadronic phase. Here we show that within the $1/N_c$ approximation, owing to the mentioned above, in path integral one should sum not over all configurations of the gauge field but only over some subclass. First we guess it and then see that in the higher order of semiclassical approximation a solution of effective field equations appears which indeed plays the role of this master field. Form the physical point of view it describes a relativistic string with bare tension coefficient (with quarks at the edges).

2. Effective action for correlation functions. Speaking of hadrons we confine ourselves to the meson case. By assumption, each meson is an extended object and is described by the colourless field operator

$$M^{\beta}_{\alpha}(\Gamma)=\bar{\Psi}^{\beta}_{c}(y')\Big[P\exp\Big(-ie\frac{\lambda^a}{2}\int_{y}^{y'}dz_\mu A^a_\mu(z)\Big)\Big]_{cc'}\Psi_{c',\alpha}(y). \quad (1)$$

Here α, β are the set of the Lorentz and internal indices, c, c' the indices of the colour gauge group SU(N), the bare charge. Integration is made over a contour Γ connecting points y, y' in Euclidean space $\mathbb{R}^4$. The Euclidean n-particle correlator $K(1, \ldots, n)$ is determined by the relation

$$K(1,\ldots,n)=N^{-1}\int d\mu(A)D\Psi D\bar{\Psi}\Big\{e^{-S_{YM}}\big[M(\Gamma_n)\ldots M(\Gamma_1)\big]\Big\}, \quad (2)$$

where $N=\int d\mu(A)D\psi D\bar{\psi}\, e^{-S_{YM}}$,

$$S_{YM}(A,\Psi)=\int d^4x\Big\{\frac{1}{4}G^a_{\mu\nu}G^a_{\mu\nu}+\bar{\Psi}(i\gamma_\mu D_\mu-m)\Psi\Big\},$$

$$G^a_{\mu\nu}=\partial_\mu A^a_\nu-\partial_\nu A^a_\mu-ef^{abc}A^b_\mu A^c_\nu, \quad (3)$$

$$D_\mu=\partial_\mu+ie\frac{\lambda^a}{2}A^a_\mu\equiv\partial_\mu+ieA_\mu.$$

It is shown in [3] for n=2 and in [4] for an arbitrary n that neglecting quark loops ($1/N_c$-expansion) and their spin one can represent the connection part of the mean value (2) as

$$K(1,\ldots,n)\approx\sum_P\delta_P\prod_{q=1}^{n}\int\tilde{D}x_q\Big\{\exp\Big[-\frac{1}{2}\int_{\gamma_1}^{\gamma_2}d\gamma(\dot{x}^2/\lambda+\lambda m^2)\Big]_q\langle O(\Gamma)\rangle_A\Big\}, \quad (4)$$

where

$$\tilde{D}x = \frac{d\lambda\,\Delta\gamma}{4\pi i} Dx_{\mu}, \quad O(\Gamma) = \mathrm{Tr}\left[P\exp\left(-ie\oint_{\Gamma} dz_{\mu} A_{\mu}\right)\right], \tag{5}$$

δ_P is the Fermi-field interchange parity P. The integration contour Γ in (5) consists of the contours Γ_i (i=1, ..., n) contained in the operators $M(\Gamma_i)$ in (2) connected with each other at the edges of quark trajectories.

In the leading in $1/N_c$ order one should take into account only gluon planar diagrams. This means that in the integral

$$\langle O(\Gamma)\rangle_A = N^{-1}\int d\mu(A)\, e^{-S^{(A)}_{YM}} \cdot \mathrm{Tr}\left[P\exp\left(-ie\oint_{\Gamma} dz_{\mu} A_{\mu}\right)\right] \tag{6}$$

one should sum over a certain subclass of fields whose contribution is most essential in the mentioned approximation. Let us introduce an arbitrary two-dimensional surface Σ in R^4 without holes and hendles and the boundary Γ. It is given by equation $x_{\mu} = z_{\mu}(\tau^i)$, where μ = 0,1,2,3; i = 0,1. Since $h_{\mu i}(\tau) = \frac{\partial z_{\mu}}{\partial \tau^i}$, we define the tangent components of the field A^a_{μ}

$$A^a_i(\tau) = \frac{\partial z_{\mu}}{\partial \tau^i} A^a_{\mu}(z(\tau)) = h_{\mu i} A^a_{\mu}$$

and reproduce an approximate calculation of (6) as follows: first, for a fixed Σ we integrate over the fields $A^a_i(\tau) \equiv A(\Sigma)$ and afterwards integrate over all the surfaces Σ with the boundary Γ, i.e.

$$\langle O(\Gamma)\rangle_A \approx \int d\mu(\Sigma)\int d\mu\left[A^a_i(\Sigma)\right] e^{-S_{Y-M}^{[A(\Sigma)]}}\, O(\Gamma). \tag{7}$$

Here

$$S_{YM}[A(\Sigma)] = \frac{1}{4}\int_{\Sigma} d^2 z\sqrt{g}\, G^a_{ik} g^{il} g^{kn} G^a_{ln}, \tag{8}$$

$$G^a_{ik} = \frac{\partial \tilde{A}^a_k}{\partial z^i} - \frac{\partial \tilde{A}^a_i}{\partial z^k} - \varepsilon f^{abc} \tilde{A}^b_i \tilde{A}^c_k \tag{9}$$

where $g = \det g_{ik}$, $g_{ik}(z) = \frac{\partial z_\mu}{\partial z^i} \frac{\partial z_\mu}{\partial z^k}$ is the matric on Σ. Renormalize $A^a_i(z)$ setting

$$eA^a_i = \left(\frac{e}{d}\right)\left(A^a_i d\right) \equiv \varepsilon \tilde{A}^a_i, \quad \varepsilon = e/d \tag{10}$$

where d is an arbitrary constant of the length dimension, ε the bare dimensional charge. The mean value (7) will be calculated in the following way. First, we represent $O(\Gamma)$ as a path integral over auxiliary Grassmannian fields $\xi_c(\gamma)$ ($c = 1, \ldots, N_c$) [3,5]:

$$O(\Gamma) = \int \Pi\left[i D\xi(\gamma) D\xi^*(\gamma)\right] e^{-S(\xi)} \xi_c(1) \xi^*_c(0), \tag{11}$$

$$S(\xi) = \oint_{\Gamma} d\gamma\, \xi^*_c(\gamma)\left[\frac{d}{d\gamma} + ieA_\mu \frac{dz_\mu}{d\gamma}\right]_{cd} \xi_d(\gamma). \tag{12}$$

Here γ runs the closed contour $\Gamma = \partial\Sigma$ so that $x_\mu = z_\mu(\gamma)$ and monotonously varies from 0 to 1, while $z_\mu(0) = z_\mu(1)$. Then each factor of the sum (4) can be written as

$$\int \prod_{q=1}^{n} \tilde{D}x_q \int d\mu(\Sigma) \int d\mu[A(\Sigma)] \int \left[iD\xi(\gamma) D\xi^*(\gamma)\right] e^{-S_{eff}(x,A,\xi)} \xi_c(1)\xi^*_c(0), \tag{13}$$

where

$$S_{eff} = S_0(x) + S_{Y\text{-}M}(A) + S(\xi, A) \equiv$$
$$\equiv \sum_{q=1}^{n} \left[-\frac{1}{2} \int_{\gamma_1}^{\gamma_2} d\gamma \, (\dot{x}^2/\lambda + \lambda m^2) \right]_q + \frac{1}{4} \int_{\Sigma} d^2 z \sqrt{g} \, G^a_{ik} G^a_{\ell n} g^{i\ell} g^{kn} + \tag{14}$$
$$+ \oint_{\Gamma} d\gamma \, \xi^*_c(\gamma) \left[\frac{d}{d\gamma} + i\varepsilon A_i \frac{dz^i}{d\gamma} \right]_{cd} \xi_d(\gamma).$$

In the last term, since Γ is the surface boundary, we write

$$A_\mu \frac{dz_\mu(z(\gamma))}{d\gamma} = A_\mu \frac{\partial z_\mu}{\partial z^i} \frac{dz^i}{d\gamma} = A_i \frac{dz^i}{d\gamma}$$

and change the normalization of fields A^a_i according to (10). The last term in S_{eff} (14) arises from the total contribution of the ordered exponents in (1) and quark propagators. An involution such that

$$(\xi_1 \xi_2)^* = \xi^*_1 \xi^*_2 \tag{15}$$

provides the reality of $S(\xi, A)$. The action (14) takes into account the quantum up-vacuum excitations generated by the field operators (1) and is invariant with respect to arbitrary variations of γ and general transformations of z. The kinetic quark part $S_0(x_q)$ is also invariant under reparameterization [6] $\gamma \to \gamma' = f(\gamma)$, $\lambda \to \lambda' = \lambda / \dot{f}(\gamma)$.

3. <u>Solution of field effective euclidean equations.</u> For semiclassical calculating (13) we expand all the fields into their classical and quantum parts. Varying S_{eff} (14) over $A^a_i(z)$ leads to an equation of motion on the surface Σ

$$\partial_i(\sqrt{g}\, G^{a,ik}) + \varepsilon f^{abc} A_i^c \sqrt{g}\, G^{b,ik} = 0 \tag{16}$$

with the boundary value on $\partial\Sigma = \Gamma$

$$\frac{dz^j}{d\gamma}\left(\sqrt{g}\, G^{a,ik} e_{ij} + \varepsilon T^a(\gamma)\delta_j^k\right) = 0 \tag{17}$$

where e_{ij} is the antisymmetric unit tensor,

$$T^a(\gamma) = i\,\xi_c^*(\gamma)\left(\frac{\lambda}{2}\right)^a_{cc'} \xi_{c'}(\gamma) \tag{18}$$

the quark colour spin vector.

By (15) T^a is real. The variables which occur in T^a satisfy the equation of motion

$$\frac{d\xi_c}{d\gamma} + i\varepsilon\left(\frac{\lambda}{2}\right)^a_{cc'} A_i^a \frac{dz^i}{d\gamma}\xi_{c'} = 0 \tag{19}$$

obtained by varying $S(\xi, A)$ over $\xi^*(\gamma)$. It follows from (18), (19) that

$$\frac{dT^a}{d\gamma} - \varepsilon f^{abc} A_k^b T^c \frac{dz^k}{d\gamma} = 0, \qquad (T^a)^2\text{-const}. \tag{20}$$

The field configuration with the strength

$$G^{a,ik}(z) = \varepsilon e^{ik} I^a(z)/\sqrt{g(z)}, \quad G^a_{ik}(z) = \varepsilon e_{ik}\sqrt{g(z)}\, I^a(z) \tag{21}$$

satisfies (16), (17) provided $I^a(z)$ is such that

$$\frac{\partial I^a}{\partial z^i} - \varepsilon f^{abc} A_i^b I^c = 0, \qquad (I^a)^2 \text{ -const}, \tag{22}$$

$$I^a(\eta^i(\gamma)) \equiv I^a(\gamma) = T^a(\gamma) \qquad \text{on } \Sigma. \tag{23}$$

The two-dimensional Bianci identity for antisymmetric tensor is always satisfied. Below we confine ourselves to embeddings of SU(2) into SU(N_c). Then the potential $A_i^a(\eta)$ corresponding to (21) can be found solving (22)

$$A_i^a(\eta) = \frac{1}{\varepsilon}\left[e^{abc} n^b \partial_i n^c + n^a a_i\right] \tag{24}$$

where $n^a(\eta) = I^a(\eta)/\sqrt{I^2}$ is the unit vector, $a_i(\eta)$ an arbitrary abelian field. Substituting (24) into (9) we get

$$G_{ik}^a(\eta) = \frac{n^a(\eta)}{\varepsilon} F_{ik}(\eta) \tag{25}$$

where

$$F_{ik}(\eta) = e^{abc} n^a \partial_i n^b \partial_k n^c + \partial_i a_k - \partial_k a_i. \tag{26}$$

Tensor (25) satisfies the Yang-Mills equation (16), i.e. coincides with (21) provided

$$^*F \equiv \frac{1}{2}\frac{e^{ik}}{\sqrt{g}} F_{ik} = \varkappa, \qquad \text{where} \quad \varkappa = \varepsilon^2 I. \tag{27}$$

Relation (27) is invariant with respect to general transformations of coordinates η^i and gauge SU(2) transformations. If one constructs the Lorentz tensor from the world tensor (21) via

$$G_{\mu\nu}^a(z(\eta)) = \frac{\partial z_\mu}{\partial \eta^i}\frac{\partial z_\nu}{\partial \eta^k} G^{a,ik}(\eta) \tag{28}$$

then one gets

$$G^a_{\mu\nu}(z(\tau)) = \varepsilon I^a(\tau)\, \sigma_{\mu\nu}(\tau) / \sqrt{g(\tau)}, \quad \text{where} \tag{29}$$

$$\sigma_{\mu\nu} = \frac{\partial z_\mu}{\partial \tau^0}\frac{\partial z_\nu}{\partial \tau^1} - \frac{\partial z_\mu}{\partial \tau^1}\frac{\partial z_\nu}{\partial \tau^0}$$

coinciding with the field strength of chromoelectric string. Eq. (29) has been earlier studied in [7, 4, 8] but its derivation from the first principles of the theory was less consistent than ours.

Relation (27) implies that the string with the strength (29) when moving produces a surface of constant scalar curvature $|R| = 2\varkappa$ in $\mathbb{R}^4$. Indeed, in the gauge $a_i(\tau) = 0$ it follows from (27) that

$$\sqrt{h(\tau)} = \varkappa \sqrt{g(\tau)}, \quad h(\tau) = \det h_{ik}(\tau) \tag{30}$$

where h_{ik} is the metric of the unit sphere $n^a n^a = 1$. Using the conformal coordinates on Σ and S^2 and the definition of scalar curvature, one can show that $|R| = 2\varkappa$ (the details see in [9]).

In the other gauge with n^a-const, we have from (24), (26)

$$A^a_i(\tau) = \frac{n^a}{\varepsilon} a_i(\tau), \quad G^a_{ik}(\tau) = \frac{n^a}{\varepsilon}\left[\partial_i a_k(\tau) - \partial_k a_i(\tau)\right]. \tag{31}$$

setting

$$a_i(\tau) = \sqrt{g(\tau)}\, e_{ik}\, \partial^k \Lambda(\tau) \tag{32}$$

(which corresponds to the gauge $a^i_{\ ;i} = 0$) we obtain from (27) an equation for $\Lambda(\tau)$:

$$\frac{1}{\sqrt{g}}\partial_i\left(\sqrt{g}\, g^{ik}\partial_k \Lambda\right) = \varkappa. \tag{33}$$

In conformal coordinates for a surface of constant curvature we find from (33) that

$$\Lambda(z) = \mp\frac{1}{2}\ln\sqrt{g_R(z)} + f(z), \quad R = \pm 2\varkappa \tag{34}$$

where $\Delta f = 0$ and f reflects the remaining freedom in the choice of conformal coordinates on Σ. If we take stereographic coordinates u, v as the conformal coordinates, then

$$\Lambda(u,v) = \pm\ln\left\{\frac{1}{2}\left[1 \pm \varkappa\left[(u-u_0)^2 + (v-v_0)^2\right]\right]\right\}, \ R = \pm 2\varkappa \tag{35}$$

where u_0, v_0 play the role of the "instanton" center (35) of dimension $\rho \sim 1/\sqrt{\varkappa}$.

4. <u>The string field contribution to path integral.</u> In the higher semiclassical approximation

$$K(1,\ldots,n) \approx \Phi\cdot\exp\left[-S_{eff}\left(A_{cl.}, \xi_{cl.}, x_{cl.}\right)\right], \tag{36}$$

where Φ accounts non-leading in $1/N_c$ contribution, S_{eff} is determined by (14). The last term in S_{eff} vanishes substituting $\xi_{cl.}$ due to the equation of motion for ξ. The quarks sitting at the edges of the string move along the surface Σ and therefore $\dot{x}^2(z(\gamma)) = g_{ik}\dfrac{dz^i}{d\gamma}\dfrac{dz^k}{d\gamma}$. The variation of the action (4) over the quark coordinates z on Σ leads to an equation of motion which in the natural parameter $\gamma = s$ takes the form

$$S_0 = \frac{1}{2}\int_{\gamma_1}^{\gamma_2} d\gamma\,(\dot{z}^2/\lambda + \lambda m_q^2) = m_q\int_{\gamma_1}^{\gamma_2} d\gamma\,\sqrt{\dot{z}^2} = m_q\, l_q\,. \tag{38}$$

The contribution of the gauge field (21) which describes the strength inside the string is expressed via $A(\Sigma) = \int_\Sigma d^2 z\sqrt{g}$

$$\frac{1}{4}\int_\Sigma d^2 z\sqrt{g}\, G^a_{ik}\, g^{il} g^{kn}\, G^a_{ln} = \frac{\varepsilon^2 I^2}{2}\int_\Sigma d^2 z\sqrt{g} = \frac{1}{2\pi\alpha'_0}\, A(\Sigma). \tag{39}$$

The quantity

$$K_0 = \frac{1}{2\pi\alpha'_0} = \frac{\varepsilon^2 I^2}{2} = \frac{e^2 C_2(F)}{2d^2} \tag{40}$$

where $C_2(F)$ is the Casimir operator of the fundamental representation stands for the bare coefficient of string tension.

For large enough contours $\Gamma = \partial\Sigma$ the semiclassical criterion $S/\hbar \sim \dfrac{A(\Sigma)}{2\pi\alpha'_0} \gg 1$ must hold Eq. (36), with (38), (39), takes the form

$$K(1,\dots,n) \approx \Phi\cdot \exp\left[-\frac{A(\Sigma)}{2\pi\alpha'_0} - \sum_{q=1}^{n} m_q\, l_q\right]$$

therefore the square law (yet only in term of the bare α'_0) holds. Account of quantum fluctuations in the vicinity of $A_{cl.}$ must result in appearance of a pre-exponential factor Φ and provide renormalization of α'. If this program is realized we get quark confinement.

According to (40), (27) α'_0 is expressed via the Gaussian curvature R/2 of Σ:

$$K_0 = \frac{\varepsilon^2 I^2}{2} = \frac{I\varkappa}{2} = \frac{I|R|}{4} = \pm\frac{IR}{4} > 0$$

and the right-hand side of (39) can be transformed with the help of the Gauss-Bonnet theorem

$$\frac{1}{2}\int_{\Sigma} R\sqrt{g}\,d^2\xi = 2\pi\chi - \oint_{\partial\Sigma} K_g dS + \sum_i (\alpha_i - \pi)$$

where χ is the Euler characteristic of Σ, K_g is the geodesic curvature of the boundary $\partial\Sigma$, α_i are the internal angles of the boundary. We get

$$S^{cl.}_{eff.} = \sum_{q=1}^{n} m_q \ell_q \pm \frac{\sqrt{C_2(F)}}{2}\Big\{2\pi\chi + \sum_i (\alpha_i - \pi) - \oint_{\partial\Sigma} K_g dS\Big\} > 0.$$

This expression is invariant with respect to bending and variation of the position and orientation of the surface in the enveloping $\mathbb{R}^4$. If $\partial\Sigma$ contains parts of geodesics, then $K_g = 0$ on them. Note that as we see from Fig.1, the field describes also string interaction when the number of edges n is > 2.

5. <u>Conclusion.</u> In Sect.2 we defined a subclass of gauge fields whose contribution into path integral was assumed to be dominant in the leading order of $1/N_c$ expansion. The above calculations a posteriori justify our choice, since the solution of the effective Yang-Mills field equations involves a master field which describes the bare hadronic string. It is supposed [1,2] that just this pattern must result from all the planar Feynman diagrams. Note also that the obtained above master-field completely rule the excitations which arise in the QCD vacuum from quarks.

References

1. 't Hooft, G., (1974). Nucl.Phys., B72, 461.

2. Witten, E. (1980). In: Recent developments in gauge theories, ed. 't Hooft et al., Plenum, N.Y.

3. Samuel, S. (1979). Nucl.Phys., B149, 517.

4. Iroshnikov, G.S. (1983). (Nucl.Phys.) Yad. Fiz. 38, 512 (Russian).

5. Ishida, J., Hosoya, A. (1979). Prog. Theor. Phys., 62, 544.

6. Borisov, N.V., Kulish, P.P. (1982). Theor. Math.Phys., 51, 311 (Russian).

7. Ryang, S., Ishida, J. (1981). Prog.Theor.Phys., 66, 685.

8. Iroshnikov, G.S. (1985). Dokl.Akad.Nauk SSSR, 280, 88 (Russian).

9. Iroshnikov, G.S. (1985). Sov. Nucl. Phys. (Russian).

Part III

SUPERALGEBRAS AND THEIR APPLICATIONS

EMBEDDINGS OF OSP (N/2) AND THE ASSOCIATED NONLINEAR SUPERSYMMETRIC EQUATIONS

D.A.Leites, M.V.Saveliev[†], V.V.Serganova

[†]Institute for High Energy Physics, Serpukhov, 142284

This paper considers in detail the results that were announced in [1] and their relation to the supersymmetried Toda lattices in the form of [2] and the KdV and mKdv equations [3].

1. Over the last several years quite a promising tendency in the elementary particle physics and cosmology has come into view. The tendency is to unify the ideas of supersymmetry (supergravity) and dimensional reduction in the spirit of Kaluza-Klein (cf. [4-5]). These ideas can be expressed in the framework of describing integrable embeddings of supermanifolds $V_{4/4N}$ with a motion supergroup (with the Lie superalgebra g') into the flat enveloping superspace $R_{A/B}$ endowed with the structure of some Lie superalgebra g. Here the nontrivial dynamics of the arising physical systems is described by the supersymmetric extension of the Gauss, Peterson-Codazzi and Ricci equations in an integrable sector and is directly related to the properties of the embeddings of g' into g [6]. Note that the criteria we have [4] to choose the dimensions A, B and the Lie superalgebra g are of model character.

Since at present there are no effective methods to integrate (in the sense of the Cauchy problem) nonlinear equations in the spaces whose even dimension is more than two, it seemes natural to impose on the systems under consideration requirement of additional symmetry. The latter may be chosen so as to reduce the number of degrees of freedom to two even (x_1, x_2) and 2N odd $(\theta_+,\theta_-)^\alpha$, $1\leq\alpha\leq N$, coordinates. This reproduces the main features of the real physical picture.

Whereas the direct sum of the simple gauge Lie algebra g and sl(2) is the complete invariance algebra for the equations of the cylindrically-symmetric Yang-Mills fields in the complexified two-dimensional Euclidean space, it is the direct sum of the simple superalgebra g and osp(N/2) that serves as the complete algebra for their N -supersymmetric extension. The equations describing the embeddings of supermanifolds $V_{2/2}$ are related to the embeddings of the subsuperalgebra osp(1/2) into the simple Lie superalgebra g.

The construction of exactly integrable supersymmetric nonlinear equations in $V_{2/2}$ related with the embeddings of osp(1/2) into g can be performed without explicit description of these embeddings [7]. They are [8] :

$$D_+(g^{-1}D_-g) = \left[Y_-, g^{-1}Y_+g\right]_+ \qquad (1)$$

Here $D_\pm \equiv \pm\partial/\partial\theta_\pm + i\theta_\pm\partial/\partial z_\pm$ are the supersymmetric covariant derivatives in the (2/2)-dimensional superspace $[z_\pm \equiv \frac{1}{2}(x_1 \mp ix_2);\ \theta_\pm]$; $Y_\pm$ are the odd generators of osp(1/2); $g(z_\pm;\theta_\pm)$ is a function on the Grassmann envelope of the Lie group Go with the Lie algebra $g_{\bar{0}}$ represented by the exponent of the bosonic superfield with values in $g_{\bar{0}}(\Lambda)$, where g is the subalgebra $g = \oplus g_a$ graded by the Cartan element $H(=[Y_+,Y_-]_+)$ of its subsuperalgebra osp(1/2).

In order to obtain physical consequences from eqs.(1), one should examine the properties of the solutions (such as regularity for the self-dual gauge systems [5]) of various subclasses of (1) related to certain embeddings of osp(1/2) into g. To do this, one should write down all these embeddings. The complete description of semisimple subalgebras of simple Lie algebras (and embeddings of sl(2) in g) see [9]. This allows us to apply the abovementioned algebraic method and classify and obtain the explicit solutions for various types of the two-dimensional exactly integrable systems [7] (associated with the embeddings of sl(2) into g including, in particular, two-dimensional generalized Toda lattices and string-type equations). Here we describe all the embeddings of osp(1/2) into a simple Lie superalgebra.

2. For the definitions and notations used below see [10,11]. We are interested in the Lie superalgebras g containing osp(1/2) with the basis

$Y_{\pm} \in osp(1/2)_{\bar{1}}$ and $(H, X_{\pm}(=\pm 2X_{\pm}^2)) \in osp(1/2)_{\bar{0}} \simeq sl(2)$

In this case the Z-grading is determined by the element H, so that $[H, \mathfrak{g}_a]_- = a\mathfrak{g}_a$.

Recall that the list of simple finite-dimensional Lie superalgebras is a part of the following list of classical Lie superalgebras (see [10,11]).

$gl(n/m) = \{\begin{pmatrix} A & B \\ C & D \end{pmatrix}\}$, A and D being nxn and mxm matrices respectively;

$gl(n/m)_{\bar{0}} = \{\begin{pmatrix} A & 0 \\ 0 & D \end{pmatrix}\}$, $gl(n/m)_{\bar{1}} = \{\begin{pmatrix} 0 & B \\ C & 0 \end{pmatrix}\}$;

$sl(n/m) = \{X \in gl(n/m) : strX = 0\}$, where $str\begin{pmatrix} A & B \\ C & D \end{pmatrix} = trA - trD$;

$Q(n) = \{X \in gl(n/n) : [X, \pi_n] = 0\}$, where $\pi_n = \begin{pmatrix} 0 & 1_n \\ 1_n & 0 \end{pmatrix}$;

$SQ(n) = \{X \in Q(n) : otrX = 0\}$, where $otr\begin{pmatrix} A & B \\ B & A \end{pmatrix} = trB$;

$osp(n/2m) = \{X \in gl(n/2m) : X^{st}B_{n,m} + B_{n,m}X = 0\}$, where $\begin{pmatrix} A & B \\ C & D \end{pmatrix}^{st} = \begin{pmatrix} A^t & -C^t \\ B^t & D^t \end{pmatrix}$

and $B_{n,m} = diag(1_n, J_{2m})$, where $J_{2m} = \begin{pmatrix} 0 & 1_m \\ -1_m & 0 \end{pmatrix}$;

$$\pi(n) = \{X \in gl(n/n) : X^{st}\pi_n + (-1)^{p(X)} \pi_n X = 0\};$$

$s\pi(n) = \pi(n) \cap sl(n/n)$;

$W(0/n) = Der\Lambda(n)$, where $\Lambda(n)$ is the Grassmann superalgebra of n variables $\xi = \xi_1 \ldots, \xi_n)$.

Clearly, $W(0/n) = \Sigma f_i \partial/\partial\xi_i$ with $f_i \in \Lambda(n)$.

$S(0/n) = \{D \in W(0/n) : L_D(V_\xi) = 0\}$, where V_ξ is the integral volume form with constant coefficients in coordinates ξ (see [10]), and L_D is the Lie derivative along the field D.

$S'_t(n) = \{D : L_D\ (1 + t\xi_1 \ldots \xi_{2N})V_\xi = 0\}$,

where $p(t) \equiv n(2)$ and $D \in W(0/n)$ if $p(t) = \bar{0}$; $D \in \Lambda(t) \otimes W(0/n)$, if $p(t)=\bar{1}$. Clearly $S'_a=S'_b$ for $ab\neq 0$; Fixing t if $p(t)=\bar{1}$ we set $S'=S'_1$.

The explicit expressions for the elements of S(o/n) and S'(n) follow from the formula

$$L_D(fV_\xi)=(D(f)+(-1)^{p(f)p(D)} f \text{ div } D)V_\xi, \text{ with}$$
$$\text{div}(\Sigma f_i \partial/\partial\xi_i)=-\Sigma(-1)^{p(f_i)} \frac{\partial f_i}{\partial \xi_i}.$$

$P_o(0/n)$: its superspace is $\Lambda(n)$; the bracket (called Poisson bracket) is

$\{f,g\}_{P.B.}=-(-1)^{p(f)}\Sigma\partial f/\partial\xi_i\cdot\partial g/\partial\xi_i$;

$$SP_o(n)=\{f\in P_o(0/n) : \int fV_\xi=0\};$$

Put $h:f\to D_f=-(-1)^{p(f)}\Sigma \frac{\partial f}{\partial\xi_i}\frac{\partial}{\partial\xi_i}$. Denote the image of $P_o(0/n)$ under h by H(0/n); let $SH(n)=h(SP_o(n))$. As can be shown, $H(0/n)=\{D\in W(0/n): L_D(\omega)=0\}$, where $\omega=\Sigma(d\xi_i)^2$.

Let G be a Lie subsuperalgebra in gl(n/m), and $D=\{\lambda 1_{n+m}:\lambda\in\mathbb{C}\}$ a Lie subalgebra of scalar matrices. Set PG=G/D if D is an ideal in G.

The Lie superalgebras: sl(n/m) for n>m; psl(n/n) for $n\neq 1$; psq(n) for n>2; osp(m/2n) for $mn\neq 0$, $(m,n)\neq(2,1)$; π(n) for n>2; W(0/n) for n>2; S(0/n) for n>3; S'(2n) for n>1; SH(0/n) for n>4 are simple and paur-wise nonisomorphic. In what follows these Lie superalgebras are condidered under these restrictions.

D(α) is the deformation of the Lie superalgebra osp(4/2). Let us identify $D(\alpha)_{\bar{0}}$ with $sl(2)\oplus sl(2)\oplus sl(2) = =\{(a_1,a_2,a_3)\}$, and $D(\alpha)_{\bar{1}}$ with $id_1\otimes id_2\otimes id_3$, id_j being the standard representation of the j-th copy of sl(2). Let ψ_j be an invariant form on id_j with the matrix J_2. Define the mapping $p_j:id_j\otimes id_j\to sl(2)_j$ by $P_j(U,V)W = =\lambda_j(\psi_j(V,W)U-\psi_j(W,U)V)$ and $[\ ,\ ]: S^2(D(\alpha)_{\bar{1}})\to D(\alpha)_{\bar{0}}$ by the formula

$$[u_1\otimes u_2\otimes u_3, V_1\otimes V_2\otimes V_3] = = \sum_{(i,j,k)=p(1,2,3),\ p\in S_3} \psi_i(u_i,V_i)\psi_j(u_j,V_j)\psi_k(u_k,V_k).$$

The Jacobi identity holds if and only if $\lambda_1+\lambda_2+\lambda_3=0$. Evidently, the numbers λ_i are defined up to a factor and permutation. Put $\alpha=\lambda_1/\lambda_2$ assuming $\lambda_2\neq 0$. At $\alpha=0,-1,\infty$, the Lie superalgebra is not simple, $D(\alpha)\simeq D(\alpha')$ if $\alpha'=\alpha^{-1}$ or $\alpha'=-\alpha/(\alpha+1)$; for all $\alpha\neq 0,-1,\infty$, the Lie superalgebra $D(\alpha)$ is simple. Below we consider only one representative of the isomorphic Lie superalgebras.

The Lie superalgebra $D(-1)$ has the ideal $psl(2/2)$, and $D(-1)/psl(2/2)\simeq sl(2)$. The action of the Lie algebra $sl(2)$ on $psl(2/2)$ is explicitly expressed via

$$\begin{pmatrix} a & b \\ c & -a \end{pmatrix}\left(\begin{pmatrix} A & B \\ C & D \end{pmatrix}\right) = \begin{pmatrix} 0 & aB+bJ_1C^tJ_1^{-1} \\ cJ_1B^tJ_1^{-1}+aC & 0 \end{pmatrix}$$

Call the above Lie superalgebras <u>classical</u> ones. Besides, there are two <u>exceptional</u> simple Lie superalgebras, AG_2 and AB_3, described in [10,11].

A loop superalgebra $\underline{g}^{(1)}=\underline{g}\otimes\mathbb{C}[t^{-1},t]$ or its Lie subsuperalgebra

$$\underline{g}^{(m)}_{\varphi} = \bigoplus_{i\in\mathbb{Z},0\leq j\leq m-1} \underline{g}_j t^{mi+j},$$

where $\underline{g}_j=\{g\in\underline{g} : \varphi(s)=\exp(2\pi ij/m)g$ (here $i=\sqrt{-1}$)$\}$ for $\varphi\in \mathrm{Aut}\, g$ such that $\mathrm{ord}\,\varphi=m$, is called a <u>Kac superalgebra</u>. A <u>Kac-Moody</u> or a <u>current superalgebra</u> is a central extension of a Kac superalgebra. Simple Kac superalgebras classified by V.Serganova are listed in [10,12].

In the sequel we need a more detailed structure of Lie superalgebras that are very alike finite-dimensional Lie algebras: a description of the bases (simple root systems). Such a description has already been published [11,12] (with omissions). The successful generalization of the Weyl group (Appendix II) allowed us to get a transparent proof of the completeness of the bases.

3. Let Q be a free Abelian group. A Lie superalgebra $\underline{g}$ presented as the sum of subsuperspaces $\bigoplus_{\alpha\in Q} \underline{g}_\alpha$ is called Q-graded if $[\underline{g}_\alpha, \underline{g}_\beta] \subset \underline{g}_{\alpha+\beta}$. Let $\underline{g}$ be a Lie superalgebra and $\underline{h}$ be its Cartan subalgebra, i.e. maximal nilpotent subsuperalgebra coinciding with its normalizor. Let $\underline{c}$ be the <u>maximal torus</u> in $\underline{h}_{\bar{0}}$, i.e. $\underline{c}$ acts on $\underline{g}$ by diagonal operators. Then $\underline{g}$ may be represented in the form $\underline{g} = \bigoplus_{\alpha\in c} \underline{g}_\alpha$, where $\underline{g}_\alpha = \{x \in \underline{g} \mid [c,x] = \alpha(c)x,\ c \in \underline{c}\}$.

In analogy with Lie algebras, the set $R = \{\alpha \in \underline{c}^* \setminus (0) \mid \dim \underline{g}_\alpha \neq 0\}$ is called the <u>root system</u> of $\underline{g}$. A Lie superalgebra $\underline{g}$ for which $\dim \underline{g}_\alpha = \dim \underline{g}_{-\alpha}$ is <u>contragredient</u>. Some contragredient Lie superalgebras can be described with analogues of Cartan matrices to which, in their turn, analogues of Dynkin graphs may sometimes be assigned.

Let $a = (a_{ij})$ be a complex $n \times n$ matrix, $(\alpha_1, \ldots, \alpha_n)$ fixed basis in Q, $\tau \subset (1,\ldots,n)$. A <u>contragredient Lie superalgebra of type (a,τ)</u> is a Q-graded Lie superalgebra $g(a,\tau)$ with the generators

$$x_i^-,\ h_i,\ x_i^+,\ 1 \le i \le n;\quad \deg x_i^{\pm} = \pm\alpha_i,\ \deg h_i = 0,$$

$$p(x_i^{\pm}) = \bar{1} \text{ for } i \in \tau \text{ and } \bar{0} \text{ for } i \notin \tau, \text{ satisfying}$$

the following conditions:

$$\begin{aligned} &\underline{g}_{\alpha_i} = \mathbb{C}x_i^+,\ \underline{g}_{-\alpha_i} = \mathbb{C}x_i^-,\ \underline{g}_0 = \langle h_1, \ldots, h_n\rangle, \\ &[x_i^+, x_j^-] = \delta_{ij}h_i \quad (\text{hence, } p(h_i) = \bar{0} \text{ for all } i) \\ &[h_i, h_j] = 0,\quad [h_i, x_j^{\pm}] = \pm a_{ij}x_j^{\pm} \end{aligned} \tag{2}$$

Each nonzero Q-graded ideal in $\underline{g}(a,\tau)$ has a nonzero intersection with $\underline{g}_0$. Denote also via $\underline{g}(a,\tau)$ the Lie superalgebra obtained by factorization with respect to the maximal Q-graded ideal; only these quotients are considered in the sequel.

The Lie superalgebra $G(a,\tau)$ may have a centre.

<u>Convention</u>. $g(a,\tau)$ and its quotient w.r.t. the centre have the same matrix a called the <u>Cartan matrix</u>. The Cartan matrices a and a' are <u>equivalent</u> if there exists a matrix D such that $\det D \neq 0$ and $a' = Da$. Multiplying by an appropriate matrix D, we normalize Cartan matrices so that $a_{ii} = 0$ or 2. Moreover, in the columns with $a_{ii} = 0$ the nonzero elements should be minimal in the absolute value while the number of integer ones should be maximal. If $a_{ii} = 0$ then $i \in \tau$. ∎

The simple Lie superalgebras of type $g(a,\tau)$ can be determined by a system of defining relations for the generators $X_i^{\pm}$, $1 \leq i \leq n$. To relations (2) we should add

$$(\mathrm{ad}\, X_i^{\pm})^{1-b_{ij}}(X_j^{\pm}) = 0, \quad \text{where } i \neq j \text{ if } a_{ii} \neq 0 \tag{3}$$

Here the matrix b is obtained from a by substituting -1 for the nonzero elements in the rows with 0 on the diagonal.

<u>Remark</u>. For Lie algebras the matrices a and b coincide!

As for the Lie algebras [13], one can easily verify that for any a and τ there exists a Lie superalgebra $g(a,\tau)$. However, different pairs (a,τ) may be associated with isomorphic Lie superalgebras.

Put $R = \{\alpha \in Q\setminus(0) \mid g_\alpha(a,\tau) \neq 0\}$. Clearly, the elements of R are <u>roots</u>, while the expansion $\underline{g} = \bigoplus_{\alpha \in R} \underline{g}_\alpha$ with $\underline{g} = \underline{g}(a,\tau)$, $\underline{c} = \underline{g}_0$ is a root expansion, The set $\{\alpha_1, \dots, \alpha_n\}$ is called a <u>simple root system</u> or a <u>base</u>. Call two bases B_1 and B_2 <u>equivalent</u> if there exists an automorphism $\varphi \in \mathrm{Aut}\,g$, such that

$$\varphi(X_{\pm\alpha_i}) = X_{\pm\beta_i}, \quad \varphi(h_{\alpha_i}) = h_{\beta_i} \quad \text{for} \quad \alpha_i \in B_1, \beta_i \in B_2 .$$

Clearly, equivalent bases are associated with equivalent Cartan matrices. The nonzero roots $\sum_{n_i>0} n_i\alpha_i$ are called positive; they form the set $R^+ \subset R$. Clearly, $R = R^+ \cup (-R^+)$.

3. Let $F = Q \otimes_{\mathbb{Z}} \mathbb{C}$. Define the linear mapping $\Lambda: F \to \underline{e}$, $\alpha \to \hat{\alpha}$, by the formula $\hat{\alpha}_j(h_i) = a_{ij}$ $(i,j = 1,2,\dots,n)$. Clearly, $[h,x] = \hat{\alpha}(h)x$ for $h \in \underline{e}$, $\alpha \in R$ and $x \in \underline{g}$, and Λ is an isomorphism if and only if a is nondegenerate (in which case a base may be defined as the minimal subset among the subsets $B = (\alpha_1,\dots,\alpha_n)$ in $\underline{e}$, such that the elements of B are linearly independent). If $N^{\pm}$ is the Lie subsuperalgebra in $\underline{g}$ generated by $\underline{g}_{\pm\alpha_i}$, then $\underline{g} = N^- \oplus \underline{C} \oplus N^+$.

To an integral Cartan matrix we assign a graph with vertices O, ● and ⊗ for the bases of type sl(2), osp(1|2) and sl(1|1) respectively. Next associate α_i with O if $i \notin \tau$, with ● if $i \in \tau$ and $a_{ii}=2$, and with ⊗ if $a_{ii} = 0$.

Now, let us join two circles associated with the roots α_i and α_j, by $\max(|a_{ij}|, |a_{ji}|)$ segments and an arrow with the circle i as the target if $|a_{ij}|<|a_{ji}|$.

Unless otherwise stated vertices of graphs are enumerated as in Table 1, where each line may be a multiple one or endowed with an arrow sign, the dotted line may be omitted.

Table 1. Numbering of vertices in Dynkin schemes

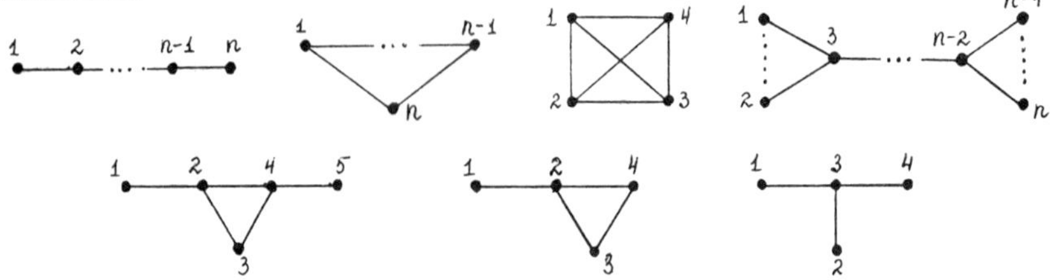

1 2 ... n-1 n

1 ... n-1, n

1 4, 2 3

1, 2, 3 ... n-2, n-1, n

1 2 4 5, 3

1 2 4, 3

1 3 4, 2

Table 2. Some elementary schemes

Schemes indicated by the same number of asterisks determine equivalent Cartan matrices (up to permutation of rows and multiplication of the row corresponding to an isotropic root by $\lambda \neq 0$). The corresponding bases are not equivalent with respect to the Weyl group W. The schemes obtained from the depicted ones by reflection with respect to the vertical axis determine bases equivalent with respect to the Weyl supergroup SW but non-equivalent in general with respect to W. They are not depicted except for sl(1/2) and osp(3/2)

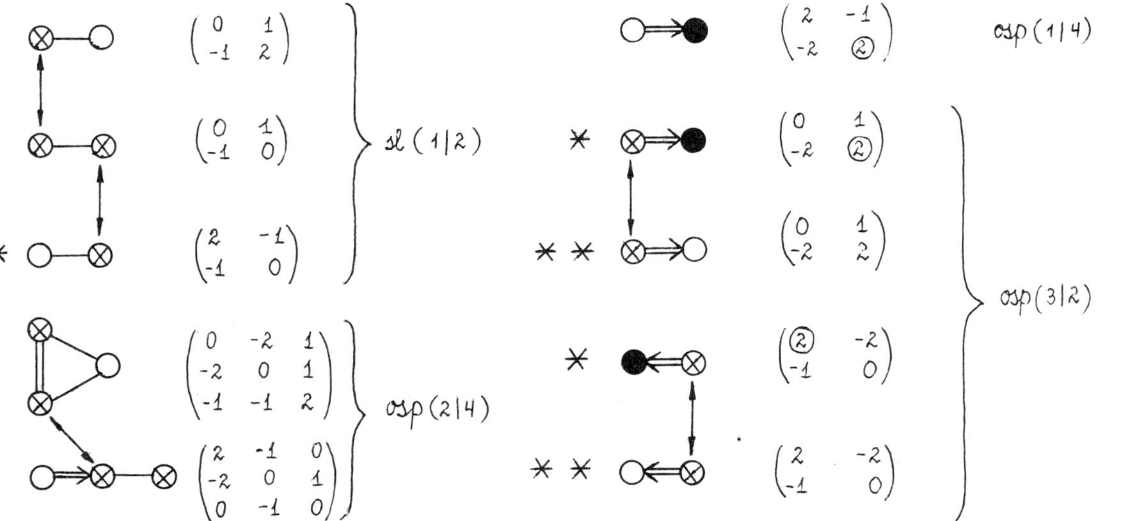

Table 2. Some elementary schemes (continued)

$$\begin{pmatrix} 2 & 0 & -1 \\ 0 & 2 & -1 \\ -1 & -1 & 0 \end{pmatrix} \quad \begin{pmatrix} 0 & -2 & 1 \\ -2 & 0 & 1 \\ -1 & -1 & 0 \end{pmatrix} \quad \begin{pmatrix} 2 & -1 & 0 \\ -2 & 0 & 1 \\ 0 & -1 & 2 \end{pmatrix} \quad \mathfrak{osp}(4|2)$$

$$\begin{pmatrix} 2 & -1 & 0 \\ -1 & 0 & 1 \\ 0 & -1 & 2 \end{pmatrix} \quad \begin{pmatrix} 0 & 1 & 0 \\ -1 & 0 & 1 \\ 0 & -1 & 0 \end{pmatrix} \quad \begin{pmatrix} 0 & 1 & 0 \\ -1 & 2 & -1 \\ 0 & -1 & 0 \end{pmatrix} \quad \mathfrak{sl}(2|2)$$

$$\begin{pmatrix} 0 & -2 & 1 & 1 \\ -2 & 0 & 1 & 1 \\ -1 & -1 & 0 & 2 \\ -1 & -1 & 2 & 0 \end{pmatrix} \qquad \begin{pmatrix} 2 & -1 & 0 & 0 \\ -2 & 0 & 1 & 1 \\ 0 & -1 & 2 & 0 \\ 0 & -1 & 0 & 2 \end{pmatrix} \quad \mathfrak{osp}(4|2)^{(1)}$$

$$\begin{pmatrix} 2 & -1 \\ -4 & \textcircled{2} \end{pmatrix} \quad \mathfrak{osp}(1|2)^{(1)} \qquad \begin{pmatrix} \textcircled{2} & -2 \\ -2 & 2 \end{pmatrix} \quad \mathfrak{sl}(1|3)^{(2)}$$

$$\begin{pmatrix} \textcircled{2} & -2 \\ -2 & \textcircled{2} \end{pmatrix} \quad \mathfrak{osp}(2|2)^{(2)} \qquad \begin{pmatrix} 2 & -1 & 0 \\ -2 & \textcircled{2} & -2 \\ 0 & -1 & 2 \end{pmatrix} \quad \mathfrak{sl}(1|4)^{(2)}$$

Table 2. Some elementary schemes (continued)

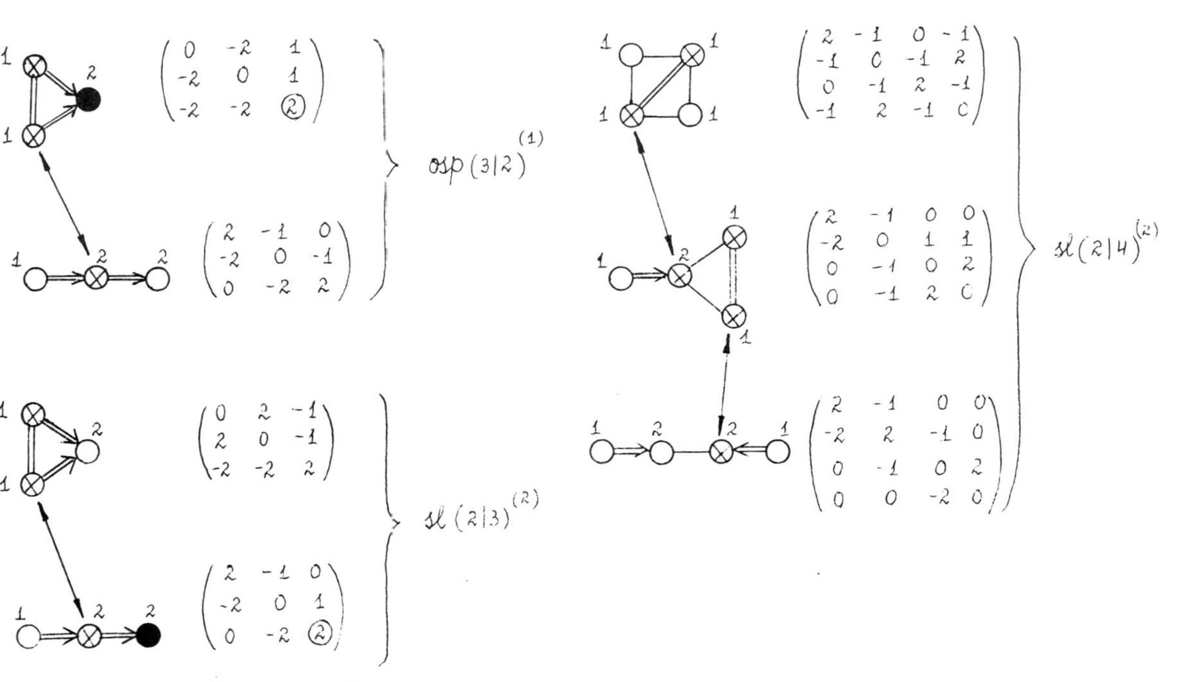

Table 3. Bases of exceptional Lie superalgebras

$D(\alpha)$: (graph for $\alpha = 3$)

$$\begin{pmatrix} 2 & -1 & 0 \\ \alpha & 0 & -1-\alpha \\ 0 & -1 & 2 \end{pmatrix} \quad \begin{pmatrix} 2 & -1 & 0 \\ -1 & 0 & \alpha \\ 0 & -1 & 2 \end{pmatrix} \quad \begin{pmatrix} 2 & -1 & 0 \\ -1-\alpha & 0 & 1 \\ 0 & -1 & 2 \end{pmatrix} \quad \begin{pmatrix} 0 & 1 & -1-\alpha \\ 1 & 0 & \alpha \\ -1-\alpha & \alpha & 0 \end{pmatrix}$$

AG_2:

$$\begin{pmatrix} 0 & 1 & 0 \\ -1 & 2 & -3 \\ 0 & -1 & 2 \end{pmatrix} \quad \begin{pmatrix} 0 & 1 & 0 \\ -1 & 0 & 3 \\ 0 & -1 & 2 \end{pmatrix} \quad \begin{pmatrix} \textcircled{2} & -2 & 0 \\ -2 & 0 & 3 \\ 0 & -1 & 2 \end{pmatrix} \quad \begin{pmatrix} 0 & -3 & -1 \\ -3 & 0 & -2 \\ -1 & -2 & 2 \end{pmatrix}$$

AB_3:

$$\begin{pmatrix} 2 & -1 & 0 & 0 \\ -3 & 0 & 1 & 0 \\ 0 & -1 & 2 & -2 \\ 0 & 0 & -1 & 2 \end{pmatrix} \quad \begin{pmatrix} 0 & 1 & 0 & 0 \\ -1 & 2 & -2 & 0 \\ 0 & -1 & 2 & -1 \\ 0 & 0 & -1 & 2 \end{pmatrix} \quad \begin{pmatrix} 0 & 1 & 0 & 0 \\ -1 & 0 & 2 & 0 \\ 0 & -1 & 2 & -1 \\ 0 & 0 & -1 & 2 \end{pmatrix}$$

$$\begin{pmatrix} 0 & 3 & 2 & 0 \\ 3 & 0 & -1 & 0 \\ -2 & -1 & 0 & 2 \\ 0 & 0 & -1 & 2 \end{pmatrix} \quad \begin{pmatrix} 0 & -2 & 0 & 1 \\ 2 & 0 & -2 & -1 \\ 0 & -1 & 2 & 0 \\ -1 & -1 & 0 & 2 \end{pmatrix} \quad \begin{pmatrix} 2 & -1 & 0 & 0 \\ -3 & 0 & 2 & 0 \\ 0 & -1 & 2 & -1 \\ 0 & 0 & -1 & 2 \end{pmatrix}$$

Table 3. Bases of exceptional Lie superalgebras (continued)

$D(\alpha)^{(1)}$: ($\alpha = 3$)

$$\begin{pmatrix} 2 & -1 & 0 & 0 \\ 1 & 0 & \alpha & -1-\alpha \\ 0 & -1 & 2 & 0 \\ 0 & -1 & 0 & 2 \end{pmatrix} \qquad \begin{pmatrix} 0 & 1 & \alpha & -1-\alpha \\ 1 & 0 & -1-\alpha & \alpha \\ \alpha & -1-\alpha & 0 & 1 \\ -1-\alpha & \alpha & \alpha & 0 \end{pmatrix}$$

$AG_2^{(1)}$:

$$\begin{pmatrix} 2 & -1 & 0 & 0 \\ -4 & 0 & 1 & 0 \\ 0 & -1 & 2 & -3 \\ 0 & 0 & -1 & 2 \end{pmatrix} \quad \begin{pmatrix} \textcircled{2} & 0 & 0 & -1 \\ 0 & 2 & -2 & 0 \\ 0 & -2 & 0 & 3 \\ -1 & 0 & -1 & 2 \end{pmatrix} \quad \begin{pmatrix} 2 & -1 & 0 & 0 \\ -4 & 0 & 3 & 0 \\ 0 & -1 & 2 & -1 \\ 0 & 0 & -1 & 2 \end{pmatrix} \quad \begin{pmatrix} 0 & -4 & 3 & 0 \\ -4 & 0 & 1 & 0 \\ -3 & -1 & 0 & 3 \\ 0 & 0 & -1 & 2 \end{pmatrix} \quad \begin{pmatrix} 0 & -3 & -2 & 0 \\ -3 & 0 & -1 & 3 \\ -2 & -1 & 2 & 0 \\ 0 & -1 & 0 & 2 \end{pmatrix}$$

$AB_3^{(1)}$:

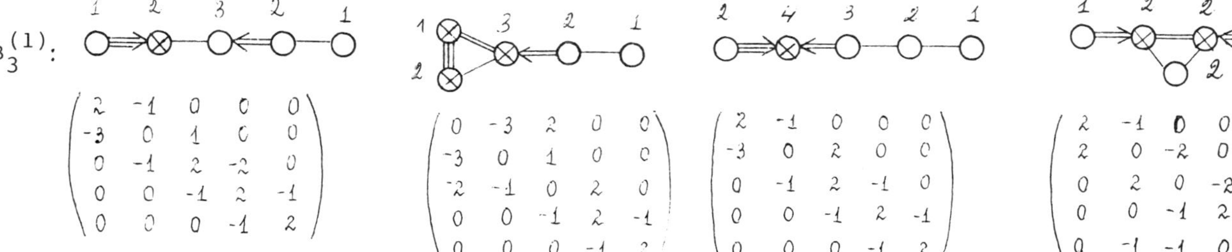

$$\begin{pmatrix} 2 & -1 & 0 & 0 & 0 \\ -3 & 0 & 1 & 0 & 0 \\ 0 & -1 & 2 & -2 & 0 \\ 0 & 0 & -1 & 2 & -1 \\ 0 & 0 & 0 & -1 & 2 \end{pmatrix} \quad \begin{pmatrix} 0 & -3 & 2 & 0 & 0 \\ -3 & 0 & 1 & 0 & 0 \\ -2 & -1 & 0 & 2 & 0 \\ 0 & 0 & -1 & 2 & -1 \\ 0 & 0 & 0 & -1 & 2 \end{pmatrix} \quad \begin{pmatrix} 2 & -1 & 0 & 0 & 0 \\ -3 & 0 & 2 & 0 & 0 \\ 0 & -1 & 2 & -1 & 0 \\ 0 & 0 & -1 & 2 & -1 \\ 0 & 0 & 0 & -1 & 2 \end{pmatrix} \quad \begin{pmatrix} 2 & -1 & 0 & 0 & 0 \\ 2 & 0 & -2 & 0 & 1 \\ 0 & 2 & 0 & -2 & -1 \\ 0 & 0 & -1 & 2 & 0 \\ 0 & -1 & -1 & 0 & 2 \end{pmatrix}$$

Table 4. Bases of simple finite dimensional Lie superalgebras

Cartan-like notation	Graph of $\mathfrak{g}$	$\mathfrak{g}$	l	$ev \geq 0$	$od \geq 1$	png	$ng \leq min$
$A_{m,n}$		$\mathfrak{sl}(m\|n)$	$m+n-1$	$m-1$ $n-1$	n m		$2m, 2n$
$\overset{\circ}{D}_{m,n}$		$\mathfrak{osp}(2m\|2n)$	$m+n$	$m-1$ $n-1$	$n-1$ $m-1$	0 1	$2m-2, 2n-1$
$D_{m,n}$				$m-2$ n	n $m-2$	1 0	$2m-3, 2n$
$C_{m,n}$				m $n-1$	$n-1$ m	0 1	$2m, 2n-1$
$B_{m,n}$		$\mathfrak{osp}(2m+1\|2n)$	$m+n$	$m-1$ n	n $m-1$	1 0	$2m-1, 2n$
$\overset{\circ}{B}_{m,n}$				m $n-1$	$n-1$ m	0 1	$2m, 2n-1$
$A_{n,n}$	 a_1 … a_{2n-1}	$\mathfrak{sl}(n\|n)$	$2n-1$	$n-1$	n		$2n-1$

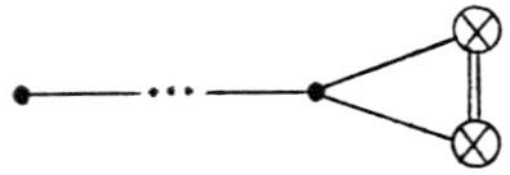

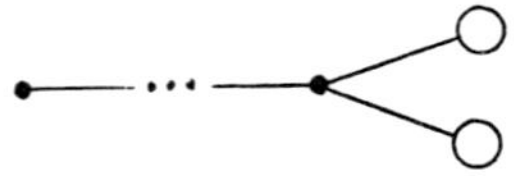

Table 5. Bases of Lie superalgebras of the form $\mathfrak{g}^{(1)}$

Graph	$\mathfrak{g}^{(1)}$	ℓ	$ev \geq 0$	$od \geq 1$	png	$ng \leq min$
	$sl(m\|n)^{(1)}$	$m+n$	m n	n m	0	$2m, 2n$
	$osp(2m\|2n)^{(1)}$	$m+n+1$	n	$m-3$	0	$2m-6, 2n$
			$m-3$	n	1	$2m-5, 2n-1$
			$m-2$	n	1	$2m-3, 2n-1$
			$m-2$	$n-1$	0	$2m-4, 2n-2$
			$m-1$	$n-1$	0	$2m-2, 2n-2$
			m	$n-1$	0	$2m, 2n-2$

Table 5. Bases of Lie superalgebras of the form $\mathfrak{g}^{(1)}$ (continued)

Graph	$\mathfrak{g}^{(1)}$	ℓ	$ev \geq 0$	$od \geq 1$	png	$ng \leq min$
			$n-1$	$m-1$	1	$2m-3,\ 2n-1$
			$m-1$	$n-1$	0	$2m-2,\ 2n-2$
	$\mathfrak{osp}(2m+1\|2n)^{(1)}$	$m+n+1$	$m-1$	n	1	$2m-1,\ 2n-1$
			n	$m-2$	0	$2m-4,\ 2n$
			$m-2$	n	1	$2m-3,\ 2n-1$
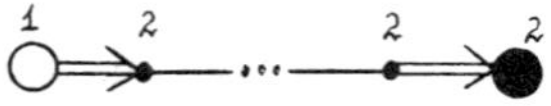			m	$n-1$	0	$2m,\ 2n-2$

Table 6. Bases of Lie superalgebras of the form $\mathfrak{g}^{(m)}$

	$\mathfrak{g}^{(m)}$	ℓ	$ev \geq 0$	$od \geq 1$	png	$ng \leq min$
	$psq(n)^{(2)}$	n	$i+j=n$		1	
			$n-2$	$m-1$	1	$2m-3,\ 2n-3$
			$m-1$ $n-1$	$n-2$ $m-2$	0	$2m-2,\ 2n-4$ $2n-2,\ 2m-4$
	$s\ell(2m\|2n)^{(2)}$	$m+n+1$	n m	$m-2$ $n-2$	0	$2m-4,\ 2n$ $2n-4.\ 2m$
			$m-2$	$n-1$	1	$2m-3,\ 2n-3$
			$n-1$ $m-1$	$m-1$ $n-1$	1	$2m-3,\ 2n-1$ $2n-3,\ 2m-1$
			$n-1$	m	1	$2m-1,\ 2n-1$

Table 6. Bases of Lie superalgebras of the form $\mathfrak{g}^{(m)}$ (continued)

	$\mathfrak{g}^{(m)}$	ℓ	$ev \geq 0$	$od \geq 1$	png	$ng \leq \min$
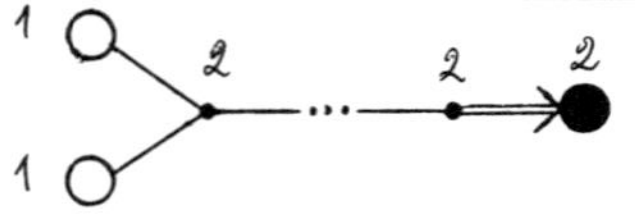	$\mathfrak{sl}(2m+1\|2n)^{(2)}$	$m+n+1$	m	$n-2$	0	$2m, 2n-4$
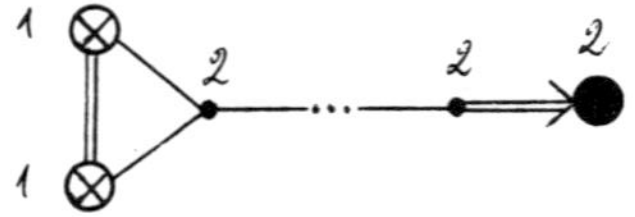			$n-2$	m	1	$2m-1, 2n-3$
			n	$m-1$	0	$2m-2, 2n$
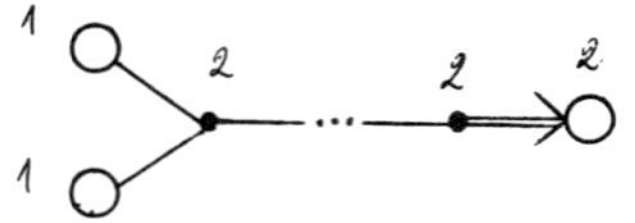			$m-1$	$n-1$	1	$2m-1, 2n-3$
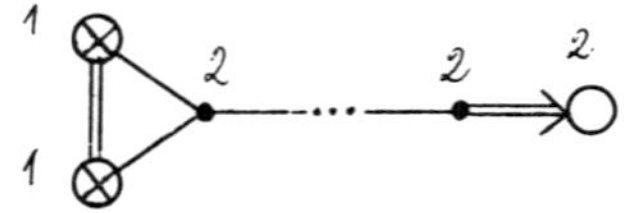			$n-1$	$m-1$	0	$2m-2, 2n-2$
			$n-1$	m	1	$2m, 2n-2$

Table 6. Bases of Lie superalgebras of the form $\mathfrak{g}^{(m)}$ (continued)

	$\mathfrak{g}^{(m)}$	ℓ	$ev \geq 0$	$od \geq 1$	png	$ng \leq min$
1 2 ... 2 2	$osp(2m\|2n)^{(2)}$	$m+n$	$m-1$	$n-1$	0	$2m-2, 2n-2$
	$sl(2m+1\|2n+1)^{(2)}$	$m+n+1$	m	$n-1$	1	$2m-1, 2n-1$
1 1 ... 1 1	$osp(2m\|2n)^{(2)}$	$m+n$	$m-2$	n	1	$2m-3, 2n-1$
	$sl(2m+1\|2n+1)^{(2)}$	$m+n+1$	m	$n-1$	0	$2m, 2n-2$
1 1 ... 1 1	$osp(2m\|2n)^{(2)}$	$m+n$	n	$m-2$	0	$2m-1, 2n$
	$sl(2m+1\|2n+1)^{(2)}$	$m+n+1$	$m-1$	n	1	$2m-1, 2n-1$

For $\underline{g}(a, \tau)$ one can uniquely recover the Cartan matrix from the graph if one decomposes it into elementary schemes (Table 2). For exceptional Lie superalgebras all the Cartan matrices and their graphs are enumerated in Table 3.

Another guide for unique recovering a matrix from a graph is as follows: if $a_{ii} = 2$, then $a_{ij} < 0$. To recover the sign of a_{ij} for $a_{ii} = 0$, we conceed to treat the graph as a 2-dimensional structure: set $a_{ij} > 0$ if the i-th vertex, ⊗ , is situated to the left of the j-th one and $a_{ij} < 0$ otherwise.

For the vertices of the form ⊗=⊗ we set $a_{ij} = a_{ji}$ and $\operatorname{sgn} a_{ij} = - \operatorname{sgn} a_{ik}$, where k is the number of the vertex joined with the i-th vertex. Thus $sl(2|2)$ and $psl(2|2)$ are associated with the graph O--⊗-- O and the Cartan matrix $\begin{pmatrix} 2 & -1 & 0 \\ -1 & 0 & 1 \\ 0 & -1 & 2 \end{pmatrix}$, while $osp(4|2)$ is associated with the graph (two vertices O joined to ⊗) vertices being enumerated as indicated and the Cartan matrix $\begin{pmatrix} 2 & 0 & -1 \\ 0 & 2 & -1 \\ -1 & -1 & 0 \end{pmatrix}$.

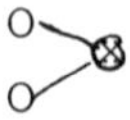

In Tables 2-6 there are listed all graphs corresponding to (in)finite dimensional contragredient Lie superalgebras of finite growth (Kac-Moody superalgebras). The numbers are coefficients of linear dependence of columns of the Cartan matrix (the coefficients of simple roots in the decomposition of the highest root).

Such coefficients for $sl(n|n)$ are defined via formulas $a_1 = 1$, $a_k = a_{k-1} + (-1)^{ngr(k-1)}$, where $ngr(k)$ is the number of "gray" vertices ⊗ among the first k ones.

4. Denote via f a Lie subsuperalgebra in $\underline{g}$, isomorphic to $osp(1|2)$. The embeddings of $osp(1|2)$ into $\underline{g}$ with the images f_1 and f_2 are equivalent, if there exists an automorphism $\varphi \in \operatorname{Aut} \underline{g}$ such that $\varphi(f_1) = f_2$.

Let ρ_1 be an irreducible representation of osp $(1|2)$ with the highest weight 1 and an even highest vector, $\pi(\rho_1)$ the same representation but with the opposite parity. We normalize the basis in osp $(1|2)$ so that all the weights of ρ_1 were integral.

Consider the decomposition $S = \bigoplus_{a \in \mathbb{Z}} S_a$ of the superspace of the standard (identical) representation of a matrix Lie superalgebra $\underline{g}$ over the field $\mathbb{C}$ and introduce the subsuperalgebras $\underline{g}'_a = \{F \in \underline{g} \mid F(S_b) = \delta_{ab} S_a\}$. Let a <u>true subsuperalgebra</u> be a superalgebra $\underline{g}' = \bigoplus_a \underline{g}'_a$ under the following restrictions on S_a :

$\underline{g}$	sl	sQ, sπ	osp
dim S_a	$m_a \mid m_a \pm 1$	$m_a \mid m_a \pm 1$	$2m_a \pm 1 \mid 2m_a$; $2m_a \pm 2 \mid 2m_a$
			$2m_a \mid 2m_a$

Here $\pi S_a = S_a$ for sQ and $S_a \perp S_b$ if $a \neq b$ for osp and sπ or π.

A Lie subsuperalgebra $R \subseteq \underline{g}$ is called <u>regular</u> (as in the case of Lie algebras [9]) if there exists a Cartan subalgebra $\underline{\mathfrak{e}}$ such that $R = R \cap (\underline{\mathfrak{e}} \oplus \bigoplus_{\alpha \in \underline{\mathfrak{e}}} \mathbb{C}X_\alpha)$, X_α being the root elements associated with α . A regular subalgebra A is called "<u>regular of type g</u>" if it is the direct sum of Lie subsuperalgebras of type g.

The embedding E : osp $(1|2) \rightarrow \underline{g}$ is called (<u>almost</u>) <u>regular</u> if the maximal regular (regular of the same type) Lie subsuperalgebra in $\underline{g}$ containing ImE coincides with $\underline{g}$.

<u>Theorem 1</u>. Table 7 describes all regular and almost regular embeddings for the classical finite-dimensional Lie superalgebras. ■

Let E : osp $(1|2) \rightarrow \underline{g}$ be an embedding, and $\underline{\mathfrak{e}}$ a Cartan subalgebra in $\underline{g}$, such that $E(H) \in \underline{\mathfrak{e}}$. Then there exists a base (α_i) such that $\alpha_i(E(H)) > 0$ for all i. The set of numbers $\alpha_i(E(H))$ is called a characteristic of the embedding E. The characteristic is naturally depicted by unperical marks on the Dynkin graph associated with the given base.

Table 7. Regular and almost regular embeddings

Regular embeddings:

$\mathfrak{g}$	$\mathfrak{sl}$		$\mathfrak{osp}$				
$\dim\mathfrak{g}$	$m+1\|m$	$m\|m+1$	$2m+1\|2m$	$2m-1\|2m$	$2m+2\|2m$	$2m\|2m$	$2m\|2m+2$
ρ	ρ_m	$\Pi(\rho_m)$	ρ_{2m}	$\Pi(\rho_{2m-1})$	$\rho_{2k}\oplus\rho_{2(m-k)}$ $2k<m$	$\Pi(\rho_{2m-2k-1})\oplus$ $\oplus\,\rho_{2k}$	$\Pi(\rho_{2k+1})\oplus$ $\oplus\,\Pi(\rho_{2m-2k-1})$ $2k<m-1$

Almost regular embeddings:

$\mathfrak{g}$	$\mathfrak{osp}$		$\mathfrak{sl}$, $\mathfrak{sq}$
$\dim\mathfrak{g}$	$4m\|4m-2$	$4m\|4m+2$	$2m+1\|2m+1$
ρ	$\rho_{2m-1}\oplus\rho_{2m-1}$	$\rho_{2m}\oplus\rho_{2m}$	$\rho_m\oplus\Pi(\rho_m)$

Reasoning again as for sl(2), cf. [9], we can prove that all the marks of a characteristic are either 1 or 0.

A superprincipal embedding for a base formed by odd elements is defined as in [9], i.e. $E(Y_+) = \Sigma X_i^+$ (here X_i is an element associated with the simple root α_i).

It is not in every Lie superalgebra one can find superprincipal embedding. They exist only for $sl(m|m\pm 1)$; $osp(2m\pm 1|2m)$; $osp(2m+2|2m)$ and $osp(2m|2m)$ for bases with $k = 0$ and $D(\alpha)$ for any α (see Table 8). All these embeddings are apparently regular.

Table 8. Superprincipal embeddings

g	$sl(n\|n\pm 1)$	$osp(2m\pm 1\|2m)$	$osp(2n\|2m)$, $n=m, m+1$	$D(\alpha)$ $(\alpha=3)$	$osp(1\|2)$
B	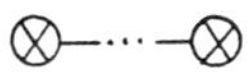 ⊗—···—⊗	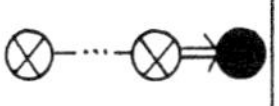 ⊗—···—⊗⇒●	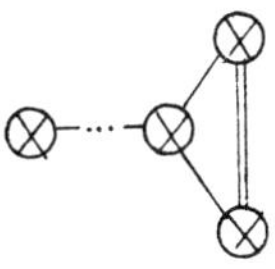 ⊗—···—⊗ with two ⊗ branches joined by a double line	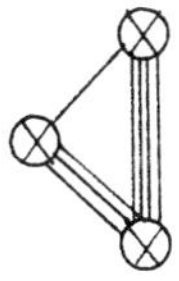 triangle of three ⊗	●

Theorem 2. All the nonequivalent embeddings of osp $(1|2)$ into the exceptional Lie superalgebras are described in Table 9.

Let $E : osp\,(1|2) \to g$ be an embedding into a classical matrix Lie superalgebra. Then there exists a true regular Lie subsuperalgebra g' of the same type $g' = \bigoplus_{a\in Z} g_a$, such that $\mathrm{Im}\ E \subset g'$ and $E = \bigoplus E_a$, $E_a : osp\,(1|2) \to g_a$ being an almost regular embedding.

Any embedding $E : osp\,(1|2) \to g$, where $g = \omega(0|n)$, $S(0|n)$ or $SH(0|n)$, is equivalent to a sum $E = \bigoplus E_a$, where E_a is one of the compositions

$$osp\,(1|2) \nearrow sl\,(1|2) \simeq \omega(0|2) \to \omega(0\ n)$$
$$osp\,(1|2) \to s\,\pi(3) \simeq S(0|3) \to S(0|n)$$
$$osp\,(1|2) \searrow psl\,(2|2) \simeq SH(0|4) \to SH(0|n)\,.\ \blacksquare$$

Conjecture. There are no embeddings $osp\,(1|2) \to S'(n)$.

Proof. Consider the most intricate case: the embeddings into AG_2. Denote via H_α the elements $H_\alpha = [X_\alpha, X_{-\alpha}]$ of the Cartan subalgebra $\underline{e}$ associated with the root α, where X_α is a root element of $\underline{g}$. Via E_o we denote the restriction of the embedding E on $sl(2)$. One can assume $E(H) \in \underline{e}$.

From ref. [9] we know the list of possible values

<u>Table 9</u>. Characteristics of embeddings of osp(1/2) into exceptional $\mathfrak{g}$ and minimal regular subalgebras (MRS)

D(α) (Graph for α = 3)		AG_2		AB_3	
MRS	Char	MRS	Char	MRS	Char
$\mathfrak{sl}(1/2)$	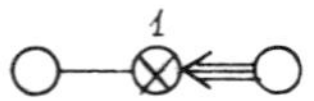	$\mathfrak{sl}(1/2)$	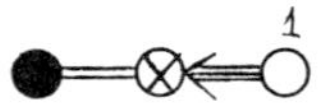	$\mathfrak{sl}(1/2)$	
$\mathfrak{sl}(1/2)'$		$\mathfrak{sl}(1/2)'$	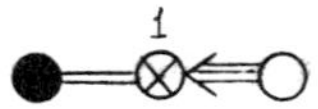	$\mathfrak{sl}(1/2)'$	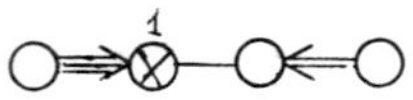
$\mathfrak{sl}(1/2)''$	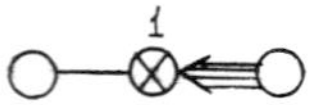	$\mathfrak{osp}(1/2)$	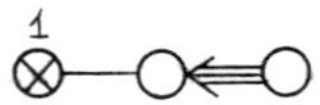	$\mathfrak{sl}(1/2)''$	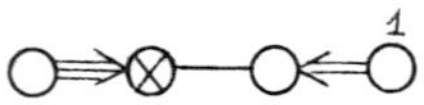
$D(3)$	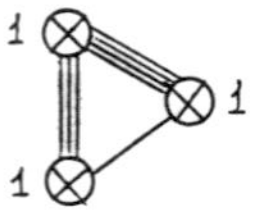	$\mathfrak{osp}(3/2)$	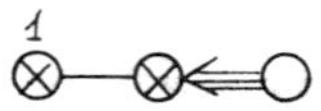	$D(2)$	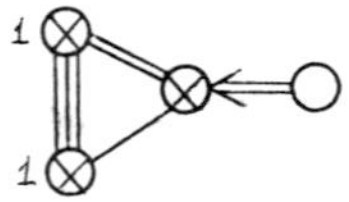
		$D(3)$	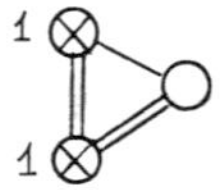		

$E_o(H) = E(H)$ can take. Up to the (Aut $\underline{g}_{\bar{o}}$)-action they are as follows. (Each of them is associated with the unique embeddings E_o and we give $E_o(H)$):

1) H_{γ_i}; 2) H_{γ_2}; 3) $H_{2\delta}$; 4) $H_{\gamma_1} + H_{2\delta}$; 5) $H_{\gamma_2} + H_{2\delta}$;

6) $2H_{\gamma_1} + 2H_{\gamma_1+3\gamma_2}$; 7) $2H_{\gamma_1} + 2H_{\gamma_1+3\gamma_2} + H_{2\delta}$;

8) $H_{\gamma_1} + H_{\gamma_1+2\gamma_2}$; 9) $H_{\gamma_1} + H_{\gamma_1+2\gamma_2} + H_{2\delta}$; 10) $H_{\gamma_1} + 6H_{\gamma_2}$;

11) $10H_{\gamma_1} + 6H_{\gamma_2} + H_{2\delta}$.

Cases 6,8,10 and 11 can be crossed out, since the restriction of the adjoint representation $E_o(sl(2))$ on $\underline{g}_{\bar{1}}$ does not contain the highest vector with weight 1, and therefore E_o does not extend to any embedding E. Let us consider the remaining cases.

1) $E(H) = H_{\gamma_1}$; $E(X^+) = X_{\gamma_1}$; then $E(Y^+) = X_{\gamma_1+\gamma_2} \otimes v + X_{-\gamma_2} \otimes \omega$. Besides, since $E(Y^+), E(Y^+) = X_{\gamma_1}$, $(v, \omega) = 1$, where $(,)$ is a nondegenerate invariant form. It follows then that there exists the unique embedding with the given $E(H)$ (all $E(Y^+)$ are transformed into each other by means of the action of $sl(2) \subset (AG_2)_{\bar{o}}$).

2) $E(H) = H_{\gamma_2}$; $E(X^+) = X_{\gamma_2}$; then $E(Y^+)$ may equal $X_{2\gamma_2+\gamma_1} \otimes v + X_{-(\gamma_2+\gamma_1)} \otimes \omega$, where $(v, \omega) = 1$. Further as in case 1).

3) $E(H) = H_{2\delta}$, $E(X^+) = X_{2\delta}$, $E(X_-) = X_{-2\delta}$. Then $E(Y^+)$ takes the form $\Sigma\, (a_\alpha X_{\alpha+\delta} + bX_\delta)$, where α runs the set of the short roots of the Lie algebra G_2.

4) $E(H) = H_{\gamma_1} + H_{2\delta}$, $E(X^+) = X_{\gamma_1} + X_{2\delta}$. Then $E(Y^+) = aX_{\gamma_1+2\gamma_2+\delta} + bX_{-\gamma_1-2\gamma_2+\delta} + cX_\delta$ since $E(Y^+)$ is

a linear combination of the highest vectors of weight 1. Clearly, $[E(Y^+), E(Y^+)] \neq X_{\gamma_1} + X_{2\delta}$ for any a,b,c. Thus, there are no embeddings E such that $E(H) = H_{\gamma_1}+H_{2\delta}$.

5) $E(H) = H_{\gamma_2}+H_{2\delta}$, $E(X^+) = X_{\gamma_2}+X_{2\delta}$. The space of the highest weight 1 vectors is one-dimensional, therefore the embedding with the given E(H) is unique. This embedding does exist and $E(Y^+) = X_{\gamma_2-\delta} + X_{2\delta}$.

7) $E(H) = 2H_{\gamma_1}+2H_{\gamma_1+3\gamma_2}+H_{2\delta}$, $E(X^+) = X_{\gamma_1}+X_{\gamma_1+3\gamma_2}+ X_{2\delta}$. Then $E(Y^+) = aX_{\gamma_1+\gamma_2-\delta} + bX_{\gamma_1+2\gamma_2-\delta}+cX_{\gamma_2+\delta}+eX_{\delta}+\alpha X_{\gamma_2+\delta}$, where $abcd \neq 0$. But then $|E(Y^+),E(Y^+)| \neq 0$, and hence there exist no embeddings with this E(H).

9) This embedding does exist, since there exists the embedding $D(3)\to AG_2$ and it is unique. The latter is true since the space of the highest vectors of weight 1 is three-dimensional and belongs to D(3). Therefore $E(osp(1|2))\subset D(3)$, and E is superprincipal.

Let $\underline{g}$ be a matrix Lie superalgebra, S the superspace of its standard representation, $S = \oplus S_a$ a decomposition into irreducible osp (1|2)-moduli.

If $\underline{g} \simeq sl$, then the image of osp (1|2) will be clearly found in $\underline{g}' = \oplus \underline{g}'_a$, where the embeddings $E_a: osp(1|2) \to \underline{g}_a$ are uniquely defined and are equivalent to the super principal ones.

If $g\simeq osp$, the decomposition $\oplus S_a$ may be represented in the form $\oplus\ \omega_b$, so that $\omega_a \perp \omega_c$, and $\omega_a = S_a$ or $\omega_b = S_b \oplus S_{b'}$, with $S_c \simeq S_b$, where the restrictions of the bilinear form onto S_b and $S_{b'}$ are zeros. Further as in the case of sl.

For $g \simeq sq$ or $s\pi$ every S_a has its own $S_{b(a)}$, such that $S_{b(a)} \simeq \pi(S_a)$ as osp (1|2)-moduli. Put $\omega_a =$

$= S_a \oplus S_{b(a)}$ and so on, as above.

The consideration of the other cases we omit. ∎

5\. The classification of embeddings of osp $(1|2)$ into a Lie superalgebra $\underline{g}$ allows one to explicitly associate each embedding with an integrable supersymmetric system. We illustrate this with the example of a supersymmetric extension of the two-dimensional Toda lattice with fixed end-points associated with the superprincipal embeddings of osp $(1|2)$ into classical Lie superalgebras. In this case the subalgebra $\underline{g}_o$ is abelian, and the elements of $\underline{g}_{\pm 1}$ are odd; $d_o = d_{\pm 1}$; the numerical coefficients $c_i^{o\pm}$ of the decomposition $H = \Sigma\, c_i^o F_i^o$ and $Y_{\pm} = \sum\limits_{l} c_i^{\pm} F_i^{\pm}$ are fixed by the embedding and constrained by the formula $c_i^o = \sum\limits_{j} a_{ji}^{-1} = c_i^+ c_i^-$. Then, parametrizing the function $g(z_{\pm}\,;\,\theta_{\pm})$ on G_o entering the general equation (1) as follows: $g = \exp \sum\limits_{i} F_i^o\, \Phi_i\, (z_{\pm}\,;\,\theta_{\pm})$; where $\Phi_i(z_{\pm}\,;\,\theta_{\pm})$ are the components of a bosonic superfield taking values in the Grassmann algebra $\Lambda\,(\theta_{\pm})$ and using commutation relations (2), (3), one comes to the system:

$$D_+ D_-\, \Phi_i = \exp\,(a\Phi)_i \tag{4}$$

For the series $sl(n|n+1)$ and $osp(2|2n)^{(2)}$ with the graph from Tables 2,10 this system reproduces the result of [6] and allows a uniform description for the supersymmetric extensions of the Liouville and sine-Gordon equations (for $n = 1$).

6\. The construction of the N-supersymmetric extension of the nonlinear systems given above is related with the consideration of a certain type of the embeddings of supermanifolds $V_{2|2N}$ with the coordinates $x^a = (z_{\pm}\,;\, \theta_{\pm}, 1 \le \alpha \le N)$ into the enveloping linear supermanifold

$V_{A|B}$ with the coordinates X^A and the structure of the Lie superalgebra $\underline{g}$ and the connection $\Gamma_A(X)$. Here Γ_A takes values is Λ-points of the Lie superalgebra $\underline{g}$ (cf. [10]) (otherwise called the Grassmann envelope $\underline{g}(\Lambda)$), while the induced connections $\gamma_a(x)$ on the supermanifold $V_{2|2N}$ are related to $\Gamma_a(X)$ at points x via the components $E_a^A = \partial X^A / \partial x^a$ of the frame: $\gamma_a(x) = \sum_A E_a^A \Gamma_A$. The zero curvature representation of the enveloping linear supermanifold

$$R_{AB} \equiv [\partial/\partial X^A + \Gamma_A, \ \partial/\partial X^B + \Gamma_B] = 0$$

induces on $V_{2|2N}$ the following system of equations:

$$[\partial/\partial x^a + \gamma_a(x), \ \partial/\partial x^b + \gamma_b(x)] = 0 \qquad (5)$$

Note that if the connections Γ_A are arbitrary functions on $\underline{g}(\Lambda)$, then they, and hence γ_a, may be turned into zero using the corresponding gauge transformation generated by $\underline{g}(\Lambda)$. It is due to the fact that the representation $R_{AB} = 0$ is form-invariant under these transformations. However, if these connections are zero on certain subsuperspaces of the graded Lie superalgebra $\underline{g} = \bigoplus_{a\in \mathbb{Z}} \underline{g}_a$, the representation $R_{AB} = 0$, where these conditions are imposed on the connection, is form-invariant under the transformations generated by the Lie superalgebra $\underline{g}_0$ only. In this case, system (5) would lead to nontrivial nonlinear equations.

In order to find the explicit form of the supersymmetric equations resulting from (5), it is convenient for $N = 2n$ to pass from $\theta_{\pm\alpha}$, $1 \leq \alpha \leq N$, to $\theta_j^{\pm}$ and $\overline{\theta}_j^{\pm}$, $1 \leq j \leq n$, and introduce the following covariant supersymmetric derivatives:

$$D_{\pm j} = i\,\theta_j^{\pm}\,\partial/\partial z_{\pm} + \partial/\partial\,\overline{\theta}_j^{\pm}\ , \quad \overline{D}_{\pm j} = i\,\overline{\theta}_j^{\pm}\,\partial/\partial z_{\pm} + \partial/\partial\theta_j^{\pm}\ ,$$

$$[D_{\pm j}\ ,\ \bar{D}_{\pm k}] = 2\delta_{jk}\partial/\partial z_{\pm}\ .$$

The case $N = 2n-1$ is considered as for an even N substituting $z_{\pm} \to z_{\pm} \mp i(\bar{\theta}^{\pm} - \theta^{\pm})$.

The construction generalizing the known $N = 1,2,4$ ([4,5]) superextensions of the Liouville equation, may be described as follows. Consider a local Lie superalgebra $\bar{g} = \bar{g}_{-1} \oplus \bar{g}_{o} \oplus \bar{g}_{1}$, where $\bar{g}_{o}$ is a vacuum invariance subsuperalgebra with $\dim \bar{g}_{o} = N$. Here the Z-grading $g = \oplus g_a$ is induced by the embedding of the Lie subsuperalgebra $\bar{g}_{o}$ in it. Put

$$\Xi_j = D_j + g_{+}^{-1}D_j g_{+} + (g_{-}^{-1}\ Y^{+} g_{-}\ )_j\ ,$$

$$\bar{\Xi}_j = \bar{D}_{-j} + g_{-}^{-1}\bar{D}_{-j}\ g_{-} + (g_{+}^{-1}Y^{-}g_{+}\ )_j\ .$$

Here $g_{\pm}\ (z_{+},z_{-};\ \theta,\ \bar{\theta}\ )$ is a function on the Grassmann envelope of the Lie supergroup G_o with Lie superalgebra $\underline{g}_o$ represented by the exponent of the superfield with values in $\underline{g}_o(\Lambda)$; $Y^{\pm}$ are the elements of $\bar{g}_{\pm 1}$. Then, commuting the elements $\Xi_j, \bar{\Xi}_{-j}$, one comes to the N-supersymmetric system

$$\begin{aligned}[][D_j + g_{+}^{-1}D_j g_{+} + (g_{-}^{-1}Y^{+}g_{-})_j\ ,\\ \bar{D}_{-k} + g_{-}^{-1}\bar{D}_{-k}\ g_{-} + (g_{+}^{-1}Y^{-}\ g_{+})_k\ ] = 0\end{aligned} \qquad (6)$$

These equations are written down in the form, which is symmetric in g_{+} and g_{-}. Its simplification is due to the use of the gauge transformations of $G_o(\Lambda)$. In particular, one can pass to the gauge where the element g_{+} (or g_{-}) turns into unit and the final equation is a matrix system for one group element of $G_o(\Lambda)$. The details of these calculations are left to the reader.

The remaining commutators of the elements Ξ_j, $\bar{\Xi}_{-j}$ as well as Ξ_{-j}, $\bar{\Xi}_{+j}$ (the latter take values in a subsuperspace dual to $\bar{g}$) guarantee together with the gauge transformations from $g_o(\Lambda)$ the irreducibility conditions for the components $g_{\pm}$ and have a "kinematic" nature.

In the simplest cases of the N-superextensions of the Liouville equation for N = 1, 2, 4 the ingredients of the system are as follows:

N	$\bar{g}$	Graph for $\bar{g}$	$\bar{g}_o$	$\bar{g}_{\pm 1} = (Y_j^{\pm})$
1	osp(1|2)	●	gl(1)	$Y_{\pm}^{\pm}$
2	sl(1|1)	⊗	gl(1)	$_{\pm}Y^{-}$
4	sl(1|2)	⊗—⊗	gl(2)	$Y_1^{\pm}$, $Y_2^{\bar{+}}$

This approach does not indicate however why cases with N >4 do not lead to equations interesting to physics. This is manifest in Ivanov-Krivonos' approach [15] and is closely related with the lack of nontrivial central extensions for superconformal Lie superalgebras of string theories, K(n) [10] .

Acknowledgements. We wish to express our gratitude for the fruitful discussions to P.P.Kulish, A.N.Leznov, Yu.I.Manin and A.O.Radul.

Appendix I. Supersymmetrization of Toda lattices, KdV and mKdV (R.Yu.Kirillova)

The requirement of supersymmetricity fixes most rigidly the choice of a base determining the superprincipal embedding which according to Theorem 2 is not realized for any type of simple Lie superalgebra. The choice of any other system of simple roots, as eg in [16] , fails to guarantee supersymmetricity,

The superextension of the sine-Gordon equation as is known from ref. [14, 6] is associated with the infinite-dimensional Lie superalgebra $osp(2|2)^{(2)}$ realized in the degenerate representation on the elements of $sl(1|2)$. A direct construction of the supersymmetric extensions of multi-component generalizations of sine-Gordon type equations (including the periodic Toda lattice) is associated with the superprincipal embeddings of $osp(1|2)$ into infinite-dimensional Lie superalgebras of finite growth (Kac-Moody superalgebras) with bases consisting of odd roots, see Table 10.

Then the arising equations can be written down as

$$D_+D_- \psi_i = \sum_j a_{ji} \exp \psi_j \quad , \tag{7}$$

Remark. System (4) for a nondegenerate Cartan matrix a may be rewritten in an equivalent form (7) with $\Phi = = a^{-1}\psi$.

Except for the case $psl(n|n)^{(1)}$ (excluded in what follows) the matrix a has the only zero eigenvalue associated with the eigenvector λ , $(\lambda\, a)_j = 0$. Hence, eqs. (4) yield $D_+D_-(\lambda\psi) = 0$. If $(\lambda\psi) = 0$, i.e. $\psi_\tau = = -\sum_{1\leq i\leq\tau-1} \psi_i\lambda_i\lambda_\tau^{-1} \equiv -\langle\psi, \mu\rangle$, $\mu_i \equiv \lambda_i \lambda_\tau^{-1}$, $\tau \equiv$ rank g, then quite analogously to the case of ordinary Lie algebras [7], we rewrite system (7) for linearly independent functions $\hat{\Phi}_i \equiv \langle a^{-1}, \psi\rangle_i$, $1 \leq i \leq \tau - 1$, in the form

$$D_+D_- \hat{\Phi}_i = \exp \langle a_o, \hat{\Phi}\rangle_i + a_{i\tau} \exp \langle\langle a_o \hat{\Phi}\rangle , -\mu\rangle \tag{8}$$

Here a_o is the Cartan matrix of the finite-dimensional Lie superalgebra of rank $\tau - 1$ obtained from the infinite-dimensional Lie superalgebra $\underline{g}$ by deleting one vertex of $\underline{g}$'s Dynkin graph.

Thus, the standard procedure of two-dimensional generalizing with the help of a cocycle deformining the Kac-Moody superalgebra, see [17], leads to (2|2)-dimensional systems only in quite rare cases of the superprincipal embedding.

Table 10. Bases associated with super KdV

$\mathfrak{g}$	$\mathfrak{sl}(n\|n)^{(1)}$, $\mathfrak{sq}(2n+1)^{(2)}$	$\mathfrak{osp}(2n+2\|2n)^{(1)}$, $\mathfrak{sl}(2n\|2n)^{(2)}$	$D(\alpha)^{(1)}$ $(\alpha=1)$
B	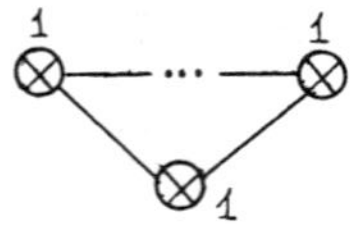	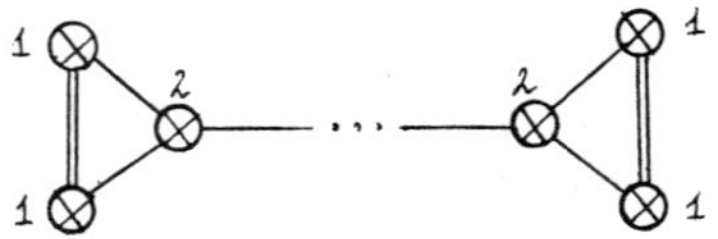	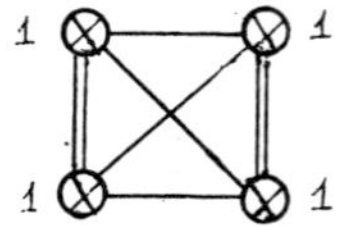

$\mathfrak{g}$	$\mathfrak{osp}(2n\|2n)^{(2)}$, $\mathfrak{sl}(2n+1\|2n+1)^{(2)}$	$\mathfrak{osp}(2\|2)^{(2)}$
B		

$\mathfrak{g}_\infty$	$L\,\mathfrak{sl}(n\|n)^{(1)}$	$L\,\mathfrak{osp}(2n+2\|2n)^{(1)}$	$L\,\mathfrak{osp}(2n\|2n)^{(2)}$
B_∞		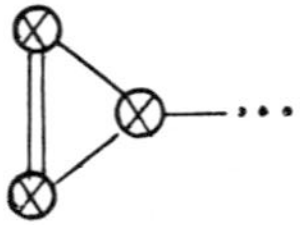	

The "true" supersymmetric versions of the KdV and mKdV equations, i.e. with (1|1)-dimensional space and time of type [18] (as opposed to ref.[19] where both time and space are one-dimensional) are also associated with the superprincipal embedding.

All written in the main text is enough to superize the KdV and mKdV equations (only for the bases admitting the superprincipal embeddings). To do this, in formulas(6.1), (6.4), (6.7), (3.11), (3.27), (10.4) in [3] one must replace $d/dx = D$ by the operator θ from [18], $\theta = \partial/\partial r + r\partial/\partial x$ and d/dt by $\theta_t = \partial/\partial\tau + \tau\partial/\partial t$. The cases when one is able to do this, are listed in Table 9. A tedious superization of the theorems from Sect. 7.1 and Chapter 9 of [3] will be given elsewhere.

Remark. sKdV considered in [18] is related with $sq(3)^{(2)}$.

Hamiltonian structures of superized KdV are constructed as in [20] with an obvious definition of residue as in [18]. Since the natural invariant scalar product in $sq^{(1)}$ and $sq^{(2)}$ is induced by odd trace otr (see [10]), then the Hamiltonian structure of sKdV associated with $sq^{(2)}$ (Table 10) is odd (i.e. carries the Buttin or anti-bracket [2,10]) and Poisson otherwise.

The "japanese" approach to KP, KdV, etc. lucidely exposed in [21] naturally superizes if in $gl(\infty,\infty)$ we choose a particular base consisting of odd roots only. The corresponding graphs are depicted in Table 10. They illustrate the misterious isomorphisms

$$Lsl(n|n)^{(1)} \simeq Lsq(2n+1)^{(2)}, \quad Losp(2n+2|2n)^{(1)} \simeq$$

$$\simeq Lsl(2n|2n)^{(1)}, \quad Losp(2n|2n)^{(2)} \simeq Lpsl(2n+1|2n+1)^{(2)},$$

where $L = \lim_{n\to\infty}$.

The soliton solutions are expressed via formulas of [21], where all the constructions are carried out in the language of Λ-points (Grassmann envelopes [10]).

Appendix II. Classification of bases of simple Lie superalgebras (V.V.Serganova)

1. Let $\underline{g}$ be a simple contragredient Lie superalgebra, $R \subset \underline{C}^*$ its root system, B a base, $R_B^+ = (\gamma \quad R | \gamma = \sum_{\alpha \quad B, n_\alpha \quad Z} + n_\alpha \alpha)$. Clearly, $R = R_B^+ \quad (-R_B^+)$. The reflection with respect to $\alpha \in B$ is $\omega_\alpha \in \mathrm{Hom}(\underline{C}^*, \underline{C}^*)$, such that for $\beta \in B$

$$\omega_\alpha(\beta) = \beta - a_{\alpha\beta}\alpha \quad , \text{ if } a_{\alpha\alpha} = 2 ;$$

$$\omega_\alpha(\beta) = \beta \quad , \text{ if } a_{\alpha\beta} = 0, \alpha \neq \beta , a_{\alpha\alpha} = 0 ;$$

$$\omega_\alpha(\beta) = \beta + \alpha , \quad \text{if } a_{\alpha\beta} \neq 0, \alpha \neq \beta , a_{\alpha\alpha} = 0 ,$$

$$\omega_\alpha(\alpha) = -\alpha , \quad \text{if } a_{\alpha\alpha} = 0 .$$

On the remaining elements of R the ω_α-action is defined by linearity.

The group generated by ω_α's for all bases is called the Weyl group of the Lie superalgebra g or the Weyl supergroup SW,

Lemma 1. $\omega_\alpha = 1$.

Proof. Obvious.

Lemma 2. Let B be a base, $\alpha \in B$, and $\gamma \in R_B^+$. Then:

1) If $a_{\alpha\alpha} = 2$, $S_\gamma = (\sum_{\{\beta \in B | \alpha \neq \beta\}} n_\beta a_{\alpha\beta}) + n_\alpha < 0$

for $\gamma \neq n_\alpha \alpha$.

2) If $a_{\alpha\alpha} = 0$, $S_\gamma = (\sum_{\{\beta \in B | a_{\alpha\beta} \neq 0\}} n_\beta a_{\alpha\beta}) - n_\alpha < 0$,

and if $[X_\alpha, X_\gamma] \neq 0$, then $S_\gamma < 0$.

Proof. 1) Let us prove this for an odd α (for the even one it is analogous). Let $g^0 = \langle X_\alpha, X_{-\alpha}, X_{2\alpha}, X_{-2\alpha}, h_\alpha \rangle$ be a Lie subsuperalgebra, isomorphic to $osp(1|2)$. Then $[X_{-\alpha}, X_\beta] = 0$ for $\beta \neq \alpha$. Therefore $(\mathrm{ad} X_{-\alpha})^{n_\alpha + 1}(X_\gamma) = 0$, and the highest weight λ of the subrepresentation generated by $g^0(X_\gamma)$ is not greater than $-(\gamma - n_\alpha \alpha)(h_\alpha) =$

$= \sum_{(\beta\in B|\ \alpha\neq\beta)} -n_\beta a_{\alpha\beta}$. Since $\gamma(h_\alpha) < \lambda$, and

$$\left(\sum_{(\beta\in B|\ \alpha\neq\beta)} n_\beta a_{\alpha\beta}\right) + 2n_\alpha < \sum_{(\beta\in B|\ \alpha\neq\beta)} -n_\beta a_{\alpha\beta} ,$$

then

$$\left(\sum_{(\beta\in B|\ \alpha\neq\beta)} n_\beta a_{\alpha\beta}\right) + n_\alpha < 0 .$$

2) Let $|\gamma| = \Sigma n_\beta$ for $\gamma = \sum_{\beta\ B} n_\beta \beta$. We prove our statement by induction in $|\gamma|$. For $|\gamma| = 1$ it is evident. If $|\gamma| = k$ then $X_\gamma = [X_{\gamma_1}, X_{\gamma_2}]$ for $|\gamma_i| < k$, $i = 1, 2$. Next, $s\gamma = s\gamma_1 + s\gamma_2$ and $[X_\alpha, X_\gamma] \neq 0$ if $[X_\alpha, X_{\gamma_i}] \neq 0$ at least for one i. If $\gamma_i \neq n_\alpha$, then the statement for γ is evident. Let $\gamma_1 = n\alpha$ then $n = 1$ since $[X_\alpha, X_\alpha] = 0$. Besides, $[X_\alpha, X_\gamma] = [X_\alpha, [X_\alpha, X_{\gamma_2}]] = 0$, Since $s\gamma = s\gamma_2 + 1$ and $s\gamma_2 < 0$, then $s\gamma < 0$.

<u>Lemma 3</u>. Let B be a base and $\alpha \in B$. Then $\omega_\alpha(B)$ is also a base and $R^+_{\omega_\alpha(B)}(-n\alpha) = R^+_B(n\alpha)$.

<u>Proof</u>. By Lemma 2 if $\gamma \in R^+_B$, $\gamma \neq n\alpha$, then

$$\gamma = \left(\sum_{\beta\neq\alpha} n_\beta \beta\right) + n_\alpha \alpha = \left(\sum_{\beta\neq\alpha} n_\beta(\beta - a_{\alpha\beta})\right) + s_\gamma \alpha =$$

$$= \left(\sum_{\beta\neq\alpha} n_\beta\, \omega_\alpha(\beta)\right) - s_\gamma \omega_\alpha(\alpha) \quad \text{for } a_{\alpha\alpha} = 2$$

and

$$\gamma = \left(\sum_{a_{\alpha\beta}\neq\alpha} n_\beta \beta\right) + \left(\sum_{\substack{a_{\alpha\beta}=0 \\ \beta\neq\alpha}} n_\beta \beta\right) + n_\alpha \alpha =$$

$$= \left(\sum_{\beta\neq\alpha} n_\beta\, \omega_\alpha(\beta) - s_\gamma \omega_\alpha(\alpha) \quad \text{for } a_{\alpha\alpha} = 0 .\right.$$

<u>Theorem</u>. Let B_1 , B_2 be two bases. Then there exists a sequence of bases $B^1, \ldots, B^n$ and roots $\alpha^i \in B^i$ such

that $\omega_{\alpha}i\,(B^i) = B^{i+1}$ and $B^1 = B_1$, $B^n = B_2$.

Proof. Let $R_B^{\bullet}$ be the set of all the roots of R_B^+ up to multiplication by an an integer. Clearly $B_1 = B_2$ if $R_{B_1}^{\bullet} = R_{B_2}^{\bullet}$. The set $\{R_{B_1}^{\bullet} \cup R_{B_2}^{\bullet}\} \setminus R_{B_2}^{\bullet}$ is finite and there exists a root $\alpha \in B_1$, such that $\alpha \in \{R_{B_1}^{\bullet} \cup R_{B_2}^{\bullet}\} \setminus R_{B_2}^{\bullet}$. It follows from Lemma 3 that in $(R_{\omega(B_1)}^{\bullet} \cup R_{B_2}^{\bullet}) \setminus R_{B_2}^{\bullet}$ there is one element less. Now the statement of the theorem is readily obtained by induction in $|\{R_{B_1}^{\bullet} \cup R_{B_2}^{\bullet}\} \setminus R_{B_2}^{\bullet}|$.

2. The following simple statement allows one to describe all bases.

Lemma 4. Let B be base and $\alpha \in B$. Then $\omega_{\alpha}(B)$ is not equivalent to B if and only if $a_{\alpha\alpha} = 0$.

Therefore if one wants to describe all the bases of a Lie superalgebra $\underline{g}$ up to equivalence, it suffices to choose any one of them and apply successively the reflections with respect to simple roots of the form ⊗. In this way one obtains all the remaining bases.

3. Let • denote the vertex ⊗ or O. Let l be the number of vertices in the graph. Clearly, any graph has three levels except $sl^{(1)}$ and $psq^{(2)}$. Decompose the maximal middle subgraph •-...-• into groups ⊗, O-...-O, O-...-O-⊗ from the left (for $sl^{(1)}$ and $psq^{(2)}$ decompose the whole graph starting from any place) so that every group contains one vertex ⊕ or possibly none if it is the last one. For instance,

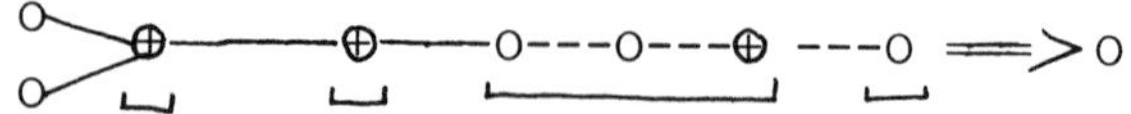

Then let ev (resp. *od*) be the number of vertices in all even (odd) groups, *png* the parity of the number of vertices ⊕ in this subgraph.

Table 11. Action of the Weyl supergroup on bases

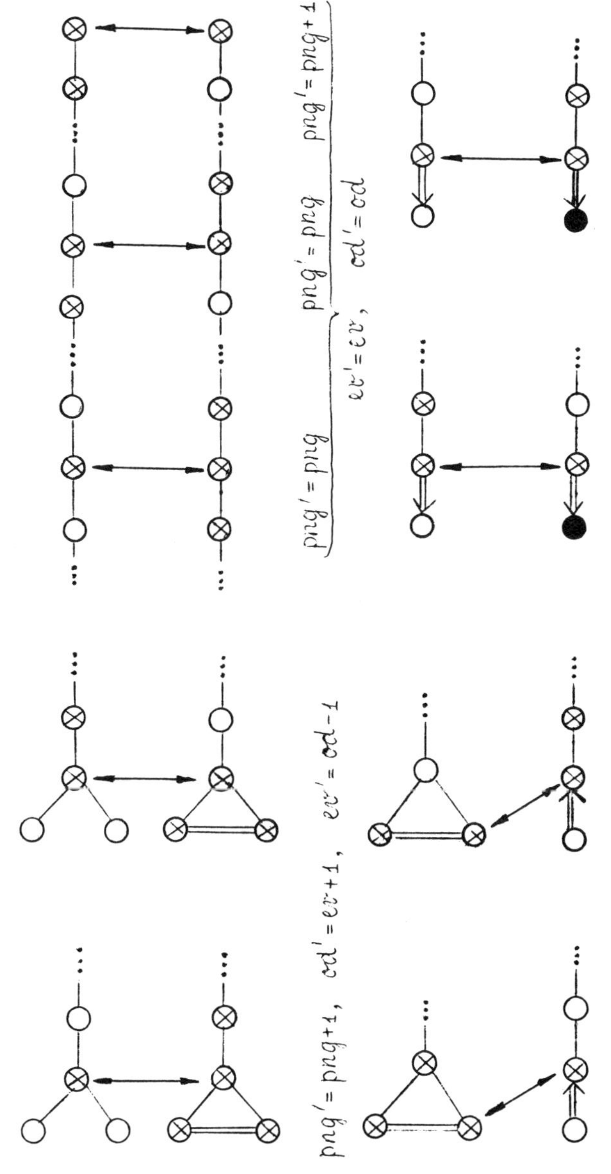

Table 11. Action of the Weyl supergroup on bases (continued)

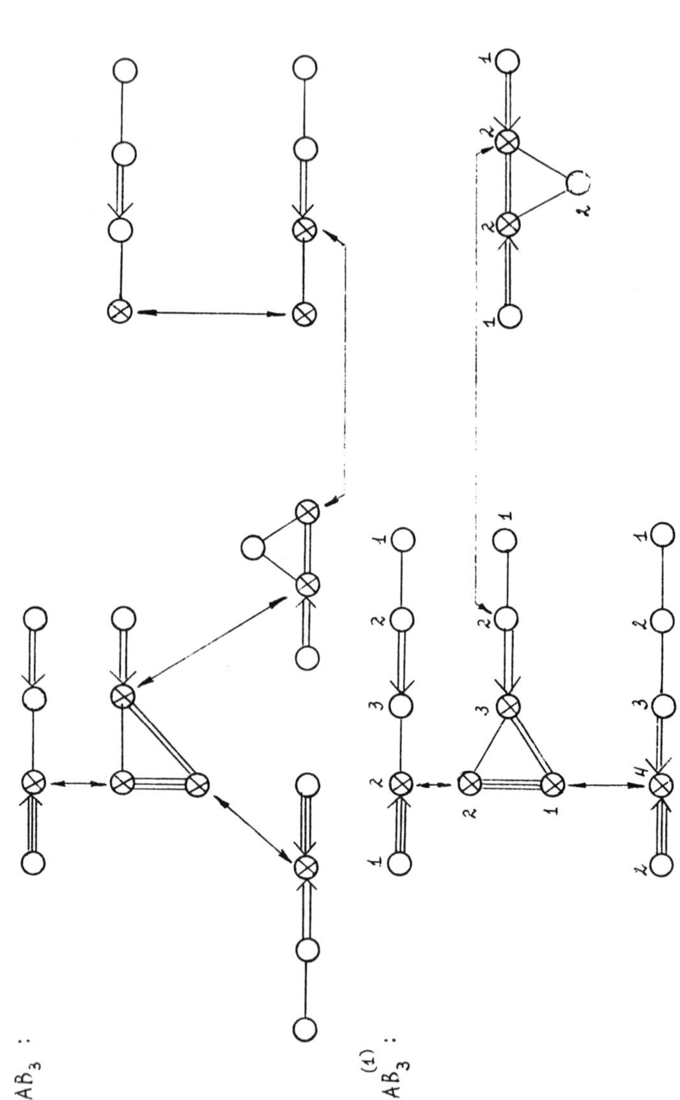

Table 11. Action of the Weyl supergroup on bases (continued)

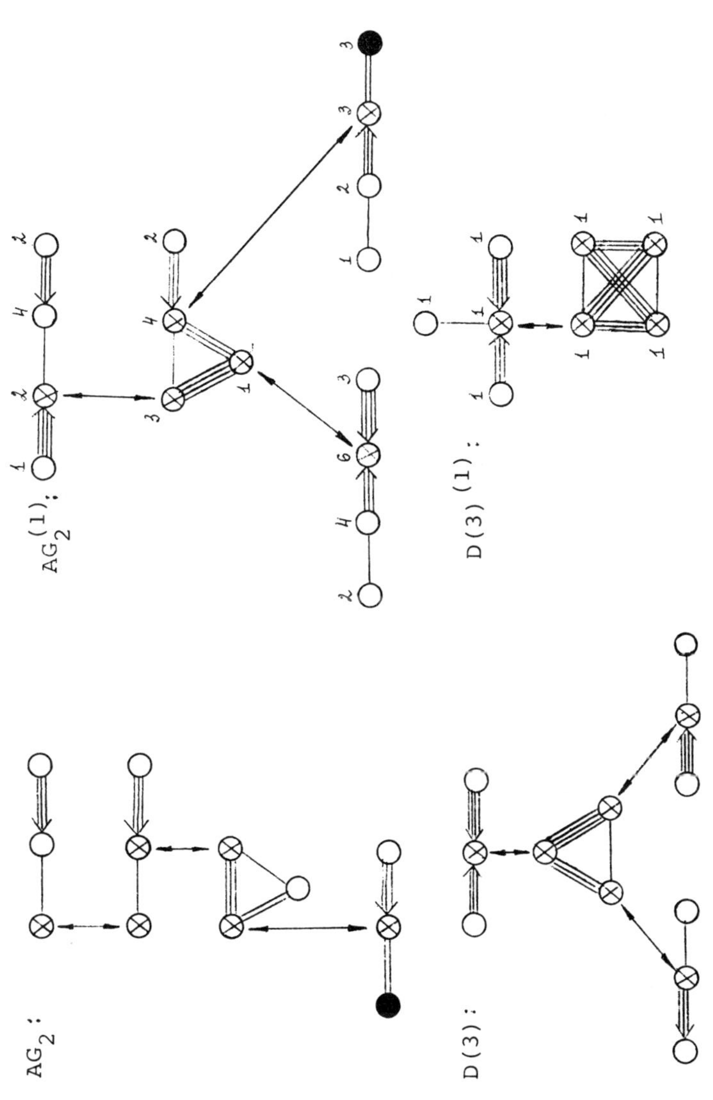

1) sl(n), $sl^{(1)}$ and $psq^{(2)}$. Take the base B with the graph

O-...-O—⊗—O-...-O, l=m+n-1, ev=m-1, od=n, png= 1 .

With the help of the reflection with respect to any internal vertex, the graph of the form •-...-• may be transformed into a similar one, in which l,ev,od and png do not change. The reflection with respect to the edge vertex changes png.

2) osp, $sl^{(m)}$, $osp^{(m)}$. The graphs for the small values of the dimension are presented in Table 2. It follows from 1) that for larger dimensions the graphs having the same schemes near edges are transformed by an appropriate reflection into each other.

Table 11 shows how the edge subgraphs are changed. There the case of $D(\alpha)$ is considered for $\alpha = 3$.

Now enumerate explicitly all the bases for exceptional Lie superalgebras.

$D(\alpha)$: The Cartan matrices for $D(\alpha)^{(1)}$ are

$$\begin{pmatrix} 2 & -1 & 0 & 0 \\ 1 & 0 & \alpha & -1-\alpha \\ 0 & -1 & 2 & 0 \\ \hline 0 & -1 & 0 & 2 \end{pmatrix} \quad \text{and} \quad \begin{pmatrix} 0 & 1 & -1 & -\alpha & \alpha \\ 1 & 0 & \alpha & & -1-\alpha \\ -1-\alpha & \alpha & 0 & & 1 \\ \hline \alpha & -1-\alpha & 1 & & 0 \end{pmatrix}$$

while those for $D(\alpha)$ are placed in the upper left box of the matrices for $D(\alpha)^{(1)}$. In all other cases the matrices are integral.

Let $\langle H_1, H_2, H_3\rangle = \mathfrak{h} \subset D(\alpha)$ and $\varepsilon_1, \varepsilon_2, \varepsilon_3$ be the standard basis of the dual space. Then all the bases are

1) $2\varepsilon_1, -\varepsilon_1+\varepsilon_2-\varepsilon_3, 2\varepsilon_3$; 2) $\varepsilon_1-\varepsilon_2-\varepsilon_3, 2\varepsilon_2, 2\varepsilon_3$;

3) $2\varepsilon_1 2\varepsilon_2, -\varepsilon_1 -\varepsilon_2 +\varepsilon_3$; 4) $\varepsilon_1-\varepsilon_2-\varepsilon_3, \varepsilon_1+\varepsilon_2+\varepsilon_3, -\varepsilon_1+\varepsilon_2-\varepsilon_3$

All the bases for $D(\alpha)^{(1)}$ are

1) $-2\varepsilon_1, 2\varepsilon_2, 2\varepsilon_3, \varepsilon_1-\varepsilon_2-\varepsilon_3$; 2) $\varepsilon_1-\varepsilon_2-\varepsilon_3, \varepsilon_1+\varepsilon_2+\varepsilon_3$,

$-\varepsilon_1+\varepsilon_2-\varepsilon_3$, $-\varepsilon_1-\varepsilon_2+\varepsilon_3$.

AG_2 : Choose a basis in $\underline{e}$ as follows: let δ be the dual to the generator H of the Cartan subalgebra of sl(2), such that $\delta(H) = 1$, γ_1 a long root in the base of the Lie algebra G_2, and γ_2 a short root. Then all the bases for AG_2 are:

1) γ_1, $-\gamma_1-\gamma_2+\delta$, $\gamma_1+2\gamma_2-\delta$; 2) γ_1,γ_2, $-\gamma_1-2\gamma_2+\delta$;

3) $-\gamma_2+\delta$, $\gamma_1+\gamma_2-\delta$,γ_2 ; 4) $\gamma_2-\delta$, γ_1,δ

All the bases for $AG_2^{(1)}$ are:

1) γ_1,γ_2, $-\gamma_1-2\gamma_2+\delta$, -2δ ; 2) $\gamma_1,-\gamma_1-\gamma_2+\delta,\gamma_1+2\gamma_2-\delta$, $-\gamma_1-2\gamma_2-\delta$; 3) $\gamma_2-\delta,\gamma_1,\delta$, $-2\gamma_1-3\gamma_2$; 4) $-\gamma_2+\delta$, $\gamma_1+\gamma_2-\delta$, $-3\gamma_2-2\gamma_1$, γ_2 ; 5) $\gamma_1,-2\gamma_1-3\gamma_2$, -2δ , $\gamma_1+2\gamma_2+\delta$.

AB_3: Let the basis in $\underline{e}$ be the natural basis $\varepsilon_1,\varepsilon_2,\varepsilon_3$ in $o(7) = B_3$, whereas δ be the same as for AG_2. All the bases for AB_3 are:

1) -2δ, $\frac{1}{2}(\varepsilon_1+\varepsilon_2+\varepsilon_3)+\delta$, $-\varepsilon_1$, $\varepsilon_1-\varepsilon_2$;

2) $\varepsilon_2+\varepsilon_3$, $\varepsilon_1-\varepsilon_2$, $-\varepsilon_1$, $\frac{1}{2}(\varepsilon_1+\varepsilon_2-\varepsilon_3)$ $-\delta$;

3) $\varepsilon_2+\varepsilon_3$, $\varepsilon_1-\varepsilon_2$, $\frac{1}{2}(-\varepsilon_1+\varepsilon_2-\varepsilon_3)-\delta$, $\frac{1}{2}(-\varepsilon_1-\varepsilon_2+\varepsilon_3)+\delta$;

4) $\frac{1}{2}(\varepsilon_1+\varepsilon_2+\varepsilon_3)-\delta$, $-\frac{1}{2}(\varepsilon_1+\varepsilon_2+\varepsilon_3)-\delta$, $\frac{1}{2}(-\varepsilon_1+\varepsilon_2+\varepsilon_3)+\delta$, $\varepsilon_1-\varepsilon_2$;

5) $-\frac{1}{2}(\varepsilon_1+\varepsilon_2+\varepsilon_3)+\delta$, -2δ, $\varepsilon_2+\varepsilon_3$, $\varepsilon_1-\varepsilon_2$;

6) $\varepsilon_2+\varepsilon_3$, $-\varepsilon_1$, $\frac{1}{2}(\varepsilon_1-\varepsilon_2+\varepsilon_3)+\delta$, $\frac{1}{2}(\varepsilon_1-\varepsilon_2-\varepsilon_3)-\delta$.

All the bases for $AB_3^{(1)}$ are:

1) -2δ, $\frac{1}{2}(\varepsilon_1+\varepsilon_2+\varepsilon_3)+\delta$, $-\varepsilon_1,\varepsilon_1-\varepsilon_2$, $\varepsilon_2-\varepsilon_3$;

2) $-\frac{1}{2}(\varepsilon_1+\varepsilon_2+\varepsilon_3)-\delta$, $\frac{1}{2}(\varepsilon_1+\varepsilon_2+\varepsilon_3)-\delta$, $\frac{1}{2}(-\varepsilon_1+\varepsilon_2+\varepsilon_3)+\delta$,

$\varepsilon_1-\varepsilon_2$, $\varepsilon_2-\varepsilon_3$;

3) $-\frac{1}{2}(\varepsilon_1+\varepsilon_2+\varepsilon_3)+\delta$, -2δ, $\varepsilon_2+\varepsilon_3$, $\varepsilon_1-\varepsilon_2$, $\varepsilon_2-\varepsilon_3$;

4) $\varepsilon_2+\varepsilon_3$, $-\varepsilon_1$, $\frac{1}{2}(\varepsilon_1-\varepsilon_2-\varepsilon_3)-\delta$, $\frac{1}{2}(\varepsilon_1-\varepsilon_2+\varepsilon_3)+\delta$, $\varepsilon_2-\varepsilon_3$.

The corresponding Dynkin graphs and Cartan matrices are presented in Tables 2-6.

References

1. Saveliev,M.V. (1984). Sov. J. Theor. Math. Phys.,59 3,367; 60, 9 (Russian).
2. Kirillova,R.Yu. (1983). Zapiski nauchnykh seminarov LOMI, 123, Nauka, Leningrad, p. 98; Leites,D.A.,Semenov-Tian-Shansky,M.A. Ibid.,p. 92 (Russian).
3. Drinfeld,V.G., Sokolov,V.V. (1984). Sovrem. Probl. Mat., (Itogi nauki), VINITI, Moscow,,p.81 (Russian).
4. Salam,A. (1984). Invited talk on the XXII Intern. Conf. on High Energy Physics, Leipzig.
5. Manin,Yu,I. (1984). Gauge fields and complex geometry. Nauka, Moscow; (1984). Sovrem. Probl. Mat., 24 (Itogi nauki) VINITI, Moscow, p. 3 (Russian).
6, Saveliev,M.V. (1984). Comm. Math. Phys., 95, 199.
7. Leznov,A.N., Saveliev,M.V. (1984). Math. Anal., 22 (Itogi nauki), VINITI, Moscow, 101 (Russian).
8. Leznov,A.N., Saveliev,M.V. Sov. J. Theor. Math. Phys., 1984, v. 61, p. 150.
9. Dynkin,E.B. (1952). Mat. Sb., 30, 349 ; (1952).Trudy MMO, 1, p. 1 (RUssian).
10. Leites, D.A. (1983). Theory of supermanifolds. Karelia Brauch of the USSR Acad. Sci., Petrozavodsk; (1984). Sovrem. Probl. Mat., v. 25 (Itogi nauki), VINITI, p. 3 (Russian).
11. Kac,V.G. (1977). Comm. Math. Phys., 53, 31; (1977). Adv. Math., 26, 8; (1978)., 30, 85.
12. Serganova,V.V. (1984). Izv. AN SSSR, ser. mat., 48, 585 (Russian).
13. Kac,V.G. (1983). Infinite-dimensional Lie algebras. Birkhauser, Boston et al.
14. Chaichian,M., Kulish,P.P. (1978). Phys. Let.,788,413.

15. Ivanov,E.A., Krivonos,S.O. (1983). Lett. Mah. Phys., 7, 523; (1984). J. Phys. A: Math. Gen., 17, L671.

16. Olshanetsky,M.A. (1983). Comm. Math. Phys., 88, 63.

17. Reyman,A.G. (1980). Zapiski nauchnykh seminarov LOMI, 95, Nauka, Leningrad, p. 3 (Russian).

18. Manin,Yu.I., Radul,A.O. (1985). Comm. Math. Phys., 98, 65-77.

19. Kupershmidt,B.A. (1984). Phys. Lett., 402, 213.

20. Reyman,A.G., Semenov -Tian-Shansky,M.A. (1984). Zapiski nauchnykh seminarov LOMI, 133, Nauka, Leningrad, p. 212 (Russian).

21. Verdier,J.-L. (1982). In: Seminaire Bourbaki, p.596.

INTEGRABLE MODELS OF FIELD THEORY AND SIMPLE LIE ALGEBRAS AND SUPERALGEBRAS

P.P.Kulish

Leningrad Branch of Steklov Mathematical Institute

The field theoretical models solved by nonpertubative approaches improve our understanding of complicated dynamical systems with infinite number of degrees of freedom. In recent years there was a considerable growth of activity in the study of the integrable two-dimensional field theoretical models. The supersymmetric nonlinear σ-models related to superstrings are of special interest. A great number of the exactly solvable models was introduced due to the quantum inverse scattering method (QISM) (cf.reviews [1-5]). A common feature of different approaches to nonlinear σ-models is an important role played by some exactly solvable one-dimensional magnetics. A systematic way to construct these magnetic models is provided by the QISM and the solutions of the Yang-Baxter equation (YBE) [2,3]. It is easy to get the solutions of YBE related to simple Lie algebras and superalgebras and their representations. Models of this type are now believed to be universal. Other integrable systems can be obtained from the above spin models in a limit as the dimension of the representation tends to infinity and the lattice spacing tends to zero. The nonlinear Schrödinger (NS) equation and the sein-Gordon model are associated with the Lie algebra SU(2). The supersymmetric (SUSY) sine-Gordon model associated with the superalgebra Sl(2/1) (the SUSY Liouville equation is related to osp(1/2) [6]) was formulated and its integrability was proved for the classical [7] and quantum approaches [3]. The many components generalization of the NSE was formulated using different approaches (e.g.

[8]). NSE associated with the superalgebras gl(m/n) contain commuting (Bose) fields as well as anticommuting (Fermi) fields [9]. The most general model with the simples quantum R-matrix related to the gl(m/n) (interacting wave system or Lee model) was solved by the QISM [10]. The application of the QISM to the model in question requirs as a starting point the construction of an auxiliary linear problem (the Lax representation) and the corresponding R-matrix that defines the commutation relations of the quantum scattering data. The authors who generalize the known integrable equations to the SUSY case [11-16] deal, as a rule, with the Lax representation omitting the discussion of the corresponding Hamiltonian structure.

In this paper we write the R-matrices invariant with respect to the simple matrix superalgebras and the corresponding integrable models. The R-matrix is a solution of the Yang-Baxter equation (YBE) with the Z_r-grading [3,9,19]:

$$_{a_1a_2}R_{c_1c_2}(u)\,_{c_1a_3}R_{c_3b_1}(u+v)\,_{c_2c_3}R_{b_2b_3}(v)\,(-1)^{p(c_2)(p(a_3)+p(c_3))} = \tag{1}$$

$$= {}_{a_2a_3}R_{c_2c_3}(v)\,_{a_1c_3}R_{c_1b_3}(u+v)\,_{c_1c_2}R_{b_1b_2}(u)\,(1)^{p(c_2)(p(c_3)+p(b_3))}$$

The gl(m/n)-invariant solution acting in the superspace $C^{m/n} = C^{m/n}$ is

$$R(n) = uI + \rho, \quad _{ab}\rho_{cd} = (-1)^{p(a)p(b)}\sigma_{ad}\sigma_{bc} \tag{2}$$

where ρ is the permutation operator and p(a) is the parity of a basic vector $e_a \in C^{m/n}$.

For Lie algebras the YBE has the form (1) without factors $(-1)^p$ [3]. Using the matrix units e_{ab} (acting in $C^{m/n}$) we can write down the permutation operator in the form

$$p = \sum_{a,b} e_{ab} = e_{ba}(-1)^{p(b)} \tag{3}$$

where $\otimes$ denotes the tensor product of superspaces.

Denote by $^{12}R(n)$ the R-matrix acting in the superspace $V_1 \otimes V_2$. The following R-metrices

$$^{i3}R(n) = uI + \sum_{a,b} e_{ab}^{(i)} p_{ba}, \quad i = 1,2 \tag{4}$$

where p_{ba} are the generators of gl(m/n) in an arbitrary representation space V_3 satisfy the YBE

$$^{12}R(u)\,^{13}R(u+v)\,^{23}R(v) = \,^{23}R(v)\,^{13}R(u+v)\,^{12}R(u) \tag{5}$$

with $^{12}R(u)$ determined by (2). It is possible to construct the gl(m/n)-invariant solution of the YBE (1),(5) corresponding to arbitrary irreducible finite-dimensional representations with the highest weights Λ_1, Λ_2 [10]. However the general expession is rather complicated and can be obtained using the generation theorem [3,10,20].

For the superalgebras osp(m/2n) there are three invariant operators in the tensor product of two fundamental representations $C^{m/2n} \times C^{m/2n}$ [20]: the identity I, the permutation operator $\rho(2)$ and a projection K of rank 1. The osp(m/2n)-invariant solution of the YBE (5) is the following linear combination of these operators:

$$R(u) = u(u-c)I + (u-c)p - uK, \text{ where } c = n+1-m/2 \tag{6}$$

Unlike the gl(m/n) case there is not osp-invariant R-matrix corresponding to arbitrary finite-dimensional representations of osp(m/2n).

The nonlinear integrable equations in classical theory are connected with the r-matrices, the quasiclassical limits of the quantum R-matrices:

$$R(u) = I + i\hbar r(u) + O(\hbar^2). \tag{7}$$

The interacting wave system can be associated with an arbitrary Lie algebra [8] or superalgebra. The corresponding auxiliary linear system is of the form

$$\partial_x f = iL(\lambda) f = i(\lambda F + Q) f \tag{8}$$

where $A = \Sigma a_i h_i$ belongs to the Cartan subalgebra and

$$Q = \sum_{\alpha \in R^+} (q_\alpha(x) e_\alpha + \rho_\alpha(x) e_{-\alpha}) \tag{9}$$

where R^+ is the subset of positive roots, $e_{\pm\alpha}$ the root vectors. If A is regular then the number of nonzero potentials equals to the number of positive roots. For a degenerate A we obtain the nonlinear equations associated with symmetric superspaces [17]:

$$\frac{SV(m/n)}{S(V(m-r/n-S) = V(r/S)};\ \frac{Osp(m/2n)}{O(2)=Osp(m-2/2n)};\ \frac{Osp(2m/2n)}{V(m:N)}.$$

The simplest new equation is the following one

$$i\partial_t q = -\partial_x q + 2q^+ qq + \psi^+\psi q - i\psi^{\partial}\ \psi,$$
$$i\partial_t\psi = -2\partial_x^2\psi + q^+ q\psi - i(2q\partial_x\psi^+ + \psi^+\partial x^{q}), \qquad (10)$$

that describes the interaction of the Bose field $q(x)$ with the one-component Fermi field $\psi(x)$.

As is mentioned above, the integrable magnetic models can be considered as regularized versions of continuous systems. The sl(2/1) magnetic on the one-dimensional chain is described by the auxiliary linear problem $f_{n+1}(n) = L_n(n) f_n$, where the L-operator in the n-th site coincides with the R-matrix (4) and $p_{ab}^{(n)}$ are the generators of the sl(2/1) representation in the space V_n. Hence the L-operator is 3x3 matrix whose entries are quantum operators. We obtain new quantum variables from the given ones $\{p^{(n)}\}$ constructing the monodromy matrix

$$T_N(u) = L_N(u) L_{N-1}(n)\ \ldots\ L_1(n) \qquad (11)$$

satisfying the relation (5). The trace of $T_N(n)$ (or supertrace in our case) is the generating function of the integrals of motion

$$t_N(u) = \mathrm{str}\ T_N(u) = \sum_{j=1,2} (T_N(u))_{jj} - (T_N(n))_{33}. \qquad (12)$$

The eigenfunctions of $t_N(u)$ are constructed with the help of the algebraic Bethe ansatz. For a finite N they are parametrised by quasimomenta satisfying the equations

$$\left(\frac{u_j+m_1}{u_j-n}\right)^N = \prod_{k=1}^{\mu 2} S_{-1}(u_j-v_k),$$

$$\left(\frac{v_j+m_2}{v_j+m_1}\right) = \prod_{k=1}^{\mu 2} S_2(v_j-v_k) \prod_{k=1}^{\mu 1} S_{-1}(v_j-u_k), \qquad (13)$$

where $S_1(u) = (u+1/2)/(u-1_2)$ and $(m_1, m_2/n)$ is the highest weight of the sl(2/1)-module in the site.

The limit as $N\to\infty$ depends crucially on the Hamiltonian in question. In the ferromagnetic system we have two types of magnons and their bounded states. The ground state for the antiferromagnetic contains an infinite number of magnons which are described by densities $\rho_a(u)$ of quasimomenta satisfying integral equations. Since the ground state's structure is nontrivial, the antiferromagnetic is rather similar to a relativistic model, where the physical vacuum is Dirac's sea filled with negative energy particles. We can use the magnetic models (11)-(13) as a regularized version of the SUSY-sigma-model.

Let us also point out the quantum analogue of the Bäcklund transformation from an eigenstate of the integrals of motion (or the transfer matrix t(u))to another one using the creation operators of the quantum scattering data. There exists the corresponding transformation for the auxiliary linear problem.

In conclusion we describe a generalization of the YBE connected with the σ-commuting algebras [21]. Consider an abelian group G and a map $\sigma: G\times G \to C$ which satisfies the relations

$$\sigma(\alpha+\beta,\gamma) = \sigma(\alpha,\gamma)\sigma(\beta,\gamma),$$
$$\sigma(\alpha,\beta)\sigma(\beta,\alpha) = 1, \quad \alpha,\beta,\gamma\in G. \tag{14}$$

The G-grading associative algebra $A = \bigoplus_{\alpha\in G} A_\alpha$ is called σ-commutative if for all $\alpha,\beta\in G$, $a\in A_\alpha$, $b\in A_\beta$ the relation

$$ab = \sigma(\alpha,\beta)ba$$

is valid. Let us define G-grading Zamolodchikov's algebra with generators Aj(u) and defined relations

$$A_i(u)A_j(v) = \sigma(p(i),p(j))\,{}_{ij}S_{kl}(u,v)A_l(v)A_k(u) \tag{15}$$

where i,j,k,l belong to a G-grading set of indeces J: $j\in J$, $p(j)\in G$. The generalized YBE for the coefficients ${}_{ij}S_{kl}(u,v)$ follows from the requirements of associativity and independence of the third power monomials in the generators Aj(n):

$$S_{a_1a_2c_1c_2}(u,v)\,S_{e_1a_3\,b_1c_3}(u,w)\,S_{c_2c_3\,b_2b_3}(v,w)\frac{\sigma(p(c_2),p(c_3))}{\sigma(p(c_2),p(a_3))} = \tag{16}$$

$$_{a_2a_3}S_{c_2c_3}(v,w)\,_{a_1a_3}S_{c_1b_3}(u,w)\,_{c_1c_2}S_{b_1b_2}(u,v)\,\frac{\sigma(p(c_1),p(c_2))}{\sigma(p(a_1),p(c_2))}.$$

The simplest nontrivial solution just as in the usual case (2) is the linear combination of identity and permutation operators

$$_{ab}P_{cd} = \sigma(p(b),p(a))\,\delta_{ad}\delta_{bc}, \tag{17}$$

$$S(u,v) = S(u-v) = (u-v)I + P.$$

References

1. Faddeev,L.D. (1980). Sov. Sci. Rev., Sec. C: Math. Phys. 1, 107-160.

2. Faddeev,L.D. (1984). In: Recent advance in field theory and statistical mechanics, Les Houches, Session 39, 563-608.

3. Kulish,P.P., Sklyanin,E.K. (1982). Lect. Notes Phys. 151, 61-119.

4. Thacker,H.B. (1981). Rev. Mod. Phys. 53, 253-285.

5. Kulish,P.P. (1985). In: Critical Phenomena, eds. V.Ceausescu, G.Costache; Birkhäuser, PPH 11, 351-374.

6. Scheunert,M., Nahm,W., Rittenberg,V. (1977). J. Math. Phys. 18, 155-162.

7. Chaichian,M., Kulish,P. (1978). Phys. Lett. 78B, 413-417.

8. Fordy,A.P., Kulish,P.P. (1983). Comm. Math. Phys. 89, 427-443.

9. Kulish,P.P. (1980). Dokl. Akad. Nauk SSSR 255, 323-327 (Russian),

10. Kulish,P.P., Reshetikhin,N.Yu. (1983). J. Phys.: Math. Gen. A16, L673-681.

11. Olshanetsky,M. (1983). Comm. Math. Phys. 88, 63-76.

12. Saveliev,M.V. (1984). Comm. Math. Phys. 95, 199-216.

13. Ivanov,E., Krivonos,S. (1984). J. Phys.: Math. Gen. A17, L671-675.

14. Leites,D.A., Saveliev,M.V., Serganova,V.V. Preprint IHEP 85-81, Serpukhov, 1985.

15. Andreev,V.A. (1985). Lebedev Inst. Reports, 2,57-60.

16. Popowicz,Z, (1985). Preprint ITP-UWr-85/633.

17. Serganova,V.V. (1983). Func. Analis i ego Pril.

(Funct. Anal. Appl.) 17, 46-54 (Russian).

18. Reshetikhin,N.Yu. (1983). Lett. Math. Phys. 7,205-213.
19. Leites,D.A., Serganova,V.V. (1984). Theor. Math. Fiz. 58, 26-37 (Russian).
20. Kulish,P.P. (1985). Zap. nauch. semin. LOMI, 145, 140-163 (Russian).
21. Kobajashi,Y., Nagamashi,S. (1984). J. Math. Phys. 25, 3367-3374.

LIE SUPERALGEBRA OSP (1/2), NEVEU-SCHWARZ SUPERALGEBRA AND SUPERIZATION OF KORTEVEG-DE VRIES EQUATION

T.G. Khovanova
Moscow State University

Relation of Virasoro and sl(2) algebras with Korteveg de Vries equation (KdV) has been repeatedly discussed in mathematical literature [1,2]. This paper proceeds with the theme. More exactly, we embed the Neveu-Schwarz superalgebra in a Lie superalgebra connected with $osp(1/2)^{\wedge}$, nontrivial central extension of $C^{\infty}(S^1, osp(1/2))$. We use notations of [3,4,5]. The "even part" of this construction related to the Virasoro algebra and $sl(2)^{\wedge}$ has not been probably published.

We perform a supervariant of Hamiltonian reduction [6] following V.G. Drinfeld and V.V. Sokolov [1] and obtain the superisation of KdV proposed by B.A. Kupershmidt [7]. The second Hamiltonian structure of sKdV is obtained reducing the Poisson-Lie (Kirillov) form in the coadjoint representation of $osp(1/2)^{\wedge}$ as in [1]; it is also the Poisson-Lie form in the coadjoint representation of the Neveu-Schwarz superalgebra [8]. Our matrix representation of sKdV is consistent with [7].

I am indebted to I.M. Gelfand, D.R. Lebedev and A.O. Radul for their useful advice and discussions.

1. Notations. Let $spo(2/1)^{\wedge}$ denote the central extension of the Lie superalgebra $C^{\infty}(S^1, spo(2/1))$ (with respect to the point-wise bracket) with coefficients in the ground field k ($\mathbb{R}$ or $\mathbb{C}$) with respect to the cocycle

$$C(X,Y)=\int_{S^1} \mathrm{str}\,(XY')\,dx, \qquad \text{where } Y'=dY/dx.$$

For brevity set $\int f=\int_{S^1} f(x)\,dx$. Recall also that spo(2/1) ≅ osp (1/2) [5].

The non-degenerate scalar product in spo(2/1)^

$$\langle (X,a),(Y,b)\rangle = ab + \int str(XY)$$

allows us to embed the space spo(2/1)^ into its dual. Denote spo (2/1)^* the image of this embedding.

Let A be an arbitrary supercommutative ring with unit over k. Set $\mathcal{G}$ and $\mathcal{G}^*$ for the A-points of spo(2/1)^ and spo(2/1)^* respectively. We will express elements of $\mathcal{G}$ in the form

$$\left(\left(\begin{array}{cc|c} a_1 & a_2 & \alpha_1 \\ a_3 & -a_1 & \alpha_3 \\ \hline -\alpha_2 & \alpha_1 & 0 \end{array}\right), b\right); \; a_i \in C^\infty(S^1, A_0), \alpha_i \in C^\infty(S^1, A_1), b \in A_0.$$

The coadjoint action of spo(2/1)^ determines the $\mathcal{G}$-action on $\mathcal{G}^*$. Since we have a pairing of $\mathcal{G}$ and $\mathcal{G}^*$ with values in A and not in k then $\mathcal{G}^*$ is not contained generally in the dual of $\mathcal{G}$ and $\mathcal{G}$-action on $\mathcal{G}^*$ is not the coadjoint one in the conventional sense. Nevertheless we will denote it ad^*:

$$ad^*_{(X,a)}(Y,b) = (-bX' + [Y,X], 0).$$

2. <u>The reduction scheme of a Poisson manifold with symmetry</u> (cf. [6, Appendix 5]). A smooth manifold M is <u>Poisson</u> if on M there is a homomorphism of the cotangent bundle into the tangent one $I: T^*M \to TM$ such that the operation

$$\{f,g\} = \langle I(df), dg\rangle \text{ for } f,g \in C^\infty(M)$$

makes $C^\infty(M)$ into a Lie algebra.

A <u>Poisson action</u> of a Lie algebra n on M is a k-linear mapping $H: \underline{n} \to C^\infty(M)$ such that

$$H_{[a,b]} = \{H_a, H_b\}.$$

The Poisson action of $\underline{n}$ determines the map $P: M \to \underline{n}^*$ which to a point x of M assigns the functional P_x on $\underline{n}$ such that $P_x(a) = H_a(x)$. The map P is called <u>moment map</u>.

Let $p \in \underline{n}^*$. Set $M_p = P^{-1}(p)$.

Lemma. Elements of $T^*_x M_p$ defining a nonzero functional on $T_x M_p$ coincide with $dH_a(x)$, where $a \in \underline{n}$.

The map $a \mapsto I(dH_a)$ determines an $\underline{n}$-action on M. Denote by $\underline{n}_p$ the stabilizer of M_p in $\underline{n}$, i.e. $a \in n_p$ if and only if $I(dH_a) \in TM_p$.

Comment. It follows from Lemma that elements of n_p define Hamiltonians locally constant on M_p, whose phase flows are tangent to M_p.

Let $N_p = \exp(\underline{n}_p)$, set $F_p = M_p/N_p$ and call F_p the reduced phase space of our system with symmetry.

Let $\pi^*: T^*F_p \to T^*M_p$ and $i^*: T^*M \to T^*M_p$ be induced by the projection $\pi: M_p \to F_p$ and embedding $i: M_p \to M$ respectively.

Lemma. There exists an operator $B: T^*M_p \to T^*M$ such that $i^*B = id$ and $IB\pi^*$ maps T^*F_p into TM_p.

Define a homomorphism $V: T^*F_p \to TF_p$ setting $V = \pi I B \pi^*$.

Theorem. 1) V is well defined, 2) V turns F_p into a Poisson manifold.

In what follows this scheme is applied to A-points of the infinite dimensional supermanifold $spo(2/1)^{\wedge *}$ leaving the details to the reader.

3. Reduced phase space. As M take the Poisson manifold G^* with the Hamiltonian operator ad^*. As $\underline{n}$ take the nilpotent subalgebra of G consisting of elements of the form

$$\left(\left(\begin{array}{cc|c} 0 & 0 & 0 \\ a & 0 & \alpha \\ \hline -\alpha & 0 & 0 \end{array}\right), b\right)$$

$ad^*|\underline{n}$ determines a Poisson $\underline{n}$-action on G^*. Then

$$M_p = \left(\left(\begin{array}{cc|c} a_1 & -1 & 0 \\ a_2 & -a_1 & \beta \\ \hline -\beta & 0 & 0 \end{array}\right), -1\right).$$

Then $\underline{n}_p = \left\{\left(\left(\begin{array}{cc|c} 0 & 0 & 0 \\ a & 0 & 0 \\ \hline 0 & 0 & 0 \end{array}\right), b\right)\right\}$. Notice that $\underline{n}_p$ is a subalgebra $((n_p)_{\bar{1}} = 0)$ in $\underline{n}$.

The standard calculations show that the orbit variety F_p can be embedded into M_p as the set of pairs

(U, -1), where

$$U=\left(\begin{array}{cc|c} 0 & -1 & 0 \\ u & 0 & \xi \\ \hline -\xi & 0 & 0 \end{array}\right).$$

Set $L=\partial+U\ (\partial=\partial_x)$. Then $ad^*_{(X,c)}(U,-1)=([L,X],0)$.

4. <u>Constructing Hamiltonian operator V</u>. Let $R=\left(\begin{array}{cc|c} 0 & 0 & 0 \\ c & 0 & \gamma \\ \hline -\gamma & 0 & 0 \end{array}\right)$ be an element of $T_U F_P$. Determine a functional ℓ on $T_U F_P$ setting $\ell(R)=\int(cy+\gamma\varphi)$, where $y\in C^\infty(S^1,A_0)$, $\varphi\in C^\infty(S^1,A_1)$.

From this functional recover a matrix Y(l) such that $\ell(R)=\langle R,Y\rangle$ and $ad^*_{(Y,b)}(U,-1)=(Y'+[U,Y],0)\in T_U F_P$.

From these conditions we uniquely recover Y :

$$Y=\left(\begin{array}{cc|c} -y'/2 & y & \varphi/2 \\ -y''/2-uy-\xi\varphi/2 & y'/2 & \varphi'/2-\xi y \\ \hline -\varphi'/2+\xi y & \varphi/2 & 0 \end{array}\right)$$

Then $V(Y)=Y'+[U,Y]=\left(\begin{array}{cc|c} 0 & 0 & 0 \\ V_1 & 0 & V_2 \\ \hline -V_2 & 0 & 0 \end{array}\right)$, where

$$\begin{pmatrix} V_1 \\ V_2 \end{pmatrix}=V(u,\xi,-1)\begin{pmatrix} y \\ \varphi \end{pmatrix}=\begin{pmatrix} -\partial^3/2-u\partial-\partial u & -\partial\xi/2-\xi\partial \\ -\partial\xi-\xi\partial/2 & (\partial^2+u)/2 \end{pmatrix}\begin{pmatrix} y \\ \varphi \end{pmatrix}.$$

The Hamiltonian operator $V(u,\xi,-1)$ is obtained by Kupershmidt as an operator for the second Hamiltonian structure of sKdV [7].

5. <u>Lie algebra G_L</u>. In this section we introduce a Lie algebra structure on the space $\mathfrak{F}$ of functionals on F_P depending on $U\in F_P$ as a differential polynomial.

A <u>monomial</u> on F_P is an element

$$\prod_{j=0}^{N}(\partial^j u)^{i_j}(\partial^j\xi)^{k_j},\qquad \text{where } i_j, k_j, N\in\{0,1,2,\dots\}.$$

Denote DPol(U) the space of finite sums of monomials with coefficients of $C^\infty(S^1,A)$ with the natural parity (Z_2-grading).

Define $\mathfrak{F}$ as the space of functionals l on F_p whose coefficients y and φ belong to $DPol(U)_0$ and $DPol(U)_1$ respectively. The entries of the matrix Y(l) also belong to DPol(U). Define the bracket of such matrices setting

$$[Y_1,Y_2]_L=[Y_1,Y_2]+\frac{\partial Y_2}{\partial L}\circ V(Y_1)-\frac{\partial Y_1}{\partial L}\circ V(Y_2)\quad (\text{Cf. }[10,11]).$$

Here $\frac{\partial Y_2}{\partial L}\circ V(Y_1)$ denotes $\frac{d}{d\varepsilon}(Y_2(u+\varepsilon V_1,\xi+\varepsilon V_2))|_{\varepsilon=0}$ for $\varepsilon\in k$, the coefficient-wise action of the vector field $V(Y_1)$ on Y_2.

Statement. 1) $[Y_1,Y_2]_L$ makes $\mathcal{F}$ into a Lie algebra; 2) $V([Y_1,Y_2]_L)=[[V(Y_1),V(Y_2)]]$, where $[[\,,]]$ is the bracket of vector fields on F_p.

Let us express $[Y_1(l_1),Y_2(l_2)]_L$ in terms of l_1 and l_2:

$$[(y_1,\varphi_1),(y_2,\varphi_2)]=(y_1y_2'-y_1'y_2+\varphi_1\varphi_2/2+\frac{\partial y_2}{\partial L}\circ V(Y_1)-\frac{\partial y_1}{\partial L}\circ V(Y_2),$$
$$-y_1'\varphi_2/2+y_1\varphi_2'+y_2'\varphi_1/2-y_2\varphi_1'+\frac{\partial\varphi_2}{\partial L}\circ V(Y_1)-\frac{\partial\varphi_1}{\partial L}\circ V(Y_2).$$

Denote $\mathcal{F}$ with the bracket $[\,,\,]_L$ by G_L.

6. Lie superalgebra K(1). Statement. 1) Pairs (y,φ) independent of U constitute a Lie superalgebra K in G_L.

2) K is isomorphic to the Lie algebra of A-points of K(1) (notations see in [3,4]).

7. Neveu-Schwarz superalgebra (NS).

Lemma. The restriction of the cocycle c(X,Y) onto matrices $Y_1(l_1)$ and $Y_2(l_2)$ determines a cocycle on G_L with values in the nontrivial module $\mathcal{M}=\{\int Q,\,Q\in DPol(U)\}$ (a point of F_p makes a differential polynomial into a function on the circle which is the integrand).

An explicit calculating $c(Y_1,Y_2)$ shows that it is cohomologic to the cocycle $\frac{1}{2}\int(y_1'y_2''+\varphi_1'\varphi_2')$ on G_L. The extension of K with respect to this cocycle constitutes the set of A-points of NS which we will denote by NS(A). Let $(y,\varphi,h)\in NS(A)$, $(u,\xi,-1)\in NS^*(A)$, then

$$ad^*_{(y,\varphi,h)}(u,\xi,-1)=(V_1,V_2,0),\text{ where }\begin{pmatrix}V_1\\V_2\end{pmatrix}=V(u,\xi,-1)\begin{pmatrix}y\\\varphi\end{pmatrix}$$

i.e. the coadjoint NS-action can be obtained reducing the coadjoint action of $spo(2/1)^\wedge$.

8. Constructing conservation laws for sKdV [8]. Set $V^1=V(u+1,\xi,-1)-V(u,\xi,-1)$, $V^2=V(u,\xi,-1)$.

Then V^1 and V^2 are the first and the second Hamiltonian structures of sKdV respectively. An infinite chain of integrals in involution $H_0, H_1, \dots$ is determined from the conditions $V^1(\delta H_0)=0,\ V^1(\delta H_{n+1}) = V^2(\delta H_n)$, where $\delta H_n = \begin{pmatrix} \delta H_n/\delta u \\ \delta H_n/\delta \xi \end{pmatrix}$. Set $H_0 = \int 2u$, then

$H_1 = \int (u^2/2 - 2\xi\xi'),\ H_2 = \int (u^3/4 - (u')^2/8 - 3u\xi\xi' - 2\xi\xi''')$, etc.

9. A superisation of Korteveg de Vries equation (cf. [7]).

Now we automatically write a sKdV:

$$\begin{pmatrix} u \\ \xi \end{pmatrix}_t = V^1(\delta H_2) = V^2(\delta H_1) = \begin{pmatrix} -u'''/2 - 3uu' + 6\xi\xi'' \\ -3u\xi' - 3\xi u'/2 - 2\xi''' \end{pmatrix}.$$

For the Lax representation $L_t = [L, M]$ of sKdV take $L = \partial + U,\ M = \Upsilon(\ell),\ \ell = \delta H_1$. Then sKdV is the compatibility condition for the system

$$L\Psi = 0,\quad \Psi_t = -M\Psi,\quad \text{where}\quad \Psi = \begin{pmatrix} \psi_0 \\ \psi_1 \\ \psi_2 \end{pmatrix}.$$

The scalar differential operator $\hat{L} = \partial^2 + u + \xi\partial^{-1}\xi$ corresponds to the first order differential operator L, namely the equation $L\Psi = 0$ is equivalent to the following system of equations for components of Ψ:

$$\psi_1 = \psi_0',\quad \psi_2 = \partial^{-1}\xi\psi_0,\quad \hat{L}\psi_0 = 0.$$

Equations $L\Psi = 0$ and $\Psi_t = -M\Psi$ allow us to write the evolution in time of ψ_0 and ψ_2 in the following form

$$\psi_{0t} = -3u\psi_0' - 3u'\psi_0/2 - 2\psi_0'''$$

$$\psi_{2t} = -2\xi''\psi_0 - 2\xi\psi_0'' - 3u\xi\psi_0 - 2\xi'\psi_0'.$$

Comment. Consider the coadjoint NS(A)-action on (u, 0, -1). Then $V(u,0,-1) = \begin{pmatrix} -\partial^3/2 - u\partial - \partial u & 0 \\ 0 & (\partial^2+u)/2 \end{pmatrix}$.

The operator V(u, 0, -1) splits into the direct sum of two operators: the Schroedinger operator $(\partial^2+u)/2$ and the operator $-\partial^3/2 - u\partial - \partial u$ of the second Hamiltonian structure for KdV [2]. This in particular implies that for $\xi=0$ equations $V^2\begin{pmatrix} y \\ y \end{pmatrix} = 0$ and $\hat{L}\psi_0 = 0$ determine the same equations for y and ψ_0. (Perhaps that is why evolutions ξ_t and Ψ_{ot} are determined by the same operator and the evolution Ψ_{2t} is symmetric with respect to ξ and Ψ_o .)

10. Miura transformation. An operator $L_1 = \partial + U_1$ is gauge equivalent $L_1 \tilde{g} L$ to L if $L = g^{-1} L_1 g$ for $g \in \exp(\underline{n})$

Statement. 1) $L_1 = \partial + \left(\begin{array}{cc|c} v & -1 & \eta \\ 0 & -v & 0 \\ \hline 0 & \eta & 0 \end{array}\right) \tilde{g}\, L,\ g = \left(\begin{array}{cc|c} 1 & 0 & 0 \\ v & 1 & \eta \\ \hline -\eta & 0 & 1 \end{array}\right).$

2) The coefficients u, ξ of L are expressed in terms of v, η as follows:

$$u = -v^2 + v' - \eta'\eta$$
$$\xi = \eta' - v\eta \tag{1}$$

3) The following scalar differential operator corresponds to L_1:

$$\hat{L}_1 = (\partial - v)(1 - \eta\partial^{-1}\eta)(\partial + v).$$

Comments. 1) Formulas (1) determine the Miura transformation for sKdV [7].

2) Note that unlike the "purely even" situation [1], $\underline{n}_p \neq \underline{n}$ and $L_1 - \partial$ does not belong to M_p.

11. spo(2/1) again. Pass to NS. Let Z be the centre of NS. Let $(y, \varphi, a) \in NS$, $(u, \xi, -1) \in NS^*$ (here $y, \varphi, u, \xi \in C^\infty(S^1), a \in k$). Denote $G(u, \xi, -1)$ the stabilizer of $(u, \xi, -1)$ in the coadjoint representation of NS. The pairs (y, φ) of solutions of the equation $V(u, \xi, -1)\begin{pmatrix} y \\ \varphi \end{pmatrix} = 0$ constitute the Lie superalgebra $G(u, \xi, -1)/z$.

Consider the Lie superalgebra $K_{\mathbb{R}}(1)$ consisting of pairs (y, φ), where $y, \varphi \in C^\infty(\mathbb{R}^1)$. Determine the bracket setting

$$[(y_1, \varphi_1), (y_2, \varphi_2)] = (y_1 y_2' - y_1' y_2 + \varphi_1 \varphi_2/2,\ -y_1' \varphi_2/2 + y_1 \varphi_2' + y_2' \varphi_1/2 - y_2 \varphi_1').$$

Clearly $K(1)$ is embedded into $K_{\mathbb{R}}(1)$. Consider the subalgebra G_0 in $K_{\mathbb{R}}(1)$ consisting of solutions of equation (2) in $K_{\mathbb{R}}(1)$.

Statement. (cf. [9]). $G_0 \cong spo(2|1)$.

References

1. Drinfeld, V.G., Sokolov, V.V. (1984). In: Results of Science and Technique. Modern Problems of Mathematics, 24, VINITI, Moscow, pp. 81-180. (Russian).

2. Kirillov, A.A. (1981). Funct. Anal. Appl., 15, 2, 75-76. (Russian).

3. Leites, D.A. (1983). Supermanifold Theory. Karelia Branch of Acad. of Sci., Petrozavodsk. (Russian).

4. Leites, D.A., Feigin, B.L. (1983). In: Group Theoretical Methods in Physics, 1. Nauka, Moscow, pp. 269-273. (Russian); (1986). Gordon & Breach, 2.

5. Manin, Yu.I. (1984). Gauge Fields and Complex Geometry. Nauka, Moscow. (Russian).

6. Arnold, V.I. (1979). Mathematical Methods of Classical Mechanics. Nauka, Moscow. (Russian).

7. Kupershmidt, B.A. (1984). Phys. Lett. 102A, 5, 213-215.

8. Kupershmidt, B.A. (1985). Phys. Lett. 109A, 9, 417-423.

9. Khovanova, T.G. (1986). Funct. Anal. Appl., 20 (To appear, Russian).

10. Gelfand, I.M., Diky, L.A. (1978). The USSR Acad. of Sci. Institute of Applied Mathematics. Preprint - 136. (Russian).

11. Lebedev, D.R., Manin, Yu.I. (1979). Funct. Anal. Appl. 13, 4, 40-46 (Russian).

12. Reyman, A.G., Semenov-Tien-Shansky, M.A. (1980). Math. USSR. Doklady, 251, N6, 1310-1313.

SUPERSYMMETRIC GENERALIZED TODA LATTICE

V.A. Andreyev
P.N. Lebedev Physical Institute, Moscow

A generalized supersymmetric Toda lattice is described in [1,2] by an algebraic method. Here we use the inverse scattering transformation to investigate this lattice as it is done in [3] with the ordinary Toda lattice.

The integrable Toda lattice is the compatibility condition of the system

$$D_1\Psi=\Big(-D_1\sum_\alpha f_\alpha K_\alpha+\lambda\sum_\alpha e^{\varphi_\alpha}E_\alpha\Big)\Psi, \qquad (1)$$

$$D_2\Psi=\Big(D_2\sum_\alpha f_\alpha K_\alpha+\frac{1}{\lambda}\sum_\alpha e^{\varphi_\alpha}E_{-\alpha}\Big)\Psi \qquad (2)$$

where $D_1=-\frac{\partial}{\partial\theta_2}+i\theta_2\frac{\partial}{\partial x}$, $D_2=\frac{\partial}{\partial\theta_1}+i\theta_1\frac{\partial}{\partial\theta_1}$ and the root vectors $E_{\pm\alpha}$, satisfy the relations (defining K_α):

$$[K_\alpha, E_\beta]=C_{\alpha\beta}E_\beta,\quad [K_\alpha, K_\beta]=0,$$

$$\{E_\alpha, E_{-\beta}\}=\delta_{\alpha\beta}K_\alpha \quad , \quad \text{for} \quad \alpha,\beta=1,\dots,n$$

where $\alpha\in R^+$ and R^+ is a base consisting of odd positive roots of some simple finite dimensional Lie superalgebra [2] . The compatibility of eqs. (1), (2) implies

$$D_1 D_2+D_2 D_1=0 \qquad (3)$$

hence

$$2D_1 D_2 f_\alpha+e^{2\varphi_\alpha}=0\ , \qquad (4)$$

$$\varphi_\alpha + \sum_\beta C_{\alpha\beta} f_\beta = 0, \quad \alpha,\beta = 1,\dots,n \tag{5}$$

Call (4), (5) equations of the first class.

Let the maximal root ω be odd. Then from (1), (2) we get

$$2D_1 D_2 (f_\alpha + a_\alpha f_\omega) + e^{2\varphi_\alpha} + a_\alpha e^{2\varphi_\omega} = 0, \tag{6}$$

$$K_\omega = \sum_{\alpha=1}^{n} a_\alpha K_\alpha, \ \varphi_\alpha + \sum_{\beta=1}^{n+1} C_{\alpha\beta} f_\beta = 0, \quad \alpha = 1,\dots,n. \tag{7}$$

Call (6), (7) equations of the second class.

The bases of simple Lie superalgebras are listed in [2,4].

Instead of the operator corresponding to the maximal root we can use another operator. We demonstrate this with an example. To find the conservation laws of eqs. (6),(7) it is convenient to use another L-A pare instead of (1), (2)

$$D_1 \Psi = \Big(D_1 \sum_\alpha 2 f_\alpha K_\alpha + \lambda \sum_\alpha E_\alpha \Big) \Psi, \tag{8}$$

$$D_2 \Psi = \Big(\frac{1}{\lambda} \sum_\alpha e^{2\varphi_2} E_{-\alpha} \Big) \Psi. \tag{9}$$

An infinite number of local conservation laws exists if the matrix $\mathcal{E} = \sum_{\alpha=1}^{n+1} E_\alpha$ is invertible, so we will verify this for all algebras under consideration.

Let R_0 be the set of even roots and R_1 the set of odd roots.

1) $SU(n|n) \cdot R_0 = \{e_i - e_j,\ i,j = 1,\dots,n \ \ e_p - e_q,$

$p,q = n+1,\dots,2n\};\ R_1 = \{\pm(e_i - e_p)\}$. As basic roots select $e_1 - e_{n+1},\ -e_1 + e_{n+2},\ e_2 - e_{n+2},$

$-e_2 + e_{n+3},\ e_3 - e_{n+3},\ \dots,\ -e_{n-1} + e_{2n},\ e_n - e_{2n}$.

The maximal root is then $e_{-\omega}=-e_n+e_{n+1}$.

The $2n\times 2n$ enlarged Cartan matrix and its Dynkin scheme [2] are:

$$\left\|\begin{matrix} 0 & -1 & 0 & & & 0 & 1 \\ 1 & 0 & -1 & & & & 0 \\ 0 & 1 & 0 & & & & \\ & & & & & & \\ & & & & 0 & -1 & 0 \\ 0 & 0 & & & 1 & 0 & -1 \\ -1 & 0 & & & 0 & 1 & 0 \end{matrix}\right\|$$

A matrix realization of the Cartan-Weyl basis is

$$E_{e_i-e_p}=E_{p,i}\,,\quad E_{-e_i+e_p}=E_{i,p}\,,\quad K_{i,p}=E_{i,i}+E_{p,p}$$

where $E_{i,p}$ is the matrix unit. The matrix $\mathcal{E}$ is invertible. The system corresponding to $SU(n|n)$ is described by the action

$$S=\int dx\,dt\,d\theta_1 d\theta_2\left(\frac{1}{2}\sum_{i=1}^{2n}(-1)^{i+1}\left(u_i D_1 D_2 u_i+e^{u_i-u_{i+1}}\right)\right). \quad (11)$$

2) $B(n\mid n)$. $R_0=\{\pm 2e_i,\pm e_i\pm e_j,\ i,j=1,\dots,n$

$\pm e_p,\pm e_p\pm e_q,\ p,q=n+1,\dots,2n\}$; $R_1=\{\pm e_1,\pm e_i\pm e_p\}$.

As base roots select $e_1,\ -e_1-e_{n+1},\ e_2+e_{n+1},\ \dots,$

$e_n+e_{2n-1},\ -e_n-e_{2n}$.

The maximal root is then $e_{-\omega}=-e_n+e_{2n}$.

A matrix realization of the Cartan-Weyl basis is

$$E_{e_1}=-E_{1,2}+E_{2n+2,1} \qquad E_{-e_1}=E_{21}+E_{1,2n+2}$$

$$E_{-e_1-e_{n+1}}=E_{2,3n+2}+E_{n+2,2n+2} \qquad E_{e_1+e_{n+1}}=E_{3n+2,2}-E_{2n+2,n+2}$$

$$E_{e_2+e_{n+1}}=E_{2n+3,n+2}-E_{3n+2,3} \qquad E_{-e_2-e_{n+1}}=E_{n+2,2n+3}+E_{3,3n+2}$$

$E_{-e_2-e_{n+2}}=E_{3,3n+3}+E_{n+3,2n+3}$ $\quad E_{e_2+e_{n+2}}=E_{3n+3,3}-E_{2n+3,n+3}$

. . . .

. . . .

$E_{-e_{n-1}-e_{2n-1}}=E_{n,4n}+E_{2n,3n}$ $\quad E_{e_{n-1}+e_{2n-1}}=E_{4n,n}-E_{3n,2n}$

$E_{e_n+e_{2n-1}}=E_{3n+1,2n}-E_{4n,n+1}$ $\quad E_{-e_n-e_{2n-1}}=E_{2n,3n+1}+E_{n+1,4n}$

$E_{-e_n-e_{2n}}=E_{n+1,4n+1}+E_{2n+1,3n+1}$ $\quad E_{e_n+e_{2n}}=E_{4n+1,n+1}-E_{3n+1,2n+1}$

The maximal root operator is $E_{-e_n+e_{2n}}=E_{n+1,2n+1}+E_{4n+1,3n+1}$.

Clearly in this case ε is degenerate and one cannot construct an integrable system. Add to the basis operators $E_{e_1},\dots,E_{-e_n-e_{2n}}$ the odd operator $E_{-\delta}=$ $=E_{4n+1,4n+2}+E_{4n+2,2n+1}$ instead of $E_{-e_n+e_{2n}}$. The enlarged matrix ε is invertible. The operator $E_{-\delta}$ does not belong to $B(n|n)$ and together they generate $OSP^{(2)}(2n|2n)$ with the Dynkin diagram

The system corresponding to this algebra is described by the action •⟸⊗ – … – ⊗⟹•

$$S=\int dx\,dt\,d\theta_1\,d\theta_2\Big(\frac{1}{2}\sum_{i=1}^{2n}(-1)^i u_i D_1 D_2 u_i+ \qquad \tag{12}$$
$$+e^{-u_1}+e^{-u_{2n}}+\sum_{i=1}^{2n-1}\exp\big((-1)^{i+1}(u_i+u_{i+1})\big)\Big).$$

For $B(n|n+1)$ we similarly construct a system with the action

$$S=\int dx\,dt\,d\theta_1\,d\theta_2\Big(\frac{1}{2}\sum_{i=1}^{2n+1}(-1)^i u_i D_1 D_2 u_i+e^{-u_1}+$$
$$+e^{u_{2n+1}}+\sum_{i=1}^{2n}\exp\big((-1)^{i+1}(u_i+u_{i+1})\big)\Big).$$

3) $D(2,1,\alpha)$. $R_0=\{\pm 2e_i,\ i=1,2,3\}$. $R_1=\{\pm e_1\pm e_2\pm e_3\}$. As basic roots select $e_1+e_2+e_3, e_1-e_2-e_3, -e_1-e_2+e_3$. Then the maximal root is $e_{-\omega}=-e_1+e_2-e_3$

and $\mathcal{E}$ is degenerate and we cannot construct an integrable system.

4) Now discuss any other possibility to construct integrable supersymmetric equations. Consider $SU(n|n+1)$. As basis roots select $e_1-e_{n+1}, -e_1+e_{n+2}, -e_2+e_{n+3}, \ldots$ $\ldots, e_n-e_{2n}$ and the maximal root $e_{-\omega}=e_{n+1}-e_{2n+1}$ is even. The matrix realization of the Cartan-Weyl basis is

$$E_{e_i-e_p}=E_{p,i},\ E_{-e_i+e_p}=E_{i,p},\ K_{e_i-e_p}=E_{i,i}+E_{p,p}.$$

Instead of the maximal root operator $E_{e_{n+1}-e_{2n+1}}=$ $=E_{2n+1,n+1}$ consider the odd operator $E_{-\delta}=$ $=-E_{2n+1,2n+2}+E_{2n+2,n+1}$.

Instead of the L-A pairs (1), (2) or (9), (10) introduce new L-A pairs

$$D_1\Psi=\Big(-D_1\sum_{\alpha=1}^{2n+1} f_\alpha H_\alpha+\lambda\sum_\alpha e^{\varphi_\alpha}E_\alpha\Big)\Psi, \tag{14}$$

$$D_2\Psi=\Big(D_2\sum_{\alpha=1}^{2n+1} f_\alpha H_\alpha+\frac{1}{\lambda}\sum_\alpha e^{\varphi_\alpha}E_{-\alpha}\Big)\Psi \tag{15}$$

and

$$D_1\Psi=\Big(-D_1\sum_\alpha 2f_\alpha H_\alpha+\lambda\sum_\alpha E_\alpha\Big)\Psi,$$

$$D_2\Psi=\Big(\frac{1}{\lambda}\sum_\alpha e^{2\varphi_\alpha}E_{-\alpha}\Big)\Psi,\quad \varphi_{2n}=\varphi_1+i\pi$$

where $H_1-K_1-E_{2n+1,2n+1}, H_2=K_2,\ldots,H_{2n-1}=K_{2n-1}$, $H_{2n}=K_{2n}-E_{n+1,n+1},\ H_{2n+1}=-E_{2n+1,2n+1}+E_{n+1,n+1}$.

Instead of the Cartan matrix introduce the matrix $H=(h_{\alpha\beta})$, where $[H_\alpha, E_\beta]=h_{\alpha\beta}E_\beta$:

$$H=\begin{Vmatrix} 0 & 1 & 0 & & & & & & 0 & 1\\ -1 & 0 & 1 & 0 & & & & & & 0\\ 0 & -1 & 0 & 1 & 0 & & & & & \\ & & & & & & & & & \\ 0 & & & & & -1 & 0 & 1 & 0 & \\ -1 & 0 & & & & & -1 & 0 & 1 & \\ 1 & 0 & & & & & 0 & 1 & -1 & \end{Vmatrix}$$

And instead of the Dynkin diagram the coupling diagram characterizing the operators of the L-A pairs and the system of equations: ⊠— · · · —⊠ (joined to ■)

An odd operator E_α corresponds to a square ⊠ and the operator $E_{-\delta}$ corresponds to the square ■. The number of lines between them equals the number of nontrivial commutators $\{E_\alpha, E_{-\delta}\}, [E_\alpha, \{E_\alpha, E_{-\delta}\}], \ldots$ For $SU(n|n+1)$ the matrix is invertible and we get the system

$$D_1 D_2 u_1 + e^{u_1} + e^{-u_1+u_2} = 0,$$

$$D_1 D_2 u_2 + e^{-u_1+u_2} + e^{-u_2+u_3} = 0,$$

$$D_1 D_2 u_3 + e^{-u_2+u_3} + e^{-u_3+u_4} = 0,$$

$$\cdots\cdots\cdots\cdots\cdots$$

$$\cdots\cdots\cdots\cdots\cdots$$

$$D_1 D_2 u_{n-2} + e^{-u_{n-3}+u_{n-2}} + e^{-u_{n-2}+u_{n-1}} = 0,$$

$$D_1 D_2 u_{n-1} + e^{-u_{n-2}+u_{n-1}} + (1)^n e^{-u_{n-1}} = 0.$$

We have enumerated all the algebras corresponding to the integrable supersymmetric Toda lattice of the second class. In [2,4] other superalgebras with odd bases and odd maximal root, (say, $D(n|n+1)$) are given but for them $\mathcal{E}$ is degenerate and we cannot construct an integrable system.

To find conservation laws of the eqs. (11)-(15), diagonalize the matrix $\mathcal{E} = \sum_\alpha E_\alpha$. Some components of the column ψ are even and some are odd; we must consider the ratio $\gamma^{i,j} = \psi^i / \psi^j$, where ψ^j is even. The quantities $\gamma^{i,j}$ satisfy the Riccaty equations. From these equations we can find in the usual way [5] an infinite number of equalities $D_1 P^j_k = D_2 R^j_k$, $k = 1, 2, \ldots$, that can be transformed into ordinary conservation laws [6,7] :

$$\frac{\partial}{\partial x}\left(\frac{\partial}{\partial \theta_1}\frac{\partial}{\partial \theta_2}\theta_2 P^j_k\right)=\frac{\partial}{\partial t}\left(\frac{\partial}{\partial \theta_1}\frac{\partial}{\partial \theta_2}\theta_1 R^j_k\right),\quad k=1,2,3,\ldots$$

The author is thankful to V.P. Karasev and V.N. Tolstoy for many informative discussions.

References

1. Saveliev, M.V. (1984). Theor. and Math. Phys., 59, 3, 367.
2. Leites D.A., Saveliev M.V., Serganova V.V. Lie superalgebra osp(N,2)... (this issue).
3. Mikhailov, A.V., Olshnetsky, M.A., Perelomov, A.M. (1981). Commun. Math. Phys., 79, 473.
4. Serganova, V.V. (1984). Math. USSR Izvestia. 48, 3, 585. (Russian).
5. Alberty, J.M., Koikawa, T., Sasaki, R. (1982). - Physica, 5D, 43.
6. Vladimirov, V.S. Volovich, I.V. (1985). Theor. Math. Phys., 62, No.1, 3. (Russian).
7. Volovich, I.V. (1985). Theor. Math. Phys., 63, No.2, 313. (Russian).

EXTREMAL PROJECTIONS AND REDUCED SUPERALGEBRAS OVER FINITE-DIMENSIONAL CONTRAGREDIENT LIE SUPERALGEBRAS

V.N. Tolstoy

Institute of Nuclear Physics, Moscow State University, Moscow 119899

The extremal projections for reductive Lie algebras [1,2] give an effective method for solving many problems of the representation theory for these algebras [3-5] and so do their super counterparts. One example of application of these projections (for $osp(1/2)$ is considered in [6]).

Here we explicitly describe the extremal projections for all the finite-dimensional contragredient complex Lie superalgebras [9] and give a complete description (in terms of generators and relations) of certain important reduced (also known as step, shift, S-, transfer,...) superalgebras (Z -superalgebras) over $gl(n|m)$ and $osp(M|2n)$.

Note that reduced superalgebras make it possible to construct bases of cyclic representations effectively, see e.g. [10].

1. Ordered root system. Let $g = g^0 \oplus g^1$ be a finite-dimensional contragredient Lie superalgebra [7]. h a Cartan subalgebra, R_+ a positive root system of the pair (g, h) . Order R_+ so that each composite root lay between its components. Such an ordering is called normal [8]. Obviously, simple roots and their multiples must lay at the ends of any normal ordered system R_+.

We say that a root γ is generated by roots α and β if it can be presented as a linear combination of α and β. Let $\{\alpha, \beta\}_+$ be the set of all the positive roots geterated by $\alpha, \beta \in R_+$. If, in a normal ordered system, all the roots of $\{\alpha, \beta\}_+$ are each other's neighbours, then placing the roots of $\{\alpha, \beta\}_+$ in the inverse order we again get a normal ordered system R_+. Such a transformation will be called an elementary transposition.

Lemma1. A normal ordering in R_+ exists for an arbitrary arrangement of simple roots. Any two normal orderings can be transformed into each other by transpositions of the form:

$$\begin{aligned}
&\alpha, \beta \rightleftarrows \beta, \alpha; \\
&\alpha, 2\alpha \rightleftarrows 2\alpha, \alpha; \\
&\alpha, \alpha+\beta, \beta \rightleftarrows \beta, \alpha+\beta, \alpha; \\
&\alpha, \alpha+\beta, \alpha+2\beta, \beta \rightleftarrows \beta, \alpha+2\beta, \alpha+\beta, \alpha; \\
&\alpha, 2\alpha, 2\alpha+\beta, \alpha+\beta, \beta \rightleftarrows \beta, \alpha+\beta, 2\alpha+\beta, 2\alpha, \alpha; \qquad (1.1) \\
&\alpha, \alpha+\beta, 2\alpha+2\beta, \alpha+2\beta, \beta \rightleftarrows \beta, \alpha+2\beta, 2\alpha+2\beta, \alpha+\beta, \alpha; \\
&\alpha, \alpha+\beta, 2\alpha+2\beta, \alpha+2\beta, 2\beta, \beta \rightleftarrows \beta, 2\beta, \alpha+2\beta, 2\alpha+2\beta, \alpha+\beta, \alpha; \\
&\alpha, \alpha+\beta, 2\alpha+3\beta, \alpha+3\beta, \alpha+2\beta, \beta \rightleftarrows \beta, \alpha+2\beta, \alpha+3\beta, 2\alpha+3\beta, \alpha+\beta, \alpha,
\end{aligned}$$

where $\alpha-\beta$ is not a root.

Proof is like that in [2].

2. Extremal projections. Let $U(\underline{g}) = U^0(\underline{g}) \oplus U^1(\underline{g})$ be the universal enveloping superalgebra of a Lie superalgebra $\underline{g}$. (The parity in $U(\underline{g})$ is induced by the parity in $\underline{g}$).

First, we extend $U(\underline{g})$ to the superalgebra

$$\tilde{U}(\underline{g}) = U(\underline{g}) \otimes_{U(\underline{h})} R(\underline{h}) \qquad (2.1)$$

where $R(\underline{h})$ is the field of fractions of the commutative algebra $U(\underline{h})$. The superalgebra $\tilde{U}(\underline{g})$ is a free $R(\underline{h})$ -module, with basis consisting of monomials of the form

$$e_{-\alpha_1'}^{r_1} \dots e_{-\alpha_n'}^{r_n} e_{-\alpha_1}^{k_1} \dots e_{-\alpha_m}^{k_m} e_{\alpha_m}^{q_m} \dots e_{\alpha_1}^{q_1} e_{\alpha_n'}^{s_n} \dots e_{\alpha_1'}^{s_1}, \quad (2.2)$$

where $R_+^0 = \{\alpha_1, \alpha_2, \dots, \alpha_m\}$ is the set of positive (even) roots of the pair $(\underline{g}^0, \underline{h})$; $R_+^1 = \{\alpha_1', \alpha_2', \dots, \alpha_n'\}$ is the set of positive (odd) weights of $\underline{g}^1$, e_α denotes a root vector corresponding to root α, where $r_i, s_i = 0,1$. The monomials (2.2) are called even (odd) if so is $\sum (s_i + r_i)$; they form a basis in $\tilde{U}^0(\underline{g})$ ($\tilde{U}^1(\underline{g})$). The superspace $\tilde{U}(\underline{g})$ is the direct sum of weight subspaces $U_\mu(\underline{g})$.

Further, we extend each $\tilde{U}_\mu(\underline{g})$ to the vector space $\mathfrak{F}_\mu(\underline{g})$ of all the formal series in the monomials (2.2) with coefficients in R(h) and fixed weight $\mu \in \underline{h}^*$. Finally, set

$$\mathfrak{F}(\underline{g}) = \bigoplus_{\mu} \mathfrak{F}_\mu^0(\underline{g}) \bigoplus_{\mu'} \mathfrak{F}_{\mu'}^1(\underline{g}). \quad (2.3)$$

Now set

$$M(\underline{g}) = \tilde{U}(\underline{g}) / \tilde{U}(\underline{g})\underline{g}_+ \quad (2.4)$$

where $\underline{g}_+$ is the linear hull of the root vectors e_α, $\alpha \in R_+$. Clearly, $M(\underline{g})$ is invariant with respect to the multiplication of its elements from the left by elements of $\mathfrak{F}(\underline{g})$, i.e. the elements of $\mathfrak{F}(\underline{g})$ can be interpreted as linear operators on $M(\underline{g})$.

Theorem 1. The vector space $\mathfrak{F}(\underline{g})$ is an associative superalgebra with respect to multiplication of formal series. The left $\mathfrak{F}(\underline{g})$ -module $M(\underline{g})$ is exact and simple.

This theorem is an analogue of Theorem 1 of [3]. In what follows we take the fixed normalization of the even root vectors $e_{\pm\alpha}$, $\alpha \in R_+^0$ given by

$$[e_\alpha, e_{-\alpha}] = h_\alpha, \quad \alpha(h_\alpha) = 2, \quad \alpha \in R_+^0. \quad (2.5)$$

For each $\alpha \in R_+$ take the following even element of $\mathfrak{F}(\underline{g})$:

$$p_\alpha = \sum_{k=0}^{\infty} \frac{(-1)^k}{k!} \varphi^{-1}_{\alpha,k} e^k_{-\alpha} e^k_\alpha \qquad \text{for} \quad \alpha \in R^0_+ ,$$
$$p_\alpha = 1 - \varphi^{-1}_{\alpha,1} e_{-\alpha} e_\alpha \qquad \text{for} \quad \alpha \in R^1_+ . \tag{2.6}$$

Here

$$\varphi_{\alpha,k} = \prod_{i=1}^{k} (\varphi_\alpha + i),$$

$$\varphi_\alpha = h_\alpha + \rho(h_\alpha) + \delta_\alpha \qquad \text{for} \quad \alpha \in R^0_+ ,$$
$$\varphi_\alpha = h_\alpha + \rho(h_\alpha) + \frac{1}{2}\alpha(h_\alpha) - 1 \quad \text{for} \quad \alpha \in R^1_+ , \tag{2.7}$$

where $\delta_\alpha = 1/2$ is $\alpha/2 \in R^1_+$ and $\delta_\alpha = 0$ if $\alpha/2 \notin R^1_+$; ρ is the difference between the half-sum ρ^0 of the positive even roots and the half-sum ρ^1 of the positive odd roots.

Let $p = p(\mathfrak{g})$ be the product of all p_α arranged in accordance with the fixed ordering of R_+. It can easily be verified that $p(\mathfrak{g}) \in \mathcal{F}(\mathfrak{g})$.

Theorem 2. The element $p(\mathfrak{g})$ is the unique up to a factor of $R(\mathfrak{h})$ element of $\mathcal{F}(\mathfrak{g})$ such that

$$e_\alpha p = p e_{-\alpha} = 0 \qquad \text{for all} \quad \alpha \in R_+ . \tag{2.8}$$

In particular, $p(\mathfrak{g})$ is independent of the choice of normal ordering in R_+ and satisfies

$$p^2 = p. \tag{2.9}$$

Proof. If $\mathfrak{g}$ is a Lie algebra or superalgebra of rank $\leqslant 2$ the condition (2.8) can be verified immediately. Further, we can show that an element of $\mathcal{F}(\mathfrak{g})$ which is a normal ordered product of p_γ's, $\gamma \in \{\alpha, \beta\}_+$, is independent of the choice of normal ordering in $\{\alpha,\beta\}_+$. Next, using Lemma 1 we get $p_\alpha p = p p_\alpha = p$ for all simple roots $\alpha \in R_+$; this implies (2.8) in the general case. Finally if $p' \in \mathcal{F}(\mathfrak{g})$ satisfies (2.8) and the constant term of p' is 1, then $(p' - p) M(\mathfrak{g}) = 0$, therefore $\rho' = \rho$.

Remark. Replacing all the odd root vectors, δ_α and ς^1 in $\underline{p}(\underline{g})$ by zero we obtain the extremal projection $p(\underline{g}^\circ)$ for the Lie algebra $\underline{g}^\circ$ [2].

Now we apply the extremal projection to description of reduced superalgebras over contragredient Lie superalgebras.

3. Reduced superalgebras. The formal construction of reduced algebras over reductive Lie algebras [3,4] can obviously be transferred to the case of Lie superalgebras.

Let $\underline{g}$ be an arbitrary finite-dimensional Lie superalgebra, $\underline{g}'$ a contragredient supersubalgebra reductive in $\underline{g}$, $\underline{h}'$ a Cartan subalgebra of $\underline{g}'$, R'_+ a positive root system of the pair $(\underline{g}', \underline{h}')$, $\underline{g}'_+$ is the linear hull of e_α for $\alpha \in R'_+$ and $\tilde{U}'(\underline{g})$ the extension of the universal enveloping superalgebra $U(\underline{g})$ over the field $R(\underline{h}')$. Set

$$M(\underline{g},\underline{g}') = \tilde{U}'(\underline{g}) / \tilde{U}'(\underline{g})\underline{g}_+ . \tag{3.1}$$

with the natural $\mathcal{F}(\underline{g}')$ module structure. Further, set

$$Z(\underline{g},\underline{g}') = \{\xi \in M(\underline{g},\underline{g}') : \underline{g}'_+ \xi = 0\}. \tag{3.2}$$

On the other hand, consider the subsuperalgebra

$$T(\underline{g},\underline{g}') = \{x \in \tilde{U}'(\underline{g}) : \underline{g}'_+ x \subset \tilde{U}'(\underline{g})\underline{g}'_+\} \subset \tilde{U}'(\underline{g}). \tag{3.3}$$

Obviously,

$$Z(\underline{g},\underline{g}') \simeq T(\underline{g},\underline{g}') \,|\, \tilde{U}'(\underline{g})\underline{g}'_+ . \tag{3.4}$$

Since $\tilde{U}(\underline{g})\,\underline{g}'_+$ is a two-sided ideal in $T(\underline{g},\underline{g}')$, then $Z(\underline{g},\underline{g}')$ is the quotient superalgebra modulo $\tilde{U}(\underline{g})\underline{g}'_+$.

Thus, the $\mathbb{Z}_2$-graded vector space (3.2) is a superalgebra (with respect to the multiplication induced in $\tilde{U}'(\underline{g})$. At the same time, $Z(\underline{g},\underline{g}')$ is a free

$R(\underline{h}')$ -bimodule whose basis is composed of weight vectors with respect to $\underline{h}'$. According to Theorem 2, the reduced superalgebra $Z(\underline{g},\underline{g}')$ can be characterized as the image of the extremal projection:

$$Z(g,g') = p(g')M(g,g'). \tag{3.5}$$

Here $Z(g,g')$ is generated by

$$z_i = pe_i, \quad i = 1,2,\dots,N, \tag{3.6}$$

where $\{e_i, 1 \leqslant i \leqslant N\}$ is a weight basis for the complementary $\underline{g}'$-module $\underline{g}'' = \underline{g} \ominus \underline{g}'$. Alternatively, $Z(\underline{g},\underline{g}')$ is the linear hull over $R(\underline{h})$ of the monomials in the generators (3.6).

The importance of the reduced superalgebra $Z(\underline{g},\underline{g}')$ in the representation theory of $\underline{g}$ is illustrated as follows. Let V be a representation space for $\underline{g}$ with cyclic vector ξ (here $\dim U(\underline{g}')\xi < \infty$) and V_+ the subspace of highest vectors with respect to $\underline{g}'$. Then,

$$V_+ = Z(\underline{g},\underline{g}')\xi. \tag{3.7}$$

Thus, the reduced superalgebra $Z(\underline{g},\underline{g}')$ is relevant to the construction of representation bases of the Lie superalgebra $\underline{g}$ in the reduction $\underline{g}\downarrow\underline{g}'$.

Now we give a complete description (in terms of generators and relations) of most important reduced superalgebras $Z(g,g')$ for $gl(n|m)$ and $osp(M|2n)$.

4. <u>The reduced superalgebra $AZ(n|m)$</u>. Let e_{ij} be the $(n+m)\times(n+m)$ matrix with 1 is in the (i,j) -th place and other entries 0. The supercommutator for such matrices be

$$[e_{ij}, e_{\ell,k}]_s := e_{ij}e_{\ell k} - (-1)^{\theta_{ij}\theta_{\ell k}} e_{\ell k}e_{ij}, \tag{4.1}$$

$$\theta_{ij} = \begin{cases} 0 & \text{if either } i,\ j \leqslant n \text{ or } i,j > n \\ 1 & \text{otherwise} \end{cases} \tag{4.2}$$

The elements e_{ij} with the supercommutator (4.1) generate the Lie superalgebra $gl(n|m)$. The basis e_{ij}, $i,j = 1,2,\dots,n+m$, is called standard. Clearly

$$[e_{ij}, e_{\ell k}]_s = \delta_{j\ell} e_{ik} - (-1)^{\theta_{ij}\theta_{\ell k}} \delta_{ik} e_{\ell j}, \tag{4.3}$$

where δ_{ij} is the Kronecker symbol. The elements $h_i = e_{ii}$, $i = 1,2,\dots, n+m$, form a basis in the Cartan subalgebra $\underline{h} \subset gl(n \mid m)$. The corresponding dual basis in $\underline{h}^*$ will be denoted as $\{\varepsilon_i\}$ with $\varepsilon_i(h_j) = \delta_{ij}$; ordering of these vectors is set by $\varepsilon_i > \varepsilon_j > 0$ for $i<j$. Then the root vectors e_{ij} for $i<j$ $(i>j)$ correspond to positive (negative) roots. The normalizing condition (2.5) is satisfied for all the even root vectors e_{ij}, where $i,j \leqslant n$ or $i,j > n$. The half-sum of the positive even roots minus the half-sum of the positive odd roots is

$$\rho = \frac{1}{2}\left\{\sum_{i=1}^{n}(n-m-2i+1)\varepsilon_i + \sum_{j=n+1}^{n+m}(3n+m-2j+1)\varepsilon_j\right\}. \tag{4.4}$$

The coefficients φ_α (2.7) in the elements p_α:

$$p_{ij} = \sum_{k=0} \frac{(-1)^k}{k!} \varphi^{-1}_{ij,k}\, e^k_{ji}\, e^k_{ij}, \tag{4.5}$$

(where $i<j$) take the form

$$\varphi_{ij} = H_i - (-1)^{\theta_{ij}} H_j, \quad H_i = h_i + n - i. \tag{4.6}$$

The upper bound for k in the r.h.s. of (4.5) is 1 if $\theta_{ij} = 1$ and ∞ if $\theta_{ij} = 0$.

Set

$$AZ(n|m) = Z(gl(n|m+1), \hat{g}l(n|m)), \tag{4.7}$$

where $\hat{g}l(n|m)$ is $gl(n|m) \oplus \mathbb{C}\cdot h_{n+m+1}$. The reduced superalgebra $AZ(n|m)$ is generated by

$$z_i = pe_i, \quad i = \pm 1, \pm 2, \dots, \pm(n+m), \tag{4.8}$$

where $e_i = e_{i,n+m+1}$, $e_{-i} = e_{n+m+1,i}$, $1 \leqslant i \leqslant n+m$. All the generators z_i are the weight ones with respect to the Cartan subalgebra of $gl(n|m+1)$:

$$[h_i, z_{\pm j}]_- = \pm \delta_{ij} z_{\pm j}, \quad i = 1,2,\ldots, n+m,$$
$$[h_{n+m+1}, z_{\pm j}]_- = \mp z_{\pm j} \quad \text{for all} \quad 1 \leq j \leq n+m. \tag{4.9}$$

Theorem 3. $AZ(n|m)$ is given by the generators (4.8) and the relations

$$z_i z_j = \alpha_{ij} z_j z_i \quad \text{for} \quad i+j \neq 0, \; i,j = \pm 1,\ldots, \pm(n+m), \tag{4.10}$$

$$z_i z_{-i} = \sum_{j=1}^{n+m} \beta_{ij} z_{-j} z_j + \gamma_i \quad \text{for} \quad i = 1,2,\ldots, n+m \tag{4.11}$$

to which one should add the weight relations (4.9). Here $N = n+m+1$ and

$$\alpha_{ij} = (-1)^{\theta_{iN}\theta_{jN}} \quad \text{if} \quad \operatorname{sgn} i \neq \operatorname{sgn} j, \tag{4.12}$$

$$\alpha_{ij} = (-1)^{\theta_{iN}\theta_{jN}} (\varphi_{ij} + (-1)^{\theta_{iN}}) / \varphi_{ij} \quad \text{for} \quad 1 \leq i < j \leq n+m, \tag{4.13}$$

$$\alpha_{-j-i} = (-1)^{\theta_{iN}\theta_{jN}} \frac{\varphi_{ij} + 2\theta_{ij} + (-1)^{\theta_{iN}}}{\varphi_{ij} + 2\theta_{ij}} \quad \text{for} \quad 1 \leq i < j \leq n+m, \tag{4.14}$$

$$\beta_{ij} = (-1)^{\theta_{iN}} (\varphi_{ji} + 1)^{-1} \delta_i^- \delta_j^+, \tag{4.15}$$

$$\gamma_i = (-1)^{\theta_{iN}+1} (\varphi_{Ni} + 1) \delta_i^-, \tag{4.16}$$

where

$$\delta_i^- = \prod_{k=i+1}^{n+m} \frac{\varphi_{ki} + 1}{\varphi_{ki}}, \qquad \delta_j^+ = \prod_{k=j+1}^{n+m} \frac{\varphi_{jk} + 1}{\varphi_{jk} + 2\theta_{jk}}. \tag{4.17}$$

Remarks. (1) The matrix $\beta = (\beta_{ij})$ is non-singular. Hence the elements of the form $z_{-j} z_j$ may be expressed unambiguously via $z_i z_{-i}$.

(2) The elements z_i, $i = \pm(n+1), \pm(n+2), \ldots, \pm(n+m)$ for $n = 0$ generate the reduced algebra AZ_m (see [4]).

(3) Unlike the AZ_m case, the elements (4.8) cannot be renormalized so that they commute for $i+j\neq 0$.

The following equations are useful

$$z_{\pm i}^{k} = p e_{\pm i}^{k}, \quad i = 1,2,\dots, n+m. \tag{4.18}$$

This and (4.3) imply, in particular, that

$$z_{\pm i}^{2} = 0, \qquad \text{for} \quad i = \pm 1, \pm 2, \dots, \pm n. \tag{4.19}$$

Moreover,

$$z_{n+m}^{k_{n+m}} z_{n+m-1}^{k_{n+m-1}} \cdots z_{1}^{k_1} = p e_{n+m}^{k_{n+m}} e_{n+m-1}^{k_{n+m-1}} \cdots e_{1}^{k_1}, \tag{4.20}$$

$$z_{-1}^{k_1} z_{-2}^{k_2} \cdots z_{-n-m}^{k_{n+m}} = p e_{-1}^{k_1} e_{-2}^{k_2} \cdots e_{-n-m}^{k_{n+m}}. \tag{4.21}$$

The elements (4.21) are used to construct directly a basis of irreducible representations for $gl(n|m)$ (see [9]).

5. <u>The reduced superalgebra $BZ(m|n)$</u>. Let e_{ij} be the standard basis of $gl(2n|M)$ where $i,j=(0), \pm 1, \pm 2, \dots, \pm(n+m)$ $(m = [M/2]$ and the index 0 is used only for odd M) and the supercommutation relations

$$[e_{ij} e_{\ell k}]_s = e_{ij} e_{\ell k} - (-1)^{\theta_{ij}\theta_{\ell k}} e_{\ell k} e_{ij} = \\ = \delta_{j\ell} e_{ik} - (-1)^{\theta_{ij}\theta_{\ell k}} \delta_{ik} e_{\ell j}, \tag{5.1}$$

where

$$\theta_{ij} = \begin{cases} 0 & \text{if either } i,j = \pm 1, \pm 2, \dots, \pm n \text{ or } i,j = (0), \pm(n+1), \dots, \pm(n+m) \\ 1 & \text{otherwise}. \end{cases}$$

The Lie superalgebra $osp(M\,|\,2n)$ is embedded in $gl(2n\,|\,M)$ as the linear hull of the following elements:

(a) the basis of the symplectic Lie algebra $sp(2n)$:

$$\bar{e}_{ij} = e_{i-j} + e_{j-i}\,\mathrm{sgn}\, j, \quad \bar{e}_{-j-i} = e_{-ji} + e_{-ij}\,\mathrm{sgn}\, j \\ e_{kk} = e_{k-k}, \qquad \bar{e}_{-k-k} = e_{-kk}, \quad h_k = e_{kk} - e_{-k-k} \tag{5.2}$$

for $1 \leq i < |j| \leq n$, $k = 1,2,\dots,n$;

(b) The basis of the orthogonal Lie algebra $O(M)$:

$$\bar{e}_{ij} = e_{i-j} - e_{j-i}, \quad \bar{e}_{-j-i} = e_{-ji} - e_{-ij}, \tag{5.3}$$

$\bar{e}_{k0} = \sqrt{2}(e_{k0} - e_{0-k})$, $e_{0-k} = \sqrt{2}(e_{0k} - e_{-k0})$ (for odd M),

$h_k = e_{kk} - e_{-k-k}$, for $n+1 \leq i < |j| \leq n+m$, $k = n+1, n+2, \dots, n+m$;

(c) The odd basis elements:

$$\bar{e}_{ij} = e_{i-j} - e_{j-i}, \quad \bar{e}_{j-i} = e_{-ji} - e_{-ij}, \tag{5.4}$$

$\bar{e}_{k0} = \sqrt{2}(e_{k0} - e_{0-k})$, $\bar{e}_{0-k} = \sqrt{2}(e_{0k} + e_{-k0})$ (for odd M)

for $1 \leq i \leq n < |j| \leq n+m$, $k = 1,2,\dots,n$.

Remark. For $M = 1$ the elements (5.3) are absent, while the odd elements are e_{k0}, e_{0-k}, $k = 1,2,\dots,n$. For $M = 2$ only h_{n+1} enters (5.3).

The elements h_i, $i = 1,2,\dots,n+m$, generate the Cartan subalgebra $\underline{h} \subset osp(M \mid 2n)$. The ordering in the dual basis ε_i, $i = 1,2,\dots,n+m$, with $\varepsilon_i(h_j) = \delta_{ij}$ is fixed by the same condition as above. The root vectors of the left-hand and right-hand columns of (5.2)-(5.4) correspond to the positive and negative roots, respectively. The normalizing condition (2.5) is satisfied by all the even root vectors. The half-sum of the positive even roots minus the half-sum of the positive odd roots is

$$\rho = \sum_{i=1}^{n}(n+1-\tfrac{1}{2}M-i)\varepsilon_i + \sum_{j=n+1}^{n+m}(n+\tfrac{1}{2}M-j)\varepsilon_j . \tag{5.5}$$

The coefficients φ_α in the elements p_α:

$$p_{i\pm j} = \sum_{k=0} \frac{(-1)^k}{k!}\varphi^{-1}_{i\pm j;k}\, e^k_{\pm j-i}\, e^k_{i\pm j} \tag{5.6}$$

take the form

$$\varphi_{i\pm j} = H_i \pm (-1)^{\theta_{ij}} H_j - \theta_{ij} \qquad \text{for } 0 < i < j \leq n+m$$

$$\varphi_{ii} = H_i + x_M \qquad \text{for } i = 1,2,\dots,n,$$

$$\varphi_{i0} = 2H_i, \quad i = 1,2,\dots,n+m, \tag{5.7}$$

where $x_M = 0$ for even M and $x_M = \frac{1}{2}$ for odd M,

$$H_i = h_i + n + \frac{1}{2} - i - \frac{1}{2}(M-1)\,\mathrm{sgn}\,(n + \frac{1}{2} - i) \tag{5.8}$$

for $i = 1, 2, \ldots, n+m$.

Set

$$BZ(m|n) = Z(osp(2m+1|2n), osp(2m|2n)). \tag{5.9}$$

The elements

$$z_i = pe_i, \quad i = \pm 1, \pm 2, \ldots, \pm n+m, \tag{5.10}$$

where $e_i = \bar{e}_{i0}, e_{-i} = \bar{e}_{0-i}$ $(i = 1, 2, \ldots, n+m)$, generate the reduced superalgebra $BZ(m|n)$.

Theorem 4. $BZ(m|n)$ is given by the generators (5.10) and the relations of the form (4.10) and (4.11). Here

$$\alpha_{ij} = (-1)^{\theta_{i0}\theta_{j0}} (\varphi_{i-j} + (-1)^{\theta_{i0}})/\varphi_{i-j} \quad \text{for } 1 \leq i < |j| \leq n+m \tag{5.11}$$

$$\alpha_{j-i} = (-1)^{\theta_{i0}\theta_{j0}} \frac{\varphi_{ij} + 2\theta_{ij} + (-1)^{\theta_{i0}}}{\varphi_{ij} + 2\theta_{ij}} \quad \text{for } 1 \leq i < |j| \leq n+m \tag{5.12}$$

$$\beta_{ij} = (-1)^{\theta_{i0}} \sigma_{ij}\,\delta_i^- \delta_j^+, \quad \gamma_i = \tau_i \delta_i^-, \tag{5.13}$$

where

$$\delta_i^- = \prod_{k=i+1}^{n+m} \frac{\varphi_{i,-k} - (-1)^{\theta_{ik}}}{\varphi_{i,-k}}, \quad \delta_j^+ = \prod_{k=j+1}^{n+m} \frac{\varphi_{j,-k} + 1}{\varphi_{j,-k} + 2\theta_{jk}} \tag{5.14}$$

$$\sigma_{ij} = \tau_i \tilde{\varphi}_{ij}^{-1} (\varphi_{j,-i} + 1)^{-1}, \quad \tau_i = 2(h_i + n + (-1)^{\theta_{i0}} m - i),$$

$$\tilde{\varphi}_{ij} = \varphi_{ij} \text{ for } i \neq j, \quad \tilde{\varphi}_{ii} = 2H_i, \quad M = 2m. \tag{5.15}$$

Remarks. (1) The matrix $\beta = (\beta_{ij})$ is nonsingular. (2) The elements z_i, $i = \pm(n+1), \pm(n+2), \ldots, \pm(n+m)$ for $n = 0$ generate the reduced algebra BZ_m [4].

6. The reduced superalgebra $DZ(m|n)$. Set

$$DZ(m|n) = Z(osp(2m+2|2n), osp(2m+1|2n)). \tag{6.1}$$

$DZ(m|n)$ is generated by $\bar{e}_{ij}$ for $i, j \neq n+m+1$ and

$$\bar{e}_{i0} = \bar{e}_{i,-n-m-1} + \bar{e}_{i,n+m+1}, \quad \bar{e}_{0,-i} = \bar{e}_{n+m+1,-i} + \bar{e}_{-n-m-1,-i}, \tag{6.2}$$

where $i = 1, 2, \ldots, n+m$. The elements

$$e_0 = \bar{e}_{n+m+1,-n-m-1},$$

$$e_i = \bar{e}_{i,-n-m-1} - \bar{e}_{i,n+m+1}, \tag{6.3}$$

$$e_{-i} = \bar{e}_{n+m+1,-i} - \bar{e}_{-n-m-1,-i},$$

where $i = 1, 2, \ldots, n+m, 0,$ form the weight basis for $osp(2m+2 \mid 2n) \ominus osp(2m+1 \mid 2n)$. Hence $DZ(m \mid n)$ is generated by

$$z_i = pe_i, \quad i = 0, \pm 1, \pm 2, \ldots, \pm (n+m). \tag{6.4}$$

Theorem 5. $DZ(m|n)$ is given by the generators (6.4) and the relations of the form (4.10) and (4.11) for $i, j = 0, \pm 1, \pm 2, \ldots, \pm (n+m)$. The coefficients are determined as in Theorem 4 , moreover

$$\alpha_{i0} = (\varphi_{i0} - 1 + 2(-1)^{\theta_{i0}})(\varphi_{i0} - 1)^{-1},$$

$$\alpha_{0-i} = (\varphi_{i0} + 1 + 2(-1)^{\theta_{i0}})(\varphi_{i0} + 1)^{-1}, \quad \beta_{i0} = -4\delta_i^-(\varphi_{i0} - 1)^{-1},$$

and σ_{ij} for $i \neq j$ are replaced by $\sigma_{ij}\chi_j$ where

$$\chi_j = (\varphi_{j0} + 1 + (-1)^{\theta_{j0}})(\varphi_{j0} + 1)^{-1}; \quad M = 2m+1. \tag{6.5}$$

Remark. The elements $z_{\pm i}$, $i = 0, \pm (n+1), \ldots \pm (n+m)$ for $n = 0$ generate the reduced algebras DZ_m . Note that the corresponding formulas in [4] are misprinted.

The same result can be obtained also for the exceptional simple contragredient Lie superalgebras. With Theorems 3-5 we can construct the bases of cyclid representations of $osp(m|2n)$ and $gl(m|n)$, see [9].

References

1. Asherova R.M., Smirnov Yu.F., Tolstoy V.V. (1971). Theor. Math. Phys., 8.

2. Asherova R.M., Smirnov Yu.F., Tolstoy V.N. (1979) Math. Notes 26.

3. Zhelobenko D.P. (1983), Soviet Math. Dokl. 28, 696-700.

4. Zhelobenko D.P. (1983), Soviet Math. Dokl. 28, 777-781.

5. Asherova R.M. Smirnov Yu.F., Tolstoy V.N. (1979). Theor. Math. Phys.,, 15.

6. Berezin F.A., Tolstoy V.N. (1981). Commun. Math. Phys., 78, 409-428.

7. Kac V.G. (1977). Commun. Math. Phys., 53, 31-64; Leites D.A., Saveliev, M.V. Serganova, V.V. in these proceedings.

8. Leznov A.N. Saveliev M.V. (1974), Function. Anal. Appl. 8.

9. Tolstoy V.N., (1985). Russian Math. Surveys, 225-226.

10. Tolstoy. V.N. Istomina I.F., Smirnov Yu.F. The Gelfand-Tsetlin bases for the Lie superalgebra GL(n/m).

THE GELFAND-TSEITLIN BASIS FOR THE LIE SUPERALGEBRA gl(n/m)

V.N.Tolstoy, I.F.Istomina, Yu.F.Smirnov
Institute of Nuclear Physics, Moscow State University, Moscow 119899

1. Introduction. The study of the representation structure of Lie superalgebras (in particular, gl(n/m)) is at present a very urgent problem in theoretical physics and mathematics. In physics the interest in such objects arose mainly from the feasibility of unified description of the boson-fermion systems, that is, the systems with mixed statistics. The supersymmetric or boson-fermion nuclear model based on gl(n/m) see e.g. in [1,2] .

The representation theory for gl(n/m) was treated in some works among which [3] gives a description of a certain class of star irreducible representational(IR's) for gl(n/m). The method applied in [3] is based on the reciprocity between the representations of gl(n/m) and those of the graded permutation group. It is important, however, to describe the representation structure for gl(n/m) in terms of gl(n/m) itself for all the star IR. The present work is devoted to this problem.

Our tool is the extremal projection [4]. Our calculations are sufficiently elementary and similar to the case of the Lie algebra g(n+m) [5] .

2. Star and superstar representations for gl(n/m) [6]. Let e_{ij} (i,j = 1,2,..., n+m) be the standart basis of gl(n/m), see [4] . It is well known that there are four types of generalised adjoint operations in gl(n/m). The extensions of the conventional adjoint operation for the Lie algebra gl(n) ⊕ gl(m) [6] are:

I. $e_{ij}^{+} = e_{ji}$, $1 \leq i, j \leq n+m$;

II. $e_{ij}^{+} = (-1)^{\theta_{ij}} e_{ji}$, $1 \leq i, j \leq n+m$;

III. $e_{ij}^{\ddagger} = e_{ji}$ for $\theta_{ij} = 0$,
$e_{ij}^{\ddagger} = e_{ji} \operatorname{sgn}(j-i)$ for $\theta_{ij} = 1$;

IV. $e_{ij}^{\ddagger} = e_{ji}$ for $\theta_{ij} = 0$,
$e_{ij}^{\ddagger} = e_{ji} \operatorname{sgn}(i-j)$ for $\theta_{ij} = 1$.

Types I and II are called the star adjoints the two others the superstar adjoints. We have

$$(e_{ij}^{+})^{+} = e_{ij}, \quad [e_{ij}, e_{kl}]_s^{+} = [e_{kl}^{+}, e_{ij}^{+}]_s \tag{2.1}$$

whereas

$$(e_{ij}^{\ddagger})^{\ddagger} = (-1)^{\theta_{ij}} e_{ij}, \quad [e_{ij}, e_{kl}]^{\ddagger} = -[e_{ij}^{\ddagger}, e_{kl}^{\ddagger}]_s. \tag{2.2}$$

A representation D such that $D^{+}(e_{ij}) = D(e_{ij}^{+})$ or $D^{\ddagger}(e_{ij}) = D(e_{ij}^{\ddagger})$ is called a star or superstar representation, respectively.

Each finite dimensional IR D^{Λ} for $gl(n/m)$ is characterized (up to isomorphism) by its highest weight $\Lambda = (\lambda_1, \lambda_2, \cdots, \lambda_{n+m})$, $\lambda_i \in \mathbb{C}$, where the differences $\lambda_i - \lambda_{i+1}$ are nonnegative integers for $i \neq n$ and the parity of the highest weight vector ζ_Λ.

By definition

$$D(e_{ij}) \zeta_\Lambda = \lambda_i \delta_{ij} \zeta_\Lambda \quad \text{for } 1 \leq i \leq j \leq n+m. \tag{2.3}$$

All the superstar IR's of $gl(n/m)$ are one-dimensional for $m > 1$. For $gl(n/1)$ the set of superstar IR's is extremly narrow. Therefore we only consider the star IR's.

3. The Gelfand-Tseitlin basis for $gl(n/m)$. The construction of the (Gelfand-Tseitlin) G-basis for $gl(n/m)$ involves the reduction of an IR D^{Λ} to the canonical

chain of subalgebras:

$$gl(n|m) \downarrow gl(n|m-1) \downarrow \ldots \downarrow gl(n) \downarrow \ldots \downarrow gl(1). \tag{3.1}$$

Because of the inductive form of the construction, it suffices to examine the first reduction, i.e.

$$gl(n|m) \downarrow gl(n|m-1). \tag{3.2}$$

The subalgebra $gl(n/m-1)$ is selected so that it is spanned by the elements e_{ij}, $1 \le i,j \le N-1 := n+m-1$.

Let $A_{ij} = D(e_{ij})$ and V^+ be the subspace of highest vectors with respect to $gl(n/m-1)$, i.e.

$$V^+ := \{ \xi \in V^\lambda : A_{ij}\xi = 0, \quad 1 \le i < j \le N-1 \}. \tag{3.3}$$

Our aim is to construct the weight basis in V^+. Obviosly, the subspace V^+ is spanned by the vectors

$$\xi_{\Lambda'} = p\Big(\prod_{i=1}^{N-1} A_{Ni}^{r_i}\Big)\xi_\Lambda, \tag{3.4}$$

where p is the extremal projection for $gl(n/m-1)$. The vectors (3.4) are weight ones with respect to A_{ii}, $i = 1,\ldots, N$. In this case

$$\lambda_i' = \lambda_i - r_i, \qquad i = 1, 2, \ldots, N-1, \tag{3.5}$$

$$\lambda_N' = \lambda_N + \sum_{i=1}^{N} r_i.$$

We see that the numbers r_i determine the weight unambiguously and, therefore, all the nonvanishing vectors $\xi_{\Lambda'}$ are linearly independent and form a basis of V^+.

To determine the restrictions on the highest weight Λ and to find the nonvanishing vectors of the form (3.4) we calculate

$$\|\xi_{\Lambda'}\|^2 = (-1)^{\varkappa \sum_{i=1}^{n} r_i} \langle \xi_\Lambda | \Big(\prod_{i=1}^{N-1} A_{iN}^{r_i}\Big) p \Big(\prod_{i=1}^{N-1} A_{Ni}^{r_i}\Big) | \xi_\Lambda \rangle \tag{3.6}$$

where

$$\varkappa = \begin{cases} 0 & \text{if} \quad D^{\Lambda} \quad \text{is of class I,} \\ 1 & \text{if} \quad D^{\Lambda} \quad \text{is of class II;} \end{cases}$$

Note that the numbers $r_i=0,1$ for $i=1,2,\ldots,n$ (see (4.18) in [4]). Using (4.10)-(4.21) of [4] and since $z_i \xi_\Lambda = 0$ for $i = 1,2,\ldots, N-1$ (where $z_i = pA_{iN}$) we get

$$\|\xi_{\Lambda'}\|^2 = \prod_{i=1}^{n}\left[(-1)^{\varkappa}(l_i+l_N+1)\prod_{j=i+1}^{n}\frac{(l_i-l_j+r_j-1)}{(l_i-l_j)}\prod_{j=n+1}^{N-1}\frac{(l_i+l_j+1)}{(l_i+l_j-r_j)}\right]^{r_i} \times \tag{3.7}$$

$$\times \prod_{j=n+1}^{N-1} r_j! \prod_{n+1\leq i<j\leq N-1}\frac{(l_i-l_j-r_i+r_j)!}{(l_i-l_j+r_j)!}\prod_{n+1\leq i<j\leq N}\frac{(l_i-l_j-1)!}{(l_i-l_j-r_j-1)!}.$$

Here,

$$l_k = \lambda_k + n - k . \tag{3.8}$$

We assume that $\|\xi_\Lambda\|^2 = 1$.

Analyzing (3.7) we get the following results.

<u>Theorem 1</u>. For $gl(n/m)$ the IR D^Λ with the highest weight $\Lambda = (\lambda_1, \lambda_2, \ldots, \lambda_{n+m})$ is a star IR if and only if all λ_i are real, the differences $\lambda_i - \lambda_{i+1}$ are non-negative integers for $i \neq n$, and, besides, one of the following conditions holds:

(a) $\lambda_n + \lambda_{n+m} - m + 1 \geq 0$;

(b) there exists $\nu \in \{1, 2, \ldots, m-1\}$ such that $1 \geq \lambda_n + \lambda_{n+\nu} - \nu + 1 \geq 0$ and then $\lambda_{n+\nu} = \lambda_{n+\nu+1} = \ldots = \lambda_{n+m}$;

(c) $\lambda_1 + \lambda_{n+1} + n - 1 \leq 0$.

Any of the star IR is of finite dimension. (Conditions (a) and (b) relate to type I representations, (c) to type II representations).

Proof. It can be easily verified that, if Λ satisfies the assumptions of the theorem, vectors of the form (4.4) for the full chain (3.1) have non-negative squared norm. If, however, Λ fails to satisfy the conditions of the theorem, a vector can be indicated whose squared norm is negative. Therefore, the respective representation is not star. The finite dimension of a star IR follows from Theorem 2. Further, since each of the finite-dimensional IR is determined (up to equivalence) by its highest weight, the respective IR is equivalent to a star representation if this weight satisfies the restrictions of the theorem.

Theorem 2. The subduction D^{Λ} gl(n/m-1), where Λ satisfies the conditions of Theorem 1, contains just once all the IR's of gl(n|m-1) whose highest weight $\Lambda' = (\lambda_1', \lambda_2', \dots, \lambda_{n+m-1}')$ obey the following restrictions:

$$\begin{array}{ccccccccc}
\lambda_1 & \geq & \lambda_2 & \geq & \lambda_3 & \geq \dots \geq & \lambda_{n-1} & \geq & \lambda_n \\
 & \searrow\!\!\geq & & \searrow\!\!\geq & & & & \searrow\!\!\geq & \\
\lambda_1' & \geq & \lambda_2' & \geq & \lambda_3' & \geq \dots \geq & \lambda_{n-1}' & \geq & \lambda_n' \\
\vee\!\!\| & & \vee\!\!\| & & \vee\!\!\| & & \vee\!\!\| & & \vee\!\!\| \\
\lambda_1-\varepsilon_1 & & \lambda_2-1 & & \lambda_3-1 & \dots & \lambda_{n-1}-1 & & \lambda_n-\varepsilon_n,
\end{array} \qquad (3.9)$$

$$\lambda_{n+1} \geq \lambda_{n+1}' \geq \lambda_{n+2} \geq \lambda_{n+2}' \geq \dots \geq \lambda_{n+m-1} \geq \lambda_{n+m-1}' \geq \lambda_{n+m}\,.$$

where $\varepsilon_1 = 0$ if $\lambda_1 + \lambda_{n+1} + n - 1 = 0$, otherwise $\varepsilon_1 = 1$; similarly, $\varepsilon_n = 0$, if $\lambda_1 + \lambda_{n+1} + n - 1 = 0$ or $\lambda_n + \lambda_{n+\nu} - \nu + 1 = 0$, but $\lambda_{n+\nu-1}' \neq \lambda_{n+\nu}$, $1 < \nu \leq m$ otherwise $\varepsilon_n = 1$. Besides, all the differences $\lambda_i - \lambda_i'$ (i = 1,2,..., n+m-1) are integers.

Proof. The explicit expression (3,7) is used to find all the restrictions imposed on r_i, i = 1,2,..., n+m-1, for which (3.7) is nonvanishing. Replacing $r_i = \lambda_i - \lambda_i'$ we get (3.9).

Remark. The restriction $\varepsilon_1 = 0$ (or $\varepsilon_n = 0$) arises in the case of of class II (class I) representations.

We write $\Lambda_N = (\lambda_{1N}, \lambda_{2N}, \dots, \lambda_{NN})$ and $\Lambda_{N-1} = (\lambda_{1\,N-1}, \lambda_{2\,N-1}, \dots, \lambda_{N-1\,N-1})$, where $N = n+m$ instead of $\Lambda = (\lambda_1, \lambda_2, \dots, \lambda_N)$ and $\Lambda' = (\lambda'_1, \dots, \lambda'_{N-1})$. Besides set $|\Lambda_N> := \xi_\Lambda$.

As a consequence of the above results, we formulate the following.

<u>Lemma</u>. Vectors of the form

$$\left| \begin{matrix} \lambda_{1N} & \lambda_{2N} & \dots & \lambda_{NN} \\ \\ \lambda_{1,N-1} & \lambda_{2,N-1} & \dots & \lambda_{N-1,N-1} \end{matrix} \right\rangle = F(\Lambda_{N-1}; \Lambda_N) | \Lambda_N > \tag{3.10}$$

where

$$F(\Lambda_{N-1}; \Lambda_N) = N^{-1}(\Lambda_{N-1}; \Lambda_N) p\Big(\prod_{i=1}^{N-1} A_{Ni}^{\lambda_{iN} - \lambda_{i,N-1}}\Big) | \Lambda_N > \tag{3.11}$$

$$N(\Lambda_{N-1}; \Lambda_N) = \Bigg\{ \prod_{i=1}^{n} \Bigg[\Bigg| (l_{iN} + l_{NN} + 1) \prod_{j=i+1}^{n} \frac{(l_{i,N-1} - l_{j,N-1})}{(l_{iN} - l_{jN})} \prod_{j=n+1}^{N-1} \frac{(l_{iN} + l_{jN} + 1)}{(l_{iN} + l_{j,N-1})} \Bigg| \Bigg]^{\lambda_{iN} - \lambda_{i,N-1}} \times$$

$$\times \prod_{j=n+1}^{N-1} (l_{jN} - l_{j,N-1})! \prod_{n+1 \le i < j \le N-1} \frac{(l_{i,N-1} - l_{j,N-1})!}{(l_{iN} - l_{j,N-1})!} \prod_{n+1 \le i < j \le N} \frac{(l_{iN} - l_{jN} - 1)!}{(l_{i,N-1} - l_{jN} - 1)!} \Bigg\}^{1/2} \tag{3.12}$$

$$l_{kN} = \lambda_{kN} + n - k, \quad l_{k,N-1} = \lambda_{k,N-1} + n - k,$$

for all the possible signatures Λ_{N-1} whose components λ_{iN-1} satisfy Theorem 2, form an orthonormal basis in the subspace (3.3) of the star IR with highest weight Λ_N of gl(n/m).

<u>Remark</u>. If $n = 0$ (and $N = m$) we get the results for representations of gl(N) subduced to the subalgebra gl(N-1) and the normalization factor $N(\lambda_{N-1}; \lambda_N)$ takes the form

$$N(\Lambda_{N-1};\Lambda_N)=\left\{\prod_{j=1}^{N-1}(l_{jN}-l_{j,N-1})!\prod_{1\le i<j\le N-1}\frac{(l_{i,N-1}-l_{j,N-1})!}{(l_{iN}-l_{j,N-1})!}\prod_{1\le i<j\le N}\frac{(l_{iN}-l_{jN}-1)!}{(l_{i,N-1}-l_{jN}-1)!}\right\}^{1/2} \tag{3.13}$$

Now, apply the lemma to each link of the chain (3.1). As a result, we get the following

Theorem 3. In the space of star IR D^{Λ_N} of gl(n/m) there exists the orthonormal basis of vectors of the form

$$e_\lambda \equiv \left| \begin{matrix} \lambda_{1N} & \lambda_{2N} & \lambda_{3N} & \cdots & \lambda_{NN} \\ \lambda_{1,N-1} & \lambda_{2,N-1} & \cdots & \lambda_{N-1,N-1} & \\ \cdots & \cdots & \cdots & \cdots & \\ \lambda_{12} & \lambda_{22} & & & \\ \lambda_{11} & & & & \end{matrix} \right\rangle \equiv \tag{3.14}$$

$$\equiv F(\Lambda_1;\Lambda_2)F(\Lambda_2;\Lambda_3)\cdots F(\Lambda_{N-1};\Lambda_N)|\Lambda_N\rangle,$$

where the numbers λ_{ij} satisfy the following betweenness conditions

$$\lambda_{i,j+1}\geqslant\lambda_{ij}\geqslant\lambda_{i+1,j+1} \qquad \text{for } j\geqslant i>n \quad \text{or } i\leqslant j<n,$$

$$\begin{matrix} \lambda_{1,j+1} & \geqslant & \lambda_{2,j+1} & \geqslant & \cdots & \geqslant & \lambda_{n-1,j+1} & \geqslant & \lambda_{n,j+1} \\ \rotatebox{90}{$\leqslant$} & & \rotatebox{90}{$\leqslant$} & & & & \rotatebox{90}{$\leqslant$} & & \rotatebox{90}{$\leqslant$} \\ \lambda_{1j} & \geqslant & \lambda_{2j} & \geqslant & \cdots & \geqslant & \lambda_{n-1,j} & \geqslant & \lambda_{nj} \\ \rotatebox{90}{$\geqslant$} & & \rotatebox{90}{$\geqslant$} & & & & \rotatebox{90}{$\geqslant$} & & \rotatebox{90}{$\geqslant$} \\ \lambda_{1,j+1}-\varepsilon_{1,j+1} & & \lambda_{2,j+1}-1 & & \cdots & & \lambda_{n-1,j+1}-1 & & \lambda_{n,j+1}-\varepsilon_{n,j+1} \end{matrix} \tag{3.15}$$

$$\text{for } j>n$$

In the last line, $\varepsilon_{1,j+1}=0$, if $\lambda_{1,j+1}+\lambda_{n+1,j+1}+n-1=0$, otherwise $\varepsilon_{1,j+1}=1$; similarly, $\varepsilon_{n,j+1}=0$ if $\lambda_{n,j+1}+\lambda_{n+1,j+1}=0$ or

$\lambda_{n,j+1} + \lambda_{n+\nu,j+1} - \nu + 1 = 0$, but $\lambda_{n+\nu-1,j} \neq \lambda_{n+\nu,j+1}$ ($1 < \nu \leq j+1$), otherwise $\varepsilon_{n,j+1} = 1$. All λ_{ij} 's are real and $\lambda_{i,j+1} - \lambda_{ij}$ are integers.

Remark. The restrictions $\varepsilon_{1,j+1} = 0$ (or $\varepsilon_{n,j+1} = 0$) arise in the case of class II (class I) representations.

The basis composed of the vectors (3.14) will be called the Gelfand-Tseitlin or G-basis.

4. The generators of gl(n/m) in the Gelfand-Tseitlin basis. We can find explicit formulas for A_{ij} in the G-basis. We immediatly have

$$A_{ii} e_\lambda = (S_i - S_{i-1}) e_\lambda \tag{4.1}$$

where $S_i = \sum_{k=1}^{i} \lambda_{ki}$ and $S_0 = 0$.

Examine now the operators A_{ij} for $i \neq j$. From the adjoint property for the operators A_{ij} it follows that it suffices to study e.g. the subsystem of the raising operators A_{ij}, $i < j$ and among them only the generators A_{ii+1}. The induction in the chain (3.1) makes it possible to treat the generator $A_{N-1,N}$.

So, let us find the action of $A_{N-1,N}$ on e_λ. We have

$$A_{N-1,N} e_\lambda = N^{-1}(\Lambda_{N-2};\Lambda_{N-1}) N^{-1}(\Lambda_{N-1};\Lambda_N) F(\Lambda_1;\Lambda_2) \ldots F(\Lambda_{N-3};\Lambda_{N-2}) \times$$
$$\times (z'_{-1})^{r'_1} (z'_{-2})^{r'_2} \ldots (z'_{-N+2})^{r'_{N-2}} A_{N-1,N} z_{-1}^{r_1} z_{-2}^{r_2} \ldots z_{-N+1}^{r_{N-1}} | \Lambda_N \rangle \tag{4.2}$$

where $r'_i = \lambda_{i,N-1} - \lambda_{i,N-2}$, $r_i = \lambda_{i,N} - \lambda_{i,N-1}$, $z'_{-i} = p' A_{N-1,i}$, $z_{-i} = p A_{Ni}$,

P' and P are the extremal projection for gl(n/m-2) and gl(n/m-1), respectively. To further transform the right-hand side of (4.2), we use the following relation (analogous to (4.11) of [4])

$$A_{N-1,N}\, p = \sum_{i=1}^{N-1} Z'_{-i}\, Z_i\, \gamma_{N-1,i} \;, \tag{4.3a}$$

where

$$\gamma_{N-1,i} = (\varphi_{i,N-1}+1)^{-1} \prod_{j=i+1}^{N-1} \frac{(\varphi_{ij}+1)}{(\varphi_{ij}+2\Theta_{ij})} \;,$$

$$\varphi_{ij} = H_i - (-1)^{\Theta_{ij}} H_j \;, \qquad H_k = A_{kk} + n - k \,. \tag{4.3b}$$

Besides put $Z'_{-N+1} \equiv 1$ in (4.3a). Now, we substitute (4.3) in (4.2) and then transpose all the raising elements Z_i to the right and the lowering elements Z'_{-i} to the left. We get

$$A_{N-1,N}\, e_\lambda = \sum_{j=1}^{N-1} a_{j,N-1}(\lambda)\, e_{\lambda+1_{j,N-1}} \;, \tag{4.4}$$

where

$$a_{j,N-1}(\lambda) = \frac{N(\Lambda_{N-2};\Lambda_{N-1}+1_{j,N-1})\, N(\Lambda_{N-1}+1_{j,N-1};\Lambda_N)}{N(\Lambda_{N-2};\Lambda_{N-1}) \quad N(\Lambda_{N-1};\Lambda_N)} \times$$

$$\frac{\prod_{i=n^0+1}^{N} (\ell_{j,N-1}-(-1)^{\Theta_{ij}}\ell_{iN}+2\Theta_{ij}) \prod_{i=n^0+1}^{N-2} (\ell_{j,N-1}-(-1)^{\Theta_{ij}}\ell_{i,N-2}+1)}{\prod_{i=n^0+1}^{N-1} (\ell_{j,N-1}-(-1)^{\Theta_{ij}}\ell_{i,N-1}+2\Theta_{ij}) \prod_{i=n^0+1}^{N-1} (\ell_{j,N-1}-(-1)^{\Theta_{ij}}\ell_{i,N-1}+1)} \times$$

$$(-1)^{\phi_{j,N-1}}(l_{jN}-l_{j,N-1})\prod_{i=j+1}^{n}\frac{(l_{j,N-1}-l_{i,N-1})(l_{j,N-1}-l_{i,N-1}+1)}{(l_{j,N-1}-l_{i,N-2})(l_{j,N-1}-l_{i,N}+1)},$$

$$\phi_{j,N-1}:=\Theta_{jN}\sum_{i=1}^{j-1}(\lambda_{iN}-\lambda_{i,N-1})+\Theta_{j,N-1}\sum_{i=j+1}^{n}(\lambda_{i,N-1}-\lambda_{i,N-2}), \tag{4.5}$$

$$n^{\circ}:=\max\{n,j\},$$

1_{jN-1} means the (N-1)-dimensional vector (0,...,0,1,0, ...,0) with 1 in the j-th place. This vector is added to the (N-1)-th line of the scheme λ of e_λ. (We assume that $\prod_{j+1}^{n}$ equals 1 if $j \geqslant n$). The action of $A_{N,N-1}$ can be found by adjointness.

Theorem 4. The generators $A_{p,p+1}$, $A_{p+1,p}$ act on vectors e_λ via

$$A_{p,p+1}e_\lambda=\sum_{j=1}^{p}a_{jp}(\lambda)e_{\lambda+1_{jp}} \tag{4.6}$$

$$A_{p+1,p}e_\lambda=\sum_{j=1}^{p}b_{jp}(\lambda)e_{\lambda-1_{jp}} \tag{4.7}$$

where

$$a_{jp}(\lambda)=(-1)^{\phi_{jp}}\left\{\frac{\prod_{i=n+1}^{p+1}(l_{i,p+1}-(-1)^{\Theta_{ij}}l_{jp}+2\Theta_{ij})}{\prod_{i=n+1}^{p}{}'(l_{ip}-(-1)^{\Theta_{ij}}l_{jp}+2\Theta_{ij})}\right. \times$$

$$\times \frac{\prod_{i=n+1}^{p-1}\left(l_{i,p-1}-(-1)^{\Theta_{ij}}l_{jp}-1+2\Theta_{ij}\right)}{\prod_{i=n+1}^{p}{}'\left(l_{ip}-(-1)^{\Theta_{ij}}l_{jp}-1+2\Theta_{ij}\right)} \times \tag{4.8}$$

$$\times \prod_{i=1}^{n}{}' \frac{\left(l_{ip}-(-1)^{\Theta_{ij}}l_{jp}+2\Theta_{ij}\right)\left(l_{ip}-(-1)^{\Theta_{ij}}l_{jp}-1+2\Theta_{ij}\right)}{\left(l_{i,p-1}-(-1)^{\Theta_{ij}}l_{jp}+2\Theta_{ij}\right)\left(l_{i,p+1}-(-1)^{\Theta_{ij}}l_{jp}-1+2\Theta_{ij}\right)}\Bigg\}_{+}^{1/2},$$

$$b_{jp}(\lambda) = (-1)^{\tilde{\Theta}_{jp}}\left\{\frac{\prod_{i=n+1}^{p+1}\left(l_{i,p+1}-(-1)^{\Theta_{ij}}l_{jp}+1\right)}{\prod_{i=n+1}^{p}{}'\left(l_{ip}-(-1)^{\Theta_{ij}}l_{jp}+1\right)} \times \right.$$

$$\frac{\prod_{i=n+1}^{p-1}\left(l_{i,p-1}-(-1)^{\Theta_{ij}}l_{jp}\right)}{\prod_{i=n+1}^{p}{}'\left(l_{ip}-(-1)^{\Theta_{ij}}l_{jp}\right)} \times \tag{4.9}$$

$$\left.\prod_{i=1}^{n}{}' \frac{\left(l_{ip}-(-1)^{\Theta_{ij}}l_{jp}+1\right)\left(l_{ip}-(-1)^{\Theta_{ij}}l_{jp}\right)}{\left(l_{i,p-1}-(-1)^{\Theta_{ij}}l_{jp}+1\right)\left(l_{i,p+1}-(-1)^{\Theta_{ij}}l_{jp}\right)}\right\}_{+}^{1/2}$$

for $N > p \geqslant n$ and

$$a_{jp}(\lambda) = \left\{\frac{\prod_{i=1}^{p+1}\left(l_{i,p+1}-l_{jp}\right)\prod_{i=1}^{p-1}\left(l_{i,p-1}-l_{jp}-1\right)}{\prod_{i=1}^{p}{}'\left(l_{ip}-l_{jp}\right)\prod_{i=1}^{p}{}'\left(l_{ip}-l_{jp}-1\right)}\right\}_{+}^{1/2} \tag{4.10}$$

$$b_{jp}(\lambda)=\left\{\frac{\prod_{i=1}^{p+1}(l_{i,p+1}-l_{jp}+1)\prod_{i=1}^{p-1}(l_{i,p-1}-l_{jp})}{\prod_{i=1}^{p}{}'(l_{ip}-l_{jp}+1)\prod_{i=1}^{p}{}'(l_{ip}-l_{jp})}\right\}_{+}^{1/2} \tag{4.11}$$

for $p < n$. Here prime means that the factor with $i = j$ is excluded, and the brackets with "+" replace the modulus sign. Moreover,

$$l_{ij}=\lambda_{ij}+n-i, \qquad \tilde{\Phi}_{jp}=\Phi_{jp}+\varkappa\delta_{p,n} \quad ,$$

$$\Phi_{jp}=\Theta_{jp+1}\sum_{i=1}^{j-1}(\lambda_{ip+1}-\lambda_{ip})+\Theta_{jp}\sum_{i=j+1}^{n}(\lambda_{ip}-\lambda_{ip-1}).$$

References

1. Balantekin,A.B., Bars,I.,Iachello,F. (1981), Phys.Rev. Lett. 47, 17; (1981). Nucl. Phys.,A370, 284.
2. Balantekin,A.B., Bars,I. (1981). J. Math. Phys., 22 1149; (1981), 22, 1810; (1982), 23, 1239.
3. Chen,J.Q., Chen,X.G., Gao,M.J. (1981). J. Math. Gen. Phys. A, 16, 3435.
4. Tolstoy,V.N. In these proceedings.
5. Asherova,R.M., Smirnov,Yu.F., Tolstoy,V.N. (1973). Theor. Math. Phys., 15, 107 (Russian).
6. Scheunert,M., Nahm,W., Rittenberg,V. (1977). J. Math. Phys., 18, 155.

ON DYNAMIC SYSTEMS WITH GRADED POISSON BRACKETS

D.V.Volkov, V.A.Soroka, V.I.Tkach

Institute of Physics and Technology, the Ukrainian Academy of Sciences, Kharkov 310108

Here we present a recipe for quantizing the Butten bracket or antibracket [1-5] the canonically conjugate variables of which have opposite Grassmann parities. These real canonical variables, with spinor properties [6,7] of a certain group (hereafter called <u>superspinor</u> components) are used to construct supergroup generators and the corresponding superspace coordinates. The proposed scheme of quantizing the antibracket on composite superspace coordinates and composite dynamical variables that describe their symmetry agrees with the conventional canonical quantization. This fact indicates that dynamic systems using the Butten bracket as a basis are perhaps not just another variant of Hamiltonian systems(in addition to the Poisson and Martin brackets) but represent a more elementary level of dynamic description and form a basis for the usual dynamic systems.

In conclusion we exemplify the general scheme of the antibracket quantization by $OS_p(1,1;2)$ which has been discussed in detail in [7].

Following refs.[6,7], consider a real superspinor with the $y_{\alpha i}$ and $\eta_{\beta k}$ components of parity g equal 0 and 1, respectively, where α and β will denote the spinor indices of the space-time symmetry, while i and k running 1 to N are related to the internal symmetry. The antibracket of functions of superspinors via

$$\{A,B\} = A\frac{\overleftarrow{\partial}}{\partial y_{\alpha i}} \frac{\overrightarrow{\partial}}{\partial \eta_{\alpha i}} B - A\frac{\overleftarrow{\partial}}{\partial \eta_{\alpha i}} \frac{\overrightarrow{\partial}}{\partial y_{\alpha i}} B. \tag{1}$$

This definition (1) leads to:

(i) the generalized Leibnitz rule

$$\{A,B\cdot C\} = \{A,B\}\cdot C + (-1)^{[g(A)+1]g(B)} B\cdot\{A,C\}, \tag{2}$$

where the point denotes the multiplication;

(ii) the parity relation

$$\{A,B\} = C, \quad g(A) + g(B) = g(C) + 1; \tag{3}$$

(iii) the permutation property

$$\{A,B\} = -(-1)^{[g(A)+1][g(B)+1]}\{B,A\}; \tag{4}$$

(iv) the Jacobi identity

$$\sum_{(ABC)} (-1)^{[g(A)+1][g(C)+1]}\{A,\{B,C\}\} = 0 \tag{5}$$

If for each value of A, in addition to the Grassmann parity $g(A)$, we introduce the antiparity $\tilde{g}(A) = g(A) +$. (mod 2), then eqs. (3)-(5) written in terms of $\tilde{g}(A)$ are Lie superalgebra relations.

The antibracket endows the space of functions of superspinors with an infinite dimensional Lie superalgebra structure. Define two sets $O = \{O_k = y^{2k+1} f(\eta)\}$ and $E = \{E_k = y^{2k} f(\eta)\}$ depending, respectively, on odd and even powers of the spinor y. Then, by (1), (2), these classes satisfy the relations

$$\{O_k,O_l\} = O_{k+l}; \quad \{O_k,E_l\} = E_{k+l}; \quad \{E_k,E_l\} = \tag{6}$$
$$= O_{k+l-1}$$

from which it is seen that the class O forms an infinite dimensional Lie subsuperalgebra containing, in its turn, a finite dimensional superalgebra O_o.

Now we write explicitly the algebraic relations for the Grassmann multiplication of the quantities belohging to the classes O and E:

$$O_k \cdot O_l = E_{k+l+1};\ O_k \cdot E_l = O_{k+l};\ E_k \cdot E_l = E_{k+l}. \quad (7)$$

By (6) and (7), O and E form a closed algebraic structure relative to the composition $*$:

$$O' * O'' = \{O', O''\},\ O' * E' = \{O', E'\},$$
$$E' * E'' = E' \cdot E'', \quad (8)$$

where $O', O'' \in O$ and $E', E'' \in E$.

Note that O_o and E_o are also alrebraically closed with respect to the composition $*$. Further the E_o-class quantities are used to build the superspace coordinates $Z^a = (x^\mu, \theta^{\alpha i})$ and to interprete functions $\Phi(Z)$ as wave functions, while O_o-class quantities to construct the dynamic variables that describe the Z^a-coordinate symmetry. Introducing the spin grading q

$$q(O^s_k) = \tilde{g}(O^s_k);\ q(E^s_k) = g(E^s_k), \quad (9)$$

where by O^s_k and E^s_k we mean a linear combination of monomials with a definite spin S belonging to the classes O_k and E_k, respectively.

The grading and symmetry properties for the multiplication $*$ follow from (3), (4), (8) and from the definition of the Grassmann superalgebra and with (9) have the form

$$A * B = C,\quad q(A) + q(B) = q(C);$$
$$O' * O'' = -(-1)^{q(O')q(O'')} O'' * O'; \quad (10)$$

$$E' * E'' = (-1)^{q(E')q(E'')} E'' * E'.$$

Eventually, we determine the quantum bracket $[A,B\}$ ((anti) commutator) for $A, B \in O \cup E$ when it acts on the wave function $\Phi(Z)$ as

$$[A,B\}\Phi(Z) = A * (B * \Phi) - (-1)^{q(A)q(B)} B * (A * \Phi). \quad (11)$$

By (10), the quantum bracket of the quantities of E gives $[E', E''\}\Phi = 0$, i.e., in particular, the superspace coordinates commute. In the cases when one of the quantities A, B or both belong to the class O, the Leibnitz rule (2) for the first case and the Jacobi identities (5) for the second give

$$[A,B\}\Phi = \{A,B\} * \Phi = (A * B) * \Phi \quad (12)$$

Note that only with this definition of the multiplication $*$ and the quantum bracket $[\,,\}$ we obtain a correct relation of spin values of all the quantities to the Fermi-Bose quantum relations, because the symmetry of the quantum bracket is controlled by the spin grading

$$[A,B\}\Phi = -(-1)^{q(A)q(B)} [B,A\}\Phi.$$

Note also that the scheme described above admits the restriction of $O(E)$ to $O_0(E_0)$.

The relation (12) describes the quantization procedure also for a particular case of the conventional quantum mechanics if one replaces the antibracket by a usual Poisson bracket, y by the momentum P, and η by the x coordinates. For example, in the x representation, the coordinates x and the wave functions $\Phi(x)$ belong to the class E_0, while the momentum and the moment belong to the class O_0.

Now we illustrate the general scheme of the antibracket quantization by group $OS_p(1,1;2)$ [7]. Its generators of the supertranslations $Q_{\alpha i}$, momenta P_μ, rotations $M_{\mu\nu}$ and internal symmetry I_{ik} are

$$Q_{\alpha i} = y_{\alpha i} - i\varkappa(\bar{\eta}, y_k)\eta_\alpha^k , \quad P_\mu = \frac{i\varkappa}{2}\bar{y}^k\gamma_\mu\eta_k .$$

$$M_{\mu\nu} = -i\bar{y}^k\sigma_{\mu\nu}\eta_k , \quad I_{ik} = -2i\bar{y}_{[i}\eta_{k]} .$$

Letting $\varkappa \to 0$ ($\varkappa$ playing the role of the inverse radius of the de Sitter curvature for O(1,2)) we come to the extended two-dimensional Poincaré supergroup.

The de Sitter superspace coordinates $\theta^{\alpha i}$ and x^μ have the form

$$\theta^{\alpha i} = \eta^{\alpha i}, \quad x^\mu = \frac{V^\mu}{\varkappa V^5} \quad (\mu = 0,1), \quad V^A = \bar{\tilde{y}}_i \Gamma^A \tilde{y}_k g^{ik} \quad (A = \mu, 5)$$

$$\tilde{y}_{\alpha i} = y_{\alpha k}\left(1 - \frac{i\varkappa}{4}\eta^2 + \frac{\varkappa^2}{16}(\eta^2)^2\right)\delta_i^k + \frac{i\varkappa}{2}(\bar{\eta}^k\eta_i) .$$

The antibracket relations for the composite generators $OS_p(1,1;2)$ and the superspace coordinates x^μ and $\theta^{\alpha i}$ as $\varkappa \to 0$ are

$$\{Q_{\alpha i}, x^\mu\} = -i(\gamma^\mu\theta_i)_\alpha , \{Q_{\alpha i}, \theta^{\beta k}\} = \delta_\alpha^\beta\delta_i^k ; \{I_{ik}, x^\mu\} = 0,$$

$$\{P_\mu, x^\nu\} = -i\delta_\mu^\nu , \{P_\mu, \theta^{\alpha i}\} = 0; \{I_{ik}, \theta^{\alpha l}\} = 2i\delta_{[i}^l\theta_{k]}^\alpha ; \tag{13}$$

$$\{M_{\mu\nu}, x^\rho\} = -2ix_{[\mu}\delta_{\nu]}^\rho , \{M_{\mu\nu}, \theta^{\alpha i}\} = -i(\bar{\theta}^i\sigma_{\mu\nu})^\alpha .$$

In view of (12) between the classical and quantum antibrackets, eqs. (13) reproduce the algebra of the usual canonical quantization of the coordinates and dynamic variables in terms of the composite coordinates $z^a = (x^\mu, \theta^{\alpha i})$ and the dynamic variables $Q_{\alpha i}$, P_μ, $M_{\mu\nu}$, describing the coordinate symmetry.

References

1. Butten,C. (1969). C.R.Acad. Sci. Paris, ser. A-B, 269, A-87.

2. Tulczyjew,W.M. (1974). Bull. Acad. Polon. Sci., ser. math. 22, 936.

3. Kirillov,A.A. (1977). Funkt. Anal. Appl. 11, 39. (Russian).

4. Leites,D.A. (1977). Dokl. Akad. Nauk SSSR 886, 804. (Russian).

5. Berezin,F.A. (1983). Introduction to algebra and analysis with anticommuting, MGU Moscow (Russian)

6. Volkov, D.V. (1983). Pis'ma Zh. ETPh. 38, 508 (Russian).

7. Volkov, D.V., Soroka, V.A., Tkach, V.I. (1985). In: Proc; VII Int. seminar on high energy phys. and quantum field theory) (July 1984, Protvino) v.1. (Russian)

8. Martin, I.L. (1959). Proc; Royal Soc. A251, 536.

Part IV
NONLINEAR INTEGRABLE EQUATIONS

SOLITONS AS PURELY ALGEBRAIC CONSTRUCTION

A.N. Leznov[+], S.M. Chumakov, V.I. Man'ko
[+]Institute for High Energy Physics, Serpukhov
P.N. Lebedev Physical Institute, Moscow

1. The inverse scattering method [1-8] as well as the dressing method [9-10] for finding soliton solutions of completely integrable systems [11,12,25] often called "the systems integrable by the inverse scattering transform" is now very popular. The intrinsic feature of these methods is that they have become rather complicated in multisoliton case and for many concrete systems explicit expressions for solutions with arbitrary number of solitons have not been found. On the other hand, before seeking soliton-type solutions of some equation one must be sure that this equation really possesses such solutions (i.e. that this equation is completely integrable). However the methods of studying this problem elaborated so far are based upon the study of the internal symmetry algebra of the considered system [13,14] . They are complicated and originally restricted to systems of special types. Here we elaborate a new method for finding soliton-type solutions of completely integrable systems. This method is purely algebraic one and does not involve such additional constructions as investigating the analytic structure of the scattering matrix or solutions of the matrix Riemann problem. We adopt the following plan: we propose a general technique for constructing soliton-type solutions. The question what equations they satisfy is secondary. Therefore we start from the solutions of appropriate type and arrive at the corresponding system

of equations since there are no general abstractly formulated criteria which allow us to single out the systems with required properties. Note that the form of solutions as well as the form of equations satisfied by these solutions are defined by the internal symmetry algebra of the system. (The information about the internal symmetry algebra is contained in "a pramordial function" Φ which is introduced in sect.3).

The method proposed here is the direct development of the ideas of [15,16] and [17-19,25,26]. We do not discuss here the generality of such an approach and the existence of the systems which possess soliton-type solutions but are beyond our consideration. All known and numerous new systems possessing soliton-type solutions are in the framework of our consideration, and in each case we have the possibility to obtain explicit expressions for solutions with arbitrary number of solitons.

2. Completely integrable systems and their soliton solutions are connected with matrix algebras whose elements are rational functions of a complex variable λ called spectral parameter. These algebras are infinite-dimensional, and the construction mentioned above is one of possible realizations of an infinite-dimensional algebra defined by its elements $L_i^{\alpha k}$. Let $\mathcal{L}_i$ be generators of finite-dimensional Lie algebra G. Let us define the new generators $L_i^{\alpha,k}$ by the relations

$$L_i^{\alpha,k} = (\lambda-\alpha)^{-k}\mathcal{L}_i$$

where α are complex constants and k integers (for $k<0$ set $\alpha=0$, i.e. $L_i^{s} = \lambda^s \mathcal{L}_i$). Then it follows:

$$[L_i^{\alpha,k}, L_j^{\beta,\ell}] = (\lambda-\alpha)^{-k}(\lambda-\beta)^{-\ell}C_{ij}^{m}\mathcal{L}_m =$$

$$=C_{ij}^{s}\Big\{\sum_{m=1}^{k} f_m(\alpha,\beta,k,\ell)(\lambda-\alpha)^{-m}\mathcal{L}_s + \sum_{m=1}^{\ell} h_m(\alpha,\beta,k,\ell)(\lambda-\beta)^{-m}\mathcal{L}_s\Big\}=$$

$$=C^{s}_{ij}\{\sum_{m=1}^{k} f_m(\alpha,\beta,k,\ell)L^{\alpha,m}_{s}+\sum_{m=1}^{\ell} h_m(\alpha,\beta,k,\ell)L^{\beta,m}_{s}\}$$

(here C^{k}_{ij} are structural constants of G). Considering the first and the last terms in these equations we get an infinite-dimensional algebra (which however possesses a finite-dimensional matrix representation) with spectral parameter λ .

In such a way for any finite-dimensional algebra G one can introduce the infinite-dimensional algebra $G(\alpha,\beta,\ldots,\gamma)$, where indices $\alpha,\beta,\ldots\gamma$ indicate positions of the poles involved in the definition of structure constants of the new algebra (1). The algebras of this type will be called the rational function algebras. The only systematically studied case is that of Kac-Moody algebras [20].

Note here that although the problem of constructing group elements (exponentiation) for infinite-dimensional Lie algebras generally has no solution, nevertheless it can be solved for some of the algebras considered below. One has to look for a group element corresponding to soliton-type solutions in the matrix form, where the matrix belongs to an appropriate finite-dimensional group representation and its elements depend on a spectral parameter in a special way (see (13) below).

3. We now intend to describe a general method for constructing completely integrable differential equations and their soliton-type solutions (actually, in two-dimensional space). Let g be an element of some finite-dimensional representation of a Lie group G, its parameters depending on variables $z,\bar{z}$. Then elements

$$\dot{g}g^{-1}=U,\quad g'g^{-1}=V \tag{2}$$

(the dot stands for d/dz , the prime stands for $d/d\bar{z}$) belong to the correspondent Lie algebra. There is a relation between U and V (due to the identity $(g^{\cdot})'=(g')^{\cdot}$

$$U' - \dot{V} + [U, V] = 0 \tag{3}$$

called the Maurer-Cartan equation (or a L-A pair).

The relation (3) is trivial until we put additional constraints on U and V. For example, one can require some specific structure of decompositions of U and V in the basis of the algebra. Then relations (2) become a system of equations for parameters of g and, respectively, (3) becomes a system of equations for parameters of U and V.

Consider separately one of the equations (2):

$$\dot{g} = Ug. \tag{4}$$

This is a system of equations for parameters of g, and in this sense, it is independent on the choice of the representation.

On the other hand, (4) may be considered as a matrix equation, where $g(U)$ is a matrix from a finite-dimensional representation of the group. Denote the basis vectors of the representation $|\alpha\rangle, \langle\beta|$ where $1 \leqslant \alpha, \beta \leqslant n$ and n is the dimension of the representation. Eq.(4) allows one to calculate derivatives of g:

$$\ddot{g} \equiv \overset{(\cdot 2)}{g} = (\dot{U} + U^2) g \equiv U_2 g,$$

$$\overset{(\cdot s)}{g} = U_s g, \quad U_{s+1} = \dot{U}_s + U_s U_1, \quad (U_0 = 1). \tag{5}$$

Calculating the matrix elements for the first n derivatives on $\langle 1|$ and $|\alpha\rangle$ we get

$$\langle 1|\overset{(\cdot s)}{g}|\alpha\rangle = \sum_{1 \leqslant \beta \leqslant n} \langle 1|U_s|\beta\rangle\langle\beta|g|\alpha\rangle =$$

$$= \langle 1|U_s|1\rangle\langle 1|g|\alpha\rangle + \sum_{2 \leqslant \beta \leqslant n} \langle 1|U_s|\beta\rangle\langle\beta|g|\alpha\rangle. \tag{6}$$

Eliminating n-1 matrix elements $\langle\beta|g|\alpha\rangle$ $(2 \leqslant \beta \leqslant n)$ from the system of n equation (6) we obtain an ordinary n-th order differential equation for the function $\psi_\alpha \equiv \langle 1|g|\alpha\rangle$:

$$\det_n \left\| \begin{matrix} \cdot & \cdot & \cdots & \cdot \\ \overset{(\cdot s)}{\Psi}_\alpha - \langle 1|U_s|1\rangle \Psi_\alpha, & \ldots & \langle 1|U_s|\beta\rangle & \ldots \\ \cdot & \cdot & \cdots & \cdot \end{matrix} \right\| = 0 \qquad (7)$$

where s indices the rows $(1 \leqslant s \leqslant n)$ and β indices the columns $(2 \leqslant \beta \leqslant n)$ of the determinant. So, all the elements of "the first row" $\langle 1|g|\alpha\rangle$ satisfy the same n-th order differential equation (7) (i.e. the functions $\langle 1|g|\alpha\rangle$ can be considered as its fundamental solutions). The coefficients in this equation are expressed in terms of matrix elements of U and their derivatives up to the (n-1)th order. We will call (7) the spectral equation. All the matrix elements $\langle\beta|g|\alpha\rangle$ $(2 \leqslant \beta \leqslant n)$ can be found from a linear system of the first (n-1) equations (6), $(1 \leqslant s \leqslant n-1)$ and therefore g is explicitly expressed in terms of n fundamental solutions of the spectral equations, the matrix elements of U and their derivatives up to the (n-1)th order. The coefficient functions in the spectral equation can be expressed in terms of its fundamental solutions by the known relations. For this rewrite (7) in the form:

$$\overset{(\cdot n)}{\psi} - \frac{\dot{v}}{v}\overset{(\cdot n-1)}{\psi} + \sum_{k=0}^{n-2} a_k \overset{(\cdot k)}{\psi} = 0. \qquad (8)$$

Denote by ψ_α $(1 \leqslant \alpha \leqslant n)$ the system of the fundamental solutions of (8) and set

$$\left\| \overset{(\cdot s_1)}{\psi} \; \overset{(\cdot s_2)}{\psi} \ldots \overset{(\cdot s_n)}{\psi} \right\| = \begin{vmatrix} \overset{(\cdot s_1)}{\psi}_1 & \overset{(\cdot s_1)}{\psi}_2 & \ldots & \overset{(\cdot s_1)}{\psi}_n \\ \overset{(\cdot s_2)}{\psi}_1 & \overset{(\cdot s_2)}{\psi}_2 & \ldots & \overset{(\cdot s_2)}{\psi}_n \\ \cdot & \cdot & \cdot & \cdot \\ \overset{(\cdot s_n)}{\psi}_1 & \overset{(\cdot s_n)}{\psi}_2 & \ldots & \overset{(\cdot s_n)}{\psi}_n \end{vmatrix} .$$

Then we obtain

$$v = \| \psi \dot{\psi} \dots \overset{(\cdot n-1)}{\psi} \|,$$

$$a_k v = - \| \psi \dot{\psi} \dots \overset{(\cdot k-1)}{\psi} \overset{(\cdot n)}{\psi} \overset{(\cdot k+1)}{\psi} \dots \overset{(\cdot n-1)}{\psi} \| . \tag{9}$$

Comparing with (7) we see that the coefficients a_k are expressed in terms of matrix elements of U_s (see (5)) by the formulas (9) where one should make a change

$$\overset{(\cdot s)}{\psi}_{\alpha} \rightarrow \langle 1 | U_s | \alpha \rangle, \quad U_0 = \hat{1}.$$

The system (2) is stable under the gauge transformation

$$g \rightarrow g_0 g, \quad U \rightarrow g_0 g^{-1} + g_0 U g_0^{-1} . \tag{10}$$

This transformation, in general, changes the system of fundamental solutions of the spectral equations. However, if the gauge transformation does not change the elements of "the first row" of the initial matrix, the system of fundamental solutions is not altered and consequently the spectral equation itself is stable under this transformation.

Now consider the simplest example: the algebra A_1 $(sl(2, R))$ most of the investigated systems are connected with (see sect.4). Then (2) takes the form

$$\dot{g} = (h U_0 + X^+ U_+ + X^- U_-) g,$$

where elements $h, X^{\pm}$ are the basis operators of the algebra such that $[X^+, X^-] = h$, $[h, X^{\pm}] = \pm 2X^{\pm}$. An arbitrary element of $SL(2, R)$ can be represented in the form (Gauss decomposition):

$$g = e^{\alpha X^-} e^{\tau h} e^{\beta X^+} .$$

Then

$$\dot{g} g^{-1} = (\dot{\tau} - \alpha \dot{\beta} e^{2\tau}) h + \dot{\beta} e^{2\tau} X^+ + (\dot{\alpha} + 2\alpha \dot{\tau} - \\ - \alpha^2 \dot{\beta} e^{2\tau}) X^- = h U_0 + X^+ U_+ + X^- U_- .$$

We obtain a system of equations independent of the representation:

$$\dot{\tau}-\alpha\dot{\beta}e^{2\tau}=U_0, \quad \dot{\beta}e^{2\tau}=U_+,$$

$$\dot{\alpha}+2\alpha\dot{\tau}-\dot{\beta}\alpha^2e^{2\tau}=U_-.$$

One can easily verify that the functions $\Psi_1\equiv e^{\tau}, \Psi_2\equiv\beta e^{\tau}$ satisfy the following 2nd-order linear differential equation

$$\ddot{\psi}-\frac{\dot{U}_+}{U_+}\dot{\psi}=\left[U_+\left(\frac{U_0}{U_+}\right)^{\cdot}+U_0^2+U_+U_-\right]\psi \tag{11}$$

which is just a spectral equation.

Let us also perform the calculation by the matrix method. The basis of the spinor representation of consists of two vectors $|1\rangle, \; X^-|1\rangle$. According to (7) we have

$$\det{}_2\begin{bmatrix}\dot{\Psi}_\alpha-U_0\Psi_\alpha & U_+\\ \ddot{\Psi}_\alpha-(\dot{U}_0+\frac{1}{2}SpU^2)\Psi_\alpha & \dot{U}_+\end{bmatrix}=$$

$$=-U_+\left\{\ddot{\Psi}_\alpha-\frac{\dot{U}_+}{U_+}\dot{\Psi}_\alpha-\left(\dot{U}_0-\frac{U_0\dot{U}_+}{U_+}+\frac{SpU^2}{2}\right)\Psi_\alpha\right\}=0,$$

which coincides with (11). Matrix elements of g is expressed in terms of Ψ_1, Ψ_2 and their derivatives

$$g=\begin{bmatrix}\Psi_1 & \Psi_2\\ (\dot{\Psi}_1-U_0\Psi_1)U_+^{-1} & (\dot{\Psi}_2-U_0\Psi_2)U_+^{-1}\end{bmatrix}. \tag{12}$$

The relations (11), (12) are the starting point for constructing soliton-type solutions connected with the rational function algebra $sl(2R)(\alpha_1,\alpha_2,\dots)$. (In fact, our treatment is valid in the case of the algebra $sl(2,C)$ instead of $sl(2,R)$).

3. In this section we carry out an explicit construction of the soliton-type solutions. We suggest that the following structure of a group element $g(z, \bar{z}; \lambda)$ leads to the soliton solutions:

$$g = g_0(\lambda)\exp(h\Phi) \tag{13}$$

where $\exp(h\Phi)$ is an element $\left(h \equiv \left|\begin{smallmatrix}1 & 0\\ 0 & -1\end{smallmatrix}\right|\right)$ of the Cartan subgroup, the matrix g_0 has the elements which are polynomial in λ, the function Φ is of the form $\Phi = \varphi(z, d) + \bar{\varphi}(\bar{z}, \lambda) + f(z, \bar{z})$. Here $\varphi(z, \lambda)$, $\bar{\varphi}(\bar{z}, \lambda)$ are arbitrary rational functions in λ, $f(z, \bar{z})$ is independent of λ.

In accordance with (12), (13) it is convenient to denote the first row matrix element of g:

$$\Psi_1 = e^{\Phi}\prod_{i=1}^{N_1}(a_i - \lambda) \equiv e^{\Phi}\tilde{\Psi}_1, \quad \Psi_2 = e^{-\Phi}\prod_{j=1}^{N_2}(b_j - \lambda) \equiv e^{-\Phi}\tilde{\Psi}_2. \tag{14}$$

Let the functions $a_i(z, \bar{z})$, $b_j(z, \bar{z})$, $f(z, \bar{z})$ be determined from the condition of linear dependence of the functions Ψ_1 and Ψ_2 at $(N_1 + N_2 + 1)$ different points of the λ-plane:

$$[\Psi_1 + C(\lambda)\Psi_2]\big|_{\lambda = \lambda_s} = 0, \quad 1 \leqslant s \leqslant N_1 + N_2 + 1 \tag{15}$$

(here $C(\lambda_s)$ is a set of arbitrary constants). It follows from (12), (15) that

$$\det g = P(\lambda) \equiv \prod_{s=1}^{N_1+N_2+1}(\lambda_s - \lambda).$$

It is convenient to consider first the following element of the group $L(2, R)$:

$$\tilde{g} = \begin{bmatrix} \Psi_1 & \Psi_2 \\ 0 & P/\Psi_1 \end{bmatrix} = \begin{bmatrix} \tilde{\Psi}_1 & \tilde{\Psi}_2 \\ 0 & P/\tilde{\Psi}_1 \end{bmatrix} e^{h\Phi}.$$

Further, perform a gauge transformation

$$g = \begin{bmatrix} 1 & 0 \\ \xi & 1 \end{bmatrix} \tilde{g} = \begin{bmatrix} \tilde{\Psi}_1 & \tilde{\Psi}_2 \\ \xi\tilde{\Psi}_1 & \xi\tilde{\Psi}_2 + P/\tilde{\Psi}_1 \end{bmatrix} e^{h\Phi} \qquad (16)$$

so that according to (13), g_{22} would have no poles at the points $\lambda = a_i$. This condition determines the part of ξ decreasing as $\lambda \to \infty$:

$$\xi = \sum_{i=1}^{N_1} \frac{A_i}{a_i - \lambda}, \quad A_i \equiv \frac{-P(a_i)}{\prod_{k=1}^{N_2}(b_k - a_i)\prod_{j\neq i}^{N_1}(a_j - a_i)} .$$

Now, define the function $U_+(\lambda)$ setting

$$P(\lambda)U_+(\lambda) = \det\begin{bmatrix} \Psi_1 & \Psi_2 \\ \dot{\Psi}_1 & \dot{\Psi}_2 \end{bmatrix} =$$

$$= \prod_{i=1}^{N_1}(a_i - \lambda)\prod_{j=1}^{N_2}(b_j - \lambda)\left\{-2\dot{\Phi} + \sum_{j=1}^{N_2}\frac{\dot{b}_j}{b_j - \lambda} - \sum_{i=1}^{N_1}\frac{\dot{a}_i}{a_i - \lambda}\right\}. \qquad (17)$$

Calculating the residues at the poles $\lambda = a_i$ we get from (17):

$$\dot{a}_i = \frac{-P(a_i)U_+(a_i)}{\prod_m (b_m - a_i)\prod_{k\neq i}(a_k - a_i)} = A_i U_+(a_i). \qquad (18)$$

(Here and below the notation F(a) for a rational function $F(\lambda)$ means that the argument λ in the corresponding series for $F(\lambda)$ is replaced by a). Note, that from (16)-(18) it follows that g depends on a_i, b_j, Φ involved in the definition of Ψ_1, Ψ_2 but does not actually depend on $\dot{a}_i$, $\dot{b}_j$.

One can further calculate g_{21} and g_{22} :

$$\xi = \sum \frac{\dot{a}_i U_+^{-1}(a_i)}{(a_i - \lambda)} = \sum \frac{\dot{a}_i}{a_i - \lambda}\left\{\frac{1}{U_+(a_i)} - \frac{1}{U_+(\lambda)}\right\} +$$

$$+ \frac{1}{U_+(\lambda)}\sum\frac{\dot{a}_i}{a_i - \lambda} \equiv \frac{1}{U_+(\lambda)}\left\{-\tilde{U}_0 + \dot{\tilde{\Psi}}_1/\tilde{\Psi}_1\right\},$$

$$\tilde{U}_0 \equiv \sum \frac{\dot{a}_i}{U_+(a_i)} \left\{ \frac{U_+(a_i) - U_+(\lambda)}{a_i - \lambda} \right\},$$

$$g_{21} = \frac{\dot{\tilde{\Psi}}_1 - \tilde{U}_0 \Psi_1}{U_+}, \quad g_{22} = \frac{\dot{\tilde{\Psi}}_2 - (U_0 + 2\dot{\Phi})\tilde{\Psi}_2}{U_+}.$$

Therefore, the matrix g takes the form (12) if we set

$$U_0 = \tilde{U}_0 + \dot{\Phi} = \dot{\Phi} + \sum_{i=1}^{N_1} \frac{\dot{a}_i}{U_+(a_i)} \left\{ \frac{U_+(a_i) - U_+(\lambda)}{a_i - \lambda} \right\}. \tag{19}$$

Consequently, g satisfies (4):

$$\dot{g} = \begin{bmatrix} U_0 & U_+ \\ U_- & -U_0 \end{bmatrix} g,$$

where U_+, U_0 are defined by (17), (19) and accordingly Ψ_1, Ψ_2 are the fundamental solutions of the spectral equation (11), and U_- is defined by the relation

$$U_- = \frac{1}{\Psi_1} \left(\frac{\dot{\Psi}_1}{U_+} \right)^{\cdot} - \left(\frac{U_0}{U_+} \right)^{\cdot} - \frac{U_0^2}{U_+}.$$

After simple manipulations (see Appendix) we get

$$U_- = -2 \sum_{i=1}^{N_1} \dot{a}_i [\dot{\Phi}(a_i) - \dot{\Phi}(\lambda)] U_+^{-1}(a_i)(a_i - \lambda)^{-1} + \\ + \sum_{i=1}^{N_1} \sum_{j=1}^{N_1} \frac{\dot{a}_i \dot{a}_j (a_i - \lambda)^{-1}}{U_+(a_i) U_+(a_j)} \left\{ \frac{U_+(a_j) - U_+(\lambda)}{a_j - \lambda} - \frac{U_+(a_j) - U_+(a_i)}{a_j - a_i} \right\}. \tag{20}$$

Note that due to the definition (15) of the functions a_i, b_j, f the element $U_+(\lambda)$ has no poles at the points λ_s. Consequently its analytic structure in the λ-plane is fully determined by that of the function $\Phi(\lambda)$. The same holds for U_0, U_-.

Note that g does not actually depend on the derivatives $\dot{a}_i$, $\dot{b}_j$ and consequently, one can perform the described above procedure with any other variable which we will denote by $\bar{z}$, i.e.

$$g' \equiv \partial g / \partial \bar{z} = \begin{bmatrix} V_0 & V_+ \\ V_- & -V_0 \end{bmatrix} g = V g \tag{21}$$

where V_+, V_0, V_- are defined by the formulas which are analogous to (17), (19), (20).

The compatibility condition for eqs. (4), (21) (i.e. the Maurer-Cartan relations (3)) leads (after equating residues in the right hand side of (3)) to a system of nonlinear equations satisfied by the soliton-type solutions defined by (17), (19), (20).

Due to the symmetry of the formulas (17), (19), (20) under all the permutations of a_i's (or all permutations of b_j's) the elements U_+, U_0, U_- as well as V_+, V_0, V_- can be expressed in terms of elementary symmetric functions constructed from a_i's (or b_j's):

$$S_1(a) = \sum_{i=1}^{N_1} a_i, \quad S_2(a) = \sum_{i<j} a_i a_j, \ldots \quad S_{N_1}(a) = \prod_{i=1}^{N_1} a_i .$$

On the other hand the algebraic system of equations (15) is the linear one with respect to $S_1(a), S_2(a), \ldots S_{N_1}(a)$, $S_1(b), \ldots S_{N_2}(b)$ and $\exp f$. Therefore the formulas (17), (19), (20) lead actually to the explicit expressions for soliton solutions in terms of solutions of the linear algebraic system of equations (15). The examples will be considered in the next section.

Here we point out that all the information on the properties of the internal symmetry algebra is concentrated in the analytic structure of the dependence of Φ on λ. It defines by means of the formulas (17), (19), (20) the analytic structure of the elements U and V which take values in the rational function algebra $sl(2,R)(\alpha_1, \alpha_2, \cdots)$ (Here $\alpha_1, \alpha_2 \ldots$ are the poles involved in the expression of Φ).

4. Interesting cases of the construction which has been developed above arise if the spectral equation (11) is

invariant under some transformation of the spectral parameter λ. Then one can set $\Psi_1 = \Psi_2(\tilde{\lambda})$, where $\tilde{\lambda}$ is a transformed value of the parameter. In particular, the invariance under the reflection $\lambda \to -\lambda$ leads to the classical Korteweg-de Vries and Sine Gordon equations. We begin with constructing their soliton solutions.

Let us consider the simplest possible form of the function Φ involved in (14): $\Phi = \lambda z + \bar{y}(\lambda)$ (where we assume $\bar{y}(-\lambda) = -\bar{y}(\lambda)$) and realize the above procedure with respect to z. The functions Ψ_1, Ψ_2 defined by (14) take the form

$$\Psi_1 = \exp(\lambda z + \bar{y}(\lambda)) \prod_{i=1}^{N} (a_i - \lambda), \quad \Psi_2 = \exp(-\lambda z - \bar{y}(\lambda)) \prod_{i=1}^{N} (a_i + \lambda).$$

Set

$$P(\lambda) = \det g = -2\lambda \prod_{s=1}^{N} (\lambda_s^2 - \lambda^2)$$

(the factor λ appears due to linear dependence of the functions Ψ_1, Ψ_2 at $\lambda = 0$, the product contains only even powers of λ since $P(\lambda)$ is stable under the transformation $\lambda \to -\lambda$, and the factor -2 is introduced for convenience).

The relation (17) which defines U_+ takes the form:

$$\prod_{i=1}^{N} (a_i^2 - \lambda^2) \left[1 + \sum_{i=1}^{N} \frac{\dot{a}_i}{a_i^2 - \lambda^2} \right] = \prod_{s=1}^{N} (\lambda_s^2 - \lambda^2) U_+ . \qquad (22)$$

In (22) tending λ to ∞ we find $V = 1$. Further, it follows from (19), (20) that $U_0 = \dot{\Phi} = \lambda$, $U_- = -2 \sum_{1 \le i \le N} \dot{a}_i$, i.e. U is of the form

$$U = \begin{bmatrix} \lambda & 1 \\ -2\sum \dot{a}_i & -\lambda \end{bmatrix} . \qquad (23)$$

Now, take $\bar{y}(\lambda) = \lambda^3 \bar{z}$ and make similar calculations with respect to $\bar{z}$. From (17) we obtain:

$$\prod_{i=1}^{N}(a_i^2-\lambda^2)\left[\lambda^2+\sum_{j=1}^{N}\frac{a_j'}{a_j^2-\lambda^2}\right]=\prod_{s=1}^{N}(\lambda_s^2-\lambda^2)\left[A+\lambda^2 B\right]. \quad (24)$$

In order to determine A and B one can consider (24) as $\lambda\to\infty$ and at $\lambda=0$. One gets accordingly $B=1$ and

$$A\prod_{s=1}^{N}\lambda_s^2=\left(\prod_{i=1}^{N}a_i^2\right)\left(\sum_{i=1}^{N}a_i'a_i^{-2}\right).$$

Moreover, calculating residues at the poles $\lambda=a_i$ in (22), (24) one gets

$$\dot{a}_i=\prod_s(\lambda_s^2-a_i^2)\Big/\prod_{j\neq i}(a_j^2-a_i^2)$$

$$a_i'=\frac{\prod_s(\lambda_s^2-a_i^2)}{\prod_{j\neq i}(a_j^2-a_i^2)}\,(A+a_i^2)=\dot{a}_i(A+a_i^2).$$

It follows from (22) (if we put $\lambda=0$) that $\left(\prod_i a_i^2\right)\left(1+\sum_i\dot{a}_i a_i^{-2}\right)=\prod_s\lambda_s^2$, therefore $A=\sum\dot{a}_i$, and $V_+=\lambda^2+\sum\dot{a}_i$.
It is convenient to introduce the function $u\equiv\sum_{i=1}^{N}\dot{a}_i$.
The calculations of V_0, V_- according to (19), (20) give

$$V_0=\lambda^3+\sum\frac{a_i'(a_i+\lambda)}{a_i^2+A}=\lambda^3+\lambda\sum\dot{a}_i+\sum\dot{a}_i a_i,$$

$$V_-=-2\sum\dot{a}_i(a_i^2+\lambda a_i+\lambda^2)-\left(\sum\dot{a}_i\right)^2.$$

Further, it is convenient to use a system of 2nd-order equations satisfied by a_i's (see Appendix):

$$\ddot{a}_i+2\dot{a}_i a_i+2\dot{a}_i\sum_{j\neq i}\frac{\dot{a}_j}{a_j-a_i}=0. \quad (25)$$

Summing over i we get

$$2\sum\dot{a}_i a_i=-\sum\ddot{a}_i=-\dot{u}.$$

Further, multiplying (25) by a_i and summing over i we get

$$-2\sum \dot{a}_i a_i^2 = \left(\sum \dot{a}_i a_i\right)^{\cdot} - \left(\sum \dot{a}_i\right)^2,$$

i.e.

$$\sum \dot{a}_i a_i^2 = \ddot{u}/4 + u^2/2.$$

An arbitrary moment $J_n \equiv \sum \dot{a}_i a_i^n$ is expressed in terms of u and its derivatives $d^k u/dz^k$, $1 \leqslant k \leqslant n$; the recurrency is presented in Appendix.

Therefore,

$$V = \begin{bmatrix} \lambda^3 + \lambda u - \dot{u}/2 & \lambda^2 + u \\ -2\lambda^2 u + \lambda \dot{u} - 2u^2 - \ddot{u}/2 & -\lambda^3 - \lambda u + \dot{u}/2 \end{bmatrix}.$$

Now, substituting U and V into (3) leads to an equation satisfied by the function $u \equiv \sum_{i=1}^{N} \dot{a}_i$:

$$2u' - \dddot{u}/2 - 6u\dot{u} = 0.$$

This is the KdV equation. Recall that in the framework of our approach the solution u is determined by condition (15) which is a linear system of algebraic equations for the elementary homogeneous symmetric polynomials $S_k(a)$ of a_i's. The system (15) in this case takes the form

$$\sum_{k=1}^{N} S_k(a)\lambda^{N-K}\left[1 + (-)^{N-K}\exp(2\Phi_s)\right] = -\lambda^N\left[1 + (-)^N \exp(2\Phi_s)\right], \tag{26}$$

where $1 \leqslant s \leqslant N$, $\Phi_s \equiv \Phi|_{\lambda=\lambda_s} = \lambda_s z + \lambda_s^3 \bar{z}$, and we get the well-known KdV soliton solutions in the form $u = \sum \dot{a}_i = \dot{S}_1(a)$, where $S_1(a)$ is the ratio of appropriate determinants (see (26)).

Since all the moments $J_n \equiv \sum \dot{a}_i a_i^n$ are expressed in terms of u and its derivatives (see Appendix), every polynomial dependence $\bar{\varphi}(\lambda)$ (such that $-\bar{\varphi}(-\lambda)=\varphi(\lambda)$) leads to an equation of KdV -type, for which the order of the highest derivative equals the power of the corresponding polynomial. Soliton solutions of each equation can be easily found in terms of solution $u=\dot{S}_1$ of the system (26), where one has to put $\Phi_s=\Phi|_{\lambda=\lambda_s}=\lambda_s z+\bar{\varphi}(\lambda_s)$. On the other hand, in order to obtain an explicit expression for the equation itself one has to make rather complicated calculations. These equations are called higher order KdV equations [2,21] .

Let us consider now the second example, when $\bar{\varphi}(\lambda)=$ $=\lambda^{-1}\bar{z}$. The basic relation (17) with respect to the variable z and the form (23) of the matrix U do not change. With respect to the variable $\bar{z}$ eq. (17) takes the form

$$\prod_{i=1}^{N}(a_i^2-\lambda^2)\Big[1+\lambda^2\sum\frac{a_i'}{a_i^2-\lambda^2}\Big]=\lambda^2\prod_{s=1}^{N}(\lambda_s^2-\lambda^2)V_+ \,.$$

Consequently, we obtain

$$V_+=\lambda^{-2}(1-\sum a_i')\equiv\lambda^{-2}e^{2\rho}$$

where we have introduced a function $\rho(z,\bar{z})$. By analogy with the first example, the constraints $\dot{a}_i=a_i'\times a^2(1-\sum a_i')^{-1}$ lead to the identities $\exp(2\rho)\equiv 1-\sum a_i'$ $=\prod_i a_i^2/\prod_s\lambda_s^2=(1+\sum\dot{a}_i a_i^{-2})^{-1}$(we have used the relation (22) at $\lambda=0$). Hereafter in this example set $\prod_{s=1}^{N}\lambda_s^2=1$. It follows from (19), (20):

$$V_0=\lambda^{-1}\exp(2\rho)-\lambda^{-2}\dot{\rho}\exp(2\rho),$$

$$V_-=2\lambda^{-1}\dot{\rho}-\lambda^{-2}(\dot{\rho}^2-2\lambda\dot{\rho}(1-e^{-2\rho}))e^{2\rho}=$$

$$=(2\lambda^{-1}\dot{\rho}-\lambda^{-2}\dot{\rho}^2)e^{2\rho}.$$

Substituting U and V into the Maurer-Cartan

equation (3) with the easily verified relation $-2\sum \dot{a}_i = \dot{\varrho}^2 + \ddot{\varrho}$ leads to the following equation:

$$(e^{4\varrho} - \dot{\varrho}' e^{2\varrho})^{\cdot} = 0,$$

which after simple manipulations leads to the Sine-Gordon equation:

$$\dot{\varrho}' = e^{2\varrho} - e^{-2\varrho}.$$

Note, that it follows from (15) which explicitly determines all the elementary symmetric functions $S_k(a)$ and, in particular, $S_N(a) = \prod a_i \equiv \exp\varrho$, that $\lim\limits_{z\to\pm\infty} \exp\varrho = \lim S_N(a) = 1$. As in the case of KdV, every polynomial dependence of the function $\bar{\varphi}(\lambda)$ (such that $\bar{\varphi}(-\lambda) = -\bar{\varphi}(\lambda)$) on λ^{-1} leads to higher order Sine-Gordon equations since all the moments $\sum \dot{a}_i a_i^{-k}$ are expressed in terms of ϱ and its derivatives.

The example which contains the KdV and Sine-Gordon equations as its limit cases arises if the function φ is chosen in the form: $\varphi = \lambda z + (\alpha\lambda^{-1} + \beta\lambda^3)\bar{z}$, ($\alpha$ and β are arbitrary constants). Then the basic relation (17) (with respect to the variables z and $\bar{z}$) is of the form:

$$\prod(a_i^2 - \lambda^2)\left[\frac{\alpha}{\lambda^2} + \beta\lambda^2 + \sum \frac{a_i'}{a_i^2 - \lambda^2}\right] =$$

$$= \left(\frac{A}{\lambda^2} + B + C\lambda^2\right) P(\lambda) =$$

$$= \left(\frac{A}{\lambda^2} + B + C\lambda^2\right) \prod(a_i^2 - \lambda^2)\left[1 + \sum \frac{\dot{a}_i}{a_i^2 - \lambda^2}\right].$$

Comparing the coefficients of the same powers of λ we get

$$C = \beta, \quad A = \alpha(1 + \sum \dot{a}_i a_i^{-2})^{-1} \equiv \alpha \exp(2\varrho),$$

$$B = \beta \sum \dot{a}_i \equiv \beta u.$$

The elements of V are

$$V_+ = \beta(\lambda^2 + u) + \alpha e^{2\rho}\lambda^{-2},$$
$$V_0 = \beta(\lambda^3 + \lambda u - \dot{u}/2) + \alpha e^{2\rho}(\lambda^{-1} - \dot{\rho}\lambda^{-2}),$$
$$V_- = \beta(-\ddot{u}/2 - 2u^2 + \lambda\dot{u} - 2\lambda^2 u) +$$
$$+ \alpha e^{2\rho}(2\dot{\rho}\lambda^{-1} - \dot{\rho}^2\lambda^{-2}),$$

(the matrix U is of the form (23)). Substituting U and V into (3) leads to the equation

$$2u' = \beta(\dddot{u}/2 + 6u\dot{u}) - 4\alpha\dot{\rho}e^{2\rho}, \quad -2u = \ddot{\rho} + \dot{\rho}^2$$

which has soliton solutions for arbitrary β and α and reduces to KdV for $\alpha = 0$ and to Sine-Gordon for $\beta = 0$.

Now apply the above technique to the case $\Phi = \lambda z + \lambda^{-1}\bar{z} + f(z, \bar{z})$ non-invariant with respect to any transformations of the spectral parameter. (This case is in some sense the nearest one to the Sine-Gordon equation). Here the function $\Phi(\lambda)$ contains a zeroth component, f. The relations (17), (20), with respect to z give: $U_+ = 1, U_0 = \lambda + \dot{f}, U_- = -2\sum_{i=1}^{N_1}\dot{a}_i$. The same relations with respect to $\bar{z}$ give

$$V_+ = \lambda^{-1}f', \; V_0 = f' + \lambda^{-1}(1 - f'\sum \dot{a}_i a_i^{-1}),$$
$$V_- = \lambda^{-1}\{2\sum \dot{a}_i a_i^{-1} - f'(\sum \dot{a}_i a_i^{-1})^2\}.$$

Due to 2nd order differential equations for the functions a_i (eqs. (A3) of Appendix; cf. similar relations in the theory of KdV and Sine-Gordon eqs.) the functions $\sum_{i=1}^{N_1}\dot{a}_i a_i^{-1} \equiv x$ and $\sum_{i=1}^{N_1}\dot{a}_i \equiv u$ are connected by the relation

$$-2u = \dot{x} + x^2 + 2\dot{f}x.$$

Substituting U and V into (3) leads to the equations:

$$2f'x = \dot{f}' - 2\dot{f}f' + 2, \quad u' + 2uf' + 2x - f'x^2 = 0.$$

Algebraically eliminating u and x from this system of three equations we get a 4th-order equation satisfied by f. (Recall that the function $\exp f$, as well as the functions $S_k(a)\ (1 \leq k \leq N_1)$, $S_m(b)\ (1 \leq m \leq N_2)$, is explicitly determined from the linear algebraic system (15)).

In order to find for this case a convenient form of equations of motion let us perform the gauge $\left(G_0 = \begin{bmatrix} e^{-f} & 0 \\ 0 & e^{f} \end{bmatrix}\right)$ transformation:

$$\tilde{U} = \dot{G}_0 G_0^{-1} + G_0 U G_0^{-1} = \begin{bmatrix} \lambda & e^{-2f} \\ -2ue^{2f} & -\lambda \end{bmatrix} \equiv \begin{bmatrix} \lambda & u_+ \\ u_- & -\lambda \end{bmatrix},$$

$$\tilde{V} = G_0' G_0^{-1} + G_0 V G_0^{-1} = \begin{bmatrix} 1 - f'x & f'e^{-2f} \\ e^{2f}(2x - f'x^2) & f'x - 1 \end{bmatrix} \equiv \lambda^{-1} \begin{pmatrix} v_0 & v_+ \\ v_- & -v_0 \end{pmatrix}.$$

The corresponding system of equations

$$\dot{v}_0 = u_+ v_- - v_+ u_-, \quad v_\pm = \mp u_\pm' / 2, \quad \dot{v}_\pm = \mp 2 v_0 u_\pm$$

reduces to the system of three equations for v_0, u_+, u_-:

$$2\dot{v}_0 = (u_+ u_-)', \quad \dot{u}_\pm' = 4 v_0 u_\pm .$$

This system is connected with the well-known Pohlmeyer-Lund-Regge model. In order to verify this let us substitute

$$u_+ = -u_-^* = (\dot{\alpha}/2 + i\tilde{u})e^{iw}, \quad v_+ = v_-^* = \sin\alpha\, e^{iw}, \quad v_0 = v_0^* = -\cos\alpha$$

($\alpha(z,\bar{z})$, $w(z,\bar{z})$, $\tilde{u}(z,\bar{z})$ are real functions). Then equations of motion take the form: $\tilde{u} = \frac{\dot{w}}{2}\,\mathrm{tg}\,\alpha$ and

$$\dot{\alpha}' - \dot{w} w' \,\mathrm{tg}\,\alpha + 4 \sin\alpha = 0$$

$$\dot{w}' + \frac{\dot{w}\alpha'}{\cos\alpha\sin\alpha} + w'\dot{\alpha}\,\mathrm{ctg}\,\alpha = 0.$$

Set $\dot{w} = \dfrac{\dot{\beta}\cos\alpha}{1+\cos\alpha}$, $w' = \dfrac{\beta'}{1+\cos\alpha}$.

We get the equations of Pohlmeyer-Lund-Regge model [22,23]:

$$\dot{\alpha}' + 4\sin\alpha - \frac{\mathrm{tg}^2(\alpha/2)}{\sin\alpha}\dot{\beta}\beta' = 0$$

$$\dot{\beta}' + \frac{\dot{\alpha}\beta' + \alpha'\dot{\beta}}{\sin\alpha} = 0.$$

Another simple example of non-symmetric case leading to the well-known nonlinear Schrodinger equation arises if one chooses $\Phi = \lambda z + \lambda^2\bar{z} + f(z, \bar{z})$. The general scheme leads to the following expressions for the matrix elements of U and V:

$$U_+ = 1,\ U_0 = \lambda + \dot{f},\ U_- = -2u \quad (u \equiv \sum \dot{a}_i)$$

$$V_+ = \lambda - \dot{f},\ V_0 = \lambda^2 + f' + u,\ V_- = -2\lambda u + \dot{u} + 2\dot{f}u.$$

The functions f and u satisfy the system of equations

$$2u = \ddot{f} - 2\dot{f}^2 - 2f',\ \ddot{u} + 4\ddot{f}u + 4\dot{f}\dot{u} + 2u' = 0$$

(or, respectively, f satisfies one 4th-order equation). Substituting $u = -q\varepsilon/2,\ \varepsilon = e^{-2f}$ we get the well-known system of equations:

$$\varepsilon' - \ddot{\varepsilon}/2 + q\varepsilon^2 = 0,\quad q' + \ddot{q}/2 - q^2\varepsilon = 0.$$

(If we put $q = \pm\varepsilon^*,\ ' \equiv i\partial t,\ ^{\cdot} \equiv i\partial x$ then we get the nonlinear Schrodinger equation: $i\varepsilon_t + \varepsilon_{xx}/2 \pm \pm|\varepsilon|^2\varepsilon = 0$.)

At the end of the section consider an arbitrary rational dependence of $\varphi(\lambda)$ and $\bar{\varphi}(\lambda)$ on λ:

$$\Phi = \varphi(\lambda) + \bar{\varphi}(\lambda) + f,$$

$$\varphi(\lambda)=\sum_{\alpha=1}^{M}\sum_{s=1}^{n_\alpha}\frac{\varphi^s_\alpha(z)}{(\lambda-\mu_\alpha)^s}+\sum_{s=1}^{n+1}\varphi_s(z)\lambda^s,$$
$$\bar\varphi(\lambda)=\sum_{\beta=1}^{\bar M}\sum_{t=1}^{\bar n_\beta}\frac{\bar\varphi^t_\beta(\bar z)}{(\lambda-\nu_\beta)^t}+\sum_{t=1}^{\bar n+1}\bar\varphi_t(\bar z)\lambda^t. \tag{26}$$

Here the functions $\varphi^s_\alpha(z), \varphi_s(z), \bar\varphi^t_\beta(\bar z), \bar\varphi_t(\bar z)$, constants μ_α and ν_β and integers $M, \bar M, n_\alpha, \bar n_\beta, n, \bar n$ are arbitrary.

The basic relation (17) with respect to z takes the form:

$$U_+=\frac{\prod(a_i-\lambda)\prod(b_j-\lambda)}{-2P(\lambda)}\left\{-2\dot\Phi+\sum\frac{\dot b_j}{(b_j-\lambda)}-\sum\frac{\dot a_i}{a_i-\lambda}\right\},$$
$$P(\lambda)\equiv\prod_{s=1}^{N_1+N_2+1}(\lambda_s-\lambda).$$

(We assume that all the constants $\lambda_s, \mu_\alpha, \nu_\beta$ are different). Consequently,

$$U_+=\sum_{\alpha=1}^{M}\sum_{s=1}^{n_\alpha}\frac{(U_+)^s_\alpha}{(\lambda-\mu_\alpha)^s}+\sum_{s=0}^{n}(U_+)_s\lambda^s. \tag{27}$$

Introducing $F(\lambda)\equiv P^{-1}(\lambda)\prod_{i=1}^{N_1}(a_i-\lambda)\prod_{j=1}^{N_2}(b_j-\lambda)$ we see that

$$(U_+)^{n_\alpha}_\alpha=F(\mu_\alpha)\dot\varphi^{n_\alpha}_\alpha,$$
$$(U_+)^{n_\alpha-1}_\alpha=F^{(1)}(\mu_\alpha)\dot\varphi^{n_\alpha}_\alpha+F(\mu_\alpha)\dot\varphi^{n_\alpha-1}_\alpha, \tag{28}$$
$$\ldots$$
$$(U_+)^{n_\alpha-s}_\alpha=\sum_{k=0}^{s}\frac{F^{(k)}(\mu_\alpha)\dot\varphi^{n_\alpha-s+k}_\alpha}{k!},$$
$$F^{(k)}(\mu)\equiv\left.\frac{d^kF}{d\lambda^k}\right|_{\lambda=\mu}.$$

To calculate $(U_+)_s$ it is convenient to rewrite $F(\lambda)$ in the form

$$F(\lambda)=\sum_{s=1}^{N_1+N_2+1}\frac{A_s}{\lambda_s-\lambda}\,,\quad A_s\equiv\frac{\prod_{i=1}^{N_1}(a_i-\lambda_s)\prod_{j=1}^{N_2}(b_j-\lambda_s)}{\prod_{k\neq s}(\lambda_k-\lambda_s)}\,.$$

Now, taking into account the form (26) of $\varphi(\lambda)$ we find

$$(U_+)_s=-\sum_{\ell=1}^{N_1+N_2+1}A_\ell\sum_{m=s+1}^{n+1}\dot{\varphi}_m\lambda_\ell^{m-s-1}\,, \tag{29}$$

where $0\leqslant s\leqslant n$.
According to (19) we get

$$U_0=\dot{\Phi}+\sum_i\frac{\dot{a}_i}{U_+(a_i)}\frac{U_+(a_i)-U_+(\lambda)}{a_i-\lambda}=$$

$$=\sum_{\alpha=1}^{M}\sum_{s=1}^{n_\alpha}\left\{\frac{\dot{\varphi}_\alpha^s}{(\lambda-\mu_\alpha)^s}+(U_+)_\alpha^s\sum_{i=1}^{N_1}\frac{\dot{a}_i\left[(a_i-\mu_\alpha)^{-s}-(\lambda-\mu_\alpha)^{-s}\right]}{U_+(a_i)(a_i-\lambda)}\right\}+$$

$$+\dot{f}+\sum_{s=1}^{n+1}\dot{\varphi}_s\lambda^s+\sum_{s=0}^{n}\sum_{i=1}^{N_1}\frac{\dot{a}_i(U_+)_s}{U_+(a_i)}\left[\frac{a_i^s-\lambda^s}{a_i-\lambda}\right]$$

and

$$(U_0)_\alpha^s=\dot{\varphi}_\alpha^s-\sum_{k=s}^{n_\alpha}(U_+)_\alpha^k J_\alpha^{k-s+1}\,,$$

$$(U_0)_s=\dot{\varphi}_s+\sum_{m=0}^{n-s-1}J_m(U_+)_{m+s+1}\,, \tag{30}$$

$$J_\alpha^s\equiv\sum_{i=1}^{N_1}\frac{\dot{a}_1}{U_+(a_i)(a_i-\mu_\alpha)^s}\,,$$

$$J_m\equiv\sum_{i=1}^{N_1}\frac{\dot{a}_i a_i^m}{U_+(a_i)}\,,\quad \varphi_0\equiv f.$$

From (20) we similarly get

$$(U_-)^s_\alpha = 2\sum_{k=s}^{n_\alpha} \dot{\varphi}^k_\alpha J^{k-s+1}_\alpha - \sum_{k=s}^{n_\alpha} (U_+)^k_\alpha \sum_{\ell=0}^{k-s} J^{k-s-\ell+1}_\alpha J^{\ell+1}_\alpha ,$$

$$(U_-)_s = -2\sum_{m=0}^{n-s} \dot{\varphi}_{m+s+1} J_m - \tag{31}$$

$$- \sum_{k=s+2}^{n} (U_+)_k \sum_{m=1}^{k-s-1} J_{m-s} J_{k-m-s-1} .$$

Note, that all the functions F, A_i, J^s_α, J_m are symmetric under all the permutations of a_i's (or b_j's) and therefore all the components of U are defined in terms of $S_k(a), S_m(b)$ determined from (15). The calculations of the components of matrix elements of V are similar. The expressions for these components $(V_{\pm,0})^t_\beta$, $(V_{\pm,0})_t$ are obtained from the corresponding expressions (28)-(31) for U replacing $U \to V$, $\dot{} \to {}'$, $\varphi \to \bar{\varphi}$, $\mu_\alpha \to \nu_\beta$, etc.

The components of U and V constructed above satisfy a system of equations obtained from (3) equating the coefficients at the poles in λ-plane. One more simple example, different from those above, is $\Phi = z/(\lambda-1) + \bar{z}/(\lambda+1)$, $U = \tilde{U}/(\lambda-1)$, $V = \tilde{V}/(\lambda+1)$. The corresponding nonlinear equations

$$\tilde{U}' = \dot{\tilde{V}} = \frac{1}{2}[U, V]$$

are called the principal chiral field equations [10, 11] Other examples will be considered elsewhere.

5. Finally, we point out that in this work we have elaborated the method for constructing soliton-type solutions and finding equations these solutions satisfy. This method is directly related to the structure and properties of the internal symmetry algebra of the physical system. As in the case of exactly integrable systems [11,12]

it turns out that the important points are the structure of the algebra itself and choosing its gradation. Note, that for the case of exactly integrable systems there is a finite number of systems corresponding to each gradation, whereas for the completely integrable systems considered here there is an infinite set of systems for each gradation. The internal symmetry algebra itself is defined by positions of the poles involved in the expression for the primordial function Φ (see sect.3). The form of the obtained systems is defined by the highest orders of the poles. The choice of the primordial function Φ (determined by symmetry considerations) defines according to (17), (19), (20) the functions V, i.e. the elements of the L-A pair of a completely integrable system which leads simultaneously to nonlinear equations and to explicit expressions of their soliton type solutions.

The authors are thankful to B.A. Arbusov, O.A. Khrustalev, A.A. Kirillov, Yu.I. Manin, M.A. Mestvirishvili, M.V. Saveliev, A.B. Shabat for helpful discussions.

<u>Appendix</u>. Here we derive the expression (20) for U_-. We will also write the 2nd-order differential equations satisfied by a_i's and b_j's. Finally, we will show that for $\Phi = \lambda z + \bar{\varphi}(\lambda)$ (leading to the KdV equation and to higher order KdV equations) it follows from the system of 2nd-order equations for a_i's that all the moments of the type $\sum_{i=1}^{N} \dot{a}_i a_i^n$ are expressed in terms of the function $u \equiv \sum_i \dot{a}_i$ and its derivatives.

Consider the derivatives of the function $\Psi_1 = \exp\Phi \prod_i (a_i - \lambda)$:

$$\dot{\Psi}_1 / \Psi_1 = \dot{\Phi} + \sum \dot{a}_i / (a_i - \lambda),$$

$$\ddot{\Psi}_1 / \Psi_1 = \ddot{\Phi} + \dot{\Phi}^2 + 2\dot{\Phi} \sum \dot{a}_i / (a_i - \lambda) + \sum_i \frac{1}{a_i - \lambda} \Big[\ddot{a}_i + 2 \sum_{j \neq i} \frac{\dot{a}_j \dot{a}_i}{a_j - a_i} \Big]. \tag{A1}$$

The function Ψ_1 satisfies the spectral equation (11)

$$\ddot{\Psi} - \frac{\dot{U}_+}{U_+}\dot{\Psi} = \left[U_+\left(\frac{U_0}{U_+}\right)^{\cdot} + U_0^2 + U_+ U_-\right]\Psi .$$

Substituting (A1) into the spectral equation and comparing the coefficients at the poles $\lambda = a_i$, we have

$$\ddot{a}_i + 2\dot{\Phi}(a_i)\dot{a}_i + 2\sum_{j\neq i}\frac{\dot{a}_i\dot{a}_j}{a_j - a_i} = \frac{\dot{U}_+(a_i)}{U_+(a_i)}\dot{a}_i . \tag{A2}$$

Similar calculations for the second fundamental solution $\Psi_2 = \exp(-\Phi)\prod(b_j - \lambda)$ lead to the equations

$$\ddot{b}_j - 2\dot{\Phi}(b_j)\dot{b}_j + 2\sum_{k\neq j}\frac{\dot{b}_k\dot{b}_j}{b_k - b_j} = \frac{\dot{U}_+(b_j)}{U_+(b_j)}\dot{b}_j . \tag{A3}$$

Eqs. (A2), (A3) are the desired second-order differential equations for a_i's and b_j's. The 1st-order eqs. (18) for a_j's and similar eqs. for b_j's:

$$\dot{b}_j = \frac{P(b_j)U_+(b_j)}{\prod\limits_{l\neq j}(b_l - b_j)\prod\limits_k (a_k - b_j)} \tag{A4}$$

may be considered as a system of the first integrals of the equations (A2), (A3). (Of course, the relations (A2), (A3) can be found directly from the system of the 1st-order equations (18) (A4) without using spectral equation (11) but it leads to rather complicated calculations).

Further, substituting (A1) into (11) and using (A2) we get:

$$U_+\left(\frac{U_0}{U_+}\right)^{\cdot} + U_0^2 + U_+U_- = \dot{\Phi}^2 + \ddot{\Phi} + 2\sum\frac{\dot{a}_i\left[\dot{\Phi}(\lambda) - \dot{\Phi}(a_i)\right]}{a_i - \lambda} -$$

$$-\frac{\dot{U}_+}{U_+}\dot{\Phi} + \sum_i \frac{\dot{a}_i}{a_i - \lambda}\left[\frac{\dot{U}_+(a_i)}{U_+(a_i)} - \frac{\dot{U}_+(\lambda)}{U_+(\lambda)}\right] =$$

$$= U_+\left\{U_+^{-1}\left[\dot{\Phi} + \sum \frac{\dot{a}_i\,[U_+(a_i) - U_+(\lambda)]}{U_+(a_i)\,(a_i-\lambda)}\right]\right\}^{\cdot} +$$

$$+\dot{\Phi}^2 + 2\sum \frac{\dot{a}_i\,[\dot{\Phi}(\lambda) - \dot{\Phi}(a_i)]}{a_i - \lambda} +$$

$$+\sum \frac{\dot{a}_i\,[\dot{U}_+(a_i) - \dot{U}_+(\lambda)]}{U_+(a_i)(a_i-\lambda)} - \left[\sum \frac{\dot{a}_i\,[U_+(a_i) - U_+(\lambda)]}{U_+(a_i)(a_i-\lambda)}\right]^{\cdot}$$

and obtain the expression (20) for U_-.

Now consider $\Phi(\lambda) = z\lambda + \bar{y}(\lambda)$. According to (17), (22) $U_+ = 1$ and (A2) takes the form (24):

$$\ddot{a}_i + 2\dot{a}_i a_i + 2\dot{a}_i \sum_{j\neq i} \frac{\dot{a}_j}{a_j - a_i} = 0.$$

Multiplying these equations by a_i^k $(k \geqslant 0)$ and summing over i we get

$$\sum_i \ddot{a}_i a_i^k + 2\sum_i \dot{a}_i a_i^{k+1} - \sum_{j\neq i} \dot{a}_j \dot{a}_i \frac{a_j^k - a_i^k}{a_j - a_i} = 0$$

or

$$\left(\sum_i \dot{a}_i a_i^k\right)^{\cdot} + 2\sum \dot{a}_i a_i^{k+1} =$$

$$= \sum_{i,j} \dot{a}_i \dot{a}_j (a_j^{k-1} + a_j^{k-2} a_i + \ldots + a_i^{k-1}),$$

and the recurrency

$$J_{k+1} \equiv \sum_{i=1}^{N} \dot{a}_i a_i^{k+1} = \frac{1}{2}\left\{\sum_{s=0}^{k-1} J_s J_{k-1-s} - \dot{J}_k\right\},$$

which enables us to express all the moments J_{k+1} in terms of $u \equiv J_0$ and its derivatives up to the k-t order. The first three moments are:

$$J_1 = -\dot{u}/2, \quad J_2 = u^2/2 + \ddot{u}/4,$$

$$J_3 = -u\dot{u} - \dddot{u}/8.$$

References

1. Gardner, C.S., Green, J.M., Kruskal, M.D., Miura, R.M. (1967). Phys. Rev. Lett., 19, 1095-1097.
2. Lax, P.D. (1968). Comm. Pure and Appl. Math., 21, 467-490.
3. Zakharov, V.E., Shabat, A.B. (1972). Sov. Phys. JETP, 34, 62-69.
4. Zakharov, V.E., Shabat, A.B. (1974). Funct. Anal. Appl. 8, 383-387.
5. Zakharov, V.E., Takhtajan, L.A., Faddeev, L.D. (1974). Math. Doklady USSR, 219, 1334-1337. (Russian).
6. Faddeev, L.D. (1974). Results of Science and Technics. Modern Problems of Math., p.93-180. (Russian).
7. Ablowitz, M.J., Kaup, D.J., Newell, A.C., Segur, M. (1974). Stud. Appl. Math., 53, 249-316.
8. Zakharov, V.E., Manakov, S.V., Novikov, S.P., Pitaevsky, L.P. (1980). Soliton theory: Inverse scattering method. Nauka, Moscow. (Russian).
9. Zakharov, V.E., Shabat, A.B. (1979). Funct. Anal. Appl., 13, 13-21.(Russian).
10. Zakharov, V.E., Mikhailov, A.V. (1978). ZhETPh, 74, 1953-1980. (Russian).
11. Leznov, A.N., Saveliev, M.V., Smirnov, V.G. (1981). Theor. Math. Phys., 48, 3-12. (Russian).
12. Leznov, A.N., Smirnov, V.G., Shabat, A.B. (1982). Theor. Math. Phys., 51, 10-21. (Russian).
13. Ibragimov, N.H., Shabat, A.B. (1979). Math. USSR Doklady, 244, 56-61;(1979), ibid, 249, 26-29. (Russian).
14. Jiber, A.B., Shabat, A.B. (1979). Math. USSR Doklady, 247, 1103. (Russian).
15. Leznov, A.N. (1984). Theor. Math. Phys., 58, 156-160. (Russian).
16. Leznov, A.N. (1984). In: Proc. of the 2nd conference on Non-Linear Processes in Physics and Turbulence, Kiev, 1983, Gordon and Breach, N.Y.

17. Leznov, A.N., Man'ko, V.I., Chumakov, S.M. (1984). Prepr. P.N. Lebedev Physical Institute No.70; (1984). Prepr. IHEPh No.84-87, Serpukhov.(Russian).

18. Leznov, A.N., Man'ko, V.I., Chumakov, S.M. (1984). Lett. Math. Phys. 8, 297-303; 413-419.

19. Leznov A.N., Man'ko, V.I., Chumakov, S.M. (1985). Theor. Math. Phys., 63, 50-63.(Russian).

20. Kac, V.G. (1983). Infinite dimensional Lie algebras. Birhauser, Boston.

21. Coudrey, P.J., Dodd, R.K., Gibbon, J.P. (1976). Proc. Roy, Soc. Lond., 351, 1666, 407-422.

22. Pohlmeyer, K. (1976). Comm. Math. Phys., 46, 207-221.

23. Lund, F., Regge, T. (1976). Phys. Rev. D., D14, p. 1524-1535.

24. Cherednik, I.V. (1981). In: Results of Science and Technics, Modern Problems of Math., VINITI, Moscow, 17, pp. 175-218. (Russian).

25. Leznov, A.N., Saveliev M.V., Man'ko, V.I. (1985). P.N. Lebedev Phys. Inst. Proc. 165. (Russian).

26. Leznov, A.N., Man'ko, V.I., Chumakov, S.M. (1985). P.N. Lebedev Phys. Inst. Proc. 167. (Russian).

NIJENHUIS OPERATORS IN CLASSICAL DYNAMICS

G.Marmo

Istituto Nazionale di Fisica Nucleare, Sezione di Napoli, Dipartmento di Fisica, Universita di Napoli, Italy

<u>Introduction.</u> In the last decade remarkable progress has been made in the construction and study of exactly solvable dynamical systems. Although many of them are unrealistic, they are still important for several reasons. As pedagogical tools they illuminate methods used in the analysis of more complicated problems, in particular, the study of completely integrable field theories. In the same vein, they serve as useful laboratories to test conjuectures and approximation techniques. Lastly, there is reason to believe that some qualitative features of these systems remain true in more realistic models.

We mention just few of them: Calogero-Moser potentials [1], Toda lattice [2] in finite dimensions, Korteweg-de Vries (KdV) [3], sine-Gordon [4], non-linear Schroedinger [5], equations for infinite dimensions.

A breakthrough in the study of these systems was the introduction of the method of Lax pairs or isospectral deformation. This technique was first applied to the KdV equation, and then to the non-linear Schroedinger and the sine-Gordon equations. There after it has been applied to many systems [6].

We recall that in the finite dimensional case this method consists of finding two nxn matrices, R and A, known as a Lax pair, such that the equations of motion are equivalent to the matrix equations

$$\frac{d}{dt}R = [R, A]. \tag{1.1}$$

This equation implies that the eigenvalues of the matrix R are conserved in time. If one can prove that these eigenvalues, as functions of the canonical coordinates and momenta, are independent and in involution, i.e. the Poisson bracket of any two vanishes, then, by Liouville's theorem, the systems are completely integrable.

If one considers equations (1.1) "à la pair" of Lagrange or Hamilton equations it is natural to consider their transformation properties. It is then important to give them a geometrical setting. In several papers there have been different suggestions [7,8]; Here we shall be concerned with only one of these proposals [9].

Starting from the idey that a dynamical system is represented by a vector field on the manifold of states M, the matrix R will be given an "invariant" meaning by associating with it a (1-1)-tensor field on M. The derivative should have an "invariant" character, so we can use the Lie derivative or the covariant derivative with respect to some connection [8]. Here we further qualify our choice by replacing the derivative $\frac{d}{dt}$ with the Lie derivative.

With these assumptions we can give Lax equations a tensorial character by setting

$$L_{\Gamma} R = 0, \tag{1.2}$$

where R is a (1-1)-tensor field on M, and L_{Γ} is the Lie derivative along a dynamical vector field Γ.

The relation between (1.2) and (1.1) can be given locally.

In local coordinates (ξ^{λ}) for M, the Lie derivative of functions, vector fields and 1-forms is given respectively by

$$L_\Gamma f = \frac{d}{dt} f = \Gamma^\lambda \frac{\partial f}{\partial \xi^\lambda} \tag{1.3a}$$

$$L_\Gamma X^\mu = \frac{dX^\mu}{dt} - X^\varrho \frac{\partial \Gamma^\mu}{\partial \xi^\varrho} \tag{1.3b}$$

$$L_\Gamma X_\mu = \frac{dX_\mu}{dt} + X_\varrho \frac{\partial \Gamma^\varrho}{\partial \xi^\mu}. \tag{1.3c}$$

Then for a (1-1)-tensor fields we have

$$L_\Gamma R^\chi_\lambda = \frac{d}{dt} R^\chi_\lambda + R^\chi_\mu \frac{\partial \Gamma^\mu}{\partial \xi^\lambda} - R^\mu_\lambda \frac{\partial \Gamma^\chi}{\partial \xi^\mu}.$$

The relation (1.2) in local coordinates reads

$$\frac{d}{dt} R^\chi_\lambda = R^\mu_\lambda \frac{\partial \Gamma^\chi}{\partial \xi^\mu} - R^\chi_\mu \frac{\partial \Gamma^\mu}{\partial \xi^\lambda}$$

which can be put in the form

$$\frac{d}{dt} R = [R, A], \text{ where } A^\chi_\mu = \frac{\partial \Gamma^\chi}{\partial \xi^\mu}.$$

Thus (1.2) provides a tensorial representation of (1.1). From (1.2) we can derive interesting results in a very simple way.

From $L_\Gamma R = 0$ we derive $L_\Gamma R^k = 0$ and $L_\Gamma \operatorname{Tr} R^k = 0$, i.e. taking the trace of R and its powers provides us with constants of motion for Γ. Interesting geometrical information is provided also by eigenspaces and eigenvalues of R.

An equation like (1.2) plays a role also in the so called "inverse problem in the calculus of variations" and provides a tool to deal with the separability of the Lagrangian function. The possibility that R might play a role for the dynamics similar to the one played by dimplectic structures, Riemannian metrics and Hermitian structures suggests to insestigate the "geometry" which comes along with a (1-1)-tensor field R on M. These aspects are dealt with in the following sections.

The paper is organized as follows: in § 1 we consider (1-1)-tensor fields on M and the so called Nijenhuis condition. In § 2 we deal with transformation properties of R, infinîte dimensional Lie algebras associated with R, relation with the dynamics and conditions leading to complete integrability.

To exhibit a situation that goes beyond the problem of complete integrability we consider an aspect of the inverse problem in the calculus of variations.

As for notations, we follow [10] as closely as possible.

<u>1. Geometry of a (1-1)-tensor field [11,12,13].</u> We consider a (1-1)-tensor field R on M, i.e. $R \in \mathcal{T}^1_1(M)$. It is possible to associate with it two $\mathcal{F}(M)$-linear maps, one form vector fields to vector fields, $\hat{R} : \mathfrak{X}(M) \to \mathfrak{X}(M)$ and the "transposed one" from 1-forms to 1-forms, $\check{R} : \mathfrak{X}^*(M) \to \mathfrak{X}^*(M)$. The extension to tensor fields is the same as for linear maps and tensors in linear algebra. We confine to antisymmetric tensor fields, to extend them to all tensor fields is then a trivial matter.

By $\mathcal{F}(M)$-linearity it suffices to define the extension on monomilas, i.e.

$$dx^i \wedge dx^j,\ dx^i \wedge dx^j \wedge dx^k, \dots,\ dx^1 \wedge dx^2 \wedge \dots \wedge dx^n,$$

where (x^i) are coordinates on M. On 2-forms we have two possibilities:

$$(1.A)\quad \langle R\, dx^i \wedge dx^j | V_1 \wedge V_2 \rangle \overset{\text{def}}{=\!=} \langle dx^i \wedge dx^j | (\hat{R} V_1) \wedge V_2 + V_1 \wedge (\hat{R} V_2) \rangle$$

or

$$(1.B)\quad \langle B^{(2)} dx^i \wedge dx^j | V_1 \wedge V_2 \rangle \overset{\text{def}}{=\!=} \langle dx^i \wedge dx^j | \hat{R} V_1 \wedge \hat{R} V_2 \rangle.$$

On higher forms, say m-forms, we have many more alternatives, each one can be characterized by an integer $k \leqslant m$,

$$\langle R^{(k)}(dx^{i_1}\wedge dx^{i_2}\wedge\ldots\wedge dx^{i_m})|V_1\wedge V_2\wedge\ldots\wedge V_m\rangle \stackrel{\text{def}}{=\!=}$$

$$=\langle dx^{i_1}\wedge dx^{i_2}\wedge\ldots\wedge dx^{i_m}|\sum_{\sigma}|\sigma|(\hat{R}V_{\sigma(1)}\wedge \quad (3)$$

$$\wedge\hat{R}V_{\sigma(2)}\wedge\ldots\wedge\hat{R}V_{\sigma(k)})\wedge V_{\sigma(k+1)}\wedge\ldots\wedge V_{\sigma(m)}\rangle$$

where σ is the permutation on m elements. The point to make here is that extension of R to m-forms is not unique.

The extension that we have called R, without qualification is the only one with the property

$$d_R\, d = -\, d\, d_R \quad (4)$$

it allows us to define the exterior derivative by setting

$$(d_R f)(X) = df(\hat{R}X) \quad (5a)$$

$$(d_R\theta)(X,Y) = (Rd\theta)(X,Y) - (dR\theta)(X,Y). \quad (5b)$$

If R is invertible, it is possible to show that d_R satisfies Poincaré's lemma, i.e. $d_R\alpha = 0$, implies that, locally there exists β such that $\alpha = d_R\beta$. This exterior derivative is very useful in discussing recursion operators as we will see later on.

Remark. We could define a derivative by using the extension $R^{(2)}$. This derivative is related to the covariant derivative. On fiber bundles the connection 1-form can be associated with a (1-1)-tensor field. When this one is extended according to $R^{(2)}$, the exterior derivative that it defines is the usual covariant exterior derivative and $d_R^{(2)}\cdot d_R^{(2)}$ is related to the curvature tensor.

Here however we only consider d_R, defined via (4). If we compute $D_R\,(d_R\, f)$ we see that

$$[d_R(d_R f)](X,Y) = df(\mathbb{N}_R(X,Y)) \tag{6}$$

where

$$\mathbb{N}_R(X,Y) = [\hat{R}X, \hat{R}Y] + \hat{R}^2[X,Y] - \hat{R}[\hat{R}X,Y] - \hat{R}[X,\hat{R}Y] \tag{7}$$

(1-1)-tensor fields satisfying $\mathbb{N}_R = 0$ define important Nijenhuis operators. Here we list few properties:

1) The image of R, i.e. $R(\mathfrak{X}(M)) \subset \mathfrak{X}(M)$, is a Lie subalgebra of $\mathfrak{X}(M)$, i.e. $[\hat{R}(X), \hat{R}(Y)] = \hat{R}(Z)$. It follows that if $\hat{R}$ is of constant rank, $\hat{R}(\mathfrak{X}(M))$ defines an involutive distribution and possibly a regular foliation.

2) Ker $\hat{R} \subset \mathfrak{X}(M)$ need not be a Lie algebra, but we can define one as follows. From the definition of $\mathbb{N}_R$ we get

$$X,Y \in \ker \hat{R} \Rightarrow [X,Y] \in \ker \hat{R}^2$$

and more general $[\ker \hat{R}, \ker \hat{R}^p] \subset \ker \hat{R}^{p+1}$. Recall that an integer p is called the Riesz index of $\hat{R}$ if

$$\mathrm{Im}\,\hat{R}^{p+1} = \mathrm{Im}\,\hat{R}^p \quad \text{and} \quad \ker \hat{R}^{p+1} = \ker \hat{R}^p .$$

It follows that ker $\hat{R}^p$ is an infinite dimensional Lie algebra.

3) $d_R \circ d_R = 0 \Rightarrow d_R k \circ d_R k = 0$.

It can be shown, more generally, that $\mathbb{N}_{R_1} = 0$, $\mathbb{N}_{R_2} = 0$ and $\mathbb{N}_{R_1+R_2} = 0$ implies that $\mathbb{N}_{R_1 \cdot R_2} = 0$.

As operators $\hat{R}$ dffine $\mathfrak{F}(M)$-linear maps in the tensor algebra of a given manifold, they allow us to set eigenvalue problems, e.g. on vector fields we have

$$\hat{R}(X) = \lambda_X X, \quad \lambda_X \in \mathfrak{F}(M), \; X \in \mathfrak{X}(M)$$

It makes sense then to diagonalize these operarors.

Nijenhuis operators defined by a diagonalizable tensor R have interesting properties. We list here few of them.

We denote by S_{λ_i} the subbundle of TM with eigenvalue λ_i (we assume that each $\lambda_i \in \mathcal{F}(M)$ has constant degeneracy). Then

$$\mathbb{N}_R = 0 \Leftrightarrow \hat{R}[e_i, e_j] \in S_{\lambda_i} \oplus S_{\lambda_j}, \; (\lambda_i - \lambda_j) L_{e_i} \lambda_j = 0,$$

where $\hat{R} e_k = \lambda_H e_k$.

Remark. A useful consequences of this result is that, at least locally, it is possible to associate foliations with eigenspaces. Thus it makes sense to speak of eigenmanifolds associated with eigenvalues.

Additional results are the following [14, 15]:

A) If R is a Nijenhuis operator then

$$\frac{1}{K} \check{R}\, d\, Tr\, \check{R}^{K} = \frac{1}{K+1}\, d\, Tr\, \check{R}^{K+1} \qquad \text{for any integer K.}$$

B) If R is diagonalizable with nowhere constant eigenvalues λ_i then $\mathbb{N}_R = 0 \Rightarrow \check{R}\, d\lambda_i = \lambda_i d\lambda_i$.

C) $\mathbb{N}_R = 0 \Rightarrow d\check{R}^k \theta = 0$ for a 1-form satisfying $d\theta = 0$, $d_R \theta = 0$.

This follows from $d_R \circ d_R = 0$.

2. Transformation properties of (1-1)-tensor fields [13].

As is well known, for any map $\varphi : M \rightarrow N$, it is possible to define an induced map on functions and forms, namely

$$\varphi^* : \mathcal{F}(N) \rightarrow \mathcal{F}(M), \quad \varphi^* : \Lambda(N) \rightarrow \Lambda(M).$$

on the other hand for vector fields we cannot induce a map from $\mathfrak{X}(N)$ to $\mathfrak{X}(M)$ or from $\mathfrak{X}(M)$ to $\mathfrak{X}(N)$. The tangent map $T\varphi : TM \rightarrow TN$ which is always available will not associate vector fields on M with vector fields on N unless φ is a diffeomorphism. Because of $\mathcal{F}$-linearity it suffices to investigate a (1-1)-tensor field $R = X \otimes \theta$, where X is any vector field and θ any 1-form. It follows that we can confine ourselves to vector fields because for 1-forms φ^* is already available.

For a vector field $X \in \mathfrak{X}(M)$ we say that it is φ-related to a vector field $Y \in \mathfrak{X}(N)$ if

$$L_X \varphi^* f = \varphi^* L_Y f \qquad \text{for any} \quad f \in \mathcal{F}(N) \ .$$

The following statement is a simple consequence of the definition. Let φ: $M \to N$ be a map from M to N and X, Y vector fields on M which are φ-related to vector fields $\tilde{X}$ and $\tilde{Y}$ on N. Then $[X, Y]$ is φ-related to $[\tilde{X}, \tilde{Y}]$.

A (1-1)-tensor field R on M is φ-related to a (1-1)-tensor field $\tilde{R}$ on N if

$$\check{R}(\varphi^* df) = \varphi^* (\check{\tilde{R}}\ df) \qquad \text{for any} \quad f \in \mathcal{F}(N).$$

It can be shown that if R is φ-related to $\tilde{R}$, then $\mathbb{N}_R$ is φ-related to $\mathbb{N}_{\tilde{R}}$. This is clear, e.g. from

$$d_R \varphi^* \alpha = \varphi^* (d_{\tilde{R}} \alpha)$$

and

$$(d_R \cdot d_R)(\varphi^* \alpha) = \varphi^* (d_{\tilde{R}} \cdot d_{\tilde{R}} \alpha).$$

Now we consider the problem of restricting (1-1)-tensor fields to submanifolds. The map φ : $M \to N$ is replaced by the map $i: M \to N$ of constant and dim N $\geq$ dim M. The tangent map $Ti: TM \to TN$ is injective and identifies $T_m M$ with a vector subspace of $T_{i(m)} N$.

Let i: M $\to$ N be a map of rank dim M and R a (1-1)-tensor field on N. There exists a (1-1)-tensor field $\tilde{R}$ on M which is i-related to R if and only if $\hat{R}$ (Ti (TM)) $\subset$ Ti (TM)

As for the extension from M to N we have the following. Suppose that i (M) $\subset$ N is a proper submanifold and N is paracompact. Then every (1-1)-tensor field on M can be extended to a (1-1)-tensor field on N is such a way that they are i-related.

For projection maps we define projectable (1-1)-tensor fields. Let $\pi : M \to N$ be a submersion onto N, then R is π-projectable onto N if

$$L_Z \check{R}(\varphi^* df) = 0, \quad \langle \check{R}(\varphi^* df) | Z \rangle = 0$$

for any vertical vector field Z, i.e. such that $T\pi(Z) = 0$. All these concepts turn out to be useful when dealing with reduction of dynamical systems.

We shall now give an example of a (1-1)-tensor field which will be useful later on.

<u>3. Example: The natural (1-1)-tensor field on TQ [10, 16-17].</u>

On M) TQ it is possible to exhibit a (1-1)-tensor field which essentially contains the "geometry" of the tangent bundle.

A second order vector field Γ on TQ is the intrinsic version of a second order differential equation, or Newtonian equations, on Q. In natural local coordinates we have

$$\Gamma = \dot{q}^i \frac{\partial}{\partial q^i} + f^i(q, \dot{q}) \frac{\partial}{\partial \dot{q}^i} .$$

A (1-1)-tensor field S on TQ is defined by requiring that

$$\check{S} \cdot L_\Gamma d\pi^* f = \pi^* df$$

for any $f \in \mathcal{F}(Q)$, and $\pi : TQ \to Q$ the projection map. It follows that

$$\check{S} L_\Gamma dq^i = \check{S} d\dot{q}^i = dq^i ,$$

so that the local form of S is given by

$$S = dq^i \otimes \frac{\partial}{\partial \dot{q}^i} .$$

It is trivial to show that $\mathrm{Im}\,\hat{S} \subset \mathfrak{X}(TQ)$ and ker S $\subset \mathfrak{X}(TQ)$ are infinite dimensional Lie subalgebras (the subalgebras of vertical vector fields). S is a Nijenhuis operator because $\mathbb{N}_S = 0$. The exterior derivative

$$d_S f = \frac{\partial f}{\partial \dot{q}^i} dq^i$$

allows to associate the Cartan 1-form with a Lagrangian function $\mathcal{L}$. Indeed

$$d_S \mathcal{L} = \frac{\partial \mathcal{L}}{\partial \dot{q}^i} dq^i = p_i\, dq^i$$

and the Lagrangian 2-form $\omega_{\mathcal{L}}$ is given by

$$\omega_{\mathcal{L}} = d d_S \mathcal{L} = d\left(\frac{\partial \mathcal{L}}{\partial \dot{q}^i}\right) \wedge dq^i = dp_i \wedge dq^i .$$

Euler-Lagrange equations read

$$\left(\frac{d}{dt}\frac{\partial \mathcal{L}}{\partial \dot{q}^i} - \frac{\partial \mathcal{L}}{\partial q^i}\right) dq^i = 0$$

and in intrinsic form, replacing $\frac{d}{dt}$ by the Lie derivative along Γ,

$$L_\Gamma d_S \mathcal{L} - d\mathcal{L} = 0 .$$

By using Cartan's identity $L_\Gamma = i_\Gamma d + d i_\Gamma$ we get the canonical version of Euler Lagrange equaitons (10)

$$i_\Gamma \omega_{\mathcal{L}} = d(\mathcal{L} - d\mathcal{L}(\hat{S}\Gamma)) .$$

<u>4. The "geometry" of R with respect to dynamical vector fields [8, 17-15].</u> From the suggested representation of Lax equations in the form $L_\Gamma R = 0$, we would like to consider the consequences on Γ if R enjoys additional properties, in particular $\mathbb{N}_R = 0$.

Some simple properties are

1) $L_\Gamma R^k = 0$ (4.1)

2) $L_\Gamma \mathrm{Tr}\, R^k = 0$ (4.2)

3) If $\check{R}\theta = \lambda_X \theta$, $\hat{R}X = \lambda_X X$ with $\langle\theta|X\rangle \in \mathbb{R}$ then

$L_\Gamma \lambda_X = 0.$ (4.3)

(This follows from 2) in the finite dimensional case. But it makes sense also in the infinite dimensional case when trace is not defined).

4) Im $\hat{R} \subset \mathfrak{X}(M)$ is invariant under dynamical evolution, i.e. $L_\Gamma \hat{R}(\mathfrak{X}(M)) \subset \hat{R}(\mathfrak{X}(M))$.

5) $L_\Gamma \ker \hat{R} \subset \ker \hat{R}$.

If the (1-1)-tensor field R is diagonizable we have sharper consequences. Let us assume that R can be expressed in the form $R = \sum_i \lambda_i P_i$ where P_i is the projection onto S_{λ_i}, the eigenspace associated with λ_i. If $\mathbb{N}_R = 0$, it is possible to find collective coordinates (ξ) such that $P_i = d\xi_i \otimes \frac{\partial}{\partial \xi_i}$, where i is a collective index to denote the i-eigenspace. The eigenmanifold associated with the eigenspace S_{λ_i} has coordinates (ξ_i). The relation $L_\Gamma R = 0$ forses Γ to split into the sum of vector fields Γ_i. Additional conditions: a) dim $S_{\lambda_i} = 2$ and b) $d\lambda_i \neq 0$, for any eigenvalue, imply the complete integrability of Γ.

From the result which provides us with Γ_i on a 2-dimensional eigenmanifold with a constant of motion λ_i, what we have to show is that on a 2-dimensional manifold a vector field without null points and admitting a constant of motion is always Hamiltonian. This result follows easily. Let Ω be any volume form on M (dim M = 2). Then

$i_\Gamma \Omega = f d\lambda$

because $i_\Gamma i_\Gamma \mathcal{R} = 0$. Then if Γ has no zeros, $f \neq 0$, we obtain

$$i_\Gamma(\mathcal{R}) = d\lambda .$$

By dimension considerations $d(\mathcal{R}/f) = 0$, thus $\mathcal{R}/f$ is a symplectic structure. It is trivial that any $F(\lambda)$ can be chosen as a Hamiltonian function if $dF(\lambda) \neq 0$. In this case the corresponding symplectic structure is given by

$$\mathcal{R}_F = \frac{dF}{d\lambda}\left(\frac{\mathcal{R}}{f}\right)$$

and we have $i_\Gamma \mathcal{R}_F = dF$.

Once a choice of Hamiltonian along each eigenmanifold has been made we construct a symplectic structure $\mathcal{R} = \sum_i \mathcal{R}_i$, $F = \sum_i F_i$ and $i_\Gamma \mathcal{R} = dF$. It also follows that $\{F_i , F_j\} = 0$, i.e. the dynamical system described by Γ is completely integrable in the Liouville sense.

Now we discuss in which sense a dynamical system completely integrable in the Liouville sense admits a Lax representation of the form we are considering in this paper.

We start with a symplectic manifold (M, ω), dim M = 2n.

a) Γ is a Hamiltonian vector field with respect to ω, i.e. $i_\Gamma \omega = dH$

b) there exist functions $f_1, f_2, \ldots, f_n$ such that $\{f_i, f_j\} = 0$,
$df_1 \wedge df_2 \wedge \ldots \wedge df_n \neq 0$ and $\{H, f_i\} = 0,$ then

$$\Gamma = \frac{\partial H}{\partial f_i}\frac{\partial}{\partial \varphi^i} = \nu^i \frac{\partial}{\partial \varphi^i}$$

where $\omega = d\varphi^i \wedge df_i$ in a neighbourhood of a torus.
Consider

$$R=\sum_i \nu^i\left(d\nu^i \otimes \frac{\partial}{\partial \nu^i} + d\varphi^i \otimes \frac{\partial}{\partial \varphi^i}\right) \qquad (4^*)$$

which has all the required properties (and is well-defined) only if

$$d\nu^1 \wedge d\nu^2 \wedge \ldots \wedge d\nu^n \neq 0 \quad \text{or} \quad \det\left\| \frac{\partial^2 H}{\partial f_i \partial f_j} \right\| \neq 0 .$$

At least in this situytion it is possible to build a (1-1)-tensor field which has all the required properties leading to complete integrability.

A complementary situation arises if $\frac{\partial H}{\partial f_i} \in \mathbb{R}$. In this case

$$R=\sum_i f_i\left(df_i \otimes \frac{\partial}{\partial f_i} + d\varphi^i \otimes \frac{\partial}{\partial \varphi^i}\right) \qquad (4^{**})$$

has all the required properties. In the intermediate situation, i.e. rank $\left\| \frac{\partial^2 H}{\partial f_i \partial f_j} \right\| = k < n$, we have to deal with a splitting, find a (1-1)-tensor which is a "combination" of type (*) and type (**).

<u>5. A remarkable class of (1-1)-tensor fields [7,11,15, 18, 19].</u> Up to now we have started with the relation $L_\Gamma \hat{R} = 0$, to discuss some consequences for Γ when $\hat{R}$ has suitable properties. It remains an open problem to know how to find such (1-1)-tensor fields. A remarkable class of them called factorizable operators which arises in the framework of Hamiltonian dynamics, can be given in terms of symplectic structures.

Let us consider a dynamical system Γ on M which is Hamiltonian (or locally Hamiltonian) with respect to two symplectic structures setting

$$\omega_1(\hat{R}X,Y) = \omega_2(X,Y).$$

It trivially follows that $L_\Gamma \omega_1 = 0$, $L_\Gamma \omega_2 = 0$ implies $L_\Gamma R = 0$. We see that for all systems admitting a biHamiltonian description we can exhibit a (1-1)-tensor field which is invariant under dynamical evolution. It is quite clear that for R tob be well-defined we only need ω_1 to be non degenerate, R turns out to be non degenerate, if ω_2 is non degenerate. In general ker R coincides with ker ω_2 . The following diagram illustrates the situation:

$$\begin{array}{ccc} \mathfrak{X}(M) & \xrightarrow{\hat{R}} & \mathfrak{X}(M) \\ \downarrow \omega_1 & \searrow^{\omega_2} & \downarrow \omega_1 \\ \mathfrak{X}^*(M) & \xrightarrow[\check{R}]{} & \mathfrak{X}^*(M) \end{array}$$

In formulas $\omega_1(\hat{R}X, Y) = \omega_1(X, \hat{R}Y)$ or $\omega_1(\hat{R}X, Y) = \omega_1(X, \hat{R}Y)$. We can write symbolically $\hat{R} = (\omega_1)^{-1} \circ \omega_2$ or $\check{R} = \omega_2 \circ (\omega_1)^{-1}$. A simple consequence is that the degeneracy of eigenvalues of $\hat{R}$ is not smaller than two.

By applying powers of $\hat{R}$ to a given vector field X, we get

$$\omega_1(\hat{R}^{2K}X, X) = \omega_1(\hat{R}^K X, \hat{R}^K X) = 0$$

and in general for any power $k+m$

$$0 = \omega_1(\hat{R}^{k+m}X, X) = \omega_1(\hat{R}^k X, \hat{R}^m X) = \omega_1(\hat{R}^m X, \hat{R}^k X).$$

Thus the iterated application of R to any given vector field X produces an isotropic subspace for ω_1 . By applying iterated power of $\hat{R}$ to Γ we get the set of vector fields that constitute an infinite Lie algebra of symmetries for Γ . This algebra is abelian if $\mathbb{N}_R = 0$.

The iterated application of R to the symplectic structure or to the Hamiltonian function gives rise to the following diagrams and definitions [19]. $\check{R}$ is a H-weak recursion if

$d\check{R}^k dH = 0$ for any $\forall k$. From $i_\Gamma \omega_0 = dH_0$ we get the diagram of Hamiltonian vector fields:

$$\begin{array}{ccccc} \Gamma_0 & \xrightarrow{\hat{R}} & \Gamma_1 & \longrightarrow \cdots \xrightarrow{\hat{R}} & \Gamma_n \\ & & \omega_0 & & \\ dH_n & \longleftarrow \cdots \longleftarrow & dH_1 & \longleftarrow & dH_0 \end{array}$$

$\check{R}$ is called ω-weak recursion operator if we get the diagram

$$\begin{array}{ccccc} & dH_0 & & & \\ \omega_0 & \xrightarrow{R} & \omega_1 & \longrightarrow \cdots \xrightarrow{R} & \omega_n \\ \downarrow & & \downarrow & & \downarrow \\ \Gamma_0 & \longrightarrow & \Gamma_1 & \longrightarrow \cdots \longrightarrow & \Gamma_n \end{array}$$

When R is recoveres from two symplectic structures and $\mathbb{N}_R = 0$, we get a strong recursion operator, i.e. a set of a symplectic structure and Hamiltonian vector fields. The proof is a simple consequences of $d_R \circ d_R = 0$, and $d_R d = -dd_R$.

When R satisfies Nijeuhuis condition we easily find that

$$\{ Tr\,\hat{R}^k, Tr\,\hat{R}^h \}_1 = 0,$$

as a result of

$$\omega_1(Z_k, Z_h) = 0 \quad \text{for} \quad i_{Z_k}\omega_1 = dTr\,\hat{R}^k.$$

6. Alternative Lagrangians and (1-1)-tensor fields [20-21];

A second order vector field Γ is said to admit alternative Lagrangian descriptions, if

$$L_\Gamma d_S \mathcal{L}_1 = d\mathcal{L}_1 \tag{6.1a}$$

$$L_\Gamma d_S \mathcal{L}_2 = d\mathcal{L}_2 \tag{6.1b}$$

or equivalently

$$i_\Gamma \omega_{\mathcal{L}_1} = -dE_{\mathcal{L}_1} \tag{6.1c}$$

$$i_\Gamma \omega_{\mathcal{L}_2} = -dE_{\mathcal{L}_2} \tag{6.1d}$$

where $\omega_{\mathcal{L}} = dd_S \mathcal{L}$, $E_{\mathcal{L}} = i_\Gamma d_S \mathcal{L} - \mathcal{L}$. In this case, since $\omega_{\mathcal{L}}\left(\frac{\partial}{\partial \dot{q}^i}, \frac{\partial}{\partial \dot{q}^j}\right) = 0$, the (1-1)-tensor we recover from two alternative Lagrangians has a special form.

Again we have

$$\omega_{\mathcal{L}_1}(\hat{R}X, Y) = \omega_{\mathcal{L}_2}(X, Y) \tag{6.2}$$

and in addition

$$\omega_{\mathcal{L}_2}(SX, Y) = -\omega_{\mathcal{L}_1}(X, SY) \tag{6.3}$$

where

$$S = dq^i \otimes \frac{\partial}{\partial \dot{q}^i} .$$

Then we find

$$S\hat{R} = \hat{R}S \tag{6.4}$$

and in local coordinates R has the form

$$R = M_i^j \left(dq^i \otimes \frac{\partial}{\partial q^j} + \partial \dot{q}^i \otimes \frac{\partial}{\partial \dot{q}^j}\right) + N_i^j \, dq^i \otimes \frac{\partial}{\partial \dot{q}^j} . \tag{6.5}$$

Moreover, $L_\Gamma \hat{R} = 0$ implies

$$L_\Gamma M = N \tag{6.6a}$$

$$L_\Gamma N = ND + [M, B] \tag{6.6b}$$

$$L_\Gamma M = \frac{1}{2}[D, M] \tag{6.6c}$$

where

$$B_j^i \overset{\text{def}}{=} \frac{\partial f^i}{\partial q^j} , \quad D_i^j \overset{\text{def}}{=} \frac{\partial f^j}{\partial \dot{q}^i} \tag{7}$$

and in local coordinates Γ is given by

$$\Gamma = \dot{q}^i \frac{\partial}{\partial q^i} + f^i \frac{\partial}{\partial \dot{q}^i} .$$

In this case if R defined by (6.2) has $\mathbb{N}_R = 0$, then all the symplectic structures defined by $R^k \omega_{\mathcal{L}_1} = \omega_{k+1}$ are Lagrangians, i.e. there exist $\mathcal{L}_{k+1}$ such that $\omega_{k+1} = \omega_{\mathcal{L}_{k+1}}$ and $\check{R}^k dE_{\mathcal{L}_1} = dE_{\mathcal{L}_{k+1}}$.

Let Γ be a second order vector field admitting two alternative Lagrangians, and let the (1-1)-tensor field R defined by (6.2) be completely integrabal, see sec.4. Then one can find, in a neighbprhood of any point $q \in Q$, a bundle chart $(\xi, \dot{\xi})$ such that Γ is completely separated, i.e. $\Gamma = \sum_{1 \leq n}^{n} \Gamma_i$ with locally $\Gamma_i = \dot{\xi}_i \frac{\partial}{\partial \xi_i} + f_i \frac{\partial}{\partial \dot{\xi}_i}$ where $f_i = f_i(\xi_i, \dot{\xi}_i)$, and every Γ_i belongs to an eigenspace of $\hat{R}$ (different eigenspaces for different i's). The converse is also true.

From here we get a sharper result: All the Lagrangians related through a functiob of $\hat{R}$ (entire function or polynomial), in the conditions of the previous proposition, are reparated in the same bundle atlas in which the separation of Γ is achieved, i.e. apart from terms which can be "gauged away" (total time derivalives) any such Lagrangian is of the form $\mathcal{L} = \sum_{1 \leq n}^{n} \mathcal{L}_i$ with $\mathcal{L}_i$ a function only of the i-th pair of bundle coordinates.

<u>7. The problem of the uniqueness of Lagrangian description [22, 23].</u> A second order dynamical system admitting more than one Lagrangian description preserves a (1-1)-tensor field R recovered from two Lagrangian symplectic structures. From equations $L_\Gamma \hat{R} = 0$ we get

$$L_\Gamma M = N, \; L_\Gamma N = ND + [M, B], \; L_\Gamma M = \frac{1}{2}[D, M]$$

which reduce to

$$L_\Gamma M = 0, \; [M, B] = 0$$

when forces do not depend on velocities in the given coordinate system.

The equation $[M,B]=0$ generates an infinite chain of algebraic equations via Lie derivative, i.e.

$$(L_\Gamma)^K[M,B]=[M,(L_\Gamma)^K B]=0 .$$

We see that a necessary condition for the existence of alternative Lagrangians is a (1-1)-tensor field R obeying above equations. For forces not depending on velocities we get: Generically the Lagrangian description for a given second order vector field is unique.

A simple way to prove this result is to start from $[M,B]=0$. We notice that if $c:\mathbb{R}\to Q$ is a solution curve for the Lagrangian dynamics described by Γ, we have $[M(c(t),\frac{d}{dt}c(t)),\ B(c(t))]=0$, and brcause of $\frac{dM}{dt}=0$, we also have $[M(c(0),c(0)),B(c(t))]=0$. Thus we get a curve $\Phi:\mathbb{R}\to Mat(n,\mathbb{R}),\ t\mapsto B(c(t))$.

Now if the curve in Mat $(n,\mathbb{R})$ is in general position, M must be a multiple of the identity.

We can visualize the situation in the two-dimensional case for potential forces where dim Symm Mat $(2,\mathbb{R})=3$. We can consider coordinate axes given by

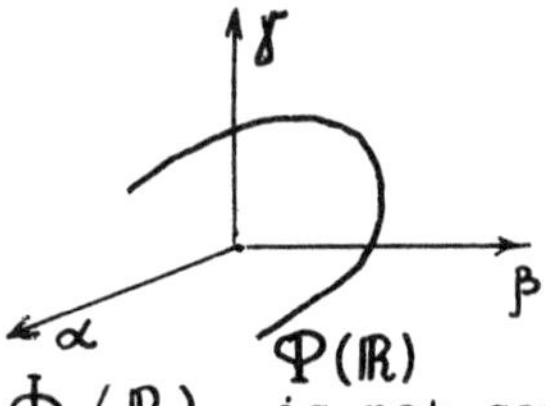

$$\alpha=\begin{Vmatrix}0&0\\0&1\end{Vmatrix},\quad \beta=\begin{Vmatrix}0&1\\1&0\end{Vmatrix},\quad \gamma=\begin{Vmatrix}1&0\\0&0\end{Vmatrix}$$

if $\Phi(\mathbb{R})$ is not contained in any plane, the matrix M must be the identity.

The point about the construction is that M is constant matrix along a solution of the Euler-Lagrange equations, while B varies from point to point. If B(c(t)) "contains" at least three independent matrices, it is not contained in a plane and M is a multiple of the identity.

It is possible to use this method also in the case of

forces depending on velocities. Some additional complications arise because we have to deal with the full tensor field R rather than the matrix representation in terms of M.

To make contact with completely integrable systems orbits coadjoint of action of Lie groups, consider the construction of (1-1)-tensor fields on cotangent bundles of Lie groups.

8. Construction of (1-1)-tensor fields on T^*G [24, 25].

Recall some well-known facts on Lie groups G and their contangent bundles T^*G [10]. Since a detailed discussion of the construction will appear elsewhere, we give here only the relevant ideas.

The group G can act on T^*G on the left and on the right. With respect to the natural symplectic structure $\omega_0 = -d\theta_0$, these two actions are symplectic and we can associate with them two momentum mappings.

If $\xi \in \mathfrak{g}_G$ set $X_\xi \in \mathfrak{X}(G)$ for the left-invariant vector fields associated with ξ, while Y_ξ will denote the right-invariant vector field associated with ξ.

From

$$[\xi_i, \xi_j] = c^k_{ij}\, \xi_k$$

we get

$$[X_{\xi_i}, X_{\xi_j}] = -c^k_{ij} X_{\xi_k}, [Y_{\xi_i}, Y_{\xi_j}] = c^k_{ij} Y_{\xi_k}.$$

Dual 1-forms will be denoted by θ^i, β^i:

$$\theta^i(X_{\xi_j}) = \delta^i_j,\ \beta^i(Y_{\xi_j}) = \delta^i_j.$$

The canonical lifting of vector fields to T^*G is defined by setting

i) $L_{\bar{X}_{\xi_i}} \pi^* f = \pi^* L_{X_{\xi_i}} f$ for any $f \in \mathcal{F}(G)$

ii) $L_{\bar{X}_{\xi_i}} \theta_0 = 0$.

From i) we get $i_{\bar{X}_{\xi_i}} d\theta_0 = -d i_{X_{\xi_i}} \theta_0$ (with same abuse of notation!).

The momentum map is defined to be $\Phi_\lambda : T^*G \to g^*_G$ such that $i_{\bar{X}_{\xi_i}} d\theta_0 = -d\Phi_\lambda(\xi)$ or $\Phi_\lambda(\xi) = i_{X_\xi}\theta_0$ (again with the same abuse of notation which consists in contracting θ_0 with X_ξ rather than $\bar{X}_\xi$). Analogously $\Phi_\varrho : T^*G \to g^*_G$ is defined by $\Phi_\varrho(\xi) = i_{Y_\xi}\theta_0$. With these two maps we get the double fibration $g^*_G \xleftarrow{\Phi_\varrho} T^*G \xrightarrow{\Phi_\lambda} g^*_G$.

The left G-action on T^*G (generated by right-invariant vector fields) defines a proncipal fiber bundle over g^*_G with projection map Φ_λ.

The main idea to construct a (1-1)-tensor field on T^*G consist of "splitting" T^*G according to Φ_λ and Φ_ϱ and use the Poisson bracket to "lift" vector fields from g^*_G to T^*G.

Given any μ on g^*_G, we have $\Phi^{-1}_\lambda(\mu) \simeq G$; on $\Phi^{-1}_\lambda(\mu)$ there is an action of a subgroup $H = G_\mu$, t isotropy group of μ under the coadjoint action.

The action of G_μ on G is Φ_ϱ-projectable onto the trivial action on g^*_G. The restriction of Φ_ϱ to $G \simeq \simeq \Phi^{-1}_\lambda(\mu)$ defines a principal bundle with G_μ as a structure group.

Our last statement is a result due to Kostant and Souriau [10]. From $\mu \in g^*_G$ we can build up a left invariant 1-form θ_μ on G. We have

$$L_{Y_\xi}\theta_\mu = 0 \qquad \text{for any } \xi \in g_G .$$

Let us consider vector fields $X_\xi \in \theta_\mu$. They integrate to a group which is the connected component of $G_\mu = \{g \in G\ (Adg)^*\mu = \mu\}$ which implies that G/Ker $(d\theta$ is a covering under Φ_ϱ of the orbit $G\cdot\mu$ in g^*_G; Note that G_μ is a closed subgroup.

The 2-form $d\theta_\mu$ defines a symplectic structure on the homogeneous space $G/\mathrm{Ker}\ d\theta_\mu$. (Let us denote by $h_{d\theta_\mu}$ the subalgebra associated with G_μ).

It is a theorem by Chu that [26]: each 2p-dimensional homogeneous space determines a left-invariant closed two-form of rank 2p on G. Conversely, an exact left-invariant two-form determines a homogeneous symplectic space.

The following theorem (Duflo-Vergne) [27] explains the relation with functions on involution: if G/G_μ has maximal dimension, the connected component of the isotropy group of μ is commutative.

Remark. For semisimple Lie group G the dual of the Lie algebra can be identified with the Lie algebra via the killing form. In this case "ker $(d\theta_\mu)$ has minimal dimension" becomes the assertion that the centralizer of the corresponding element $\mu \in g_G$ has minimal dimension and previous theorem reduces to the classical assertion that for regular elements the centralizer is abelian. Then for regular elements $h_{d\theta_\mu}$ is a Cartan subalgebra.

<u>Canonical basis on g^*.</u> If O_μ is a maximal orbit, then g_μ (the Lie algebra of G) is locally constant on a neighborhood of μ, then these g_μ fit together to form a smooth vector bundle on such neighborhood. All this implies that we can select Casimir functions (or neutral elements) $C_1, \ldots, C_{n-2p}$ and 2p functions on each orbit, so that in the given neighborhood P_i, Q_i, C_j is a coordinate system such that

$$\{P_i, Q_j\} = \delta_{ij}, \{C_j, P\} = \{C_j, Q\} = 0.$$

Thus, on g^*_G, we select a neighborhood of a regular point which allows a symplectic chart and is made up of regular elements.

Now we consider the pull back of this neighborhood to T^*G (a satured neighbourhood along Φ_λ). The restriction of Φ_ϱ to this open subset of T^*G provides again with a smooth foliation and other functions $\bar{P}_i$, $\bar{Q}_i$, C_i such that

$$\{\bar{P}_i, \bar{Q}_j\} = \delta_{ij} \quad , \{C_j, \bar{P}\} = \{C_i, \bar{Q}\} = 0.$$

Now by using the symplectic structure on T^*G, we can define vector fields Z_i, W_i, $\bar{Z}_i$, $\bar{W}_i$, given by

$$i_{Z_i}\omega_0 = -d\Phi^*_\lambda P_i, \qquad i_{\bar{Z}_i}\omega_0 = -d\Phi^*_\varrho \bar{P}_i$$

$$i_{W_i}\omega_0 = -d\Phi^*_\lambda Q_i, \qquad i_{\bar{W}_i}\omega_0 = -d\Phi^*_\varrho \bar{Q}_i$$

and "angle-variables" conjugate to $C_1, \ldots, C_{n-2p}$ according to Nehorosev's theorem [28], say $\varphi_1, \varphi_2, \ldots, \varphi_{n-2p}$, so that

$$R = R^\lambda_i (Z_i \otimes dP_i + W_i \otimes dQ_i) + R^\varrho_i (\bar{Z}_i \otimes d\bar{P}_i + \bar{W}_i \otimes d\bar{Q}_i) +$$

$$+ R_i \left(d\varphi_i \otimes \frac{\partial}{\partial \varphi_i} + dC_i \otimes \frac{\partial}{\partial C_i} \right)$$

where $R^\lambda_i = R^\lambda_i(P,Q)$, $R^\varrho_i(P,O)$, $R_i(C)$ implies that R is invariant under the Lie derivative associated with any Hamiltonian vector field generated by $H(C_1 \cdot C_2 \cdot \ldots \cdot C_{n-2a})$.

Remark. 1) It is possible to restrict the analysis to $\Phi^{-1}_\lambda(\Theta_\mu) \times U$, $\quad U \subset \mathbb{R}^{n-2p}$ so that it is made up of regular values, then it is possible to "lift" the space of Hamiltonian vector fields $\mathfrak{X}_H(\Theta_\mu)$ to Hamiltonian vector fields on $\Phi^{-1}_\lambda(\Theta_\mu)$. Such a possibility allows for the construction of (1-1)-tensor fields on $\Phi^{-1}_\lambda(\Theta_\mu)$ which are Φ_λ prjectable onto (1-1)-tensor fields on Θ_μ . Then we can build up a (1-1)-tensor field on T^*G by using

$$R_i \left(d\varphi_i \otimes \frac{\partial}{\partial \varphi_i} + dC_i \otimes \frac{\partial}{\partial C_i} \right)$$

and the other two lifted via Φ_λ^{-1} and Φ_ς^{-1}.

2) For projectable (1-1)-tensor fields we have $\varphi_* \mathbb{N}_R = \mathbb{N}_{\varphi_* R}$. In our case, due to the way we built our (1-1)-tensor field, we can show that starting with Nijenhuis operators on Θ_M we get Nijenhuis ones on T^*G.

3) It is possible to add other vector fields to the ones generated by $H(C_1, \ldots, C_{n-2p})$ if we quality R_1^λ and R_i^ς. For instance $H(C_1, \ldots, C_{n-2p}) + \Phi_\varsigma^*(f_2(\bar{P}_i)) + \Phi_\lambda^*(f_1(P_i))$ defines a completely integrable vector field (perhaps not complete) with (1-1)-tensor field constructed by qualifying the coordinate dependence of R_i^λ and R_i^ς to be $R_i^\lambda = R_i^\lambda(P_i)$, $T_i^\varsigma = T_i^\varsigma(\bar{P}_i)$. This construction provides a possible link between complete integrability through a (1-1)-tensor field and completely integrable systems on Lie groups. This construction and its relation to the decomposition of $\mathfrak{g}^*_G$ provided by subalgebras of $\mathfrak{g}_G$ is yet to be investigated. This analysis will appear elsewhere.

Aknowledgements. Results outlined in section 8 have been obtained with the collaboration of S. De Filippo and G. Vilasi. I would like to thank Professor Nijenhuis for illumination discussions.

References

1. Calogero, F. (1971). Solution of one-dimensional n-body problems with quadratic and/or inversely quadratic pair potentials, J. Math. Phys. 12, 419-436;

Moser, J.K. (1975). Three integrable Hamiltonian systems connected with isospectral deformations, Adv. in Math. 16, 197-220;

id. (1980). Various aspects of integrable Hamiltonian systems, C.I.M.E. Lectures - Bressanone 1978, Progress in Math. 8, Birkhäuser, Boston, pp;233-289;

id. (1980). Geometry of quadratics and spectral theory, in: The Chern Sympozium 1979, Springer, Berlin et.al. pp.147-188;

2. Toda, M. (1974). Studies in nonlinear lattice, Arkiv for det Fysiske Seminar i Trondheim, 2

id. (1976). Development of the Theory of Nonlinear Lattice, Progr. Theor. Phys. Suppl; 59, 1-35

Kostant, B. (1980). The solution to a generalized Toda lattice and representation theory, Adv. in Math. 34, 195-338

Flaschka, H. (1974). On the Toda lattice II: Inverse scattering solution, Progr. Theor. Phys. 51, 703-716

Goodman, R. , Wallach, N.R. (1982). Classical and quantum-mechanical systems of Toda lattice type - I. Comm. Math. Phys. 84, 355-386

(1984). II. Solutions of the classical flows, 94, 177-217

3. Gardner, C.S. (1971). Korteweg-de Vries equation and generalizations. IV. The Korteweg-de Vries equation as a Hamiltonian system, J. Math. Phys. 12, 1548-1551

Gardner, C.S., Green, J.M., Kruskal, M.D., Miura, R.M. (1974). Korteveg-de Vries equation and generalizations. VI, Methods for exact solution, Comm. Pure Appl. Math. 27, 97-133

Dubrovin, B.A., Matveev, V.B., Novikov, S.P. (1976). Non-linear equations of Korteveg-de Vries type, finite band, linear operators, and Abelian varieties, Russ. Math. Surveys 31, 59-146

Zakharov, V.E., Faddeev, L.D. (1971). Korteveg-de Vries Equations: A completely Integrable Hamiltonian system, Funct. Anal. Appl. 5, 280-287

4. Coleman, S. (1975). Quantum sine-Gordon equations as the massive Thirring Model, Phys. Rev. D11, 2088-2097

Kaup, D.J., Newell, A.C. (1978). The Cauchy and Goursat problems for the sine-Gordon equation, Siam J. Appl. Math. 34, 37-54

5. Nikolaenko, N.V. (1976). The complete integrability of the non-linear Schrödinger equation, Funct. Anal; Appl. 10, 209-220

6. Lax, P.D. (1976). Almost periodic solutions of the KdV equation, SIAM Review 18, 351-375

id. (1968). Integrals of non-linear equations of evolution and solitary waves, Comm. Pure & Appl. Math. 21, 467-490

7 Eilenberg, G. (1981). Politons. Mathematical methods for physicists, Springer Berlin et.al.

Berezin, F.A.,Perelomov, A.M. (1980). Group-theoretical interpretation of the Korteveg-de Vries type equations, Comm. Math. Phys. 74, 129-140

7. Gel'fand, I.M., Dorfman, I.Ya. (1979). Hamiltonian operators and associated algebraic structures, Funct. Anal. Appl. 13, 248-262

Gel'fand, I.M., Dikii, L.A. (1976). Fractional powers of operators and Hamiltonian systems, Funct. Anal. Appl. 10, 259-273

Magri, F. (1978). A simple model of the integrable Hamiltonian equation, J.Math. Phys. 19, 1156-1162

Fokas, A., Fuchssteiner B. (1981). Symplectic Structures, their Backlund transformations and hereditary symmetries, Physica 4D, 47

Fuchssteiner, B. (1982). The Lie algebra structure of degenerate Hamiltonian and bi-Hamiltonian systems. Progress of Theoretical Physics 68, 1082-1104

8. Filippo, G. De, Marmo, G., Vilasi, G. (1982). A Geometrical setting for the Lax representation, Phys. Letters 117B, 418

Carinena, J.F., Ibort, L.A. (1984). On Lax equations arising from Lagrangian foliations, Lett. Math. Phys. 8, 21-26

id. (1984). Bilagrangian connections and the reduction of Lax equations, Lett. Math; Phys. 8, 359-365

9. Marmo, G. (1982). A geometrical characterization of completely integrable systems. In: Proceedings "Geometry and Physics" Florence, October 1982, pp. 257-262

Filippo, De, Marmo, G., Salerno, M., Vilasi, G. (1984). A new characterization of completely integrable systems, Nuovo Cimento 83B, 97

10. Abraham, R., Marsden, J.M. (1978). Foundations of Mechanics, Benjamin, N.Y. 2nd Ed.

11. Magri, F., Morosi, C. (1984). A geometrical characterization of Integrable Hamiltonian systems through the theory of Poisson-Nijenhuis manifolds, Quaderno S/19 Dipartimento di Matematica Università di Milano

Magri, F., Morosi, C., Ragnisco, O. (1985). Reduction

Techniques for infinite dimensional Hamiltonian systems: Some ideas and Applications, Comm. Math. Phys. 99, 115-140

12. Eiseman, P.R., Stone, A.P. (1975). A note on a differential concomitant, Proceedings of the Am. Math. Soc. 53, 179-185

Stone, A.P. (1983). The Frobenius theorem associated with a certain derivation, Tensor, N.S. 40, 35-41

13. Frölicher, A., Nijenhuis, A. (1960). Invariance of Vector form operations under mappings, Comm. Math. Helsinki 34, 227-248

id. (1966). Theory of vector valued differential forms, Indag. Math. 18, 338-359

14. Stone, A.P. (1966). Analytic conservation laws, Ann. Inst. Fourier (Grenoble) 16, 319-327

15. Magri, F. Alcune osservazioni sui sistemi Hamiltoniani Integrabili, Unpublished

16. Crampin, M. (1983). Tangent bundle geometry for Lagrangian dynamics, J. Phys. A: Math; Gen; 16, 3755

17. Klein, J. (1962). Espaces variationelles et Mécanique - Thèse - Ann. Inst. Fourier, Grenoble 12, 1-124

Godbillon, C. (1969). Géométrie différentielle et mécanique analytique, Hermann, Paros

Trautmann, C. (1984). Differential geometry for physi; cists, Bibliopolis, Napoli

18. Gutkin, D. (1985). Varietées bi-structurées et operateur de recursion, Pub; IRMA - Lille, 7, Fasc. 2 n°3

19. Zakharov, V.E., Konopelchenko, B.G. (1984). On the theory of recursion operator, Comm. Math. Phys. 94, 483-509

20. Bampi, F., Morro, A. (1982). The inverse problem of the calculus of variations applied to continuum physics, J. Math. Phys. 23, 2312-2321

Sarlet, W., Cantrijn, F., Crampin, M. (1984). A new look at second oreder equations and Lagrangian mechanics, J. Phys. A: Math. Gen. 17, 1999

21. Marmo, G., Saletan, E.J. (1977). Ambiguities in the Lagrangian and Hamiltonian formalism: transformation properties, Nuovo Cimento 40B, 67

Sarlet, W., Cantrijn, F. (1983). Symmetries, first integrals, and the inverse problem of Lagrangian mechanics: II. J. Phys. A: Math. and Gen. 16, 1383

22. Hennaux, M. (1982). Equations of motions, commutation relations and ambiguities in the Lagrangian formalism, Annals of Physics 140, 45

Antonini, P., Marmo, G., Rubano, C. Alternative Lagrangians and complete integrability: some remarks, Nuovo Cimento 86B, 17-30

23. Marmo, G., Rubano, C. (1983). Equivalent Lagrangians and Lax representation, Nuovo Cimento 78B, 70

Ferrario, C., Vecchio, G.Lo, Marmo, G., Morandi, G. (1985). Separability of Completely integrable dynamical system admitting alternative Lagrangian descriptions, Lett. Math. Phys. 9, 140

24. Henneaux, M., Shepley, L.C. (1982). Lagrangians for spherically symmetric potentials, J. Math. Phys. 23, 2101-2107

25. Guillemin, V., Sternberg, S. (1984). Symplectic techniques in physics. Cambridge Univ. Press, Camnridge

26. Marsden, J.E., Ratiu, T., Weinstein, A. (1982). Semidirect products and reduction in Mechanics, Preprint CPAM - Univ. of California, Berkeley

Marle, C.M. (1976). Symplectic manifolds, dynamical groups, and Hamiltonian mechanics. In: Differential Geometry and relativity, Cahen, M., Flato, M.(eds.) Reidel, pp;249-269

27. Chu, Bon-Yao, (1974). Symplectic homogeneous space, Trans. Am. Math. Soc. 197, 145

28. Duflo, M., Vergne, M. (1969). Une propriété de la representation coadjointe d"une algèbre de Lie, C.R.Acad. Sci. Paris A-B 268, 583-588

29. Nekhoroshev, N.N. (1972). Action-angle variables and their generalizations, Trans. Moscow Math. Soc. 26, 180-198

MULTIDIMENSIONAL NONLINEAR DYNAMICAL SYSTEMS

M.Saveliev

Institute for High Energy Physics, Protvino, Serpukhov, SU-142284 Moscow Region

1. The modern methods for investigating nonlinear partial differential equations and their systems bring about constructive integrability criteria for one- and two-dimensional spaces only [1-5] . Here the theory of graded Lie algebras and their representations is of the decisive significance for both constructing integrable systems with a nontrivial internal symmetry algebra and finding an explicit form of their exact solutions.

At the same time, there are no effective methods for the systems in n-dimensional spaces. (Possibly, the only exception is some three-dimensional equations,e.g. the Kadomtsev-Petviashvili equation and the three wave problem,)

The main cause might be the following one. In the conventional two-dimensional case one takes a nilpotent structure of the defining parts of the Lax-type operators taking values in a Z-graded Lie algebra G. But it seems that for n-dimensional systems this structure fails to reflect the properties of the corresponding internal symmetry algebra adequately.

Thus the problem is to find an appropriate modification of this construction for spaces of an arbitrary dimension. The formulation [6] of the Gauss, Peterson-Codazzi and Ricci equations (obviously containing an integrable subclass) as an n-dimensional representation of the zero curvature type:

$$[\,\partial/\partial x_i + A_i(x),\quad \partial/\partial x_j + A_j(x)\,] = 0 \qquad (1)$$

for the induced connections of V_n may serve as a guide. These equations describe the manifolds $V_n(x_1,\dots,x_n)$ embedded into the flat enveloping space R_{n+p} endowed with the structure of a Lie algebra G. Naturally, representation (1) with no extra conditions on the operators A_i with values in G has only a trivial solution. This is obvious since in this case (1) is form-invariant under the gauge transformations generated by the elements of G. Therefore let us introduce the notion of the grading spectrum of the operators A_i generalizing the similar notion of the two-dimensional case [3] .

By the grading spectrum $Q_i \in Z$ of an operator A_i for a fixed grading of G ($G = \bigoplus_{a\in Z} G_a$, $[G_a, G_b] \subset G_{a+b}$) we mean a set of the indices of the subspaces G_a, where A_i or their certain linear combinations (with numerical coefficients) take values: $A_i \in \bigoplus_{a\in Q_i} G_a$. Besides,

$$[A_i, A_j] \subset \bigoplus_{a\in Q_i \cup Q_j} G_a \quad \text{for all } j.$$

Consider gauge transformations which leave (1) form-invariant and preserve the grading spectrum of the operators A_i . Since $[G_o, G_a] \subset G_a$, these transformations are generated by the subalgebra G_o only, and representation (1) produces nontrivial nonlinear systems.

However, all the attempts to use Z-grading of G reduce automatically the dimension of the arising systems to two. Therefore it seems natural to consider as internal symmetry algebras of n-dimensional systems Lie algebras whose grading is performed by formal (spectral) parameters. The number of the parameters should correspond to the dimension of the space.

The final and most essential step is to choose the grading spectrum of the operators A_i , so that representation (1) would generate integrable systems. For the two-dimensional case [3] the integrable systems are known to be associated with the grading spectrum of the operators $A_{\pm} \equiv -iA_2 \mp iA_1$ of the form $Q_{\pm} = \pm(0,1,\ldots, m_{\pm})$. Then operators $\hat{A}_{\pm} \in \bigoplus_{1 \le a \le m_{\pm}} G_{\pm a}$ $(A_{\pm} \equiv A_{\pm}^{(0)} + \hat{A}_{\pm})$ defining the integrability take values in the nilpotent parts of G. The latter circumstance provides us with the possibility to construct explicitly the solutions of the corresponding nonlinear systems as absolutely convergent series (finite polynomials for finite-dimensional algebras G) in the functions

$$\Phi_{\pm\alpha_1,\ldots,\alpha_n}(z_{\pm}) \equiv \int^{z_{\pm}} dz_1^{\pm} \ldots \int^{z_1^{\pm}} dz_2^{\pm} \ldots \\ \ldots \int^{z_{n-1}^{\pm}} dz_n^{\pm} \varphi_{\pm\alpha_1}(z_1^{\pm}) \ldots \varphi_{\pm\alpha_n}(z_n^{\pm})$$

Here $z_{\pm} \equiv x_1 \pm x_2$, $\varphi_{\pm\alpha}(z_{\pm})$ are arbitrary functions determining the Cauchy and Goursat problem. Probably these results can be interpreted in the framework of the Ritt-Kolchin formal theory [7]. We mean the theorem on the connection between the Liouvillean extension of a differential field (without new constants and realized by integrals, exponentials of integrals, and algebraic functions) and the Picard-Vessiot extension in terms of solvability of the corresponding differential Galois group.

As a whole this is the general programme of the investigation proposed. In this report we discuss some of the above questions for the systems of a special type.

2. In what follows we need infinite-dimensional Lie algebras $G^{(m)}$, which are direct modifications of the centre-free loop algebras $G^{(1)}$ in the principal grading. The latter ones are determined [8] in a degenerate representation of $\mathbb{G}^{(1)}$ by elements F_A^a, $[F_A^a, F_B^b] = c_{ABC} F_C^{a+b}$,

of the algebra $G^{(1)}$ corresponding to the finite-dimensional algebra $G(\equiv G^{(o)})$ with an orthonormal basis

$F_A(\equiv F_A^o)$. To do this introduce a formal (spectral) parameter λ and set $F_A^a = \lambda^a F_A$. Clearly, the algebra $G^{(1)}$ is of finite growth, and the numbers $a \in Z$ are grading indices of $G^{(1)}$.

Similarly, in the n-dimensional case introduce the parameters $\lambda_1, \ldots, \lambda_m$. Then the elements $F_A^{a_1, \ldots a_m} \equiv \equiv \lambda_1^{a_1} \lambda_2^{a_2} \ldots \lambda_m^{a_m} F_A$ form a filtrated infinite-dimensional algebra $G^{(m)} = \cup \bar{G}_k$ of finite growth with

$$\bar{G}_k = \{F_A^{a_1 \ldots a_m}, \max \{ |a_1|, \ldots, |a_m| \} \leq k\},$$

$$[F_A^{a_1 \ldots a_m}, F_B^{b_1 \ldots b_m}] = c_{ABC} F_C^{a_1+b_1, \ldots, a_m+b_m}$$

As an example consider a special case when the operators A_i are realized on the space of skew matrices $M_{IJ}(= -M_{JI}, 1 \leq I, J \leq n+1)$. Then $G^{(m)}$ is associated with the finite-dimensional Lie algebra so(n+1) whose space is decomposed into direct sum of the subalgebra $G_o \simeq so(n)$ and the vector space r_1 spanned by the elements M_{in+1}, $1 \leq i \leq n$; $[G_o, G_o] \subset G_o$, $[G_o, r_1] \subset r_1$, $[r_1, r_1] \subset G_o$. Accordingly, write the operators A_i as $A_i = A_i^{(o)} + A_i^{(1)}$ with

$$A_i^{(o)}(x) = \sum_{1 \leq k, l \leq n} v_i^{kl}(x) M_{kl},$$

$$A_i^{(1)}(x) = \sum_{1 \leq k \leq n} f_{ik}(x) M_{kn+1}$$

Notice that imposing some conditions on the functions

f_{ik}, which agree with the equations resulting from (1) is in a way equivalent to a consistent choice of a grading spectrum for the operators A_i taking values in (generally infinite-dimensional) Lie algebra.

To obtain a nontrivial system from representation (1), let us require that it be form-invariant under G_o. For this introduce the spectral parameters λ_j, e.g. in the following way: $f_{ik}(x) \rightarrow f_{ik}(x)\, m_i^k(\lambda)$. Here m_i^k are matrix elements of the complexi'hull of SO(n), i.e.

$$\sum_k m_i^k m_j^k = \sum_k m_k^i m_k^j = \delta_{ij} \quad ;$$

the generalized Euler angles in the first fiber of the element of SO(n) are expressed via $\ln \lambda_j$. Then the operators A_i turn out to be actually spanned by the elements of the Lie algebra $G^{(m)}$ in a degenerate representation. Then, up to the gauge transformation

$$A_i\,(x) \longrightarrow g^{-1} A_i(x) g + g^{-1}\, \partial g/\,\partial x_i \qquad (2)$$

with $g(x)$ being a function on SO(n) we get $f_{ik}(x) = f_i(x)\, \delta_{ik}$.

Thus, the nonlinear systems resulting from (1) are formally produced by the operators $A_i(x)$ of the form:

$$A_i(x) = \sum_{k,l} \upsilon_i^{kl} M_{kl} + f_i \sum_k n_i^k M_{kn+1}\,, \quad 1 \leq i \leq n \qquad (3.)$$

Here $\upsilon_i^{kl}\,(= -\upsilon_i^{lk})$, f_i and n_i^k are functions of the variables $x_1,\ldots,x_n$; n_i^k are matrix element of SO(n). Substituting (3) into (1) we obtain for the functions υ_i^{kl}, f_i and n_i^k the following system of equations

$$\partial\, \upsilon_j^{kl}/\,\partial x_i - \partial\, \upsilon_i^{kl}/\,\partial x_j - 4\sum_m \upsilon_i^{km}\, \upsilon_j^{ml} + \tfrac{1}{2} f_i f_j (n_i^k n_j^l - n_i^l n_j^k) = 0, \quad i \neq j \; ; \qquad (4.1)$$

$$\partial(f_j n_j^k)/\partial x_i - \partial(f_i n_i^k)/\partial x_j - 2f_j \sum_m \upsilon_i^{km} n_j^m +$$

$$+ 2f_i \sum_m \upsilon_j^{km} n_i^m = 0 \qquad (4.2)$$

The solution of (4.2) is given by the formula

$$-2\,\upsilon_i^{kl} = \sum_j (n_i^k n_j^l - n_j^k n_i^l) f_j^{-1} \frac{\partial f_i}{\partial x_j} + \sum_j n_j^k \, \partial n_j^l / x_i \qquad (5)$$

Then (4.1) are rewritten in terms of the variables $\rho_i \equiv \ln f_i$ as

$$e^{\rho_i - \rho_j} D_{i,jj} + e^{\rho_j - \rho_i} D_{j,ii} + \sum_k e^{\rho_i + \rho_j - 2\rho_k} \frac{\partial \rho_i}{\partial x_k} \frac{\partial \rho_j}{\partial x_k} +$$

$$+ e^{\rho_i + \rho_j} = 0, \quad i \neq j\,; \qquad (6.1)$$

$$D_{j,ik} = 0, \quad i \neq j \neq k \qquad (6.2)$$

Here

$$D_{j,ik} \equiv \frac{\partial^2 \rho_j}{\partial x_i \partial x_k} + \frac{\partial \rho_j}{\partial x_i} \frac{\partial \rho_j}{\partial x_k} - \frac{\partial \rho_j}{\partial x_k} \frac{\partial \rho_k}{\partial x_i} - \frac{\partial \rho_j}{\partial x_i} \frac{\partial \rho_i}{\partial x_k}.$$

System (6) contains no dependences on the functions n_i^k and follows from (1) when

$$A_i(x) = e^{\rho_i} \left[- \sum_k e^{-\rho_k} \frac{\partial \rho_i}{\partial x_k} M_{ik} + M_{in+1}\right] \qquad (7)$$

which may be obtained from (1) by gauge transformation (2) with

$$g^{-1} \left(\sum_k n_i^k M_{kn+1} \right) g = M_{in+1} \quad ,$$

i.e. when (1) is form-invariant.

As is easy to verify, equations (6) for real e^{ρ_i} describe the Riemannian spaces V_n of constant curvature (or the principal second-order hypersurfaces of the flat space R_{n+1}) with n orthogonal hypersurfaces. Indeed, choosing as these the coordinate hypersurfaces x_i=const, for the metric tensor g_{ij} one has $g_{ij}=0$, $i\neq j$. From here substituting $g_{ii} = e^{2\rho_i}$ we obtain the following evident equations [9]

$$R_{ijkl} = 0, \quad i \neq j \neq k \neq l;$$

$$R_{ijjk} = e^{2\rho_j} D_{j,ik}(\rho) \ , \quad i \neq j \neq k;$$

$$R_{jiij} = e^{2\rho_i} D_{i,jj} + e^{2\rho_j} D_{j,ii} + \sum_k e^{2(\rho_i+\rho_j-\rho_k)} \frac{\partial \rho_i}{\partial x_k} \frac{\partial \rho_j}{\partial x_k}$$

$$i \neq j \ .$$

Then, by (6), the components of the first-type Riemannian tensor satisfy

$$R_{ijkl} = K(g_{ik}g_{jl} - g_{il}g_{jk})$$

with the constant curvature K (=1) and $g_{ij} = \delta_{ij} e^{2\rho_i}$; while the Ricci tensor is

$$R_{ij} = (n-1)\ \delta_{ij} e^{2\rho_j} = \frac{R}{n}\ \delta_{ij} e^{2\rho_j} \quad .$$

In the same time, if $\sum_i e^{2\rho_i} = 1$ equations (6) result in a realization of the Gauss equations for hyperbolic (constant sectional curvature -1) n-dimensional submanifolds of R_{2n-1} obtained in [10] .

Let us illustrate the above considerations by the simplest case n=2. The initial system of equations for the functions f_{ij} and v_i is:

$$\frac{\partial v_2}{\partial x_1} - \frac{\partial v_1}{\partial x_2} + \cos^2\theta\ f_{11} f_{22} + \sin^2\theta\ f_{12} f_{21} = 0 \ ,$$

$$\sin\Theta \left(\frac{\partial f_{21}}{\partial x_1} - v_2 f_{12} \right) - \cos\Theta \left(\frac{\partial f_{11}}{\partial x_2} + v_1 f_{22} \right) = 0 ,$$

$$\cos\Theta \left(\frac{\partial f_{22}}{\partial x_1} - v_2 f_{11} \right) + \sin\Theta \left(\frac{\partial f_{12}}{\partial x_2} + v_1 f_{21} \right) = 0 ,$$

where $\Theta \equiv -i \ln \lambda$. Equating to zero the terms containing the same power of λ we get

$$\frac{\partial}{\partial x_1} \left(f_1^{-1} \frac{\partial f_2}{\partial x_1}\right) + \frac{\partial}{\partial x_2} \left(f_2^{-1} \frac{\partial f_1}{\partial x_2}\right) + f_1 f_2 = 0 .$$

If $f(x) = (f_1 f_2)$ is a vector of constant length, i.e. $\Sigma f_i^2 = c$, this equation coincides with the Liouville equation for $c = 0$ and with the Sine-Gordon one for $c \neq 0$. (It is well-known that both the Liouville and Sine-Gordon equations are integrable).

In general if the length of the vector f is constant we can analyse the structure of system (6) and obtain various formulations of the corresponding equations. In particular, when $n = 3$ and $\sum_j \varepsilon_j f_j^2 = c$, $\varepsilon_j = \pm 1$ introducing the "modified" Laplacians $\Delta_{(j)} \equiv \sum_{i \neq j} \varepsilon_i \partial^2/\partial x_i^2 - \varepsilon_j \partial^2/\partial x_j^2$ and substituting $\varepsilon_i^{\frac{1}{2}} \partial/\partial x_i \rightarrow \partial/\partial x_i$, $\varepsilon_i^{\frac{1}{2}} f_i \rightarrow f_i$ we can rewrite (6) in the form:

$$\Delta_{(i)} \rho_i = \sum_{j \neq i} \left(2 \frac{\partial \rho_i}{\partial x_j} \frac{\partial \rho_j}{\partial x_j} + \left[e^{2(\rho_i - \rho_j)} - 1\right] \left(\frac{\partial \rho_i}{\partial x_j}\right)^2 + e^{2(\rho_j - \rho_i)} \left(\frac{\partial \rho_j}{\partial x_i}\right)^2\right) + \left(\frac{\partial \rho_i}{\partial x_i}\right)^2 + e^{2\rho_i} - c; \quad (8.1)$$

$$\frac{\partial}{\partial x_j}\left[e^{\rho_j-\rho_i}\frac{\partial \rho_j}{\partial x_i}\right] + \frac{\partial}{\partial x_i}\left[e^{\rho_i-\rho_j}\frac{\partial \rho_i}{\partial x_j}\right] =$$

$$= -e^{\rho_i-\rho_j}\sum_{k\neq i,j} e^{2\rho_k}\frac{\partial \rho_k}{\partial x_i}\frac{\partial \rho_k}{\partial x_j}, \quad i \neq j \qquad (8.2)$$

3. The above construction shows that in the n-dimensional case at least for the systems under consideration, the conjecture of [3] is valid in the form that a two-dimensional system possesses a formal LA-pair (in our case, n operators A_i), provided its internal symmetry algebra has a finite number of generators. To divide further these systems into exactly and completely integrable (in the sense of [3]), one has to investigate in detail the properties of this algebra associated with the restrictions of the functions f_i entering the expressions for A_i .

Representation (1) yields the gradient substitution for the vector $A = \{A_i(x),\ 1 \leq i \leq n\}$:

$$A_i = G^{-1}_{(n)}\frac{\partial G_{(n)}}{\partial x_i}. \qquad (9)$$

Here $G_{(n)}$ is an element of the complexhull of the group with the Lie algebra G, where A_i take values.

Assume that we know the solution $\bar{\rho}_\alpha\ (x_1,\dots,x_{n-1})$, $1\leq\alpha \leq n-1$, of a system like (6) generated via (1) by the operators

$$A_\alpha^{(n-1)} = e^{\bar{\rho}_\alpha}\left[-\sum_{\beta=1}^{n-1} e^{-\bar{\rho}_\beta} \cdot \frac{\partial\bar{\rho}_\alpha}{\partial x_\beta} M_{\alpha\beta} + M_{\alpha n}\right].$$

Take the gradient form of the operators $A_\alpha^{(n-1)}$:

$A_\alpha^{(n-1)} = G^{-1}_{(n-1)}\partial G_{(n-1)}/\partial x_\alpha$. In this case a regular element $G_{(n)}$ of $SO(n+1)$ parametrizing the operators $A_i^{(n)}$ $(\equiv A_i)$ of the form (7) by the formula (9) is obtained from $G_{(n-1)}$ via multiplying by the matrix $\xi \in SO(n+1)/SO(n)$, $G_{(n)} = \xi g^{-1}$, $g \equiv G^{-1}_{(n-1)}$. Then the parameters of ξ themselves turn out to be the solutions of the nonlinear system. Indeed, let us parametrize ξ with the generalized Euler angles:

$\xi = \prod_{i=1}^{n} e^{\varphi_i M_{ii+1}}$, ordered w.r.t. the increasing index i, i.e.

$$\xi^I_J = (-1)^{n-I+1} \prod_{j=I+1}^{n} \cos^{-1}\varphi_j \prod_{k=I}^{n} \frac{\partial}{\partial \varphi_k} \xi^{n+1}_j ,$$

$$1 \le I, J \le n+1 ,$$

$$\xi^{n+1}_J = (-1)^{n-J+1} \cos\varphi_{J-1} \prod_{k=J}^{n} \sin\varphi_k , \quad \varphi_o \equiv 0. \qquad (10)$$

(Here the lower index of ξ numbers the row the upper one the column). Next, substituting $G_{(n)}$ into the rhs of (9) and equating it to (7), one gets the equations for the angles $\varphi_{i(x)}$

$$\frac{\partial \varphi_k}{\partial x_i} = (-1)^{n-k} n^k_i e^{-\tau_i} \frac{\partial \upsilon}{\partial x_i} \prod_{j=k+1}^{n-1} \sin^{-1}\varphi_j, \quad 1 \le k \le n-1 ,$$

$$\sum_j ((n^1_i n^k_j - n^k_i n^1_j) \frac{\partial e^{\tau_j}}{\partial x_i} - e^{\tau_i} n^k_j \frac{\partial n^1_j}{\partial x_i}) =$$

$$= (-1)^{n-1+1} n^k_i \frac{\partial \upsilon}{\partial x_i} \operatorname{ctg}\varphi_1 \prod_{q=1+1}^{n-1} \sin^{-1}\varphi_q', \quad 1 \le k \le n-1 \qquad (11.1)$$

$$\frac{\partial^2 \upsilon}{\partial x_i \partial x_j} = \frac{\partial \upsilon}{\partial x_i} \frac{\partial \tau_i}{\partial x_j} + \frac{\partial \upsilon}{\partial x_j} \frac{\partial \tau_j}{\partial x_i} , \quad i \neq j , \qquad (11.2)$$

where $\upsilon \equiv \ln \operatorname{tg} \frac{\varphi_n}{2}$, $\tau_i \equiv \ln n^n_i$, $\sum_i e^{2\tau_i} \equiv 1$.

The expression for the solutions of system (6) is

$$f_i \equiv e^{\rho_i} = (-1)^n (e^{\tau_i} ch\upsilon)^{-1} \frac{\partial \upsilon}{\partial x_i} \qquad (12)$$

Recall that we assume here that the functions n_i^k, elements of the matrix g, depend on $x_1,\ldots,x_{n-1}$. Thus, the functions f_i are defined on a class of the solutions of equations (11.2) for one function υ.

For instance, for $n = 2$ and $f_1^2+f_2^2 = 0$, it follows from (11) that υ satisfies the Laplace equation $(\partial^2/\partial x_1^2-\partial^2/\partial x_2^2)\upsilon = 0$. Then (12) reproduces automatically the known complete solution to the Liouville equation.

Certainly, our derivation with no requirements of the functions e^{ρ_i} cannot provide final solution for the problem of interest but reveals the reduction structure of the solutions.

If the length of f is constant one can obtain an interesting recurrent formula for the operators $\hat{A}_i \equiv \equiv \Sigma f_i n_i^k M_{kn+1}$ (or, more precisely, for their combinations $L_I \equiv \sum_{1\le i\le n-I} \hat{A}_i$, $0\le I\le n-1$, which define effectively the structure of (6) by means of (1). Set $f_i = n_i^n$ and parametrize the column n_i^n by an expression like (10) with the angles θ_α, $1\le \alpha \le n-1$. Then, introducing the quantities

$$G_j \equiv g_j(\theta_j)\ldots g_1(\theta_1);\ g_j(\theta_j) \equiv e^{-(\theta_j+\frac{\pi}{2})M_{n-j,n}}, \qquad (13)$$

one may easily verify the validity of the formula

$$L_j\ (\theta_1\ldots\theta_j) = (-1)^j \prod_{k=1}^{j} \sin\theta_k G_j^{-1} M_{nn+1} G_j\ ,\ 1\le j\le n-1$$

This gives the following recurrence

$$L_j = -\sin\theta_j (G_{j-1}^{-1} g_j G_{j-1})^{-1}\ L_{j-1}(G_{j-1}^{-1} g_j G_{j-1}). \qquad (14).$$

The latter realizes the transformation which generates the operator $L_j(\Theta_1,\ldots,\Theta_j)$ starting from $L_{j-1}(\Theta_1,\ldots,\Theta_{j-1})$ by adding the dependence on a new unknown function $\Theta_j(x_1,\ldots,x_n)$. This function is determined by the solution of the corresponding system.

4. The essence of the present paper is an explicit realization of the zero-curvature-type representation for n-dimensional systems like (6). Some properties of their solutions and of the corresponding operators A_i entering (1) have been established. We hope that the discovered properties will enable us to investigate the integrability problem, at least in terms of Lie pseudogroups.

Concluding the report, I would like to present a generalization of system (6) for the case of $G^{(m)}$ associated with the Lie algebra so(N) whose space is decomposed into the direct sum of the subalgebra $G_o \simeq so(n) \oplus so(N-n)$ and the vector space r_{N-n} spanned by the elements $M_{in+\alpha}$, $1 \leq i \leq n$, $1 \leq \alpha \leq N-n$. The operators A_i can be written as $A_i = A_i^{(o)} + A_i^{(1)}$ with

$$A_i^{(o)} = \sum_{k,l=1}^{n} V_i^{kl}(x)M_{kl} + \sum_{\alpha,\beta=1}^{N-n} t_i^{\alpha\beta}(x)M_{n+\alpha\ n+\beta},$$

$$A_i^{(1)} = \sum_{k=1}^{n} \sum_{\alpha=1}^{N-n} f_{ik}^{\alpha}(x)M_{kn+\alpha}$$

Repeating the reasonings similar to those of n° 2 one obtains

$$f_{ik}^{\alpha}(x) = \sum_{\beta=1}^{N-n} f_i^{\beta}(x)m_{\alpha}^{\beta}(x)\ \delta_{ik},$$

where $m^{\beta}{}_{\alpha}(x)$ are matrix elements of SO(N-n). Then, due to the form-invariance of (1) under the gauge transformation (2) generated by SO(N-n), the terms containing the functions $t_i^{\alpha\beta}(x)$ can be omitted. Thus

$$A_i(x) = \sum_{j=1}^{n} \upsilon_{ij}(x)M_{ij} + \sum_{\alpha,\beta=1}^{N-n} f_i^{\beta}(x)m_{\alpha}^{\beta}(x)M_{in+\alpha} \equiv$$

$$\equiv \hat{g}^{-1}\Big(\sum_j \upsilon_{ij}M_{ij} + \sum_\alpha f_i^\alpha M_{in+\alpha}\Big)\hat{g}, \quad g \in SO(N-n),$$

in representation (1) yield the required generalization of the nonlinear system (6):

$$e^{\rho_i-\rho_j}\Big(D_{i,jj} - \frac{\partial\rho_i}{\partial x_j}\frac{\partial\sigma_{ij}}{\partial x_j}\Big) + e^{\rho_j-\rho_i}\Big(D_{j,ii} - \frac{\partial\rho_j}{\partial x_i}\frac{\partial\sigma_{ij}}{\partial x_i}\Big) + \sum_{1\le k\le n} e^{\rho_i+\rho_j-2\rho_k}\frac{\partial\rho_i}{\partial x_k}\frac{\partial\rho_i}{\partial x_k}\cdot$$

$$\cdot\, e^{\sigma_{ij}-\sigma_{kj}-\sigma_{ki}} + e^{\rho_i+\rho_j+2\sigma_{ij}} = 0, \quad i \neq j\,;$$

$$D_{j,ik} + (1-e^{\sigma_{kj}-\sigma_{ij}-\sigma_{ki}})\frac{\partial\rho_j}{\partial x_i}\frac{\partial\rho_i}{\partial x_k} - \frac{\partial\rho_j}{\partial x_k}\frac{\partial\sigma_{kj}}{\partial x_i} = 0,$$

$$i \neq j \neq k\,;$$

with

$$\upsilon_{ji}(x) = -e^{-\sigma_{ij}}\Big(e^{\rho_j-\rho_i}\frac{\partial\rho_j}{\partial x_i}\Big)\,.$$

Here the following notations are used

$$\sum_{1\le\alpha\le N-n} f_i^\alpha f_j^\alpha = e^{\rho_i+\rho_j+\sigma_{ij}}$$

where $\sigma_{ij} = \sigma_{ji}$, and $\sigma_{ij} = 0$ for $i=j$. The functions $D_{i,jk}(x)$ are exactly as in (6). Note that the functions $m_\alpha^\beta(x)$ are determined by the equations

$$\sum_{1\leq\alpha,\beta\leq N-n} (m_\alpha^\gamma \frac{\partial m_\alpha^\beta}{\partial x_i}) f_j^\beta + \frac{\partial f_j^\gamma}{\partial x_i} -$$

$$- e^{-\sigma_{ij}} (e^{\rho_j - \rho_i} \frac{\partial \rho_j}{\partial x_i}) f_i^\gamma = 0, \quad i \neq j.$$

This system reduces to the form (6) for $N=n+1$ because then $\sigma_{ij} \equiv 0$ for all $1<i, j<n$.

Addendum. After the manuscript of this paper had been written, the author knew of [11], where a consistent description of the n-dimensional Zakharov-Shabat "dressing procedure" is presented. It is related both to the nonlocal Riemann problem and its generalization, the nonlocal "$\bar{\partial}$-problem". It would be interesting to establish a correspondence between the classes of n-dimensional equations obtained via their purely analytical technique and those investigated here in the algebraic approach.

Reference

1. Manakov,S.V., Novikov,S.P., Pitaevsky,L.P., Zakharov,V.E. (1980). The soluton theory: Inverse scattering method. Nauka, Moscow.

2. Solitons. (1980). Ed. Bullough,R.K., Caudrey,P.J. Springer, N.Y.

3. Leznov,A.N., Saveliev,M.V. (1985). Group methods for integration of nonlinear dynamical systems. Nauka, Moscow.

4. Jimbo,M., Miwa,T. (1983). Publ. RIMS. Kyoto Univ.,19, 943-1001.

5. Leznov,A.N. (1984). In: Proc. of the Internat. Workshop on Nonlinear and Turbulent Processes. Gordon and Breach. N.Y.

6. Saveliev,M.V. (1984). Teor. Mat. Fiz. 60, 9-23; (1984). Comm. Math. Phys. 95, 199-216.

7. Ritt,J.F. (1950). Differential Algebra. Amer. Math. Coll. Publ. 33, N.Y.; Kolchin,E.R. (1973). Differential Algebra and Algebraic Groups. Acad. Press,N.Y.

8. Kac,V. (1983). Infinite-dimensional Lie Algebras. Boston, Birkhauser.

9. Eisenhart,L.P. (1926). Riemannian Geometry. Princeton, Princeton Univ. Press.

10. Tenenblat,K., Chuu-Lian Terng (1980). Ann. Math. 111 477-490; Aminov Yu.A.(1980).Math.USSR Sbornik(Russian)

11. Zakharov,V.E., Manakov,S.V. (1985). Funkt. Anal. Appl. 19, 11-25 (Russian).

HIDDEN SYMMETRY OF SELF-DUAL EQUATIONS FOR STATIC AXIAL-SYMMETRIC GAUGE FIELDS

E.V. Doktorov

Institute of Physics of the BeSSR Academy of Sciences, Minsk

1. The discovery of infinite set of nonlocal conserved currents in two-dimensional chiral models [1] reflected the existence of an essentially nontrivial symmetry inherent in these models. The algebraic structure for this hidden symmetry has been elucidated in [2] where an infinite-dimensional subalgebra of the affine Lie algebra (the Kac-Moody algebra [3, 4]) is shown to be responsible for such symmetry. Lately this algebraic structure has been revealed in a number of other completely integrable field theoretical models (the self-dual Yang-Mills fields, superchiral and Kaluza-Klein models) and in a wider class of soliton equations. The results obtained in this area are systematized in [5,6].

We will consider the hidden symmetry of self-duality equations for static axial-symmetric SU(N) gauge fields. A detailed analysis of these equations is performed in [7]. Note that the self-duality equations for the fields under considerations are also of interest in gravity, because the SU(2) case is strictly equivalent to the Ernst equation [8], while the Einstein-Maxwell equations represent a particular case of the SU(3) self-duality equations [9].

2. Let $A_\nu(\rho, z)$ where $\nu = t, \rho = (x^2+y^2)^{\frac{1}{2}}, z$ and φ be gauge potentials for static axial symmetric SU(N) gauge fields. In terms of complex coordinates $y = \rho + iz$ and $\bar{y} = \rho - iz$ the self-duality equations

$F_{\mu\nu} = \frac{1}{2}\varepsilon_{\mu\nu\alpha\beta}F_{\alpha\beta}$ can be written as in [7]:

$$\partial_y A_{\bar{y}} - \partial_{\bar{y}} A_y - i[A_y, A_{\bar{y}}] + i[\Phi_y, \Phi_{\bar{y}}] = 0,$$
$$\partial_y \Phi_{\bar{y}} - i[A_y, \Phi_{\bar{y}}] + \frac{1}{2}(y+\bar{y})^{-1}(\Phi_y + \Phi_{\bar{y}}) = 0 \qquad (1)$$
$$\partial_{\bar{y}} \Phi_y - i[A_{\bar{y}}, \Phi_y] + \frac{1}{2}(y+\bar{y})^{-1}(\Phi_y + \Phi_{\bar{y}}) = 0$$

Here

$A_y = \frac{1}{2}(A_\rho - iA_z)$, $A_{\bar{y}} = \frac{1}{2}(A_\rho + iA_z)$, $\Phi_y = \frac{1}{2}(A_\varphi - iA_t)$, $\Phi_{\bar{y}} = \frac{1}{2}(A_\varphi + iA_t)$.

The existence of an additional symmetry, such as the axial one, gives a possibility to reduce the number of independent functions which define the potentials A_ν. Such a reduction is achieved in the following way. Let us decompose the algebra $g = su(N)$ into some maximal subalgebra k and its complementary subspace p, $g = k + p$, with respect to an involution θ $(\theta^2 = 1)$, where k and p are the subspaces with eigenvalues +1 and -1 respectively. The commutation relations are as follows:

$$[k,k] \subset ik, \; [k,p] \subset ip, \; [p,p] \subset ik.$$

The above reduction is expressed as an ansatz of the form $A_y, A_{\bar{y}} \in k$, $\Phi_y, \Phi_{\bar{y}} \in p$; for all this the equations (1) are invariant under the group $K \subset G$ whose Lie algebra is k.

We introduce now the linear combinations $a_\varkappa = A_\varkappa + i\Phi_\varkappa$, $\varkappa = y, \bar{y}$, multiplying the elements of p by i. Thereby $a_\varkappa$ represent elements of the real noncompact Lie algebra $g^* = k + ip$. The latter is $sl(N,R)$ or $su(p,q)$, $p + q = N$, depending on θ. Due to two equations of (1), the field strength $f_{y\bar{y}} = \partial_y a_{\bar{y}} - \partial_{\bar{y}} a_y - i[a_y, a_{\bar{y}}]$ vanishes identically, $f_{y\bar{y}} = 0$. In other words, the potentials $a_\varkappa$ are pure gauge,

$$a_\varkappa = -i\partial_\varkappa J \cdot J^{-1}, \; J(y,\bar{y}) \in G^*$$

where G^* is a real noncompact Lie group with Lie algebra g^*. The remaining equation in (1) is conveniently expressed in terms of hermitian matrix $\mathcal{M}=J^+J$:

$$\partial_y[(y+\bar{y})\partial_{\bar{y}}\mathcal{M}\cdot\mathcal{M}^{-1}]+\partial_{\bar{y}}[(y+\bar{y})\partial_y\mathcal{M}\cdot\mathcal{M}^{-1}]=0 \quad (2)$$

Here $\mathcal{M}$ is K-invariant and belongs to the irreducible globally symmetric Riemannian space of the noncompact type G^*/K. The equation (2) is the object for the subsequent analysis.

3. It was shown in [7] that there is an infinite number of nonlocal conservation laws in the system described by (2). As is well-known, the existence of infinite number of conserved currents is connected with complete integrability of the corresponding equation. As a consequence, we have the Belinsky-Zakharov-type representation [10] for the equation (2):

$$D_y\Psi(\lambda)=B_y(\lambda)\Psi(\lambda),\quad D_{\bar{y}}\Psi(\lambda)=B_{\bar{y}}(\lambda)\Psi(\lambda) \quad (3)$$

where λ is a complex parameter,

$$B_y=\frac{i\rho}{\lambda+i\rho}\partial_y\mathcal{M}\cdot\mathcal{M}^{-1},\quad B_{\bar{y}}=\frac{-i\rho}{\lambda-i\rho}\partial_{\bar{y}}\mathcal{M}\cdot\mathcal{M}^{-1},$$

$$D_y=\partial_y+\frac{i\lambda}{\lambda+i\rho}\partial_\lambda,\quad D_{\bar{y}}=\partial_{\bar{y}}-\frac{i\lambda}{\lambda-i\rho}\partial_\lambda$$

The integrability condition for the system of the linear equations (3) is fulfilled for any λ if and only if $\mathcal{M}$ is a solution of (2). If $\Psi(\lambda,y,\bar{y})$ is a solution of (3) holomorphic at $\lambda=0$ then $\mathcal{M}(y,\bar{y})=$ $=\Psi(\lambda=0,y,\bar{y})$ is a solution of (2).

Let C be a closed contour in the complex λ-plane which encircles the origin, and define a piecewise holomorphic gauge transformation for the potentials $B_\varkappa$:

$$B_\varkappa^{(\pm)}(\lambda)=\chi_\pm(\lambda)B_\varkappa(\lambda)\chi_\pm(\lambda)^{-1}+D_\varkappa\chi_\pm(\lambda)\cdot\chi_\pm(\lambda)^{-1},\ \varkappa=y,\bar{y}. \quad (4)$$

Here the matrix functions $\chi_{\pm}(\lambda)$ are analytic in $C \cup C_{\pm}$, where $C_{+}(C_{-})$ denotes the interior (exterior) of C. The new potential (4) satisfies the same linear equations (3):

$$D_{\varkappa}\Psi_{\pm}(\lambda)=B_{\varkappa}^{(\pm)}(\lambda)\Psi_{\pm}(\lambda),\quad \Psi_{\pm}(\lambda)=\chi_{\pm}(\lambda)\Psi(\lambda)$$

We have $B_{\varkappa}^{(+)}(\lambda)=B_{\varkappa}^{(-)}(\lambda)$ on C therefore we expect to obtain some relation between $\Psi_{+}(\lambda)$ and $\Psi_{-}(\lambda)$. Namely,

$$D_{\varkappa}(\Psi_{-}(\lambda)^{-1}\Psi_{+}(\lambda))=0,\quad \lambda\in C$$

We introduce now the matrix function $G_0(\lambda)=\Psi_{-}(\lambda)^{-1}\Psi_{+}(\lambda)$ such that $D_{\varkappa}G_0(\lambda)=0$ on C. Therefore, to find the functions $\chi_{\pm}(\lambda)$ and, thereby, the new solutions $\Psi_{\pm}(\lambda)$, we have to solve the homogeneous matrix Riemann-Hilbert problem

$$\chi_{+}(\lambda)=\chi_{-}(\lambda)G(\lambda),\quad \lambda\in C \tag{5}$$

where $G(\lambda)=\Psi(\lambda)G_0(\lambda)\Psi(\lambda)^{-1}$. The interrelation between solving the Riemann-Hilbert problem and solving nonlinear equations was established in [11]. It was shown in [12] for the chiral fields that the infinitesimal Riemann-Hilbert transformations possess the structure typical for the affine Lie algebras.

The matrix $G(\lambda)$ satisfies the equation

$$D_{\varkappa}G=[B_{\varkappa},G] \tag{6}$$

The existence of the infinite number of continuity equations follows from an asymptotic expansion of $G(\lambda)$ at $\lambda=\infty$ [7]:

$$G(\lambda)=\sum_{n=0}^{\infty}G^{(n)}\lambda^{-n} \tag{7}$$

Substituting (7) in (6), we get recurrences

$$\partial_y G^{(0)} = \partial_{\bar y} G^{(0)} = 0,$$
$$\partial_y G^{(n+1)} = -i\rho^{2n+1} \mathcal{M} \partial_y \left(\mathcal{M}^{-1} G^{(n)} \mathcal{M} \rho^{-2n}\right) \mathcal{M}^{-1}, \qquad (8)$$
$$\partial_{\bar y} G^{(n+1)} = i\rho^{2n+1} \mathcal{M} \partial_{\bar y} \left(\mathcal{M}^{-1} G^{(n)} \mathcal{M} \rho^{-2n}\right) \mathcal{M}^{-1}.$$

Let us choose $G^{(0)} = P$, where P is a constant non-singular matrix. Then the recurrences (8) give an infinite set of non-local continuity equations:

$$\partial_{\bar y} K_y^{(n)} + \partial_y K_{\bar y}^{(n)} = 0, \quad n = 0, 1, 2, \ldots \qquad (9)$$

where

$$K_{\text{æ}}^{(n)} = i\rho^{2n+1} \mathcal{M} \partial_{\text{æ}} \left(\mathcal{M}^{-1} G^{(n)} \mathcal{M} \rho^{-2n}\right) \mathcal{M}^{-1}$$

4. The Riemann-Hilbert transformation will be considered as a transformation $\psi \to \tilde\Psi$ expressed by

$$\tilde\Psi(\lambda) = \begin{cases} \chi_-(\lambda)\Psi(\lambda), & \lambda \in C_- \\ \chi_+(\lambda)\Psi(\lambda)G(\lambda)^{-1}, & \lambda \in C_+ \end{cases}$$

with the normalization condition $\chi_-(\infty) = 1$. The Riemann-Hilbert problem can be formulated in terms of singular integral equations [13] of the form

$$\chi_-(\lambda) = 1 - \frac{1}{2\pi i} \oint_C \frac{dt}{t-\lambda} \chi_-(t)\left[G(t)G(\lambda)^{-1} - 1\right]$$

In what follows $G_0(\lambda)$ is very close to the unit matrix, i.e. $G_0(\lambda) = \exp(\delta V(\lambda)) \approx 1 + \delta V(\lambda)$, $\delta \ll 1$. Then the infinitesimal Riemann-Hilbert transformation is represented by

$$\chi_-(\lambda) = 1 - \frac{\delta}{2\pi i} \oint_C \frac{dt}{t-\lambda} R_V(t) \equiv 1 + \delta M_V(\lambda) \qquad (10)$$

where

$$R_V(t) = \Psi(t)V(t)\Psi(t)^{-1}, \quad \mathcal{M}_V(\lambda) = -\frac{1}{2\pi i} \oint_C \frac{dt}{t-\lambda} R_V(t)$$

Hence, we obtain the transformation law for the matrix $\Psi(\lambda)$:

$$\Psi(\lambda) \to \tilde{\Psi}(\lambda) = (1 + \delta M_v)\,\Psi(\lambda) \qquad (11)$$

inducing that for $\partial_{\ae} M \cdot M^{-1}$. For finding the transformation law for the latter, we proceed as follows. We get from (5) for $\lambda \in C$

$$D_{\ae}\chi_+ \cdot \chi_+^{-1} + \chi_+ B_{\ae} \chi_+^{-1} = D_{\ae}\chi_- \cdot \chi_-^{-1} + \chi_- B_{\ae}\chi_-^{-1} = \tilde{B}_{\ae}$$

Then, due to (10, 11) and the definition of $B_{\ae}$, we find

$$\partial_{\ae} M \cdot M^{-1} \to \widetilde{\partial_{\ae} M \cdot M^{-1}} = (1 + \delta M_v)\,\partial_{\ae} M \cdot M^{-1} \quad (12)$$

where

$$M_V(\partial_y M \cdot M^{-1}) = \frac{1}{2\pi i}\oint_C \frac{dt}{t+i\rho}\left\{ i\partial_t R_V(t) + [\partial_y M \cdot M^{-1}, R_V(t)] \right\},$$

$$\mathcal{M}_V(\partial_{\bar{y}} M \cdot M^{-1}) = \frac{1}{2\pi i}\oint_C \frac{dt}{t-i\rho}\left\{ -i\partial_t R_V(t) + [\partial_{\bar{y}} M \cdot M^{-1}, R_V(t)] \right\}$$

Now it is easy to show by direct calculation that the transformed quantities $\widetilde{\partial_{\ae} M \cdot M^{-1}}$ and $\tilde{\Psi}$ satisfy up to terms of higher order in δ, the equations (3) ensuring the recovery of the solution $\tilde{M}$ on the self-duality equation (2) from $\tilde{\Psi}$: $\tilde{M} = \tilde{\Psi}(\lambda = 0)$.

Following [14], we can show that M_V realize a representation of the Lie algebra: $[M_V, M_W] = M_{[V,W]}$. Owing to its construction, $V(\lambda)$ admits the asymptotic series expansion at $\lambda = \infty$, similar to (7):

$$V(\lambda) = \sum_{n=0}^{\infty} V^{(n)} \lambda^{-n}$$

where $V^{(n)} = \sum_{a=1}^{\dim g^*} V_a^{(n)} T_a^* \in g^*$. Here T_a^* is a basis element of g^* and $V_a^{(n)}$ real functions. Hence, an arbitrary element $V(\lambda)$ admits a decomposition with respect to basis elements $T_a^* \lambda^{-n}$ which satisfy the commutation relations

$$[T_a^* \lambda^{-n}, T_b^* \lambda^{-m}] = C_{ab}^c T_c^* \lambda^{-(n+m)},$$

typical for the Kac-Moody algebras. Since (11) is the infinitesimal symmetry transformation for the Belinsky-Zakharov representation (3), then (12) gives the infinitesimal symmetry transformation for the self-duality equation (2). Hence, the algebra $g^* \otimes R[\lambda]$ is responsible for the existence of the infinite set of nonlocal conserved currents (9).

References

1. Luscher M., Pohlmeyer K. (1978) Nucl. Phys. B 137, 46.
2. Dolan L. (1981), Phys. Rev. Lett. 47, 1371.
3. Kac V.G. (1968), Math. USSR Izv. 2, 1311.
4. Moody R.V. (1968), J. Algebra 10, 211.
5. Leznov A.N., Saveliev M.V. (1981) Particles and Nuclei 12, 125.
6. Dolan L. (1984), Phys. Rep. 109, 1.
7. Bais F.A., Sasaki R., (1983), Nucl. Phys. B 227, 75.
8. Forgacs P., Horvath Z., Palla L. (1980), Phys. Rev. Lett. 45, 505.
9. Gurses M., Xantophoulos B. (1982), Phys. Rev. D 26, 1912.
10. Belinsky V.A., Zakharov V.E., (1978), JETP 75, 1953.
11. Zakharov V.E., Shabat A.B., (1979), Funct. Anal. Appl. 13, 13 (Russian).
12. Ueno K., Nakamura Y. (1982) Phys. Lett. 117 B, 208.
13. Muskhelishvili N.I. (1962). Singular Integral Equations, GIFML, Moscow.
14. Eichenherr H. (1982) Preprint CERN TH. 3299.

τ- SCHEME IN THEORY OF INTEGRABLE EQUATIONS

I.Ya.Dorfman

Institute of Chemical Physics, the USSR Academy of Sciences, Moscow

From the algebraic point of view, an integrable equation is an element of a Lie algebra $\mathfrak{A}$ belonging to its ample commutative subalgebra (the Lie algebra of commuting symmetries). The origin of integrability in many cases lies in the deformation theory of the Lie algebra $\mathfrak{A}$. An integrable system develops if there is given a one-parameter family of Lie brackets

$$[a,b]_\lambda = [a,b] + \lambda\omega(a,b), \quad a,b \in \mathfrak{A},$$

and there exists a linear operator $A: \mathfrak{A} \to \mathfrak{A}$ such that $id + \lambda A$ is an isomorphism of the $[\,,\,]_\lambda$ - structure to the $[\,,\,]$ - structure. Such a deformation of $\mathfrak{A}$ is called trivial.

Trivial deformations are in one-to-one correspondence with linear operators $A: \mathfrak{A} \to \mathfrak{A}$ with zero Nijenhuis torsion [1] :

$$[Aa, Ab] - A[Aa, b] - A[a, Ab] + A^2[a,b] = 0\,, \quad a,b \in \mathfrak{A}.$$

The role of such operators (called recursion [2] , Nijenhuis [3-5] , hereditary [6] , etc.) in the theory of integrable equations is based on the following property. Let A be a Nijenhuis operator and let its symmetry be given, i.e. an element $a_o \in \mathfrak{A}$ such that $[a_o, Ab] = A[a_o, b]$ for all $b \in \mathfrak{A}$, Then all the elements $A^n a_o$, $n = 0,1,\ldots,$ commute. The equation corresponding to $a_o \in \mathfrak{A}$ in nonformal theory is then integrable in some sence.

Nijenhuis operators are closely connected with the Hamiltonian theory of integrable equations, see [3-9]. To construct this theory we need in addition a complex (Ω, d) over $\mathfrak{A}$ [10] . Then it is possible to introduce algebraically all the notions of the usual Hamiltonian formalism; among which that of Hamiltonian operator is principal.

In the frame work of Lenart's scheme [3] the Nijenhuis operator A is presented in the form $A = HK^{-1}$, where H and K are two Hamiltonian operators forming a Hamiltonian pair [3]; the initial element a_o of Lenart's scheme is Hamiltonian with respect to both H- and K-structures. These assumptions guarantee the existence of infinite sets of symmetries, conservation laws in involution and Hamiltonian structures.

Some integrable equations, such as the Benjamin-Ono and Kadomtsev-Petviashvili equations, however, seem to have single nontrivial Hamiltonian structure. Nevertheless all the beautiful properties of integrability are present. The consideration of the results of [11,12] (see also [13]) led us to the conclusion that there exists another algebraic scheme of integrability, which we call the τ-scheme [1] and which is not manifestly connected with deformations. This scheme acompanied by some examples is considered below.

Note that in some cases (such as KdV and sine-Gordon equations) both the schemes of integrability act simultaneously. The coincidence of symmetries obtained by the two different methods is nontrivial. This phenomenon is also considered below.

<u>τ-scheme in a Lie algebra</u>. Let $\mathfrak{A}$ be a Lie algebra, $a_o, \tau \in \mathfrak{A}$. We say that τ -scheme with initial element a_o is in action if all the elements

$$a_{n+1} = [\tau, a_n], \tag{1}$$

commute.

The following theorem gives sufficient conditions for a realization of τ-scheme.

Theorem 1. Suppose for some $N, M \geq 1$ there exist elements $\tau_{-N}, \tau_{-N+1}, \dots, \tau_0, \tau_1$ such that

(a) $\tau_{s+1} = [\tau, \tau_s]$, $s = -N, \dots, 0$; $\tau_1 = \lambda\tau$, $\lambda \in \mathbb{R}$;

(b) $[\tau_{-N}, a_0] = [\tau_{-N+1}, a_0] = \dots = [\tau_{-1}, a_0] = 0$;

$[\tau_0, a_0] = \mu a_0$, $\mu \in \mathbb{R}$, $\lambda\mu < 0$;

(c) $[a_{i-1}, a_i] = 0$, $i \leq M$ and if $[p, a_{M-1}] = [p, \tau_{-N}] = 0$ then $p = \sum_0^{N-1} \nu_\alpha a_\alpha$, $\nu_\alpha \in \mathbb{R}$.

Then the procedure (1) gives a set of commuting $a_n \in \mathfrak{A}$, i.e. the τ-scheme with initial element a_0 is in action.

Proof 1. First prove that

$$[\tau_s, a_n] = \mu_{sn} a_{s+n} , \tag{2}$$

where $\mu_{sn} \in \mathbb{R}$ and $\mu_{sn} = 0$ for $s+n < 0$.

We use the induction in n. If $n = 0$ then (2) holds owing to the condition (b) of the theorem; from (a) and (b) it follows that $\mu_{00} = \mu$, $\mu_{10} = \lambda$, $\mu_{s0} = 0$ for $s < 0$, $s > 1$. We also have

$$[\tau_s, a_{n+1}] = [\tau_s, [\tau, a_n]] = [[\tau_s, \tau], a_n] + [\tau, [\tau_s, a_n]] = -[\tau_{s+1}, a_n] + [\tau, [\tau_s, a_n]].$$

The inductive hypothesis implies

$$[\tau_s, a_{n+1}] = -\mu_{s+1,n} a_{s+n+1} + [\tau, \mu_{sn} a_{s+n}] = (-\mu_{s+1,n} + \mu_{sn}) a_{s+n+1}$$

and we get (2) with

$$\mu_{s,n+1} = -\mu_{s+1,n} + \mu_{sn} . \tag{3}$$

Using (3) all μ_{sn}'s can be calculated; we need only

$\mu_{on} = \mu - n\lambda$.

2. Let us prove now that if $[a_{n-1}, a_n] = 0$ for all $n \leqslant K$ then $[a_m, a_n] = 0$ for all m, n such that $m+n \leqslant 2K$. For $K = 0$ it is true. Let it be true for some K and suppose $[a_{n-1}, a_n] = 0$, $n \leqslant K+1$. Then, if there is a pair m, n such that $m+n \leqslant 2(K+1)$ there are three possibilities: $m+n \leqslant 2K$, $m+n = 2K+1$ and $m+n = 2K+2$. In the first case there is nothing to prove; in the second case we have for $m < n$

$$[a_m, a_n] = [a_m, [\tau, a_{n-1}]] = [[a_m, \tau], a_{n-1}] + [\tau, [a_m, a_{n-1}]] =$$
$$= -[a_{m+1}, a_{n-1}] + [\tau, [a_m, a_{n-1}]]. \tag{4}$$

As $m+n-1 \leqslant 2K$, it follows that $[a_m, a_n] = -[a_{m+1}, a_{n-1}]$. Repeating this reasoning several times we get $[a_K, a_{K+1}]$ which is 0 by inductive hypothesis. In the third case we again use formula (4) noting that $[a_m, a_{n-1}] = 0$ since $m+n-1 = 2K+1$ (the situation of the second case). So $[a_m, a_n] = -[a_{m+1}, a_{n-1}]$ and we may iterate the procedure obtaining $[a_{K+1}, a_{K+1}] = 0$.

3. It is left to prove that $[a_{n-1}, a_n] = 0$ for all n. To begin the induction note that this is true for $n = 1, 2, \ldots, M$ due to (c). Let this be true for $n \leqslant K$, $K \geqslant M$. Then

$$[[a_K, a_{K+1}], a_{M-1}] = [[a_K, a_{M-1}], a_{K+1}] + [a_K, [a_{K+1}, a_{M-1}]].$$

As $K+M \leqslant 2K$ we may use the result of the preceding section to get $[[a_K, a_{K+1}], a_{M-1}] = 0$. We also have

$$[[a_K, a_{K+1}], \tau_{-N}] = [[a_K, \tau_{-N}], a_{K+1}] + [a_K, [a_{K+1}, \tau_{-N}]] = [[a_K, \tau_{-N}],$$
$$a_{K+1}] + [a_K, [a_{K+1}, \tau_{-N}]] = \mu_{-N,K}[a_{K-N}, a_{K+1}] - \mu_{-N,K+1}[a_K, a_{K+1-N}].$$

Since $2K+1-N \leqslant 2K$ then $[[a_K, a_{K+1}], \tau_{-N}] = 0$.

Using the condition (c) of our theorem, we conclude that

$$[a_K, a_{K+1}] = \sum_0^{N-1} \nu_\alpha a_\alpha, \quad \nu_\alpha \in \mathbb{R}. \tag{5}$$

Applying $\operatorname{ad} \tau_0$ to (5) we get

$$[\tau_0, [a_K, a_{K+1}]] = \sum_0^{N-1} \nu_\alpha [\tau_0, a_\alpha].$$

Using (2) we obtain

$$(\mu_{0K} + \mu_{0,K+1})[a_K, a_{K+1}] = \sum_0^{N-1} \nu_\alpha \mu_{0\alpha} a_\alpha,$$

or taking (5) into account

$$\sum_0^{N-1} (\mu_{0K} + \mu_{0,K+1}) \nu_\alpha a_\alpha = \sum_0^{N-1} \nu_\alpha \mu_{0\alpha} a_\alpha.$$

Substituting $\mu_{0n} = \mu - n\lambda$, we get

$$\sum_0^{N-1} (2\mu - (2K+1)\lambda) \nu_\alpha a_\alpha = \sum_0^{N-1} (\mu - \alpha\lambda) \nu_\alpha a_\alpha.$$

If there is an element among a_m, $m = 0, \dots, N-1$, which is linearly independent with a_0 then we immediately have $\mu = \lambda = 0$. If all a_m are proportional to a_0 then $\mu = (2K+1)\lambda$, that contradicts $\lambda\mu < 0$. Theorem is proved.

<u>Example 1</u>. The KdV equation $u_t = 6uu_x - u_{xxx}$ is included in τ-scheme if we take

$$a_0 = u_x,$$

$$\tau = x(6uu_x + u_{xxx}) + 4u_{xx} + 8u^2 + 2u_x \partial^{-1} u; \quad \partial^{-1} = \frac{1}{2}\left(\int_{-\infty}^{x} - \int_{x}^{\infty}\right).$$

Then $a_0 = u_x$, $a_1 = 6uu_x + u_{xxx}$, $a_2 = 3(u^{(5)} + 10uu_{xxx} + 20u_x u_{xx} + 30u^2 u_x), \dots$ If we take $\tau_{-1} = 1$ then

$$\tau_0=[\tau,\tau_{-1}]=-8(xu_x+2u), \qquad [\tau_{-1},a_0]=0,$$

$$\tau_1=[\tau,\tau_0]=16\tau, \qquad [\tau_0,a_0]=-8a_0,$$

so the conditions (a) and (b) of the theorem are valid (N=1). It is easy to show that the condition (c) is also satisfied for M=2. The hierarchy of equations given by the τ-scheme coincides with the well known Lenart hierarchy, but this is not trivial (Theorem 4).

<u>Example 2</u>. The Benjamin-Ono equation $u_t=2uu_x+\mathcal{H}u_{xx}$, where $\mathcal{H}$ is the Hilbert transformation $(\mathcal{H}\varphi)(x) = = \frac{1}{\pi}\int\varphi(\xi)(\xi-x)^{-1}dx$, is included into the τ-scheme with

$$a_0=u_x$$

$$\tau=x(2uu_x+\mathcal{H}u_{xx})+\frac{3}{2}\mathcal{H}u_x+u^2.$$

Then $a_0=u_x$, $a_1=2uu_x+\mathcal{H}u_{xx}$, $a_2=6u^2u_x+3u_x\mathcal{H}u_x+3u\mathcal{H}u_{xx}+ +3\mathcal{H}(u_x^2)+3\mathcal{H}(uu_{xx})-2u_{xxx}, \ldots$ The conditions (a) and (b) of the theorem are satisfied (N=1) if $\tau_{-1}=1$. Then

$$\tau_0=[\tau,\tau_{-1}]=-2(xu_x+u), \qquad [\tau_{-1},a_0]=0,$$

$$\tau_1=[\tau,\tau_0]=2\tau, \qquad [\tau_0,a_0]=-2a_0.$$

It is possible to prove that (c) is also satisfied for M=2.

<u>Example 3</u>. The Kadomtsev-Petviashvili I equation $u_t= = -2\alpha^2(6uu_x-u_{xxx}-3\alpha^2\partial^{-1}u_{yy})$ is included into the τ-scheme with

$$a_0=u_x$$

$$\tau=y(6uu_x-u_{xxx}-3\alpha^2\partial^{-1}u_{yy})-2\alpha^2xu_y-4\alpha^2\partial^{-1}u_y.$$

Then $a_0=u_x$, $a_1=-2\alpha^2u_y$, $a_2=-2\alpha^2(6uu_x-u_{xxx}-3\alpha^2\partial^{-1}u_{yy})$,

$a_3=-24\alpha^2(u_{xxy}-4uu_y-2u_x\,\partial^{-1}u_y+\alpha^2\,\partial^{-1}u_{yyy}),\ldots$. If we take $\tau_{-2}=1$, then

$$\tau_{-1}=[\tau,\tau_{-2}]=-6yu_x, \qquad [\tau_{-2},a_0]=0,$$

$$\tau_0=[\tau,\tau_{-1}]=-12\alpha^2(xu_x+2yu_y+2u), \quad [\tau_{-1},a_0]=0,$$

$$\tau_1=[\tau,\tau_0]=12\alpha^2\tau, \qquad [\tau_0,a_0]=-12\alpha^2 a_0,$$

and thus the conditions (a) and (b) of the theorem are satisfied (N=2). It is possible to prove that (c) is also satisfied for M=3 .

Example 4 (Counterexample showing that condition (c) is essential).

Take

$$a_0=u_x,$$

$$\tau=x(uu_x+u_{xxx})+u^2+2u_{xx}.$$

Then for $\tau_{-1}=1$ we have

$$\tau_0=[\tau,\tau_{-1}]=-(xu_x+2u), \qquad [\tau_{-1},a_0]=0,$$

$$\tau_1=[\tau,\tau_0]=2\tau, \qquad [\tau_0,a_0]=-a_0.$$

For the obtained hierarchy $a_0=u_x$, $a_1=uu_x+u_{xxx}$, $a_2 = =3u^{(5)}+4uu_{xxx}+11u_xu_{xx}+2u^2u_x,\ldots$ all the conditions of the theorem are satisfied, except the condition $[a_1,a_2]=0$.

τ-scheme in the frame of Hamiltonian formalism. Suppose $(\mathfrak{R},d)$ is a complex over a Lie algebra $\mathfrak{a}$ [10]. It means that to every $a\in\mathfrak{a}$ there corresponds a linear operator $i_a:\mathfrak{R}\to\mathfrak{R}$ such that

$$i_a i_b=-i_b i_a, \quad [i_a d+d i_a, i_b]=i_{[a,b]}\,.$$

Then it is possible to develop a Hamiltonian formalism (see [1], [10]). If there is given a Hamiltonian operator

K: $\Omega^1 \to \mathfrak{a}$ then the elements $a \in \mathfrak{a}$ represented as $a = Kdf$, $f \in \Omega^0$, are called Hamiltonian elements of $\mathfrak{a}$. Under what conditions does the τ-scheme give Hamiltonian elements?

Recall that the Lie derivative L_a along $a \in \mathfrak{a}$ is defined for basic objects of the theory by the following rule [1] : $L_a = i_a d + d i_a$ in the spaces Ω^q, $q \geq 0$; $L_a b = [a,b]$ for b in the Lie algebra $\mathfrak{a}$; $L_a Q = [L_a, Q]$ in the spaces of linear operators. An element $a \in \mathfrak{a}$ is called a symmetry of an object s if $L_a s = 0$. The answer to is question is given by the following simple but important theorem.

Theorem 2. Let $a_0 \in \mathfrak{a}$ be a Hamiltonian element, i.e. $a_0 = Kdf_0$ with the Hamiltonian operator K and the Hamiltonian f_0. If τ is a symmetry of K then all the a_n's obtained by (1) are also Hamiltonian elements with respect to K. Their Hamiltonians f_n are calculated by the rule

$$f_{n+1} = (\tau, df_n). \tag{6}$$

(Here[after we set (ξ, a), where $\xi \in \Omega^1$, $a \in \mathfrak{a}$, for the element $i_a \xi \in \Omega^0$. In the nonformal theory $(\xi, a) = \int \xi a \, dx$.)

Proof. It follows from definition that

$$[\tau, K\xi] \equiv L_\tau (K\xi) = (L_\tau K)\xi + KL_\tau \xi, \quad \xi \in \Omega^1.$$

By hypothesis $L_\tau K = 0$

$$[\tau, Kdf] = Kd(\tau, df).$$

Let the statement of the theorem be proved for $n \leq N$, i.e. $a_N = Kdf_N$. Then

$$a_{N+1} = [\tau, a_N] = Kd(\tau, df_N)$$

and it follows that a_{N+1} is also a Hamiltonian element with the Hamiltonian $f_{N+1} = (\tau, df_N)$. The induction concludes the proof.

In the examples considered above the initial element $a_o = u_x$ is a Hamiltonian element with $K=\partial$, $f_o = \int \frac{u^2}{2} dx$. Let us calculate $L_\tau \partial$ for each case.

Example 1 (KdV) $\tau = x(6uu_x + u_{xxx}) + 4u_{xx} + 8u^2 + 2u_x \partial^{-1} u$.

A straightforward calculation gives

$$L_\tau \partial = -4(\partial^3 + 4u\partial + 2u_x).$$

Example 2 (BO) $\tau = x(2uu_x + \mathcal{H} u_{xx}) + \frac{3}{2} \mathcal{H} u_x + u^2$.

Note that $\tau = \partial \frac{\delta}{\delta u} \int x\left(\frac{u^3}{3} + \frac{u \mathcal{H} u_x}{2}\right) dx$, i.e. τ is a Hamiltonian element with respect to ∂. We can use the following general statement: $L_a K = 0$ if K and $a = Kdf$ are Hamiltonian [1]. Thus we have

$$L_\tau \partial = 0.$$

Example 3 (KPI). $\tau = y(6uu_x - u_{xxx} - 3\alpha^2 \partial^{-1} u_{yy}) - 2\alpha^2 x u_y - 4\alpha^2 \partial^{-1} u_y$.

Note that $\tau = \partial \frac{\delta}{\delta u} \int \left(u^3 - \frac{uu_{xx}}{2} + \frac{3}{2}\alpha^2 \partial^{-1} u \, \partial^{-1} u_{yy}\right) dx$.

As in the previous case we have

$$L_\tau \partial = 0.$$

We see therefore that procedure (6) is applicable for the consequent calculation of conservation laws for both BO and KPI equations. The KdV equation is another case; the obtained formula for $L_\tau \partial$ gives just the second Hamiltonian operator of Lenart's scheme. Furthermore, as it has been mentioned above, the equations given by both schemes at their first steps, coincide (up to constant factors). Is it true for all the steps? The answer is yes. This fact is nontrivial because the generating operator ad τ of the τ-scheme has nothing to do with the

Nijenhuis operator $\partial^2+4u+2u_x\partial^{-1}$ which generates Lenart's scheme. Some elucidations are given below.

Simultaneous actions of the τ and Lenart's schemes. We need the following preliminary proposition.

Theorem 3. Suppose K and $L_\tau K$, $\tau \in \mathfrak{a}$, are Hamiltonian operators. Then K and $L_\tau K$ form a Hamiltonian pair.

This theorem is demonstrated by applying L_τ to the explicit expression [4] of the Schouten bracket $[K,K]=0$. The following theorem describes interrelations between the two schemes.

Theorem 4. Suppose K and $L_\tau K$, $\tau \in \mathfrak{a}$ are Hamiltonian operators. Let be given some Lenart's scheme generated by the Hamiltonian pair consisting of K and $H=\lambda L_\tau K$, $\lambda \in \mathbb{R}$:

$$
\begin{aligned}
a_0 &= H df_0 = K df_1,\\
a_1 &= H df_1 = K df_2,\\
&\dots\\
a_n &= H df_n = K df_{n+1},\\
&\dots
\end{aligned}
\tag{7}
$$

Suppose furthermore that $L_\tau H=\mu HK^{-1}H$, $\mu \in \mathbb{R}$ (for brevity we assume K invertible) and $(\tau, df_0)=\nu f_1$, $\nu \in \mathbb{R}$.

Then 1) the τ-scheme with initial element a_0 gives the same elements as the Lenart's scheme (up to constant factors),

2) the conservation laws f_N are given by recurrence

$$
f_{N+1}=\frac{1}{N(\mu-\lambda)+\nu}(\tau, df_N). \tag{8}
$$

Proof. First let us prove that for all n the following formula is valid:

$$d(\tau, df_n) = \nu_n df_{n+1}, \ \nu_n = n(\mu-\lambda) + \nu \in \mathbb{R}. \tag{9}$$

For $n=0$ (9) holds by hypothesis. Let (9) be true for all $n \leqslant N$. Applying L_τ to $Kdf_{N+1} = Hdf_N$ we obtain

$$\lambda H df_{N+1} + Kd(\tau, df_{N+1}) = \mu HK^{-1}H df_N + Hd(\tau, df_N).$$

Using (9) we get from the preceding formula

$$\lambda K df_{N+2} + Kd(\tau, df_{N+1}) = \mu K df_{N+2} + \nu_n K df_{N+2}.$$

So

$$Kd(\tau, df_{N+1}) = (\nu_N + \mu - \lambda) K df_{N+2}.$$

It follows that

$$d(\tau, df_{N+1}) = \nu_{N+1} df_{N+2}, \ \nu_{N+1} = \mu - \lambda + \nu_N,$$

So (9) is proved by induction. Formula (8) and the second part of the theorem is thus proved.

Assume also that the first part of the theorem is verified for $n \leqslant N$. Then due to (9) we have

$$[\tau, a_N] = [\tau, Hdf_N] = \mu HK^{-1}H df_N + Hd(\tau, df_N) =$$

$$= \mu K df_{N+2} + \nu_N H df_{N+1} = (\mu + \nu_N) a_{N+1}.$$

Thus the statement of the theorem is valid also for $n = N+1$. This concludes the proof.

Example 4. The KdV equation can be included in Lenart's scheme (7) with $K = \partial$, $H = \partial^3 + 4u\partial + 2u_x$; and $a_0 = u_x$, $a_1 = 6uu_x + u_{xxx}$, ... is Lenart's hierarchy; $f_0 = \int \frac{u}{2} dx$, $f_1 = \int \frac{u^2}{2} dx$, $f_2 = \int (u^3 + uu_{xx}/2) dx, \ldots$ is the hierarchy of conservation laws.

We have already seen that the τ-scheme is realized with

$$\tau = x(6uu_x + u_{xxx}) + 4u_{xx} + 8u^2 + 2u_x\partial^{-1}u$$

and initial element $a_0 = u_x$. It has been calculated

$$L_\tau \partial = -4(\partial^3 + 4u\partial + 2u_x),$$

and it is easy to calculate also

$$L_\tau(\partial^3 + 4u\partial + 2u_x) = -2(\partial^3 + 4u\partial + 2u_x)\partial^{-1}(\partial^3 + 4u\partial + 2u_x).$$

Note that $(\tau, df_0) = 3f_1$. We see that all the assumptions of Theorem 4 are satisfied ($\lambda = -4$, $\mu = -2$, $\nu = 3$). Therefore both schemes give us the same hierarchies (up to constant factors). It follows that there is a recurrence for the conservation laws:

$$f_{N+1} = \frac{1}{2N+3}(\tau, df_N).$$

<u>Example 5</u>. The sine-Gordon equation, $u_{xt} = \sin u$, can be included into Lenart's scheme (7) with $K = \partial^{-1}$, $H = \partial + u_x\partial^{-1} \circ u_x$ forming a Hamiltonian pair of operators (we do not consider foundation problems for nonlocal Hamiltonian operators). $a_0 = u_x$, $a_1 = u_{xxx} + \frac{u_x^3}{2}$, $a_2 = u^{(5)} + \frac{5}{2}u_x^2 u_{xxx} + \frac{5}{2}u_x u_{xx}^2 + \frac{3}{8}u_x^5, \ldots$ is Lenart's hierarchy of equations; $f_0 = -\int \cos u \, dx$, $f_1 = \int \frac{1}{2} uu_{xx} \, dx$, $f_2 = \int \left(\frac{uu^{(4)}}{2} + \frac{3}{8} uu_x^2 u_{xx} \right) dx, \ldots$ is the hierarchy of conservation laws. (Note that the sine-Gordon equation itself is included in the hierarchy setting $a_{-1} = \partial^{-1} \sin u$). Set

$$\tau = x(u_{xxx} + \frac{1}{2}u_x^3) + 2u_{xx} + \frac{1}{2}u_x\partial^{-1}(u_x^2).$$

A straightforward calculation gives

$$L_\tau\partial^{-1} = -2(\partial + u_x\partial^{-1}\circ u_x).$$

We also have

$$\tau = (\partial + u_x\partial^{-1}\circ u_x)\frac{\delta}{\delta u}\int\frac{1}{2}xuu_{xx}\,dx$$

and thus τ is a Hamiltonian element with respect to the Hamiltonian operator $\partial + u_x\partial^{-1}\circ u_x$. As it has been mentioned above it follows that

$$L_\tau(\partial + u_x\partial^{-1}\circ u_x) = 0.$$

Note that $(\tau, df_1) = 3f_2$. We see that all the assumptions of Theorem 4 are satisfied ($\lambda=2$, $\mu=0$, $\nu=1$). As in the previous case both schemes give us the same hierarchy (up to constant factors). Then the recurrence for the conservation laws of the sine-Gordon equation is

$$f_{N+1} = \frac{1}{2N+1}(\tau, df_N).$$

Any Lenart's scheme is very litely equivalent (in the above sence) to some τ-scheme. However, we have no algorithm for recovering τ from H and K.

References

1. Dorfman, I.Ya. (1984). In: Proc. 2nd Intern. Workshop Nonlin. Phenom. Kiev, 1983. Gordon & Breach, p.1313.
2. Olver, P. (1977). J. Math. Phys. 18, 1212.
3. Gelfand, I.M., Dorfman. I.Ya. (1979). Funct. anal. appl. 13, 4, 13.
4. Gelfand, I.M., Dorfman, I.Ya. (1980). Funct. anal.appl. 14, 3, 71.
5. Magri, F. (1980). LN Phys. 120, 233.
6. Fuchssteiner, B. (1979). Nonlin. Anal. TMA, 3, 849.
7. Fuchssteiner, B. (1982). Progr. Theor. Phys. 68, 1082.

8. Magri,F. (1978). J. Math. Phys. 19, 1156.

9. Gelfand,I.M., Dikii,L.A. (1978). Preprint Inst. Appl. Math. the USSR Acad. Sci. 78-136.

10. Gelfand,I.M., Dorfman,I.Ya. (1981). Funct. anal. appl. 15, 3, 23.

11. Focas,A.S., Fuchssteiner,B. (1981). Phys. Lett. 86A, 341.

12. Oevel,W., Fuchsseiner,B. (1982). Phys. Lett. 88A,323.

13. Chen,H.H., Lee,Y.C. (1983). Phisica D, 9, 439.

ABSOLUTE INVARIANTS AND KORTEWEG DE VRIES EQUATION

L.M.Berkovich
Kuibyshev State University

1. Introduction. Two happenings of 1834 were seemingly not in the least connected: Kummer published his paper [1] and Russel observed a physical phenomenon that served a basis of soliton theory [2] . Revealing an internal connection between them is our aim.

Kummer's paper is a ground-stone in transformation theory of ordinary 2nd order linear differential equations (OLDE). These are transformations of equations

$$y''+2a_1(x)y'+a_2(x)y=0, \quad (')=d/dx \tag{1.1}$$

$$\ddot{z}+2b_1(t)\dot{z}+b_2(t)z=0, \quad (^\cdot)=d/dt \tag{1.2}$$

by means of the change of variables

$$z=f(t)y(\varphi(t)), \; x=\varphi(t). \tag{1.3}$$

Note that for $t=x$ equations (1.1) and (1.2) are also transformed into each other by setting $z=\alpha(x)y+\beta(x)y'$ (the Darboux transformation), but this is not used here.

Further development in that direction in the XIX century is connected with the dissemination of Kummer's discussion on OLDE of higher orders, and especially on two Halphen's problems on equivalence and classification of OLDE's [3] . The former consists in determining necessary and sufficient conditions for mutual transformation of the n-th order equations

$$y^{(n)}(x)+\sum a_k y^{(n-k)}(x)=0, \quad k=\overline{1,n}, \tag{1.4}$$

$$z^{(n)}(t)+\sum b_k z^{(n-k)}(t)=0, \quad k=\overline{1,n}, \tag{1.5}$$

by means of substitution (1.3) and is associated with invariant theory of OLDE. The latter is constructing canonical forms of equations (1.4) and (1.5).

From the beginning of the inverse scattering method (1967) a deep connection is known between the soliton theory and the spectral theory of the Schrödinger (Stum-Liouville) 2 order linear differential operators. Moreover, integrating KdV equation

$$u_t = u''' + \mu u u' + \nu u', \quad \text{where } \mu, \nu = \text{const}, \tag{1.6}$$

an important role of the 3d order antiself - adjoint (skew-symmetric, iterative) operator

$$D^3 + 3A_2 D + 3A_2'/2. \tag{1.7}$$

was established.

The operator (1.7) also arises in the transformation problem of 2 order OLDE (Kummer's problem) [4]; an explanation of the fact with an appropriate proof is given in [5] . We show how the transformation theory of 3 order OLDE leads to the nonlinear stationary KdV equation (SKdV), i.e. to the equation corresponding to (1.6) for $u_t=0$. This is a consequence of the solutions of two Halphen's problems obtained by the author. The solutions themselves are obtained with a constructive approach to OLDE theory based on combined application of the algebraic method of factorization of ordinary linear differential operators (OLDO) and transformations of variables (1965-1967) [6] .

Notice that in solving the equivalence problem invariants of OLDE are found as conditions of compatibility of overdetermined nonlinear OLDEs.

Solving the problem of OLDE classification constructing canonical Halphen's forms provides a new insight of conditions of commuting for differential operators and integrability conditions of OLDE. In this approach a common eigenfunction for the system Lamé and Halphen equations is found.

2. The Kummer problem. Setting. Given eqs. (1.1) and (1.2) where $a_1(x) \in C(i)$; $a_2(x) \in C(i)$; $b_1(t) \in C^1(j)$, $b_2(t) \in C(j)$, i and j are open intervals of the real axes x and t respectively, there is the set of transformations

$$y = v(x)z, \quad dt = u(x)dx, \tag{2.1}$$

$u(x)v(x) \neq 0$ for any $x \in i$ reducing locally (1.1) to (1.2).

The substitution (2.1) inverse to (1.3), is called the Kummer-Liouville transformation (KL) and denoted by $(v(x), \int u(x)dx)$.

An effective solving of the local Kummer problem was given in [5,7] on the base of factorization of (1.1) and (1,2).

Theorem 1. The set of transformations $(v(x), \int u(x)dx)$ reducing (1.1) to (1,2) is described by $v(x) = |u|^{-1/2} \exp(-1/2 \int a_1 dx + 1/2 \int b_1\, dt)$, where $t(x) = \int u(x)dx$ is the general solution of the 3 order Kummer-Schwarz equation

$$\{t, x\} + B_2(t)t'^2 = A_2(x), \quad A_2 = a_2 - a_1^2 - a_1', \quad B_2 = b_2 - b_1^2 - \dot{b}_1, \tag{2.2}$$

$\{t, x\} = t'''/2t' - 3\,(t''/t')^2/4$ is the Schwarz derivative.

The general solution of (2.2) may be presented in the form $w_o(t) = (c_1 + c_2\, \omega_o(x))/(c_3 + c_4\, \omega_o(x))$, where $w_o(t) = \tau$

is a solution of the equation $\{\tau, t\} = B_2(t)$ and $\omega_0(x) = \xi$ is a solution of the equation $\{\xi, x\} = A_2(x)$, i.e. in the form

$$\int \exp(-2\int b_1 dt) z_1^{-2} dt = \frac{c_1 + c_2 \int \exp(-2\int a_1 dx) y_1^{-2} dx}{c_3 + c_4 \int \exp(-2\int a_1 dx) y_1^{-2} dx}, \quad c_1 c_4 - c_2 c_3 \neq 0.$$

Here $y_1(x)$ and $z_1(t)$ are particular solutions of (1.1) and (1.2) respectively.

Thus, under sufficiently general assumptions every 2nd order OLDE reduces to the given form by the KL transformation.

3. Two Halphen's problems. Let us consider eq. (1.4) and (1.5), where $a_k \in C^{n-k}(i)$, $b_k \in C^{n-k}(j)$ and introduce the group G of transformations of the KL type:

$$G : (x, y) \to (t, z) = (v(x), \int u(x) dx),$$

where $v(x) \in C^n(i)$, $u(x) \in C^n(i)$, $u(x)v(x) \neq 0$ for any $x \in i$.

Natural subgroups of G are $G_1 = \{(v(x), id)\}$ and $G_2 = \{(id, \int u(x)\}^2$.

Eq. (1.4) is equivalent to (1.5) if there is a substitution $g \in G$ reducing (1.4) to (1.5).

A rational differential function $I(a, a', \ldots)$, $a = (a_1, \ldots, a_n)$ with $n \geqslant 3$ of the coefficients of (1.4) such that $I(a, a', \ldots) = \lambda(u)\, I(b, b, \ldots)$ (with $t = \int u(x) dx$) is called an invariant of (1.4) with respect to G. If $\lambda(u) = 1$ then I is an absolute invariant, and if $\lambda(u) \neq$ const then I as a relative invariant.

The notions of absolute and relative semi-invariants of equation (1.4) are similarly introduced for G_1 and G_2 (with $t =$ udx).

For equations of order $n \geqslant 4$ it is expedient to introduce also the notions of pseudoinvariant and conditional invariant.

We call a rational differential function $J(a,a',\dots)$ such that

$$J(a,a',\dots) = \mu(u) J(b,\dot{b},\dots) + \lambda(u,u') I(a,a',\dots) \ (\text{with } t=\int u dx)$$

a pseudoinvariant of equation (1.4). A restriction $J(a,a',\dots)\Big|_{I=0} = I_1(a,a',\dots)$ such that $I_1(a,a',\dots) = \mu(u)\, I_1(b,\dot{b},\dots)$ $(t = \int udx)$ is called a conditional invariant.

To simplify the following calculations we consider equations in the semicanonical form

$$Ly = D^n y + \sum \binom{n}{k} A_k(x) D^{n-k} y = 0,\ D = d/dx,\ k=\overline{2,n},\ A_k \in C^{n-k}_k(i), \tag{3.1}$$

$$Mz = D_t^n z + \sum \binom{n}{k} B_k(t) D_t^{n-k} z = 0,\ D_t = d/dt,\ k=\overline{2,n},\ B_k \in C^{n-k}(j)$$

(3.2)

where A_k and B_k are absolute semiinvariants of (3.1) and (3.2) with respect to G_1.

The following problems goes back to Halphen.

Problem 1. Find necessary and sufficient conditions of equivalence of (3.1) and (3.2).

Problem 2. Give a classification of equations (3.1) reducing to canonical forms.

3.1. Solving the equivalence problem

Theorem 2. $(3.1) \Leftrightarrow (3.2)$ if and only if

$$L = \prod [D - v'/v - (k-1)u'/u - \beta_k(t(x))u],\ k = \overline{n,1}$$, where $\beta_k(t)$

satisfies

$$M = \prod [D_t - \beta_k(t)],\ k = \overline{n,1}.$$

Remark 1. Generally L and M admit noncommutative

factorizations in domains where functions u,v, β_k have no singularities.

Corrolary 1. (3.1) $\Leftrightarrow$ (3.2) if and only if the transformation (v, $\int$ udx) is of the form

$$y=|u|^{-(n-1)/2}z,\quad dt=u\,dx,$$

$$\{t,x\}+\frac{3}{n+1}B_2(t)t'^2=\frac{3}{n+1}A_2(x).$$

Lemma 1. (3.1) $\Leftrightarrow$ (3.2) if and only if the following overdetermined system of nonlinear in v(x) ODE is consistent:

$$v''-\frac{n-2}{n-1}v'^2/v+3\frac{n-1}{n+1}A_2v-3\frac{n-1}{n+1}B_2v^{(n-5)/(n-1)}=0, \qquad (3.3^{I})$$

$$v'''-\frac{3(n-3)}{n-1}v'v''/v+\frac{2(n-2)(n-3)}{(n-1)^2}v'^3/v^2+\frac{12}{n+1}A_2v'+$$
$$+\frac{2(n-1)}{n+1}A_3v-\frac{2(n-1)}{n+1}B_3v^{(n-7)/(n-1)}=0, \qquad (3.3^{2})$$

$$v^{IV}-\frac{4(n-4)}{n-1}v'v'''/v-\frac{22(n-4)}{9(n-1)}v''^2/v+\frac{2(49n-125)(n-4)}{9(n-1)^2}v'^2v''/v^2-$$
$$-\frac{(49n-125)(n-2)(n-4)}{9(n-1)^3}v'^4/v^3-\frac{10(n-4)(n+7)}{3(n-1)(n+1)}A_2v'^2/v+ \qquad (3.3^{3})$$
$$+\frac{10(n+5)}{3(n+1)}A_2v''+\frac{20}{n+1}A_3v'+\frac{5(n-1)}{3(n+1)}A_4v-$$
$$-\frac{5(n-1)}{3(n+1)}B_4v^{(n-9)/(n-1)}=0$$

. . . .

$$v^{(n)}+\sum_{k=2}^{n}\binom{n}{k}A_kv^{(n-k)}-B_nv^{(n-3)/(n-1)}=0. \qquad (3.3^{n-1})$$

Theorem 3. (3.1)$\Leftrightarrow$(3.2) (the system (3.3) is consistent) if and only if (n-2) relations on the invariants of (3.1) and (3.2) hold:

$$I_0(A)=u^3I_0(B),\ J_{n,I}(A)=u^4J_{n,I}(B)+6u'/uI_0(A),\dots,$$

$$J_{n,n-3}(A)=u^nJ_{n,n-3}(B)+\lambda_0(u'/u)^{n-3}I_0(A)+\sum_{k=1}^{n-4}\lambda_k(u'/u)^{n-3-k}J_{n,k}\ (A)$$

(λ_0, λ_k are numerical parameters), where

$$I_0(A)=A_3-3/2A_2',\ J_{n,I}(A)=A_4-2A_3'+6/5A_2''-5/3(5n+7)/(n+I)A_2^2,$$

$$J_{n,2}(A)=A_5-5/2A_4'+15/7A_3''-5/7A_2'''-10/7(7n+13)/(n+I)A_2I_0(A),\dots$$

I_0 is called the Laguerre relative invariant.

If $I_0 = 0$ then (3.3) can be shortened, since (3.3^2) is a differential consequence of (3.3^1). Then the pseudo-invariant $J_{n,I}$ becomes a conditional one: $I_{n,I}(A) = J_{n,I}\big|_{I_0=0}$.

If $I_0(A) = I_{n,I}(A) = 0$ then one can also withdraw equation (3.3^3) from (3.3). The conditional invariants $I_{n,I}(A)$ and $I_{n,2}(A)$ hace the form

$$I_{n,I}(A)=A_4-9/5A_2''-3/5(5n+7)/(n+I)A_2^2,\ I_{n,2}(A)=A_5-5/2A_4'+5/2A_2'''.$$

3.2. Solving the classification problem

Theorem 4 . The set of equations (3.1) is divided into n-1 classes according to Table 1 :

Table 1

Class	Invariants	Transformation ($u^{-(n-1)/2}$, $\int u dx$)	Halphen's canonical form
Y_0	$I_0 \neq 0$	$u_0 = \sqrt[3]{I_0}$	H_0 is principal, depends on n-2 parameters
Y_k $k=\overline{1,n-3}$	$I_0=J_{n,1}=\ldots$ $\ldots=J_{n,k-1}=0$ $J_{n,k}=I_{n,k}\neq 0$	$u_k=\sqrt[k+3]{I_{n,k}}$	H_k is degenerate, depends on n-k-2 parameters
Y_{n-2}	$I_0=J_{n,1}=\ldots$ $\ldots=J_{n,n-3}=0$	$1/2u''/u-$ $-3/4(u'/u)^2=$ $=3/(n+1)A_2$	H_{n-2} is the simplest degenerate: $z^{(n)}(t)=0$

The coefficients of the canonical forms are absolute (Halphen) invariants , where

$$z^{(n)}(t)+\sum\binom{n}{k}h_{k,0}\,z^{(n-k)}(t)=0,\quad k=\overline{2,n}, \tag{H_0}$$

$$h_{2,0}=h_0,\ h_{3,0}3/2\dot{h}_0+I,\ h_{4,0}=9/5\ddot{h}_0+3/5(5n+7)/(n+1)h_0^2-$$

$$-6u_0'u_0^{-2}+I_{n,I}u_0^{-4},\ldots$$

$$z^{(n)}(t)+\sum\binom{n}{k}h_{k,I}\,z^{(n-k)}(t)=0,\quad k=\overline{2,n} \tag{H_1}$$

$$h_{2,I}=h_I,\ h_{3,I}=3/2\dot{h}_I,\ h_{4,I}=9/5\ddot{h}_I+3/5(5n+7)/(n+1)h_I^2+I_{n,I}u_I^{-4},$$

. . . .

4. Halphen's equations and commutative operators

The obtained classification of OLDE allows to look the problem of finding integrable cases in a new way.

We are interested in the third order equation (not self-joint):

$$y''' + 3A_2 y' + (3/2 A_2' + a)y = 0, \quad a = const \neq 0. \tag{4.1}$$

Let us denote the first and the second absolute invariants of Halphen by $h=3\dot{A}_2$, and $l=3A_2'$ respectively. He has formulated the following theorem.

Theorem 5 [3] . For (4.1) being integrable in elliptic functions it suffices that

$$h^3 = Pl^2 + Q, \; P = (I - n^2)/4, \; n \in Z, \; n \neq 0 \pmod 3), \; Q = const; \tag{4.2}$$

then (4,1) becomes

$$y''' + (I - n^2)P(x)y' + \left(\frac{(I-n^2)}{4}P'(x) + a\right)y = 0, \tag{4.3}$$

where $P(x)$ is Weierstrass' elliptic P-function.

Henceforth the special case of (4.3) with n=2

$$y''' - 3Py' - (3/2 P' - a)y = 0 \tag{4.4}$$

is needed. The equation (4.3) (and also (4.4)) is Halphen's equation,

By Theorem 5 (4.4) is integrable and its coefficients parametrize an elliptic curve which, unlike (4.2), we write in the form

$$l^2 + 4/3h^3 - 3g_2 h + 9g_3 = 0, \quad g_2, g_3 = const \tag{4.5}$$

admitting various degenerate cases.

The parametrization $h=3P$, $l=-3P'$ of the curve (4.5) leads to the nonlinear autonomous differential equation

$h''' + 4hh' = 0$ which belongs to the SKdV type.

As it is known, commutation conditions of the operators

$$L = D^2 + u(x), \quad A = D^3 + 3A_2(x)D + A_3(x) \tag{4.6}$$

also lead to equations of the SKdV type.

Theorem 6. Operators (4.6) commute if and only if any of the following equivalent conditions holds:

1) A has to be presented in the Halphen canonical form, the relative invariant I_0=const, and the absolute invariants h and l have to be related by (4.5);

2) the overdetermined system

$$(L-\lambda)y = 0, \quad (A-\mu)y = 0 \tag{4.7}$$

must be consistent (have a common eigenfunction), moreover, the algebraic relation between the spectral parameters λ and μ (between operators L and A respectively)

$$4\mu^2 = 4\lambda^3 - g_2\lambda - g_3$$

must hold.

The second part of Theorem 6 is known [8], and the first one is a consequence of the knew approach to the commutative operator problem, based on the OLDE invariants.

Theorem 7. System (4.7) has the explicit form

$$\begin{aligned} &y'' - (2P(x) + \lambda)y = 0, \quad \lambda = P(\alpha), \\ &y''' - 3P(x)y' - (3/2\,P' + \mu)y = 0, \quad \mu = 1/2\,P'(\alpha), \end{aligned} \tag{4.8}$$

moreover, the common eigenfunction $\psi(x; \lambda, \mu) = \tilde{\psi}(x, \alpha)$ of system (4.8) is found in Table 2 :

Table 2

$\wp(x)$	$4\mu^2=4\lambda^3-g_2\lambda-g_3$ $\Delta=g_2^3-27g_3^2$	$\psi(x;\lambda,\mu)$	$\tilde{\Psi}(x,\alpha)=\frac{\sigma(x+\alpha)}{\sigma(x)}$ $\exp(-x\zeta(\alpha))$
$\wp(x)$	$\mu=0,\ \Delta\neq 0$	$\psi(x;e_1,0)=\sqrt{\wp(x)-e_1}$	$\sqrt{\wp(x)-\wp(\omega)}$, $\alpha=\omega$
$\frac{1}{x^2}$	$\mu^2=\lambda^3,\ \Delta=0$	$(1+\frac{1}{\sqrt{\lambda}x})\exp(-x\sqrt{\lambda})$	$\frac{x+\alpha}{x}\exp(-\frac{x}{\alpha})$
$q\,\mathrm{cosec}^2\sqrt{q}\,x-q/3$ $q=3e_1/2$	$\mu^2=(\lambda-e_1)(\lambda+e_1/2)^2,$ $e_1>0$	$(\sqrt{\lambda-e_1}+\sqrt{q}\,\mathrm{ctg}\sqrt{q}\,x)$ $\cdot\exp(-x\sqrt{\lambda-e_1})$	$\frac{\sin\sqrt{q}\,(x+\alpha)}{\sin\sqrt{q}\,x}$ $\exp(-x\sqrt{q}\,\mathrm{ctg}\sqrt{q}\,\alpha)$
* $q\,\mathrm{sech}^2\sqrt{-q}\,x-q/3$ $q=3e_1/2$	$\mu^2=(\lambda-e_1)(\lambda+e_1/2)^2,$ $e_1<0$	$(\sqrt{\lambda-e_1}+\sqrt{-q}\,\mathrm{th}\sqrt{-q}\,x)\cdot$ $\cdot\exp(-x\sqrt{\lambda-e_1})$	** $\frac{\mathrm{sh}\sqrt{-q}\,(x+\alpha)}{\mathrm{ch}\sqrt{-q}\,x}\cdot$ $\cdot\exp(-x\sqrt{-q}\,\mathrm{th}\sqrt{-q}\,\alpha)$

<u>Remark</u>. $\sigma(x)$ and $\zeta(x)$ are Weierstrass σ- and ζ-functions ($\wp(x)=-\zeta'(x)$, $\zeta(x)=\frac{d}{dx}(\ln\sigma(x))$ and satisfy the addition theorems [9]:

$$\zeta(x+\alpha)=\zeta(x)+\zeta(\alpha)+\frac{1}{2}\,\frac{\wp'(x)-\wp'(x)}{\wp(x)-\wp(\alpha)},\quad \wp(x+\alpha)+\wp(x)+\wp(\alpha)=\frac{1}{4}\left[\frac{\wp'(x)-\wp'(\alpha)}{\wp(x)-\wp(\alpha)}\right]^2.$$

In the case (*) the Lamé equation is the Sröedinger equation with potential ($2q\,\mathrm{sech}^2\sqrt{-q}x - e_1$). Its solution for the special case $q=-1$ was obtained in various ways in [10,11] in forms different from (**).

Although some results of Table 2 were known, the eigenfunction problem for the Lamé - Halphen system has not been considered (see Introduction to the review [12]).

References

1. Kummer,E.E. (1887). Abdruck aus Program des evangelischen Konigl una Stadtgymnasiums in Liegnitz von Jahre 1834 (reprint: J.Reine Angew. Math.,100,1-9.
2. Solitons et al. (1980). Ed. R.K.Bullough, P.J.Caudrey. Springer. Berlin.
3. Halphen,G.-H. (1884). Mémoires présentés par divers savants a l'Acad. des Sci. dé l'inst. Mat. de France, 28, 1, 301.
4. Gelfand,I.M., Diky,L.A. (1975). Russian Math. Surveys, 30, 5, 50-100.
5. Berkovich,L.M. (1982). Funct. Anal. Appl. 16, 3, 42-44.
6. Berkovich,L.M. (1965). Izv. Vuzov. Math. 4, 8-16; (1967), 12, 3-14 (Russian).
7. Berkovich,L.M. (1978). Transformation of ordinary linear differential equations. Kuibyshev Univ., Kuibyshev (Russian).
8. Burchnall,J.L., Chaundy,T.W. Proceedings of the London Mathematical Society, 1923, 21, 420-440.
9. Handbook of mathematical functions (1964). Ed. Milton Abramowitz and Irene A. Stegun. National Bureau of Standards. Applied Mathematical Series. June.
10. Lamb,G.L.,Jr. (1980). Elements of Soliton theory. John Willey & Sons. N.Y.
11. Landau,L.D., Lifshits,E.M. (1974). Quantum mechanics. Nonrelative theory. Nauka, Moscow (Russian).
12. Dubrovin,B.A. (1983). In: Modern problems of math. 23, VINITI, Moscow, pp. 33-78 (Russian).

ON A METHOD OF FINDING EXACT SOLUTIONS OF EQUATIONS OF MATHEMATICAL PHYSICS

Yu.A.Danilov
Kurchatov Institute of Atomic Energy, Moscow

One of the main aims of modern mathematical physics appears to be elaboration of regular methods for solving large classes of nonlinear equations of different types (integral, ordinary and partial differential, difference, etc.) describing important and subtle physical phenomena having escaped from linear theory.

One of the most effective methods of integrating nonlinear partial differential equations with solutions localized in space (the so-called solutions) has been discovered by Gardner, Green, Kriskal and Miura [1]. They established a nontrivial relation among solutions of Korteweg-de Vries equation

$$u_t + u u_x + u_{xxx} = 0 \qquad (1)$$

and those of the linear Schrödinger equation. The proposed method is known as inverse scattering method (using the so-called scattering data one can restore the potential of the Schrödinger equation satisfying the Korteweg-de Vries equation). Later the inverse scattering method was successfully applied by Zakharov and Shabat [2,3] to the integration of the nonlinear Schrödinger equation

$$i u_t + u_{xx} + |u|^2 u = 0 \qquad (2)$$

A long (but, nevertheless, by far noncomplete) list of nonlinear equations integrated by the inverse scattering method is given in the [4 , Ch. 7] by Zakharov. The most part of these equations are of physical importance. The general scheme for integrating equations of mathematical physics by inverse scattering method is proposed by Zakharov and Shabat [5]. It is worth mentioning that the inverse scattering method allows group-theoretical interpretation.

In this paper an exposition of the group-theoretical approach is given which enables us to indicate for each (both linear and nonlinear) equation for a set of such equations, $L_1 u = 0$, another set, $L_2 v = 0$, such that the solutions of the latter can be mapped into the solutions of the former by an explicit transformation. When $L_2 v = \phi$ is a linear equation this approach enables one to reduce solving a nonlinear equation to that of a linear one.

The idea of solving an equation using ample information about the set of solutions of another one can be traced back to Darboux [6] who has proved the following theorem. Let $y_0(x)$ be a particular solution of the equation

$$-y'' + q(x)y = \lambda y \tag{3}$$

corresponding to $\lambda = \lambda_c$ and nonvanishing in a neighbourhood of $x = a$ and let $y(x, \lambda)$ be a general solution of the same equation for any $\lambda \neq \lambda_c$. Then

$$y_1(x,\lambda) = WRON(y(x,\lambda); y_0(x))/(\lambda-\lambda_0)\, y_c(x) \tag{4}$$

satisfies the equation

$$-y_1''(x,\lambda) + \left[q - 2\frac{d^2}{dx^2}\ln y_0(x)\right] y_1 = \lambda y_1 \tag{5}$$

Here WRON is a Wronscian of the solutions in parentheses.

Darboux's theorem means that using the Darboux transformation (4,5) one can construct from an integrable potential of Schrödinger equation (3) an infinite series of integrable potentials producing for each of then an explicit analytic solution (4) of the new equation. Darboux's method like many other methods of integrating has a group-theoretical foundation [7] and is widely used for constructing explicit solutions of completely integrable systems [8], in particular, in the inverse scattering method [9].

In the simple case, when the solutions of the equation $L_2v = 0$ can be mapped to the solutions of the initial equation $L_1u = 0$ by the transformation in the dependent variables only (unknown functions to find), the proposed method is founded on the coincidence of the truncations of the symmetry groups (in the sense of S.Lie) of equations $L_1u = 0$ and $L_2v = 0$ for independent variables.

The proposed method of finding the "source of solutions", $L_2v = 0$, can be actually generalized for the case when the solutions of $L_2v = 0$ are mapped in the solutions of $L_1u = 0$ by the transformation not only in functions but also in independent variables. We skip it here.

Let us illustrate the proposed method of finding the equation $L_2v = 0$ ("source of solutions") with the Burgers equation (BE)

$$L_1 u \equiv u_t + u u_x - \mu u_{xx} = 0 \tag{6}$$

and the linear heat equation (HE)

$$L_2 v \equiv v_t - \mu v_{xx} = 0. \tag{7}$$

The basis of Lie algebra of the transformation group of (6) is

$$\begin{gathered} X_1 = \frac{\partial}{\partial x}, \quad X_2 = \frac{\partial}{\partial t}, \; X_3 = x\frac{\partial}{\partial x} + 2t\frac{\partial}{\partial t} - u\frac{\partial}{\partial u}, \\ X_4 = t\frac{\partial}{\partial x} + \frac{\partial}{\partial u}, \quad X_5 = tx\frac{\partial}{\partial x} + t^2\frac{\partial}{\partial t} + (x - tu)\frac{\partial}{\partial u}. \end{gathered} \tag{8}$$

Here X_1 generates the space translations; X_2 time translations; X_3 dilations; X_4 Galilean translations; X_5 inversions in the space of independent variables accompanied by compensating transformations of the dependent variable u that preserves the form of (6).

The transformation group of (7) is infinite-dimensional and modulo the "linearity ideal" (i.e. after taking into account that together with any two solutions u_1 and u_2 their linear combination $\alpha u_1 + \beta u_2$ satisfies (7)) turns into a transformation group whose Lie algebra is spanned by

$$Y_1 = \frac{\partial}{\partial x}, \quad Y_2 = \frac{\partial}{\partial t}, \quad Y_3 = x\frac{\partial}{\partial x} + 2t\frac{\partial}{\partial t}, \tag{9}$$
$$Y_4 = u\frac{\partial}{\partial u}, \quad Y_5 = t\frac{\partial}{\partial x} - \frac{1}{2\mu}xu\frac{\partial}{\partial u},$$
$$Y_6 = tx\frac{\partial}{\partial t} + t^2\frac{\partial}{\partial t} - \frac{1}{4\mu}(x^2 + \mu t)u\frac{\partial}{\partial u}.$$

where Y_1 generates space translations; Y_2 time translations; Y_3 dilations of independent variables; Y_4 dilations of dependent variable; Y_5 Galilean translations; Y_6 inversions of independent variables accompanied by compensating transformations of the unknown function.

Lie algebras with basises (8) and (9) are different. But the truncations of these algebras on the space of independent variables coincide:

$$W_1 = \frac{\partial}{\partial x}, \quad W_2 = \frac{\partial}{\partial t}, \quad W_3 = x\frac{\partial}{\partial x} + 2t\frac{\partial}{\partial t}, \tag{10}$$
$$W_4 = t\frac{\partial}{\partial x}, \quad W_5 = tx\frac{\partial}{\partial x} + t^2\frac{\partial}{\partial t}.$$

This is due to the fact that transformation groups of (6) and (7) act in the space of independent variables in the same way. Lie algebra (8) is distinguished from Lie algebra (9) since the dependent variables in eqs. (6) and (7) transform differently (they have to compensate the changes in the "appearence" of the different equations caused by the same transformations of independent variables).

It is easy to verify that

$$u(x,t) = -2\mu(\ln v)_x\,, \tag{11}$$

where v is a solution of (7), satisfying the nonlinear BE. One can find the "source of solutions" for BE and the transformation (11) as follows.

Let us take the truncated operators (10). They do not constitute a basis of any Lie algebra because they are not closed under bracket. Operators W_1 and W_2 belong to basises of Lie algebras of equations (6) and (7) because the coefficients of (6) and (7) do not involve x and t explicitly. Therefore, if one wants to find the "source of solutions" among equations whose coefficients do not explicitly depend on space and time variables, it is neccessary to include the operators W_1 and W_2 "per se" in the basis of the sought equation, i.e. to put

$$\widetilde{W}_1 \equiv W_1\,, \qquad \widetilde{W}_2 \equiv W_2\,, \tag{12}$$

to "extend" all the other operators with unknown functions $\varphi_i\ (x,t,u)$:

$$\widetilde{W}_i \equiv W_i + \varphi_i\,(x,t,u)\frac{\partial}{\partial u} \tag{13}$$

(i = 3,4,5) and to add one more operator

$$\widetilde{W}_6 = \varphi_6\,(x,t,u)\frac{\partial}{\partial u}, \tag{14}$$

which could disappear completely under truncation on the space of independent variables. Operators W_1 - W_6 close under bracket if φ_i (i = 3,...,6) satisfy some equations. Solving them one can find all the Lie algebras with truncation (10) on the space of independent variables. Knowing a basis of Lie algebra and the type of sought equation (for example, requiring the "source of solutions" to be a partial 2nd order differential equation involving the order time derivative) one can find such an equation (if it exists). It may be a potential "source of solutions" for (6).

Comparison of Lie algebras (8) and (9) shows that the transformation mapping a solution of HE in the solution of BE can be found in the form

$$u(x,t) = F(v(x,t)). \tag{15}$$

The unknown function F only depends on one argument v . Therefore to find F it is necessary to solve an ordinary differential equation. The transformation (11) was discovered by Hopf [10] and Cole [11].

The proposed method of "extending" the elements of a truncated Lie algebra allows us to find "bilinear forms" using somewhat exotic Hirota method [12] for finding explicit solutions of many well-known completely integrable systems.

A manifest resemblance of BE (6) and KdV (1) caused many authors to find a transformation similar to (11) which would reduce integrating (1) to integrating some known linear equation. Our method shows the frustration of all these attempts was not due to the "bad luck" but because a linear "source of solutions" for KdV (1) does not exist. Neither does the counterpart of Hopf-Cole transformation.

References

1. Gardner,C.S., Green,J.M., Kriskal,M., Miura,R.M.(1967). Phys. Rev. Lett., 19, 1095.

2. Zakharov,V.E., Shabat,A.B. (1971). ZhETPh, 61, 118 (Russian).

3. Zakharov,V.E., Shabat,A.B. (1973). ZhETPh, 6, 1627 (Russian).

4. Kunin,Y.A. (1974). Theory of Elasticity of Media with

Microstructure. Nauka, Moscow (Russian).

5. Zakharov,V.E., Shabat,A.B. (1974). Funk. Anal. Appl., 8, 3, 43 (Russian).
6. Darboux,J.G. (1882). C.R. Acad. Sci. Paris, 94, 1465.
7. Chudnovsky,D.V. (1983). In: Group Theoretical Methods in Physics. L.N. in Phys. 180. Springer-Berlin, 65.
8. Matveev,V.B. (1979). Lett. Math. Phys.,3, 213.
9. Faddeev,L.D. (1952). Russian Math. Surveys, 14, 4,57 (Russian).
10. Hopf,E. (1950). Comm. Pure Appl. Math., 3, 201.
11. Cole,J.D. (1951). Quart. Appl. Math., 9, 225.
12. Hirota,R. (1981). J. Phys. Soc. Japan, 50, 3785.

SOME EXACT SOLUTIONS OF FIVE-DIMENSIONAL NONLINEAR WAVE EQUATION

V.M.Fedorchuk

The Ukrainian SSR Academy of Sciences Institute of Ap plied Problems of Mechanics and Mathematics, Lvov

A well-known method of finding exact solutions of differential equations is to use the subgroup structure of the symmetry groups.

Five-dimensional linear wave equations are used in the quantum theory to describe the particles with variable mass [1] . Nonlinear wave equations in the Minkowsky space M(1,n) have been considered in [2-4] .

Here the equations' symmetry has been used to construct multiparameter families of exact solutions. Equationsof the same type were also investigated in [5,6] .

Here we construct some classes of exact solutions of the following nonlinear five-dimensional wave equation:

$$\Box u = F(u) \tag{1}$$

where

$$\Box = \frac{\partial^2}{\partial x_0^2} - \frac{\partial^2}{\partial x_1^2} - \frac{\partial^2}{\partial x_2^2} - \frac{\partial^2}{\partial x_3^2} - \frac{\partial^2}{\partial x_4^2},$$

u(x) is a scalar twice-differentiable function of a point $x \in M(1,4)$, F(u) is an arbitrary function. The symmetry group of (1) is P(1,4), i.e. the group of the motions of M(1,4). To find exact solutions of this equation we use the subgroup structure of the group P(1,4) [7,8] ; we have only considered splitting subalgebras of the Lie algebra of P(1,4) [7,8] . Using these subalgebras we can reduce (1) to the equations which have less number of variables.

Let the symmetry group G of (1) be generated by operators

$$X=\xi^{\mu}(x,u)\frac{\partial}{\partial x^{\mu}}+\eta(x,u)\frac{\partial}{\partial u}\ ,\ \mu=0,1,2,3,4\ . \qquad (2)$$

Assume that in general the rank of the matrix $M = \| \xi^{\mu}, \eta \|$ equals r_* . Then the rank of an invariant solution of (1) is [9]: $\rho = n-r_*$, n=5. We give here the solutions of rank $\rho = 1$ only, i.e. only those that satisfy the ordinary differential equations derived from the wave equation:

$$\frac{d^2u}{d\omega^2}+\frac{du}{d\omega}k\omega^{-1}=eF(u), \qquad (3)$$

where $k = 0,1,2,3,4$; $l = \pm 1$, ω is one of the invariants of the splitting subalgebras with $r_* = 4$. If $k \geqslant 1$ then direct integration of (3) is impossible. However if $F(u) = \lambda u^n$ eq. (3) takes the form

$$\frac{d^2u}{d\omega^2}+\frac{du}{d\omega}k\omega^{-1}=e\lambda u^n\ (n\neq 1). \qquad (4)$$

We seek solutions of this equation in the form

$$u=\alpha\omega^{\nu}. \qquad (5)$$

Substituting (5) into (4) we get

$$\alpha=\left[\frac{2\{1+k+n(1-k)\}}{e\lambda(1-n)^2}\right]^{\frac{1}{n-1}},\ \nu=\frac{2}{1-n}\ . \qquad (6)$$

Hence particular solutions of (1) with the right hand side $F(u) = \lambda u^n$ $(n \neq 1)$ are given by (5), (6) and the following data

N	Variables	k	e
1.	$(x_1^2 + x_2^2)^{\frac{1}{2}}$	1	-1
2.	$(x_0^2 - x_4^2)^{\frac{1}{2}}$	1	1
3.	$(x_3^2 + x_4^2)^{\frac{1}{2}}$	1	-1
4.	$(x_1^2 + x_2^2 + x_3^2)^{\frac{1}{2}}$	2	-1
5.	$(x_4^2 + x_3^2 - x_0^2)^{\frac{1}{2}}$	2	-1
6.	$(x_1^2 + x_2^2 + x_3^2 + x_4^2)^{\frac{1}{2}}$	3	-1
7.	$(x_3^2 + x_2^2 + x_1^2 - x_0^2)^{\frac{1}{2}}$	3	-1
8.	$(x_4^2 + x_2^2 + x_1^2 - x_0^2)^{\frac{1}{2}}$	3	-1
9.	$(x_4^2 + x_3^2 + x_2^2 + x_1^2 - x_0^2)^{\frac{1}{2}}$	4	-1

If $k = 0$ then eq. (3) is integrated in quadratures and for eq. (4) we get:

$$\omega + C_0 = \int \frac{du}{\sqrt{2e\lambda u^{n+1}/n+1+C_1}}, \tag{7}$$

where C_0 and C_1 are arbitrary constants, ω is one of the $x_1's$ and for the first of them $l = 1$, whereas for the others $l = -1$. For $\omega_1 = x_0+x_4$ and $\omega_2 = x_0-x_4$ we get $\Box u(\omega) \equiv 0$.

Note that using the invariants of subalgebras of the Lie algebra of $P(1,4)$ with $r_* = 3$ also reduces (1) to ODE. Here we do not consider this. The same question for other groups important from the point of view of mathematical physics has been studied in detail in [6].

References

1. Fushchich,W.I. (1970). Teor. Math. Phys., 4, 360.
2. Fushchich,W.I. In: Algebraic-Theoretical Studies in

Mathematical Physics. Kiev, 1981, p. 6.

3. Fushchich,W.I., Serov,N.I., Shtelen,W.M. (1983). In: Group-theoretical Methods in Physics. 2. Nauka, Moscow, p. 407.
4. Fushchich,W.I., Serov,N.I. (1983). J. Phys. A, 16, 3645.
5. Segeda,Yu.N. (1981). In: Algebraic-Theoretical Studies in Mathematical Physics. Kiev, p. 54.
6. Grundland,A.M., Harnad,J., Winternitz,P. (1984). J. Math. Phys. 25, 791.
7. Fedorchuk,V.M. (1978). Preprint, IM-78.18, Kiev.
8. Fedorchuk,V.M. (1979). Ukr. Mat. Zh. 31, 717 (Ukrainian).
9. Ovsyannikov,L.V. (1978), The Group Analysis of Differential Equations, Nauka, Moscow.

ON THE CONFORMAL GENERALIZATION OF THE DIRAC-HEISENBERG EQUATIONS FOR SPINOR FIELD AND SOME EXACT SOLUTIONS

W.I.Fushchich, W.M.Shtelen, R.Z.Zhdanov
Mathematical Institute, Kiev

Consider the following Poincaré-invariant nonlinear generalization of the Dirac equation

$$\gamma^{\mu}[i\partial_{\mu}+F_1\bar{\Psi}\gamma_{\mu}\Psi+F_2\bar{\Psi}\gamma_4\gamma_{\mu}\Psi+F_3(\bar{\Psi}\gamma_{\mu}\Psi)\gamma_4+F_4(\bar{\Psi}\gamma_4\gamma_{\mu}\Psi)\gamma_4]\Psi+$$
$$+F_5(\bar{\Psi}\sigma_{\mu\nu}\Psi)\sigma^{\mu\nu}\Psi+F_6(\bar{\Psi}\sigma_{\mu\nu}\Psi)\gamma_4\sigma^{\mu\nu}\Psi=(F_7+F_8\gamma_4)\Psi \qquad (1)$$

where $F_1,\dots,F_8$ are arbitrary functions of $\bar{\Psi}\Psi$ and $\bar{\Psi}\gamma_4\Psi$

$\gamma_4=i\gamma_0\gamma_1\gamma_3$, $\sigma_{\mu\nu}=\frac{i}{4}(\gamma_{\mu}\gamma_{\nu}-\gamma_{\nu}\gamma_{\mu})$.

The well-known Dirac-Heisenberg [1] and Dirac-Gürsey [2] equations belong to this class.

From (1) select the equations stable under the scale transformation

$$x'_{\mu}=x_{\mu}e^{\theta},\quad \Psi'(x')=e^{k\theta}\Psi(x),\quad k,\theta - \text{const} \qquad (2)$$

and under the conformal transformations [3]

$$x'_{\mu}=\frac{x_{\mu}-c_{\mu}x^2}{\sigma(x)},\quad \Psi'(x')=\sigma(x)(1-\gamma\cdot c\cdot\gamma\cdot x)\Psi(x)$$

$$\sigma(x)=1-2c\cdot x+c^2x^2,\ c\cdot x\equiv c^{\nu}x_{\nu},\ c^2=c^{\nu}c_{\nu},\ \nu=0,1,2,3.$$

Theorem 1. Eq. (1) is stable under (2) if and only if

$$F_1=\Phi_1[(\bar{\Psi}\gamma_{\mu}\Psi)(\bar{\Psi}\gamma^{\mu}\Psi)]^{-\frac{1+2k}{4k}},\quad F_2=\Phi_2[(\bar{\Psi}\gamma_4\gamma_{\mu}\Psi)(\bar{\Psi}\gamma_4\gamma^{\mu}\Psi)]^{-\frac{1+2k}{4k}}$$

$$F_3=\Phi_3\left[(\bar\Psi\gamma_\mu\Psi)(\bar\Psi\gamma^\mu\Psi)\right]^{-\frac{1+2k}{4k}}, \quad F_4=\Phi_4\left[(\bar\Psi\gamma_4\gamma_\mu\Psi)(\bar\Psi\gamma_4\gamma_\mu\Psi)\right]^{-\frac{1+2k}{4k}}$$

$$F_5=\Phi_5\left[(\bar\Psi\sigma_{\mu\nu}\Psi)(\bar\Psi\sigma^{\mu\nu}\Psi)\right]^{-\frac{1+2k}{4k}}, \quad F_6=\Phi_6\left[(\bar\Psi\sigma_{\mu\nu}\Psi)(\bar\Psi\sigma^{\mu\nu}\Psi)\right]^{-\frac{1+2k}{4k}} \tag{4}$$

$$F_7=\Phi_7\left[\bar\Psi\Psi\right]^{-1/2k}, \quad F_8=\Phi_8\left[\bar\Psi\Psi\right]^{-1/2k}$$

where $\Phi_1, \dots, \Phi_8$ are arbitrary functions of $\bar\Psi\Psi/\bar\Psi\gamma_4\Psi$.

<u>Proof</u>. It easy to see that transformations (2) leave eq. (1) invariant if

$$e^{\theta(2k+1)}F_b(\bar\Psi\Psi e^{2k\theta}, \bar\Psi\gamma_4\Psi e^{2k\theta}) = F_b(\bar\Psi\Psi, \bar\Psi\gamma_4\Psi), \quad b=\overline{1,6}$$

$$e^{\theta(k+1)}F_c(\bar\Psi\Psi e^{2k\theta}, \bar\Psi\gamma_4\Psi e^{2k\theta}) = F_c(\bar\Psi\Psi, \bar\Psi\gamma_4\Psi), \quad c=7,8. \tag{5}$$

Taking into account the well-known identities [4]

$$(\bar\Psi\Psi)^2+(\bar\Psi\gamma_4\Psi)^2-(\bar\Psi\sigma_{\mu\nu}\Psi)(\bar\Psi\sigma_{\mu\nu}\Psi)=0$$

$$(\bar\Psi\Psi)^2-(\bar\Psi\gamma_4\Psi)^2-(\bar\Psi\gamma_4\gamma_\nu)(\bar\Psi\gamma_4\gamma^\nu\Psi)=0 \tag{6}$$

$$(\bar\Psi\gamma_\mu\Psi)(\bar\Psi\gamma^\mu\Psi)-(\bar\Psi\gamma_4\gamma_\mu\Psi)(\bar\Psi\gamma_4\gamma^\mu\Psi)=0$$

the general solution of (5) can be written in the form (4). One can directly verify that eq. (1) with functions (4) is stable under transformations (2).

<u>Theorem 2</u>. Eq. (1) is stable under the conformal group C(1,3) if and only if functions $F_1, \dots, F_8$ are of the form (4) with $k = -3/2$.

<u>Proof</u>. Since C(1,3) contains the extended Poincaré group $\tilde P(1,3) = (P(1,3), D)$, we can use the result of Theorem 1. Then one can verify that transformations (3) preserve eq. (1) with functions (4) for $k = -3/2$.

<u>Corollary</u>. If $F_7 = \lambda(\bar{\Psi}\Psi)^{1/3}$ and $F_A = 0$, $A = 1,\ldots,6,8$, then eq. (1) coincides with Dirac-Gürsey's equation [2] :

$$\left[i\gamma\cdot\partial - \lambda(\bar{\Psi}\Psi)^{1/3}\right]\Psi = 0, \quad \lambda = \mathrm{const}. \tag{7}$$

In another case when $F_4 = \lambda\left[(\bar{\Psi}\gamma_\mu\Psi)(\bar{\Psi}\gamma^\mu\Psi)\right]^{-1/3}$, $F_B = 0$, $B = 1, 2, 3, 5, 6, 7, 8$ we obtain a conformal invariant version of Dirac-Heisenberg equation

$$\left[i\gamma\cdot\partial + \lambda\frac{(\bar{\Psi}\gamma_\mu\Psi)\gamma^\mu}{\left[(\bar{\Psi}\gamma_\mu\Psi)(\bar{\Psi}\gamma^\mu\Psi)\right]^{1/3}}\right]\Psi = 0. \tag{8}$$

As it is well known the original Dirac-Heisenberg equation [1] is not invariant with respect to the conformal transformations.

Now we use the symmetry properties of eq. (8) to construct its exact solutions. Following [5,3] we take the anzatz

$$\Psi(x) = \varphi(\beta\cdot x), \quad \beta\cdot x \equiv \beta^\nu x_\nu, \quad \beta^\nu - \mathrm{const} \tag{9}$$

$$\Psi(x) = \frac{\gamma\cdot x}{(x^\nu x_\nu)^2}\,\Phi\left(\frac{\beta\cdot x}{x^\nu x_\nu}\right). \tag{10}$$

Eqs. (9) and (10) are translationally and conformally invariant respectively. The substitution of (9), (10) into (8) gives the following system of ODE

$$i\gamma\cdot\beta\,\frac{du}{d\omega} + \nu\,\frac{(\bar{u}\gamma_\nu u)\gamma^\nu u}{\left[(\bar{u}\gamma_\nu u)(\bar{u}\gamma^\nu u)\right]^{1/3}} = 0 \tag{11}$$

where $u = \left\{\varphi(\omega),\ \omega = \beta\cdot x \ \text{ or } \ \Phi(\omega),\ \omega = \dfrac{\beta\cdot x}{x^\nu x_\nu}\right\}$,

$\nu = \lambda$ for φ and $\nu = -\lambda$ for Φ. There are three different cases (χ is a constant spinor,

$$\beta_\mu = \frac{\bar{\chi}\gamma_\mu \chi}{[(\bar{\chi}\gamma_\nu\chi)(\bar{\chi}\gamma^\nu\chi)]^{1/3}}):$$

a) $\mathrm{Im}\,\nu = 0$, $u = e^{i\nu\omega}\chi$;

b) $\mathrm{Re}\,\nu = 0$, $u = (C + \frac{2}{3}\mu\omega)^{-3/2}\chi$, $\mu = \mathrm{Im}\,\nu$;

c) $\mathrm{Im}\,\nu$ $\mathrm{Re}\,\nu \neq 0$, $u = (f_1 + if_2)\chi$, $\nu = \nu_1 + i\nu_2$ (12)

$$f_1 = \pm\left[(w-2v)^{1/2} + (w+2v)^{1/2}\right],\ f_2 = \mp\left[(w-2v)^{1/2} - (w+2v)^{1/2}\right]$$

$$\int \frac{dv}{\left(C_1 - 2\frac{\nu_2}{\nu_1}v^2\right)^{2/3}} = 2\nu_1\omega + C_2,\ w = \left(C_1 - 2\frac{\nu_2}{\nu_1}v^2\right)^{1/2}.$$

Remark. Let us show that conformally invariant ansatz (10) can be obtained from (9) applying the generating solutions procedure with the use of conformal transformations. As it is shown in [3], the formula generating solutions is:

$$\psi_{new}(x) = \frac{1-\gamma\cdot x\gamma\cdot c}{\sigma^2(x)}\psi_{old}(x'),\ x'_\mu = \frac{x_\mu - c_\mu x^2}{\sigma(x)} \tag{13}$$

$$\sigma(x) = 1 - 2c\cdot x + c^2x^2,\ c^2 \equiv c^\nu c_\nu,\ x^2 \equiv x^\nu x_\nu$$

applying (13) with $c_0 = 1$, $c_1=c_2=c_3=0$ to (9) and then changing x_0 to x_0+1 (violating transformation invariance) we obtain (10).

Now let us apply the generating solutions procedure to the (10), (12a)

$$\psi(x) = \frac{\gamma\cdot x}{(x^\nu x_\nu)^2}\exp\left\{-i\lambda\frac{\beta\cdot x}{x^\nu x_\nu}\right\}\chi,\ \beta_\mu = \frac{\bar{\chi}\gamma_\mu\chi}{[(\bar{\chi}\gamma_\nu\chi)(\bar{\chi}\gamma^\nu\chi)]^{1/3}}. \tag{14}$$

Performing translations we get from (14) another family

of solutions of eq. (8)

$$\psi(x)=\frac{\gamma\cdot x+\gamma\cdot a}{(x^\nu x_\nu+2a\cdot x+a^\nu a_\nu)^2}\exp\left\{-i\lambda\frac{\beta\cdot x+\beta\cdot a}{(x^\nu x_\nu+2a\cdot x+a^\nu a_\nu)}\right\}\chi . \tag{15}$$

It is a remarkable family of solutions, because the transformations of its parameters are invariant with respect to the full 15-parameter conformal group.

In conclusion note that using symmetry we have obtained exact solutions of the nonlinear Dirac equation [3], nonlinear equations of quantum electrodynamics [6], Yang-Mills equations [7] and some scalar nonlinear equations [8,9].

References

1. Heisenberg,W. (1966). Introduction to the unified field theory of elementary particles. Interscience Publ. London.

2. Gürsey,F. (1956). Nuovo Cim. 988.

3. Fushchich,W.I., Shtelen,W.M. (1883). J. Phys. 16A,271.

4. Finkelstein,R., Fronsdal,C. ,Kaust,P. (1956). Phys. Rev. 103, 5.

5. Fushchich,W.I. (1981). In: Algebraic-theoretical studies in mathematical physics, UkSSR Acad. Sci. Math. Institute, Kiev, p. 6 (Russian).

6. Fushchich,W.I., Shtelen,W.M. (1983). Phys. Lett. 128B, 215.

7. Fushchich,W.I., Shtelen, W.M. (1983). Lett. Nuovo Cim. 38, 37.

8. Fushchich,W.I., Shtelen,W.M. (1982). Lett. Nuovo Cim. 34, 498.

9. Fushchich,W.I., Serov,N.I. (1983). J. Phys. 16A,3445.

Part V

GROUP REPRESENTATIONS AND SPECIAL FUNCTIONS

ZEROS OF POLYNOMIALS AND GENERALIZED CHEBYSHEV POLYNOMIALS

Henri Bacry

Centre de Physique Theorique, CNRS - Luminy, Case 907, 13288 - Marseille Cedex 9

1. Introduction. A very large number of theorems have been discovered during the nineteenth and twentieth centuries concerning the location of the zeros of a polynomial. The methods used were diversified and it is not easy to classify them. If one wants to have an idea of the main results obtained in that domain until 1963, one may read the book of Marden [1].

Before giving some new insight about the zeros of polynomials, I would like to make some general remarks on that kind of problems.

1. The question of location of points in the complex plane is essentially a geometrical question and, accordir to Klein's famous programme, geometry is related with group theory. However, until now, group theory has been of little use in that kind of problem. Obviously, there is a group naturally involved (when one looks for instance for the number of zeros contained in a given disk) it is the Moebius group, that is the group of linear transformations $z \longmapsto \frac{az+b}{cz+d}$. One of the pioneers in that direction is Laguerre [2] who introduced the projective generalization of the derivative which he calls emanant but which is sometimes referred to as the polar derivative [3]. An interesting fact is that the emanant has a group theoretical interpretation in the framework of constellation theory [4] introduced by the author some years ago, when one reduces the Kronecker product of a

representation of spin j by a representation of spin 1/2.

2. In fact, the Moebius group is not the only group which appears when one deals with the problem of zeros. in reference 4 , it was shown that the projective orthogonal, symplectic and unitary groups in any finite dimension are also of interest. People having worked in apolar polynomials were quite close to group theory but in our knowledge did not use it.

3. The set of all zeros of a complex polynomial of degree n is of cardinality less than or equal to n. But it is no longer a set if we take into account the multiplicity of the zeros. That is why we prefer to use a special word, the word constellation. Moreover if we want to use the Moebius transformations, a root can be infinite and we are led to define the polynomials of degree less than or equal to n. In other words, we arrive at the following projective notions:

Let $[a_0 : a_1 : a_2 : \ldots : a_n]$ an element of the projective space $P_n(\mathbb{C})$. We associate with it the following ray-polynomial:

$$P([x:y]) = a_0 x^n - a_1 x^{n-1} y + a_2 x^{n-2} y - \ldots + (-)^n y^n \quad (1)$$

where $[x:y] \in P_1(\mathbb{C})$.

One can identify any ray-polynomial of degree n with the corresponding ray of $P_n(C)$. It maps $P_1(C)$ on 0 (zero) if $[x : y]$ is a root and on $\bar{0}$ (non zero) if it is not.

It follows from the fundamental theorem of the algebra that the ray-polynomial (1) can be factorized as follows:

$$P([x:y]) = (\beta_1 x - \alpha_1 y)(\beta_2 x - \alpha_2 y)\ldots(\beta_n x - \alpha_n y) \quad (2)$$

and $[[\alpha_1 : \beta_1], [\alpha_2 : \beta_2], \ldots, [\alpha_n : \beta_n]]$ is the constellation of roots of the ray-polynomial. Since there are one-to-one correspondences between the constellations, the ray-polynomials of degree n and $P_n(C)$ we have the following property [6-7].

$$P_n(C) = \frac{P_1(C) \times P_1(C) \times \ldots \times P_1(C)}{S_n} \tag{3}$$

where S_n denotes the n-th symmetric group. We also note that a constellation could be considered as an "algebraic variety" of dimension zero.

In the next section, we will give an alternative definition of constellations (in fact the dual of the above one, in the sense of projective geometry). It will permit us to define in $P_n(C)$ a canonical variety which will play an important role in the description of our generalized Chebyshev polynomials [8-9]. In the same section we will give the definition of the Laguerre emanant. It will lead us to a symmetric matrix we will call the discriminant matrix B. Its rank is related with the number of distinct roots (i.e. the apparent order of the constellation). The determinant of B is the discriminant of the polynomial.

In section III, we will interpret the characters of SL(n,C) as polynomials in n-1 variables (those variables being the fundamental characters of the group, that is the coefficients of the Cayley-Hamilton polynomial associated with a matrix of SL(n,C)). The results of Section II are applied to the C-H equation. Its discriminant matrix permits us to obtain the partial differential equation obeyed by the characters. Since in the case n = 2, they are Chebyshev polynomials of the second kind, we call them generalized Chebyshev polynomials. Moreover, the orthogonality of characters is obtained with the aid of an integration, the domain of which being defined in using the canonical variety introduced in Section II.

II. The Laguerre emanant and the discriminant matrix.

1. Constellations and the canonical variety of $P_n(C)$.

The equation

$$a_0 x^n - a_1 x^{n-1} y + \ldots \pm a_n y^n = 0 \tag{4}$$

where $[x : y] \in P_1(C)$ and $[a_0 : a_1 \ldots a_n] \in P_n(C)$, can be considered either as a constellation (that is an "algebraic variety" of zero dimension and degree n) or, in a dual way, as the equation of a family F of (n-1) dimensional linear varieties (hyperplanes) in $P_n(C)$, parametrized by $[x : y]$; i.e. by points of the Riemann sphere.

From the fundamental theorem of algebra, through each point $[a_0 : a_1 : \ldots a_n]$, one can draw n (non necessarily distinct) hyperplanes of F, i.e. a constellation of hyperplanes. We will denote by V the canonical variety of $P_n(C)$ where V is the envelope of the family F. In other words, from each point not on V one can draw exactly n hyperplanes tangent to V. It is clear that each such hyperplane has a linear variety of dimension n-2 in common with V and that V is a fibered space with $P_1(C)$ as a base. It is also clear that the equation of V is $\delta = 0$ where δ is the discriminant of the polynomial (4).

The most trivial example is for n = 2. V is just the Riemann sphere itself. If we restrict ourselves to the real subspace of $P_2(C)$, Eq.(4) becomes

$$a_0 x^2 - a_1 xy + a_2 y^2 = 0 \tag{5}$$

where a_0, a_1, a_2 are real. The equation of V is the well known discriminant $a_1^2 - 4a_0a_2 = 0$. By making $a_0 = 1$, one removes the point at infinity and we get a parabola. From each point outside the parabola one can draw a constellation of two (real) lines associated with the two roots. From a point inside the parabola, the two tangents are imaginary (see Fig.1).

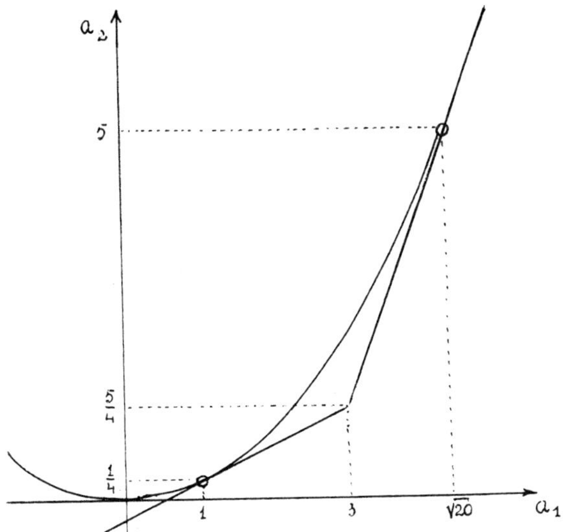

Fig.1. Each tangent to the parabola is labelled by a real number. For $(a_1, a_2)=(3, \frac{5}{4})$ the two roots are $\frac{1}{2}$ and $\frac{5}{2}$.

2. The Laguerre emanant. The concept of derivative has been generalized by Laguerre [2] in order to have at his disposal a projective tool. He introduced the emanant. It is an operator which maps $P_n(C)$ on $P_{n-1}(C)$ or the polynomial $P([x:y])$ on $P'_{[x_0:y_0]}([x:y])$ as follows

$$P'_{[x_0:y_0]}([x:y]) = \left(x_0\frac{\partial}{\partial x} + y_0\frac{\partial}{\partial y}\right)P([x:y]). \qquad (6)$$

The emanant is also known as the polar derivative of pole $[x_0 : y_0]$. For $[x_0 : y_0] = [1:0]$, one gets the usual derivative [10].

Let us now introduce the following homogeneous polynomial of degree n-2

$$B([x:y],[x':y']) = \frac{P'_{[x_0:y_0]}([x:y])P'_{[x'_0:y'_0]}([x':y']) - P'_{[x_0:y_0]}([x':y'])P'_{[x'_0:y'_0]}([x:y])}{(x_0y'_0 - x'_0y_0)(xy' - x'y)}$$

$$= \sum_{k=0}^{n-2}\sum_{l=0}^{n-2}(-)^{k+l+1} B_{kl}\, x^{n-2-k} y^{k} x'^{\,n-2-l} y'^{\,l} \qquad (7)$$

The coefficient $B_{k\ell}$ are entries of a symmetric $(n-1)\times(n-1)$-matrix B; they are homogeneous polynomials of degree two in the coefficients $a_0, a_1, \dots, a_n$ of $P([x:y])$. For n = 2, B is just the quantity $a_1^2 - 4a_0 a_2$; for n = 3, B is the matrix

$$B = \begin{vmatrix} 2a_1^2 - 6a_0 a_2 & a_1 a_2 - 9a_0 a_3 \\ a_1 a_2 - 9a_0 a_3 & 2a_2^2 - 6a_1 a_3 \end{vmatrix} \tag{8}$$

This matrix has interesting properties, for instance:
a) Det B is the discriminant of P ([x : y])
b) The rank of B equals the number of distinct roots minus one.

We will refer to B as the discriminant matrix. We will give in the next section another important property of it.

III. Generalized Chebyshev Polynomials of the second kind and the groups SU(n), SL(n,R), SL(n,C).

1. The restricted canonical manifolds. We intend to apply the above results to the Cayley-Hamilton polynomials of the matrices of SL(n,C). Let X be such a matrix and

$$a_0 X^n - a_1 X^{n-1} + a_2 X^{n-2} \dots \pm a_n I = 0 \tag{9}$$

its C-H equation. We suppose $a_0 = 1$; it follows that a_n = Det X = 1 or, equivalently, $a_0 - a_n = 0$, $\frac{1}{2}(a_0 + a_n) = 1$.

The condition $a_0 - a_n = 0$ reduces the $P_n(C)$ space introduced in Section II into a $P_{n-1}(C)$ space. Moreover, the condition $\frac{1}{2}(a_0 + a_n) = 1$ discards the "points at infinity" of $P_{n-1}(C)$. Consequently, the (n-1)-dimensional manifold M reduces to a (n-2)-dimensional submanifold M' without points at infinity. It is clear that M' is the envelope of a family of (n-2)-dimensional linear spaces parametrized by a complex number ($z = \infty$ discarded). For n = 2, the equation $a_1^2 - 4a_0 a_2 = 0$ (equation of M) reduces to two points $a_1 = \pm 2$ on the complex axis a_1 (R^2 topology). Obviously, we have kept the property that from each point a_1 one can draw two "tangents" through those two points.

Let us now turn towards the case n = 3. The restricted canonical manifold M' in the complex plane (a_1, a_2) is given by det B = 0 when B is given by Eq.(8) with $a_0 = a_3 = 1$. One gets the equation

$$a_1^2 a_2^2 - 4a_1^3 - 4a_2^3 + 18a_1 a_2 - 27 = 0. \qquad (10)$$

It is an algebraic curve of degree 4.

A further restriction of M' is obtained by restricting SL(n,C) to its subgroup SU(n) or SL(n,R). For SL(n,R), the coefficients a_k are real; for SU(n) one has the conditions

$$a_k = \bar{a}_{n-k} . \qquad (11)$$

For SL(2,R), a_1 lies on the real line; for SU(2), a_1 lies in the closed interval [-2,2]. For SL(3,R), the restricted M_r is represented on Fig.2. One verifies that from

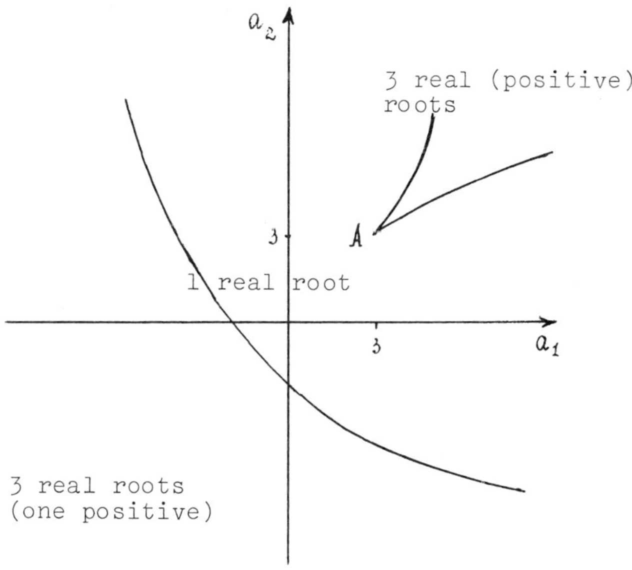

Fig.2. Canonical curve M_r for SL(3,R).

any point not on the real curve (10), one can draw 1 or 3 tangents to M_r depending on the number of real roots (1 or 3). Any point lying on M_r corresponds to a double real root except at point $A(a_1 = a_2 = 3)$ corresponding to the triple root 1.

For SU(3), the curve M_u (Eq. 10 with $a_2 = \bar{a}_1$) is represented on Fig.3. It is a hypocycloid in the (Re a_1, Ima_1)- plane. All points corresponding to SU(3) are those located <u>inside</u> the hypocycloid. From each such point, one can draw three tangents to M_u. The three cusps correspond to triple roots (the three cubic roots of 1).

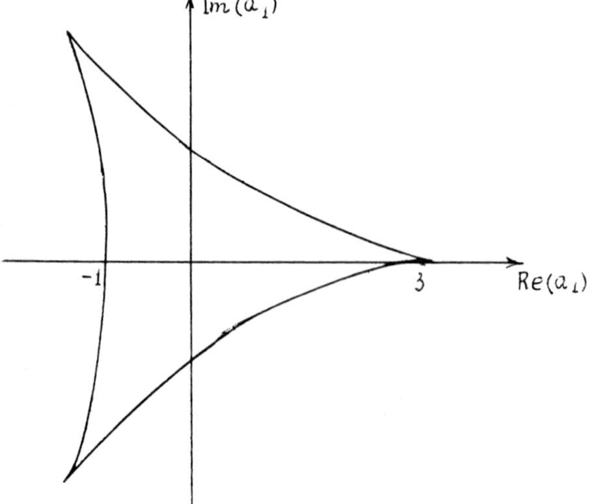

Fig.3. Canonical curve M_u for SU(3): $Re(a_1) = 2\cos\alpha + \cos 2\alpha$, $Im(a_1) = 2\sin\alpha - \sin 2\alpha$. From a point inside the hypocycloid, one can draw three tangents to M_u.

For SU(4), one could easily see that we are concerned with the points inside a "curved tetrahedron" S_4 in the (Rea_i, Ima_i, a_2) space with a_2 real. The faces of S_4 correspond to a double root, two of the edges to two double roots, the other edges to one triple root and the four vertices to a quadruple root. From each point <u>inside</u> S_4, one can draw four planes tangent to S_4.

2. The characters of SL(n,C) as polynomials. In order to avoid the use of multiple indices we will only consider the case of SL(3,C) because it presents almost all features common to all SL(n,C) for $n \geqslant 2$.

Let us write the C-H equation for $X \in SL(3,C)$ as follows

$$X^3 - tX^2 + uX - I = 0 . \tag{12}$$

It is clear that any X can be written in the form

$$X^m = f_{m-1} X^2 + g_{m-1} X + h_{m-1} I \tag{13}$$

where f_{m-1}, g_{m-1}, h_{m-1} are polynomials in t and u. These polynomials satisfy the same recurrence relation:

$$f_{m+1} = t f_m - u f_{m-1} + f_{m-2} \tag{14}$$

with

$$\begin{array}{lll} f_0 = 0 & f_1 = 1 & f_2 = t \\ g_0 = 1 & g_1 = 0 & g_2 = -u \end{array} \tag{15}$$

$$(h_m = f_{m-1})$$

We note that m can be chosen as a negative integer and

$$\begin{aligned} f_{-m}(t,u) &= h_m(u,t) \\ g_{-m}(t,u) &= g_m(u,t) \end{aligned} \tag{16}$$

which justify the labelling.

Their generating functions are

$$F = \sum_{m=0}^{\infty} f_m z^m = \frac{z}{\det(I - zX)} \tag{17}$$

$$G = \sum_{m=0}^{\infty} g_m z^m = \frac{1 - tz}{\det(I - zX)} \tag{18}$$

with $\det(1 - zX) = 1 - tz + uz^2 - z^3$.

Let a,b,c be the eigenvalues of X. They satisfy the C-H-equation. One readily obtains

$$f_{m-1} = \det\left(\begin{vmatrix} a^m & a & 1 \\ b^m & b & 1 \\ c^m & c & 1 \end{vmatrix} \begin{vmatrix} a^2 & a & 1 \\ b^2 & b & 1 \\ c^2 & c & 1 \end{vmatrix}^{-1}\right) \tag{19}$$

$$g_{m-1} = \det\left(\begin{vmatrix} a^2 & a^m & 1 \\ b^2 & b^m & 1 \\ c^2 & c^m & 1 \end{vmatrix} \begin{vmatrix} a^2 & a & 1 \\ b^2 & b & 1 \\ c^2 & c & 1 \end{vmatrix}^{-1}\right). \tag{20}$$

It readily follows that f_m, g_m are irreducible characters of the group SL(n,C). Let us write the most general irreducible character as

$$\chi_{m_1,m_2,m_3} = \det\left(\begin{vmatrix} a^{m_1} & a^{m_2} & a^{m_3} \\ b^{m_1} & b^{m_2} & b^{m_3} \\ c^{m_1} & c^{m_2} & c^{m_3} \end{vmatrix} \begin{vmatrix} a^2 & a & 1 \\ b^2 & b & 1 \\ c^2 & c & 1 \end{vmatrix}^{-1}\right) \tag{21}$$

where $m_1 > m_2 > m_3 = 0$. Therefore

$$f_m = \chi_{m+1,1,0} \qquad g_m = -\chi_{m+1,2,0}\,. \tag{22}$$

It is not difficult to prove the following generalized Cayley-Hamilton equation

$$\chi_{m_2,m_3,0}\,X^{m_1} - \chi_{m_1,m_3,0}\,X^{m_2} + \chi_{m_1,m_2,0}\,X^{m_3} = \chi_{m_1,m_2,m_3}I\,. \tag{23}$$

The χ 's satisfy the recurrence relation (13) for each index $\chi_{m_1+1,m_2,m_3} = t\,\chi_{m_1,m_2,m_3} - u\,\chi_{m_1-1,m_2,m_3} + \chi_{m_1-2,m_2,m_3}$ there, we did not impose the condition $m_1 > m_2 > m_3$ on indices; under a permutation of two of its indices χ changes its sign. With that convention (and redundancy), we easily get the following simple generating function for the characters of SL(n,C)

$$\sum_{m_1=0}^{\infty}\sum_{m_2=0}^{\infty}\sum_{m_3=0}^{\infty} \chi_{m_1,m_2,m_3}\, x^{m_1} y^{m_2} z^{m_3} = \frac{(x-y)(x-z)(y-z)}{\det[(I-xX)(I-yX)(I-zX)]} \tag{24}$$

(the main cause of redundancy disappears by setting z = o). The polynomials $\chi_{m_1,m_2,0}$ with $m_1 > m_2 > o$ will be called

<u>generalized Chebyshev polynomials of the second kind</u> because the case n = 2 refers explicitly to the standard ones.

(More precisely [11], $f_m(2\cos\theta) = \frac{\sin m\theta}{\sin\theta}$).

3. <u>Characters as orthogonal polynomials</u>. It is well known that the characters χ of SL(n,C) are orthogonal but it can be shown that they are in the usual sense of orthogonal polynomials. In fact, the corresponding measure is given by $(2\pi)^{1-n}\sqrt{|\det B|}$ and the domain of integration is the interior of the restricted canonical manifold S_u associated with SU(n). More explicitly, we have

$$\frac{1}{(2\pi)^{n-1}}\int_{\mathrm{Int}\,S_u} \sqrt{|\det B|}\,\overline{\chi_{m_1,m_2,\dots,0}(a_1,a_2,\dots,a_{n-1})}\,\chi_{m'_1,m'_2,\dots,0}(a_1,a_2,\dots,a_n) \tag{25}$$

$$d\Omega_n = \prod_i \delta_{m_i m'_i} \tag{11}$$

where

$$a_k = \bar{a}_{n-k}$$

$$d\Omega_n = \sqrt{\prod_k d(\mathrm{Re}\,a_k)\,d(\mathrm{Im}\,a_k)}\,. \tag{26}$$

The proof of this result is too long to be given here. It consists in replacing the eigenvalues of χ by their implicit functions of the variables a_k.

4. <u>Partial Differential Equation obeyed by the characters</u>. It was not obvious that the characters obey a partial differential equation. I do not have a short proof of it and I will only give the result*. Define the following operator

$$\Delta_n = \sum_{k=1}^{n-1}\sum_{\ell=1}^{n-1} b_{k\ell}\frac{\partial}{\partial a_k}\frac{\partial}{\partial a_\ell} \tag{27}$$

$$D_n = \sum_{k=1}^{n-1} k(n-k)\,a_k\frac{\partial}{\partial a_k}\,.$$

The $\chi_{m_1,m_2,\dots,m_n}$ obey the following equation

* After this work had been finished, I became aware of previous works on the subject: see [6] and references therein.

$$\left[\Delta_n+(n+1)D_n-\frac{1}{2}\sum_{i=1}^{n}\sum_{j=1}^{n}(m_i-m_j)^2+\frac{n^2(n^2-1)}{12}\right]\chi_{m_1,m_2,\dots,m_n}=0. \tag{29}$$

IV. Conclusion. As a conclusion, we only want to mention some open problems.

1. Generalized Chebyshev polynomials of the first kind. If $X \in SL(2,C)$, one wasily proves that $Tr(X^m)$ is a polynomial in $t = Tr\ X$. This polynomial $\ell_m(t)$ is such that

$$\ell_m(2\cos\theta) = 2\cos m\theta .$$

One recognizes, up to a labelling, the well known Chebyshev polynomials of the first kind. It is easy to show that if $X \in SL(n,C)$, the polynomials $\ell_m = Tr(X^m)$ satisfy the recurrence relation of the characters:

$$\ell_{m+1} = a_1\ell_m - a_2\ell_{m-1} + a_3\ell_{m-2} \dots \pm a_n\ell_{m-n+1} .$$

It can be shown that they obey the following equation

$$(\Delta_n + D_n - m^2)\ell_m = 0$$

where Δ_n and D_n are defined by Eqs. (27) and (28).

The polynomials ℓ_m can be called generalized Chebyshev polynomials of the first kind but they do not form a complete set. It is natural to look for an orthonormal complete set.

The most natural generalization of these polynomials is the following one. If $\lambda_1, \lambda_2, \dots, \lambda_n$ are the eigenvalues of $X \in SL(n,C)$, we define

$$T_{m_1,m_2,\dots,m_n} = \mathrm{Perm}\begin{vmatrix} \lambda_1^{m_1} & \lambda_1^{m_2} & \dots & \lambda_1^{m_n} \\ \lambda_2^{m_1} & \lambda_2^{m_2} & \dots & \lambda_2^{m_n} \\ \vdots & \vdots & & \vdots \\ \lambda_n^{m_1} & \lambda_n^{m_2} & \dots & \lambda_n^{m_n} \end{vmatrix}$$

where Perm denotes the permanent of the matrix but

expressed in terms of the elementary symmetric functions $a_1, a_2, \ldots, a_n$ of the λ's. Clearly a permutation of the m_i's does not change the polynomial T. Moreover, since det X = 1, we have, as for the χ's

$$T_{m_1+k, m_2+k, \ldots, m_n+k} = T_{m_1, m_2, \ldots, m_n}$$

It follows that to avoid redundancy, we can impose the condition $m_1 \geq m_2 \geq \ldots \geq m_n = 0$. We also note that they obey the same recurrence relation as the χ's and that they are eigenfunctions of $\Delta_n + \mathcal{D}_n$. Unfortunately, we were not able to interpret the T's if we except those of type $T_{m,m,\ldots m,0,0,\ldots 0}$ which are easily related with the fundamental invariants of X^m.

2. The zeros of the f_m's and l_m's. In the case n = 2, it is well known that the zeros of the f_m's and l_m's are in the interval $]-2, +2[$. It is interesting to mention that for n = 3, f_m and l_m have exactly $(m-1)^2$ zeros in the plane $Re\, a_1, Im\, a_1$. It seems a good conjecture that all these zeros are inside the hypocycloid (this has been checked for m = 1 to 6).

References and notes

1. Marden, M. (1966). Geometry of Polynomials, Am. Phys. Soc., Providence, Rhode Island.
2. Laguerre, E. Oeuvres, 1, p.48.
3. Marden, M. Oeuvres, 1, p.44.
4. Bacry, H. (1980). Comm. Math. Phys. 72, 119.
5. Marden, M. - ibid, 60.
   Since $P_1(C)$ is the Riemann sphere, a constellation of order n is described by n points (not necessarily distinct) on this sphere. The apparent order of a constellation is less than or equal to n; it is the cardinality of the set associated with the constellation. When the sphere is the celestial sphere, the Moebius group can be interpreted as the Lorentz group and the word constellation (proposed by A. Grossmann) is quite natural.
6. Bacry, H. J. math. Phys. 15 (1974), p. 1686.

7. Bacry, H. - In: Lecture Notes in Physics, 201, p.483 (1984). Springer-Verlag.

8. Bacry, H. - In: Proceedings of the Symposium International sur les polynomes orthogonaux et leurs applications. Bar-Le-Duc (1984), (to appear).

9. Two remarks are of interest: 1) The word "polar" is justified here by its use in Algebraic Geometry, 2) the emanant has a group theoretical interpretation: we consider the irreducible projective representation of SU(2) on $P_n(C)$ and $P_1(C)$. The tensor product $P_n(C) \times P_1(C)$ is a representation space which is reducible in two irreducible representation spaces corresponding to $P_{n+1}(C)$ and $P_{n-1}(C)$. The emanant takes out the P (C) component. The $P_{n+1}(C)$ component is obtained by multiplying the polynomial $P([x:y])$ by $x_0 y - y_0 x$. We note the commutator:

$$\left[x_0 \frac{\partial}{\partial x} + y_0 \frac{\partial}{\partial y}, x'_0 y - y'_0 x\right] \neq 0 \text{ except iff } [x_0 : y_0] = [x'_0 : y'_0].$$

10. We have already justified the labeling for n = 3. Here we have $f_m(t) = (-)^m f_{-m}(t)$ as a first justification. The second one is that f_m divides f_n iff m divides n.

11. Unfortunately, there are some errors due to misprints in reference [9]. They concerned the expressions (27), (28) and (29) of the present paper.

12. After this work had been finished I became aware of previous works on this subject: see Eier, R. and Lidl, R. - Math. Ann., (1982). 260, 93.

IRREDUCIBLE REPRESENTATIONS OF SO(n), SO(p,q), SU(n), S_4 AND THEIR MATRIX ELEMENTS

N.Ya.Vilenkin[+], G.V.Denisova, T.G.Kudryashova, A.A.Rudnev
[+]State Extra-mural Pedagogic Institute, Moscow

1. N.Ya.Vilenkin and G.V;Denisova studied the problem of the separation of variables for the operator $\square_{p,q} = \sum_{1\le k\le p} \partial^2/\partial x_k^2 - \sum_{1\le k\le q} \partial^2/\partial x_{p+k}^2$ on the hyperboloid $[\underline{x}, \underline{x}] = 1$, where $[\underline{x}, \underline{y}] = \sum_{1\le k\le p} x_k y_k - \sum_{1\le k\le q} x_{p+k} y_{p+k}$ in coordinate systems corresponding to a given tree. This problem was solved for q = 0 and for p = 1 in [1] (see also [2]); in [3] it was solved for arbitrary p and q for Euclidean and hyperbolic rotations. G.V.Denisova has considered the case when horospheric rotations (see [1], p;540),i.e. reductions of the form $SO(p,q) \supset JSO(p-1, q-1) \supset SO(p-1, q-1)$, are also taken into account. In the coordinates corresponding to the tree from Fig.1 we have, on $[\underline{x}, \underline{x}]_{2,\bar{4}} = -1$:

$$x_1 = \sinh \varphi_1 - \frac{1}{2} e^{i\varphi_1} \varphi_{11}^2, \quad x_2 = e^{\varphi_1} \varphi_{11} \cosh \theta_1 \cosh \theta_2,$$

$$x_3 = \cosh \varphi_1 + \frac{1}{2} e^{\varphi_1} \varphi_{11}^2, \quad x_4 = e^{\varphi_1} \varphi_{11} \cosh \theta_1 \sinh \theta_2,$$

$$x_5 = e^{\varphi_1} \varphi_{11} \sin \theta_1 \sin \theta_3, \quad x_6 = e^{\varphi_1} \varphi_{11} \sinh\theta_1 \cos\theta_3$$

(horospheric rotation in x_1, x_2, x_3). To the forks drawn on Fig.2,the angular part of the operator $\square_{p,q}$ corresponds:

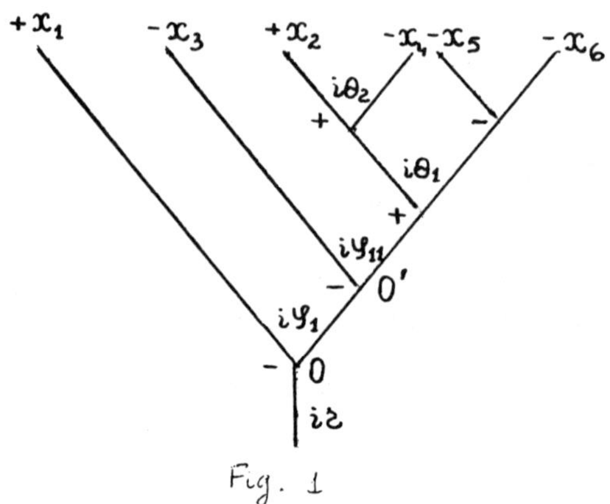

Fig. 1

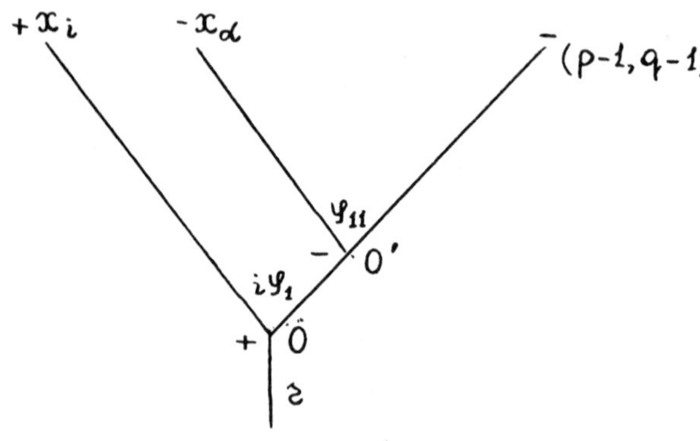

Fig. 2

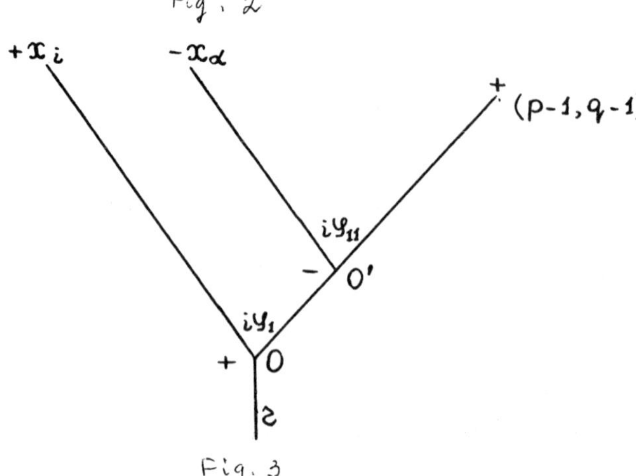

Fig. 3

$$\Delta^{\mathfrak{R}}_{p,q} = -e^{-m\varphi_1}\frac{\partial}{\partial\varphi_1}e^{m\varphi_1}\frac{\partial}{\partial\varphi_1} - e^{2\varphi_1}\left(\varphi_{11}^{-m+1}\frac{\partial}{\partial\varphi_{11}}\varphi_{11}^{m-1}\frac{\partial}{\partial\varphi_{11}} + \varphi_{11}^{-2}\Delta^{\mathfrak{R}}_{p-1,q-1}\right), \qquad (1)$$

while to those on Fig.3

$$\Delta^{\mathfrak{R}}_{p,q} = -e^{-m\varphi_1}\frac{\partial}{\partial\varphi_1}e^{m\varphi_1}\frac{\partial}{\partial\varphi_1} + e^{-2\varphi_1}\left(\varphi_{11}^{-m+1}\frac{\partial}{\partial\varphi_{11}}\varphi_{11}^{m-1}\frac{\partial}{\partial\varphi_{11}} + \varphi_{11}^{-2}\Delta^{\mathfrak{R}}_{p-1,q-1}\right),$$

where $m=p+q-2$, corresponds. When $m=1$, the corresponding operator $\Delta^{\mathfrak{R}}_{p-1,q-1}$ should be omitted. On the other hand, if $m>1$, then the form of the operator $\Delta^{\mathfrak{R}}_{p-1,q-1}$ depends on the mode of reduction to subgroups and is determined by the rules indicated in [3].

For each of the operators (1) and (2), after separation of variables, we obtain two equations:

$$\left[\varphi_{11}^{-m+1}\frac{\partial}{\partial\varphi_{11}}\varphi_{11}^{m-1}\frac{\partial}{\partial\varphi_{11}} - \mathcal{L}_{11}(\mathcal{L}_{11}+m-2)\varphi_{11}^{-2} - \lambda\right]u_{\mathcal{L}_{11},\ell}(\varphi_{11}) = 0,$$

$$\left[e^{-m\varphi_1}\frac{\partial}{\partial\varphi_{11}}e^{m\varphi_1}\frac{\partial}{\partial\varphi_1} - \lambda e^{-2\varphi_1} + \mathcal{L}(\mathcal{L}+m)\right]v_{\mathcal{L}\ell}(\varphi_1) = 0$$

and respectively

$$\left[\varphi_{11}^{-m+1}\frac{\partial}{\partial\varphi_{11}}\varphi_1^{m-1}\frac{\partial}{\partial\varphi_{11}} - \mathcal{L}_{11}(\mathcal{L}_{11}+m-2) - \lambda\right]u_{\mathcal{L}_{11},\ell}(\varphi_{11}) = 0,$$

$$\left[e^{-m\varphi_1}\frac{\partial}{\partial\varphi_1}e^{m\varphi_1}\frac{\partial}{\partial\varphi_1} + \lambda e^{-2\lambda_1} + \mathcal{L}(\mathcal{L}+m)\right]v_{\mathcal{L}\ell}(\varphi_1) = 0,$$

whose solutions can be expressed in terms of Bessel functions and modified Bessel functions.

N.Ya.Vilenkin and G.V.Denisova have considered the behaviour of the harmonics corresponding to trees while approaching manifolds that separate domains of applicability of the given systems of coordinates. These harmonics have equal limits as they approach the separating manifolds from different sides, and hence define continuous functions on the whole hyperboloid; these functions have different analytic expressions in the different domains. The transition from one of these expressions to another one is made by analytic continuation in angular variables.

Furthermore it was shown that the harmonics for complex and quaternion hyperboloids (see [4]) are reduced to harmonics for real hyperboloids for an appropriate choice of trees (e.g., for complex hyperboloids one has to join the vertices in pairs and then construct a tree from the newly obtained vertices).

G.V.Denisova has also considered the overlapping coefficients for basis on cones and hyperboloids, including bases connected with parabolic sections, found expressions for these coefficients in terms of Meyer's G-functions and obtained similar results for $SU(n,1)$. N.Ya.Vilenkin and G.V.Denisova have also considered the overlapping coefficients for bases of representation spaces of discrete series for $SO(n,1)$. With the help of these coefficients, the matrix elements of representations in mixed bases have been computed.

2. N.Ya.Vilenkin and T.G.Kudryashova have studied the irreducible component expansion problem for the restriction to $SO(n)$ of irreducible representations of $SU(n)$. F.Murnaghan [5] indicated an algorithm for finding all the components of this expansion, but he gave neither projection operators nor a method for separation of multiplicities. A number of papers by physicists treat the cace $n=3$. We used the realization of irreducible representation of $SU(n)$ of weight $(p,p,0,\ldots,0)$ by right translations in the space $F_{p,p}$ of p-th degree po-

lynomials of 2nd order minors of the matrix $\begin{pmatrix} \underline{x} \\ \underline{y} \end{pmatrix}$, where $\underline{x} = (x_1, \dots, x_n)$, $\underline{y} = (y_1, \dots, y_n)$. Introduce the following notations:

$$\underline{x}^2 = \sum_{1 \leqslant k \leqslant n} x_k^2, \ \underline{x}\underline{y} = \sum_{1 \leqslant k \leqslant n} x_k y_k, \ J = \underline{x}^2 \underline{y}^2 - (\underline{x}\underline{y})^2,$$

$$\Delta_{\underline{x}^2} = \sum_{1 \leqslant k \leqslant n} \frac{\partial^2}{\partial x_k^2}, \ \Delta_{\underline{x}\underline{y}} = \sum_{1 \leqslant k \leqslant n} \frac{\partial^2}{\partial x_k \partial y_k}, \ \delta = \Delta_{\underline{x}^2} \Delta_{y^2} - \Delta_{\underline{x}\underline{y}}^2,$$

$$R_1 = \underline{x}^2 \Delta_{\underline{x}^2} + 2\underline{x}\underline{y}\, \Delta_{\underline{x}\underline{y}} + \underline{y}^2 \Delta_{y^2}, \ \mathcal{L}_{\underline{x}} = \sum_{k=1}^{n} x_k \frac{\partial}{\partial x_k},$$

$$\mathfrak{R}_{\underline{x}\underline{y}} = R_1 - \mathcal{L}_{\underline{x}}(\mathcal{L}_{\underline{x}} + n - 2) - \mathcal{L}_{\underline{y}}(\mathcal{L}_{\underline{y}} + n - 4), \ \vartheta = J\delta - \mathcal{L}_x(\mathcal{L}_x + n - 2)\mathfrak{R}_{\underline{x}\underline{y}}$$

$$k_m = 2m(n + 2p - 2m - 4), \ \sigma_p = p(n + p - 2), \ \beta_m = 2(m + 2p)(2mp + 2p^2 + 3m - 3).$$

Then $F_{p,p} = \sum_{0 \leqslant \ell \leqslant [p/2]} F^{\ell}_{p,p}$, where

$$F^{\ell}_{p,p} = \prod_{0 \leqslant i \leqslant [p/2],\, i \neq \ell} \Big(\vartheta + \sigma_{p-2i} \mathfrak{R}_{\underline{x}\underline{y}} - \sum_{1 \leqslant j \leqslant [p/2]} \beta_{p-2j}\Big) F_{p,p}.$$

Note that $F^{\ell}_{p,p} = y^{\ell} H^{\delta}_{p-2\ell,\, p-2\ell}$, where H^{δ}_{ss} is the subspace of all $f_{ss} \in F_{ss}$ such that $\delta f_{ss} = 0$. The subspaces $F^{\ell}_{p,p}$ are invariant under right translations, therefore some representations of $SU(n)$ are realized in these subspaces. It turns out that their restrictions to $SO(n)$ contain at most one irreducible component. Subspaces which support these components are defined by the formula

$$H^{i\delta}_{p,p} = \prod_{0 \leqslant j \leqslant [p/2],\, j \neq i} \frac{R_1 - k_j}{k_i - k_j} H^{\delta}_{p,p}.$$

The representation of $SO(n)$ with weight $(p, p-2i, 0, \ldots, 0)$ is realized in $H^{i\delta}_{p,p}$.

T.G.Kudryashova obtained the following expression for the 4th order Laplace operator Γ_4 restricted to f_{pp}:

$$\Gamma_4 f_{pp} = \left\{\frac{1}{2}\Gamma_2^2 + \left(\frac{1}{2}(n-2)(n-3) - 4(p+1)(n+p-1) + 3(p+1)\right]\Gamma_2 - \right.$$

$$\left. -4J\delta + 4p(p+1)(n+p-1)(n+p-4)\right\} f_{p,p},$$

where Γ_2 is the 2nd order Laplacian. She has also found an expression for spherical functions on the Stiefel manifold of 2-frames and deduced differential equations for them; she obtained as well certain results generalizing [7]. Some of these results are generalized to representations with weight $(p, p, \ldots, p, 0, \ldots, 0)$.

3. N.Ya.Vilenkin [6] introduced the functions $P^{nl}_{km_j}(\cos\theta)$ using which the matrix elements of irreducible representations of class 1 for $SO(n)$ may be expressed. A.A.Rudnev obtained relations involving these functions and Hahn polynomials $h^{(\alpha,\beta)}_n(x, N)$ (see [9]). He proved that

$$P^{nl}_{km_j}(\cos\theta) = i^{k-m} d^{ml}_{m_j} d^{nl}_{k_j} \sum_{0\leq j\leq l-j} \varrho^l_j(s) h^{p,p}_{m-j}(s, l-s+1) h^{p,p}_{k-j}(s, l-j+1), \tag{3}$$

where $p = j + (n-4)/2$, $\varrho^l_j(s)$ is the weight, and $d^{nl}_{k_j}$ and $d^{nl}_{m_j}$ are the norms of the Hahn polynomials $h^{p,p}_{k_j}$ and $h^{p,p}_{m_j}$ respectively:

$$\varrho^l_j(s) = \frac{\Gamma(s+p+1)\Gamma(l-s+(n-2)/2)}{s!\,(l-j-s)!},$$

$$\left(d^{n\ell}_{mj}\right)^2 = \frac{[(\ell-j)!]^2 (m+j+n-2)!\,(2m+p-3)}{\Gamma(p+1)(\ell-m)!\,(m-j)!\,(\ell+m+n-3)!}.$$

From (3) one gets an expansion of $P^{n\ell}_{kmj}$ in $\sin k\theta$ and $\cos k\theta$

A similar formula is deduced for the functions $\mathfrak{P}^{n\sigma}_{kmj}$ (see [6]). The matrix elements of irreducible representations of class 1 for $SO(n-1,1)$ are expressed by means of these functions:

$$\mathfrak{P}^{n\sigma}_{kmj}(\cosh\theta) = C^{n}_{kmj}\,\frac{e^{(j-\sigma)\theta}}{\Gamma(m-\sigma)\Gamma(\sigma+k+2q)}\times$$

$$\times\int_{a-i\infty}^{a+i\infty} e^{-2\theta t}\,\Gamma(-\sigma-t+j)\Gamma(\sigma+q+t)\Gamma(t+m-j)\Gamma(k+q-t)\,\times$$

$$\times\,{}_3F_2\left(\begin{matrix}-m+j,\,-m-q+1,\,-\sigma-t+j\\ j+q,\,-t-m+j+1\end{matrix}\,\middle|\,1\right){}_3F_2\left(\begin{matrix}-k+j,\,-k-q+1,\,\sigma+q+t\\ j+q,\,t-k-q+1\end{matrix}\,\middle|\,1\right)dt,$$

where $q=(n-2)/2$ and

$$C^{n}_{kmj} = \frac{\left[\Gamma(m+j+2q-1)(2m+2q-1)\Gamma(k+j+2q-1)(2k+2q-1)\right]^{1/2}}{2\pi i\,\Gamma^2(j+q)\left[\Gamma(m-j+1)\Gamma(k-j+1)\right]^{1/2}}.$$

Using residues, this yields an expansion of $\mathfrak{P}^{n\sigma}_{kmj}(\cosh\theta)$ in powers of e^{θ}. The obtained formulas lead to various recurrent relations for Hahn polynomials.

N.Ya.Vilenkin and A.A.Rudnev using Miller's results ([8]sec; 4.15) found an expression for Racah's coefficients related to the group S_4 of matrices of the form

$$\begin{pmatrix} 1 & \frac{1}{2} e^{-i\alpha} \bar{w} & i\delta - \frac{w\bar{w}}{8} \\ 0 & e^{-i\alpha} & -\frac{w}{2} \\ 0 & 0 & 1 \end{pmatrix}$$

where $w \in \mathbb{C}$, $0 \leq \alpha < 2\pi$, and δ is real. It is proved that

$$\begin{Bmatrix} (\lambda_1, \ell_1) & (\lambda_2, \ell_2) & (\lambda_{12}-n, \ell_{12}) \\ (\lambda_3, \ell_3) & (\lambda-n-p, \ell) & (\lambda_{23}-j, \ell_{23}) \end{Bmatrix} =$$

$$= (n!p!(n+p-j)!)^{1/2} \left(\frac{(\ell_1 \ell_3)^{p-j} (\ell_2 \ell)^{j-n}}{\ell_{12}^{n+p} \ell_{23}^{n+p}} \right)^{1/2} (-\ell_1 \ell_3)^n \times$$

$$\times \sum_{\alpha} \frac{1}{\alpha!(n-\alpha)!(n+p-j-\alpha)!(j-n+\alpha)!} \left(-\frac{\ell_2 \ell}{\ell_1 \ell_3} \right)^{\alpha}.$$

Using the expressions of the Clebsch-Gordon coefficients for S_4 presented in [8], certain relations for hypergeometric functions are obtained; these can also be expressed as addition theorems for Kravchuk polynomials.

References

1. Vilenkin, N.Ya. (1965). Special functions and the theory of representations of groups. Nauka, Moscow (Russian)

2. Vilenkin, N.Ya. (1965). Matem;Sbornik **68**:3, p.432. (Russian)

3. Smirnov, Yu.F., Shustov, A.P. (1981). Nucl. Phys. 34, p.626.

4. Faraut , J.J. (1979). Math. pures et appl. 58, p.**369**.

5. Murnaghan, F.D. (1938). The theory of group representations.

6. Vilenkin, N.Ya. (1963). Trans. Mosc. Math. Soc. 12, p.185 (Russuian).

7. Bargmann, V., Todorov, I.T. (1977). J. of Math. Phys. 18, p.1141.

8. Miller, W.Jr. (1968). Lie Theory and Special Fucntions. Academic Press

9. Nikiforov, A.F., Uvarov, V.B. (1984). The special functions of the mathematical physics, Nauka, Moscow (Russian)

CLASSICAL ORTHOGONAL POLYNOMIALS OF A DISCRETE VARIABLE ON NONUNIFORM LATTICES

A.F.Nikiforov, S.K.Suslov [+]

Keldysh Institute of Applied Mathematics, Moscow 125047

[+]Kurchatov Institute of Atomic Energy, Moscow 123182

Special functions of mathematical physics usually arise as solutions of second-order differential equations. These solutions have been studied thoroughly, there are detailed tables and efficient computational algorithms for many of them. Up to now there was no need in replacing the differential equations with the difference ones to find approximate values of special functions. But recently it has become clear [1-5,8] that particular solutions of difference equations may be easily obtained in the explicit form and their properties can be studied by generalizing the method earlier used in [6,7] for original differential equations. It has been proved that some solutions obtained in this way are of interest in itself and have been for a long time used implicitly in quantum mechanics, theory of group representations and computational physics. Examples are the Clebsch-Gordan and Racah coefficients widely used in atomic and nuclear spectroscopy.

This paper is concerned with classical orthogonal polynomials of discrete variable that are the difference analogues of the Jacobi, Laguerre and Hermite polynomials on uniform and nonuniform lattices. Though the study of classical orthogonal polynomials of a discrete variable began in the last century [9] it is not yet clear that orthogonal polynomials introduced by different authors belong to this class of special functions.

The classical (Jacobi, Laguerre and Hermite) orthogonal polynomials are the simplest solutions of the differential equations of hypergeometric type

$$\sigma(x)y'' + \tau(x)y' + \lambda y = 0, \qquad (1)$$

where $\sigma(x)$ and $\tau(x)$ are polynomials of degree $\leqslant 2$ or $\leqslant 1$ respectively and λ a constant.

Let us consider the difference equation on a lattice $x = x(s)$ with nonuniform step $\Delta x(s) = x(s+h)-x(s)$:

$$\frac{\sigma[x(s)]}{x(s+h/2)-x(s-h/2)}\left[\frac{y(s+h)-y(s)}{x(s+h)-x(s)} - \frac{y(s)-y(s-h)}{x(s)-x(s-h)}\right] + \qquad (2)$$

$$+ \frac{\tau[x(s)]}{2}\left[\frac{y(s+h)-y(s)}{x(s+h)-x(s)} + \frac{y(s)-y(s-h)}{x(s)-x(s-h)}\right] + \lambda y(s) = 0.$$

Note that all the fundamentals of the theory of classical orthogonal polynomials remain valid if (1) is replaced by difference equation (2) on certain classes of uniform and nonuniform lattices.

1. After the change of variable $s \to hs$ (2) becomes an equation of the same form with step $h = 1$ in s:

$$\sigma[x(s)]\frac{\Delta}{\Delta x(s-1/2)}\left[\frac{\nabla y(s)}{\nabla x(s)}\right] + \frac{\tau[x(s)]}{2}\left[\frac{\Delta y(s)}{\Delta x(s)} + \frac{\nabla y(s)}{\nabla x(s)}\right] + \lambda y(s) = 0, \qquad (3)$$

where $\Delta y(s) = y(s+1) - y(s)$, $\nabla y(s) = y(s-1)$. Some classes of nonuniform lattices $x(s)$ have the fundamental property, i.e. the difference differentiation of the original equation (3) on nonuniform lattice gives an equation of the same type.

Theorem 1. Let $y(s)$ be a solution of (3), $x(s)$ satisfy the equation

$$\frac{1}{2}\left[x(s+1) + x(s)\right] = \alpha x\left(s + \frac{1}{2}\right) + \beta \qquad (4)$$

(α and β are constants) and $x^2(s+1)+x^2(s)$ is a polynomial of degree $\leqslant 2$ in $x(s+1/2)$. Then the functions

$$v_k(s)=\frac{\Delta v_{k-1}(s)}{\Delta x_{k-1}(s)},\quad v_0(s)=y(s),\quad x_k(s)=x(s+k/2)$$

($k = 1,2,\ldots$) satisfy equations of type (3):

$$\sigma_k[x_k(s)]\frac{\Delta}{\Delta x_k(s-1/2)}\left[\frac{\nabla v_k(s)}{\nabla x_k(s)}\right]+ \tag{5}$$

$$+\frac{\tau_k[x_k(s)]}{2}\left[\frac{\Delta v_k(s)}{\Delta x_k(s)}+\frac{\nabla v_k(s)}{\nabla x_k(s)}\right]+\mu_k v_k(s)=0,$$

where $\sigma_k(x_k)$ and $\tau_k(x_k)$ are polynomials of degree $\leqslant 2$ and $\leqslant 1$ in x_k respectively, and μ_k are constants. Besides

$$\sigma_k[x_k(s)]-\tfrac{1}{2}\tau_k[x_k(s)]\Delta x_k(s-\tfrac{1}{2})= \tag{6}$$

$$=\sigma[x(s)]-\tfrac{1}{2}\tau[x(s)]\Delta x(s-\tfrac{1}{2})\equiv\bar{\sigma}(s).$$

The theorem holds for the following functions

$$x(s)=\begin{cases} C_1 q_1^{2s}+C_2 q_2^{2s}+\beta/(1-\alpha), & \alpha\neq\pm 1 \qquad (7)\\ 4\beta s^2+C_1 s+C_2, & \alpha=1 \qquad (8)\end{cases}$$

where q_1 and q_2 are the roots of the equation $q^2 - -2\alpha q + 1 = 0$, C_1 and C_2 are constants. The case $\alpha = = -1$ is usually not interesting in practice.

Theorem 2. Let the conditions of Theorem 1 be satisfied and let there be a constant $\lambda = \lambda_n$ in (3) such that $\mu_k = \mu_k(\lambda)$ in (5) do not vanish for $k = 0,1,2,\ldots$

..., n-1, while $\mu_n(\lambda) = 0$. Then equation (3) has a particular solution in the form $y(s) = y_n[x(s)]$, where $y_n(x)$ is a polynomial of degree n in x.

$$\lambda_n = -\omega(n)\varkappa(n-1)\tau'(x) - \frac{1}{2}\omega(n)\omega(n-1)\sigma''(x)$$

for lattices (7), where $\omega(n) = (q_1^n - q_2^n)/(q_1 - q_2)$, $\varkappa(n) = (q_1^n + q_2^n)/2$, and

$$\lambda_n = -n\tau'(x) - \frac{1}{2}n(n-1)\sigma''(x)$$

for lattices (8).

2. The polynomial solutions of equation (3), $y(s) = y_n(x)$, $x = x(s)$, can be defined by the Rodrigues type formula

$$y_n(x) = \frac{B_n}{\rho(s)} D^{(n)}[\rho_n(s)], \tag{9}$$

where B_n is a constant, $\rho(s)$ is a solution of the equation

$$\frac{\Delta}{\Delta x(s-1/2)}\left[\bar{\sigma}(s)\rho(s)\right] = \bar{\tau}(s)\rho(s), \quad \bar{\tau}(s) = \tau[x(s)];$$

$$\rho_n(s) = \rho(s+n)\prod_{k=1}^{n}\bar{\sigma}(s+k); \tag{10}$$

$D^{(n)}$ is a divided difference operator:

$$D^{(n)} = \left(\frac{\nabla}{\nabla x_1}\right)\left(\frac{\nabla}{\nabla x_2}\right)\cdots\left(\frac{\nabla}{\nabla x_n}\right), \qquad x_k = x(s+k/2).$$

3. If $\bar{\sigma}(s)\rho(s)x^k(s-1/2)\big|_{s=a,b} = 0$ $(k = 0,1,\ldots)$, the polynomial solutions $y_m(x)$ and $y_n(x)$ of equation (3) are orthogonal for $\lambda_m \neq \lambda_n$:

$$\sum_{s_i=a}^{b-1} y_m(x_i)\, y_n(x_i)\, \rho_i\, \Delta x_{i-1/2} = 0 \tag{11}$$

Here $x_i = x(s_i)$, $\rho_i = \rho(s_i)$, $\Delta x_{i-1/2} = \Delta x(s_i - 1/2)$. Solutions (9) for which (11) is satisfied with constant sign of the weight function ρ_i and monotonous function x_i on $[a, b-1]$ will be called classical orthogonal polynomials of discrete variable on nonuniform lattices.

With conditions $\int_C \Delta\left[\bar{\sigma}(s)\rho(s)x^k(s-1/2)\right] ds = 0$ $(k = 0, 1, \dots)$ on a contour C in the complex plane the polynomials (9) satisfy the orthogonality relation

$$\int_C y_m[x(s)]\, y_n[x(s)]\, \rho(s)\, \Delta x(s-1/2)\, ds = 0 \tag{12}$$

4. These properties allow us to study classical orthogonal polynomials of discrete variable using the simplest mathematical technique. Let the function $x = x(s)$ be real-valued. With the help of the linear transformations $x(s) \to C_1 x(s) + C_2$, $s \to s + C_3$ which do not change the type of the equation (3) the lattices $x(s)$ can be reduced to the following canonical forms

$$x(s) = \begin{cases} s & (\alpha = 1,\ \beta = 0), \\ s(s+1) & (\alpha = 1,\ \beta = 1/4), \\ e^{2\gamma s} & \\ \operatorname{ch} 2\gamma s,\ \operatorname{sh} 2\gamma s & (\alpha > 1,\ \alpha = \operatorname{ch}\gamma,\ \beta = 0), \\ \cos 2\gamma s & (0 < \alpha < 1,\ \alpha = \cos\gamma,\ \beta = 0). \end{cases}$$

Basic systems of orthogonal polynomials on these lattices are considered in [4,8]. Some of these polynomials had been previously introduced from other view point [9-25]. Using (12) and the analytic continuation one can include some other important families [25-28] in the general theory of classical orthogonal polynomials of a discrete variable.

Classical orthogonal polynomials of a discrete variable arise in the representation theory of the three-dimensional rotation group [29,30,8], Lorentz group [27] and Chevalley groups [31], in the nonrelativistic Coulomb problem [32], while solving multidimensional Laplace equation [33], in combinatorics and coding theory [19,34,35], in the theory of partitions [20] and in some other cases.

References

1. Nikiforov,A.F., Suslov,S.K., Uvarov,V.B. (1982). Preprint 165, Keldysh Inst. Appl. Math. (Russian).

2. Nikiforov,A.F., Uvarov,V.B. (1983). Preprint 17, Keldysh Inst. Appl. Math. (Russian).

3. Nikiforov,A.F., Suslov,S.K., Uvarov,V.B. (1984). Preprint 142, Keldysh Inst. Appl. Math. (Russian).

4. Nikiforov,A.F., Suslov,S.K. (1985). Preprint 8, Keldysh Inst. Appl. Math. (Russian).

5. Nikiforov,A.F., Suslov,S.K., Uvarov,V.B. (1985). Funct. Anal. Appl., 19(3), 22 (Russian).

6. Nikiforov,A., Ouvarov,V. (1983). Fonctions Speciales de la Physique Mathematique, Mir, Moscow.

7. Nikiforov,A.F., Uvarov,V.B. (1984). Special Functions of Mathematical Physics, 2nd ed., Nauka, Moscow (Russian).

8. Nikiforov,A.F., Suslov,S.K., Uvarov,V.B. (1985). Classical Orthogonal Polynomials of a Discrete Variable, Nauka, Moscow (Russian).

9. Tchebychef,P.L. (19). Oeuvres, Chelsea, New York, t. 1, p. 203, p. 381; t.II, p. 219.

10. Markoff,A. (1884). On Some Applications of Algebraic Continued Fractions, Thesis, St.Petersburg, p. 131 (Russian).

11. Hahn,W. (1949). Math. Nachr., 2, 4.

12. Szegö,G. (1975). Orthogonal Polynomials, Amer. Math. Soc. Colloq. Publ. 23, Amer. Math. Soc. Providence, R I.

13. Erdelyi,A. et al. (1952). Higher Transcendental Func-

tions, vol. 2, McGraw Hill, New York; (1981), reprinted Krieger, Malabar, Florida.

14. Weber,M., Erdelyi,A. (1952). Amer. Math. Month. 59, 163.
15. Karlin,S., McGregor,J.L. (1961). Scripta Math. 26, 33.
16. Lesky,P. (1961). Monatsh. Math. 65, 1.
17. Lambina,E.N. (1970). Dokl. Acad. Nauk Bel. SSR 14, 786(Russian).
18. Rozenblum,A.V. (1975). Dokl. Acad. Nauk Bel. SSR 19, 297 (Russian).
19. Delsarte,Ph. (1976). SIAM J. Appl. Math. 31, 262; (1978), 34, 157.
20. Andrews,G., Askey,R. (1977). In: Higher Combinatorics, M.Aigner and D.Reidel eds, Dordrecht, Boston,p. 3.
21. Dunkl,C.F. (1978). Monat. Math. 85, 5.
22. Chihara,T.S. (1978). An Introduction to Orthogonal Polynomials, Gordon and Breach, N.Y.
23. Askey,R., Wilson,J.A. (1979). SIAM J. Math. Anal. 10, 1008.
24. Stanton,D. (1980). Amer. J. Math. 102, 625.
25. Wilson,J.A. (1980). SIAM J. Math. Anal. 11, 690.
26. Askey,R., Wilson,J.A. (1982). SIAM J. Math. Anal. 13, 651.
27. Atakishiyev,N.M., Suslov,S.K. (1985). J. Phys. A: Math. Gen. 18, 1583.
28. Askey,R., Wilson,J.A. (1985). Mem. Amer. Math. Soc. 319.
29. Koornwinder,T.H. (1981). Nieuw Arch. voor Wisk.(3), 29, 140; (1982), SIAM J. Math. Anal. 13, 1011.
30. Smorodinsky,Ya.A., Suslov,S.K. (1982). Sov. J. Nucl. Phys. 35, 192; (1982), 36, 1066.
31. Stanton,D. (1984).. In: Special functions: Group Theoretical Aspects and Applications, R.Askey et al. eds, D.Reidel, Dordrecht et al., p. 87.
32. Suslov,S.K. (1984). Sov. J. Nucl. Phys. 40, 126.
33. Suslov,S.K. (1983). Sov. J. Nucl. Phys. 38, 1367.
34. Delsarte,Ph., Goethals,J.M. (1975). J. Comb. Th. A19, 26.
35. Delsarte,Ph. (1976). J. Comb.Th. A20, 230; (1978), A25, 226.

REPRESENTATIONS OF THE ROTATION GROUP AND DISCRETE ORTHOGONAL POLYNOMIALS

A.V.Rozenblyum, L.V.Rozenblyum

Grodno State University

1. Recently a considerable interest has become apparent for special functions of discrete variables connected with a number of quantities important in the quantum field theory, in the atomic and nuclear spectroscopy such as Clebsch-Gordan coefficients, Racah coefficients, matrix elements (m. e.) of the representations of groups, etc. [1-4] . Some general methods of investigating discrete properties of m e from different points of view are given in this paper with SO(3) as an example.

The description of m e irreducible unitary representations (IUR) of SO(3) in terms of Jacobi [5] and Krawtchouk polynomials [4] is known. Moreover, expressing these polynomials via the hypergeometric function one can connect them with Meixner polynomials [6] . We also consider the connection between the m e and Krawtchouk and Meixner polynomials and also discrete analogues of Bessel functions. However, other methods are used in this case. These methods are based on obtaining some finite-difference equations for m.e. by means of group theoretical considerations. The suggested methods may be extended to the rotation groups of greater dimensions and other Lie groups. Here less investigated orthogonal polynomials of many discrete variables appear.

As is known [5] , in the space of IUR T of SO(3) with the highest weight l one can select a canonical basis $\{f_m\}$, $m = -l,\ldots,l$, in which the matrix elements of T have a form (in Euler angles)

$$T^{l}_{mn}(\varphi_1, \theta, \varphi_2) = \exp[-i(m\varphi_2 + n\varphi_1)]\, P^{l}_{mn}(\cos\theta);$$

$$m,n = -l,\dots,l$$

where $P^{l}_{mn}(\cos\theta)$ is a generalized spherical function. For convenience of describing the dependence of m e on the parameters under investigation redenote some indices. Let

$$\mu = \cos\theta;\ r = l-m;\ s = l-n;\ t = 2l;\ P^{l}_{mn}(\cos\theta) =$$

$$= M(\mu; r; s; t).$$

Subsequently under m e of IUR of SO(3) group we mean the induced complex-valued function M depending on four arguments: continuous μ and discrete (integer) r, s,t such that $-1 \leq \mu \leq 1$; $0 \leq r,s \leq t < \infty$.

Further denote $M = M(x)$, $x = \mu,r,s,t$, i.e. we consider m e $M(\mu; r,s;t)$ as a function of only one variable x, assuming the other arguments to be fixed parameters. Sometimes we use the latter as indices. Then $M = M(\mu)$ as is konown [5], is reduced to Jacobi polynomial. We show that functions $M = M(r)$ and $M = M(s)$ are reduced to Krawtchouk polynomials, and $M = M(t)$ to Meixner polinomials. Jacobi polynomials is a fixed matrix element of an IUR; Krawtchouk polynomial is a row or a column of a IUR; and Meixner polynomial is generated from m e belonging to inequivalent IUR's. Furthermore, it is shown here that $P^{l}_{mn}(\cos\theta)$, as a function of l with fixed m,n and θ, may be treated as a finite-difference analogue of Bessel functions.

2. First consider the dependence of m e $M = M(r)$ on parameter r. Let us describe the general scheme of our method. Let T be a representation of N-dimensional linear Lie group G, $\{a_{mn}(g)\}$ the matrix of adjoint representation of G in a basis $a_1,\dots,a_N$ of a Lie

algebra X, $g \in G$. Then $g a_m g^{-1} = \sum_{n=1}^{N} a_{mn}(g) a_n$. Passing to the operators of the representation T we get

$$\sum_{n=1}^{N} a_{mn}(g) A^n T(g) - T(g) A^m = 0 \qquad (1)$$

where $A^m = D(a_m)$, D is the differential of T; $m,n = 1,\ldots,N$. Further, let there be some basis in the representation space. Let matrices $\{A^n_{ij}\}$ of infinitesimal operators A^n be known in this basis, me $T_{ij}(g)$ to be found in the same basis. Let

$$\tilde{A}^m(g) = \sum_{n=1}^{N} a_{mn}(g) A^n \quad , \qquad (2)$$

and $\tilde{A}^m_{ij}(g)$ denote its me in the chosen basis. Then rewriting the identity (1) in these elements we get

$$\sum_k \left[\tilde{A}^m_{ik}(g) T_{kj}(g) - A^m_{kj} T_{ik}(g)\right] = 0; \quad m = 1,\ldots, N \quad (3)$$

Eq. (3) gives a set of recurrences between me $T_{ij}(g)$ that can be used for different expressions of me in finite differences. In particular, if A^m is diagonal with eigenvalues λ_j, then Eq. (3) has the form

$$\sum_k \tilde{A}^m_{ik}(g) T_{kj}(g) = \lambda_j T_{ij}(g). \qquad (4)$$

It follows that the columns of the matrix $\{T_{ij}(g)\}$ may be considered as eigenvectors of the matrix $\tilde{A}^m(g)$. Thus, if in some basis the infinitesimal operator A^m is given by a diagonal matrix, then we get the following method of estimating me $T_{ij}(g)$ in this basis: using formula (2) we generate $\tilde{A}^m(g)$ and make it diagonal by solving Eq. (4), the spectrum $\tilde{A}^m(g)$ coinsiding with the spectrum of the operator A^m. Note that functions $a_{mn}(g)$ giving the decomposition of $\tilde{A}^m(g)$ into basic infinitesimal operators A^n do not depend on the representation T and are only defined by the structural constants of G. In a number of cases Eq. (4) transforms into an equation with finite differences (ordinary or partial) for eigenvalues.

Let us realize the described scheme and illustrate it with SO(3). Take infinitesimal operators A^n corresponding to subgroups $g_n(\varphi)$ of the rotations about the axes Ox_n, n = 1,2,3, as basis. These operators change the canonical basis $\{\xi_r\}$ in the space of IUR with the highest weight $l = t/2$ by the formulas [5]

$$\begin{aligned} A^1 \xi_r &= -\frac{i}{2}\alpha_{r-1}\xi_{r-1} - \frac{i}{2}\alpha_r \xi_{r+1} \\ A^2 \xi_r &= -\frac{1}{2}\alpha_{r-1}\xi_{r-1} + \frac{1}{2}\alpha_r \xi_{r+1} \\ A^3 \xi_r &= (r - \frac{t}{2})\xi_r \ ; \ 0 \le r \le t \ ; \quad \alpha_r = \sqrt{(r+1)(t-r)} \end{aligned} \tag{5}$$

Since A^3 is diagonal in the canonical basis, set m=3 and estimate the obtained operator $\tilde{A}^3(g)$ using formula (2). To find $a_{3n}(g)$, that operators Ad(g) change the basis elements a_n, n = 1,2,3, just as operators $g \in SO(3)$ change the Descartes basis in the standard representation. Hence, assuming $g = g_1(\theta)$ we get $a_{31}(g) \equiv 0$; $a_{32}(g) = -\sin\theta$; $a_{33}(g) = \cos\theta$. Thus,

$$A^3[g_1(\theta)] = -\sin\theta\, A^2 + \cos\theta\, A^3 \tag{6}$$

Considering eigenvectors of $\tilde{A}^3$ as functions of the discrete variable numbering its components we get from (4)-(6) an equation for eigenvalues

$$\sqrt{p(1-p)}\,\alpha_{r-1}M_s(r-1) + i(2p-1)(\frac{t}{2} - r)M_s(r) - \sqrt{p(1-p)}\,\alpha_r M_s(r+1) = \lambda_s M_s(r) \tag{7}$$

where $p = \frac{1+\mu}{2} = \cos^2\frac{\theta}{2}$; $\lambda_s = i(\frac{t}{2} - s)$. Substituting

$$M_s(r) = i^r \left[\frac{p^r(1-p)^{t-r}}{r!\,(t-r)!}\right]^{1/2} \mathcal{K}_s(r) ,$$

in (7) we get

$$p(t-r)\mathcal{K}_s(r+1) + (1-p)r\mathcal{K}_s(r-1) + [r(2p-1) - tp+ + t-s]\mathcal{K}_s(r) = 0 .$$

The latter can be written in terms of finite differences

$$r\,\Delta\nabla\mathcal{K}_s(r) + \frac{tp-r}{1-p}\Delta\mathcal{K}_s(r) + \frac{t-s}{1-p}\mathcal{K}_s(r) = 0$$

or in a self-adjoint form

$$\Delta\,[C_t^r p^r (1-p)^{t-r}\,\nabla\mathcal{K}_s(r)] + C_t^r p^r (1-p)^{t-r-1}(t- - s)\mathcal{K}_s(r) = 0 \qquad (8)$$

where $\Delta\mathcal{K}(r) = \mathcal{K}(r+1) - \mathcal{K}(r)$, $\nabla\mathcal{K}(r) = \Delta\mathcal{K}(r-1)$. Eq. (8) describes Krawtchouk polynomials $\mathcal{K}_n^{t,p}(r)$ of degree $n = t-s$. The latter are orthogonal in the interval $[0;t]$ with weight $C_t^r p^r (1-p)^{t-r}$, the square of the norm of standartized polynomials $\mathcal{K}_n^{t,p}(r)$ is $C_t^n p^n (1-p)^n$ [3] . Thus, it follows that $\mathcal{K}_s(r)$ coincides with the Krawtchouk polynomial $\mathcal{K}_{t-s}^{t,p}(r)$ up to a constant (independent of r) factor L. To find L (up to a phase), make use of unitarinty of T , i.e. the orthonorming of $M_s(r)$. Estimating, we get

$$M_s^{t,\mu}(r) = i^{r+s}\left[C_t^r (C_t^s)^{-1} p^{r+s-t}(1-p)^{s-r}\right]^{1/2}\mathcal{K}_{t-s}^{t,p}(r), \quad p = \frac{1+\mu}{2} \qquad (9)$$

The relation (9) leads to a connection between the generalized spherical functions and Krawtchouk polynomials:

$$P^{l}_{mn}(\cos\theta) = i^{2l+m+n}\left[\frac{(l+n)!\,(l-n)!}{(l+m)!\,(l-m)!}\, p^{n+m}(1 - p)^{n-m}\right]^{1/2} \mathcal{K}^{2l,p}_{l-n}(l+m),$$

where $p = \cos^2\frac{\theta}{2}$.

3. Determine now the connections between me of IUR of SO(3) and other classes of functions in a discrete variable. For this it is necessary to consider me belonging to inequivalent representations in total. A general method of jointly investigating such me be tensor products. Considering SO(3) it is convenient to use the tensor product of an IUR T^{l} with an arbitrary highest weight l by an IUR $T^{l'}$ with the highest weight $l' = 1/2$ or $l' = 1$. Here consider the case $l' = 1/2$ that will lead to expressing me $M(t) = M^{r,\mu}_{s}(t)$ via Meixner polynomials. Expanding the considered tensor product into two irreducible: $T^{l-1/2}$ and $T^{l+1/2}$, we get

$$\left\{\begin{aligned} M^{r,\mu}_{s}(t) &= \frac{1}{t+1}\left[\sqrt{\frac{(r+1)(s+1)}{p}}\, M^{r+1,\mu}_{s+1}(t+1) + \sqrt{\frac{(t-r)(t-s)}{p}}\, M^{r,\mu}_{s}(t-1)\right], \\ M^{r+1,\mu}_{s}(t) &= \frac{1}{t+1}\left[\sqrt{\frac{(t-r)(s+1)}{p-1}}\, M^{r+1,\mu}_{s+1}(t+1) - \sqrt{\frac{(r+1)(t-s)}{p-1}}\, M^{r,\mu}_{s}(t-1)\right]. \end{aligned}\right. \tag{10}$$

where $p = (1+\mu)/2$. Excluding the function $M^{r+1,\mu}_{s+1}(t+1)$ we get

$$\sqrt{(1-p)(r+1)}\, M^{r+1,\mu}_{s}(t) = \sqrt{s-t}\, M^{r,\mu}_{s}(t-1) - \sqrt{p(r-t)}\, M^{r,\mu}_{s}(t) \tag{11}$$

Making use of the symmetry relation $M_s^{r,\mu}(t) = M_{t-s}^{t-r,\mu}(t)$, we have

$$\sqrt{(1-p)r}\, M_s^{r-1,\mu}(t) = \sqrt{p(r-t-1)}\, M_s^{r,\mu}(t) - \\ - \sqrt{s-t-1}\, M_s^{r,\mu}(t+1) \qquad (12)$$

Substituting the expressions for $M_s^{r+1,\mu}(t)$ and $M_s^{r-1,\mu}(t)$ from (11) and (12) into (7) we get

$$\sqrt{p(t-r+1)(t-s+1)}\, M_s(t+1) + \sqrt{p(t-r)(t-s)}\, M_s(t-1) + \\ + (r+s-p-t-pt)\, M_s(t) = 0$$

Substituting here

$$M_s^{r,\mu}(t) = \left[p^t \frac{(t-s)!}{(t-r)!}\right]^{1/2} L_s(t)$$

We see that $L_s(t)$ satisfies

$$p(t-s+1)L_s(t+1) + (t-r)L_s(t-1) + (r+s-p-t-pt)\, L_s(t) = 0$$

The latter can be written in terms of finite differences

$$\Delta \left[p^t \frac{(t-s)!}{(t-r-1)!} \nabla L_s(t)\right] + s(1-p)p^t \frac{(t-s)!}{(t-r)} L_s(t) = 0 \qquad (13)$$

By the symmetry of me in r and s we may assume without loss of generality that $r \geq s$. If we set $z=t-r$; $\gamma = r-s+1$; $L_s(t) = \tilde{m}_s^{\gamma,p}(z)$, then (13) transforms into an equation for Meixner polynomials $\tilde{m}_s^{\gamma,p}(z)$ of degree s of discrete variable z .These polynomials are orthogonal on $\mathbb{R}_+$ with weight $p^z \Gamma(\gamma+z)\Gamma^{-1}(z+1)$. One comes to the expression

$$M_s^{r,\mu}(t) = i^{s-r} \left[\frac{(t-s)!\, p^{t-r+s}(1-p)^{r-s}}{s!\, r!\, (t-r)!}\right]^{1/2} \times \\ \times \tilde{m}_s^{r-s+1,p}(t-r) \; ; \quad r \geq s .$$

Hence the connections between generalized spherical functions and Meixner polynomials follow

$$P^{l}_{mn}(\cos\theta) = i^{m-n}\left[\frac{(l+n)!\ p^{2l+m-n}(1-p)^{n-m}}{(l-n)!\ (l+m)!\ (l-m)!}\right]^{1/2}\cdot$$

$$\cdot\ \tilde{m}^{n-m+1,p}_{l-n}(l+m);\quad p = \cos^2\frac{\theta}{2}\ .$$

4. Now let us consider the tensor product of IUR of SO(3) with the highest weights l and 1. That gives a possibility to describe the dependence of $P^{l}_{mn}(\cos\theta)$ on the parameter l with fixed m,n and θ. Expanding the mentioned tensor product into irreducible ones we obtain the recurrence [5] :

$$\frac{[(l+m+1)(l-m+1)(l+n+1)(l-n+1)]^{1/2}}{(l+1)(2l+1)} P^{l+1}_{mn}(\mu) +$$

$$+\frac{[(l+m)(l-m)(l+n)(l-n)]^{1/2}}{l(2l+1)} P^{l-1}_{mn}(\mu) + \frac{mn}{l(l+1)} P^{l}_{mn}(\mu) =$$

$$= \mu P^{l}_{mn}(\mu) \qquad (14)$$

Substituting

$$P^{l}_{mn}(\cos\theta) = i^{2l}\left[\frac{(l+m)!(l+n)}{(l-m)!(l-n)!}\right]^{1/2} R^{\theta}_{mn}(l)$$

to (14) we get a discrete equation for R^{θ}_{mn} :

$$\frac{(l+m+1)(l+n+1)}{(l+1)(2l+1)} R(l+1) + \frac{(l-m)(l-n)}{l(2l+1)} R(l-1) -$$

$$- \frac{mn}{l(l+1)} R(l) = -\mu R(l)$$

The latter can be written in terms of finite differences

$$(l-m)(l-n)\ \Delta\nabla R(l) + l\left[2m+2n+l - \frac{mn}{l(l+1)}\right]\Delta R(l) +$$

$$+ 2l(2l+1)pR(l) = 0 \qquad (15)$$

The solution of this equation is no longer the polynomial

of l. Let us show that under a particular limited transition (15) transforms into some differential equation connected with Bessel functions. For this reduce (15) written with step 1 to an equation with an arbitrary step h. Set

$$l = x/h \ ; \ \theta = \pi + \rho h \ ; \ R^{\theta}_{mn}(l) = I^{h}(x) \quad .$$

Then (15) becomes

$$(x-hm)(x-hn)\,\frac{\Delta_h \nabla_h}{h^2}\,I^h(x) + x\,\Big|2m+2n+1 - \frac{h^2 mn}{x(x+h)}\Big)\frac{\Delta_h}{h} I^h(x) =$$

$$+\ \rho^2 x\left(x+\frac{h}{2}\right)\sin^2\frac{\rho h}{2}\left(\frac{\rho h}{2}\right)^{-2} I^h(x) = 0 \ , \qquad (16)$$

where $\Delta_h I(x) = I(x+h) - I(x)$; $\nabla_h I(x) = \Delta_h I(x-h)$. Now let $h \to 0$. Then (16) transforms into the following differential equation

$$x^2 y'' + x(2m+2n+1)y' + \rho^2 x^2 y = 0 \qquad (17)$$

This transformation of (15) shows that this equation may be treated as discrete analogue of (17). It is not difficult to show that with the abovementioned limited transition the function $R(l)$ transforms into a solution of (17). The general (finite at 0) solution of the latter is

$$y(x) = cx^{-m-n} Y_{m+n}(\rho\, x),$$

where $Y_\nu(x)$ is the order ν Bessel function c an arbitrary constant. Thus, the function $R^{\theta}_{mn}(l)$ satisfying (15) can be treated as a finite-difference analogue of Bessel functions.

Noted that (15) is also connected with IUR of the group M(3) of movements of 3-dimensional Euclidean space.The equation for eigenvectors of an infinitesimal operator of IUR of M(3) (the operator corresponding to a one-parameter subgroup of sheafs along the x_3-axis) is reduced to (15). Also, having made the spectral analysis of boosts generators of M(n) we obtain manydimensional finite-difference analogues of the Bessel functions.

References

1. Smorodinsky,Ya.A., Suslov,S.K. (1982). Sov. J. Nucl. Phys., 35(1), 192-201.
2. Suslov,S.K. (1982). Sov. J. Nucl. Phys., 36(4), 1063-1065.
3. Smorodinsky, Ya.A., Suslov,S.K. (1982). Sov. J. Nucl. Phys., 36(4), 1066-1071.
4. Koornwinder,T.H. (1982). SIAM J. Math. Anal., 13, N 6, 1011-1023.
5. Gel'fand,I.M., Minlos,R.A., Shapiro,Z.Ya. (1958). Representations of the rotation group and of the Lorentz group. Fizmatgiz, Moscow (Russian).
6. Nikiforov,A.F., Uvarov,V.B. (1978). Special functions of mathematical physics, Nauka, Moscow, 320 p. (Russian).

REPRESENTATIONS OF SO(3,1) AND THE HAHN POLYNOMIALS

N.M.Atakishiyev, S.K.Suslov [+]
Physics Institute, Baku 370143
[+] Kurchatov Institut of Atomic Energy, Moscow 123182

In [1] an analytic continuation of the Hahn polynomials, leading to a real system of polynomials orthogonal with respect to a continuous measure is considered. Here we discuss connection between Hahn's polynomials of an imaginary argument thus obtained and unitary irreducible representations (UIR) of the Lorentz group SO(3,1).

1. Hahn's polynomials of an imaginary argument [1]

$$p_n(x) = p_n^{(\alpha,\beta)}(x,\gamma) = i^{-n} h_n^{(\alpha,\beta)}(z,N) \qquad (1)$$

are defined as the analitic continuation of the Hahn polynomials $h_n^{(\alpha,\beta)}(z,N)$ [2] in $z = i(x+\gamma)/2 - (\beta+1)/2$ and parameter $N = i\gamma - (\alpha+\beta)/2$. They satisfy the recurrence with real coefficients:

$$x p_n(x) = \frac{2(n+1)(\alpha+\beta+n+1)}{(\alpha+\beta+2n+1)(\alpha+\beta+2n+2)} p_{n+1}(x) +$$

$$+ \frac{\gamma(\beta^2-\alpha^2)}{(\alpha+\beta+2n)(\alpha+\beta+2n+2)} p_n(x) + \frac{2(\alpha+n)(\beta+n)\left[\gamma^2+\left(n+\frac{\alpha+\beta}{2}\right)^2\right]}{(\alpha+\beta+2n)(\alpha+\beta+2n+1)} p_{n-1}(x) \qquad (2)$$

i.e. for real values of α, β and γ they have real coefficients in the expansion in powers of x. The polynomials $p_n(x)$ are orthogonal on the infinite interval

$$\int_{-\infty}^{\infty} p_n(x)\, p_m(x)\, \rho(x)\, dx = d_n^2\, \delta_{nm} \quad (\alpha>-1, \beta>-1) \qquad (3)$$

with the weight $\rho(x)=(4\pi)^{-1}\left|\Gamma\left(i\frac{x-\gamma}{2}+\frac{\alpha+1}{2}\right)\Gamma\left(i\frac{x+\gamma}{2}+\frac{\beta+1}{2}\right)\right|^2.$

A proof of the orthogonality relation (3) and the valuating d_n^2 are given in [1] .

It is possible to introduce analogously the polynomials [3,1]

$$q_n(x) = q_n^{(\alpha)}(x,\delta) = i^{-n}\, h_n^{(\alpha,\alpha)}(z,N) \qquad (4)$$

where $z = ix/2 - (\alpha-\delta+1)/2$, $N = \delta-\alpha$. They take real values for real α, δ and x. The orthogonality relations for $q_n(x)$ are of the form (3), where $\rho(x) = (4\pi)^{-1}\times \left|\Gamma(i\frac{x}{2}+\frac{\alpha+\delta+1}{2})\,\Gamma(i\frac{x}{2}+\frac{\alpha-\delta+1}{2})\right|^2$.

The Hahn polynomials of imaginary argument, $p_n^{(\alpha,\beta)}(x,\gamma)$ and $q_n^{(\alpha)}(x,\delta)$, turn out to be useful for describing UIR of the Lorentz group SO(3,1) for the principal and complementary series, respectively. In this way the basic properties of the Lorentz group representations can be stated by analogy with the well-known to physicists problem of the coupling of two angular momenta.

2. As it is known the infinitesimal rotations operators $\vec{J}$ and boosts $\vec{K}$ for the Lorentz group SO(3,1) satisfy

$$[J_p, J_q] = i e_{pqr} J_r, \quad [J_p, K_q] = i e_{pqr} K_r,$$
$$[K_p, K_q] = -i e_{pqr} J_r \qquad (p,q,r = 1,2,3) \qquad (5)$$

where e_{pqr} is the Levi-Civita symbol.

The transformation $\vec{A} = \frac{1}{2}(\vec{J}+i\vec{K})$, $\vec{B} = \frac{1}{2}(\vec{J}-i\vec{K})$ yields

$$[A_p, A_q] = i e_{pqr} A_r, \quad [B_p, B_q] = i e_{pqr} B_r, \tag{5a}$$
$$[A_p, B_q] = 0$$

Therefore constructing irreducible representations of SO(3,1) is closely connected with the coupling problem two "complex conjugate" momenta $\vec{A}$ and $\vec{B}$ into one "real" vector $\vec{J} = \vec{A} + \vec{B}$.

According to (5a), in the space of an irreducible representation of SO(3,1) it is possible to construct the basis $\Phi_{m_1 m_2}$ on which the operators $A_{\pm} = A_1 \pm iA_2$, A_3 and $B_{\pm} = B_1 \pm iB_2$, B_3 act via

$$A_{\pm}\Phi_{m_1 m_2} = [(j_1 \mp m_1)(j_1 \pm m_1 + 1)]^{1/2}\Phi_{m_1 \pm 1, m_2},\ A_3\Phi_{m_1 m_2} = m_1\Phi_{m_1, m_2};$$
$$B_{\pm}\Phi_{m_1 m_2} = [(j_2 \mp m_2)(j_2 \pm m_2 + 1)]^{1/2}\Phi_{m_1, m_2 \pm 1},\ B_3\Phi_{m_1 m_2} = m_2\Phi_{m_1 m_2}. \tag{6}$$

The constants j_1, j_2, m_1, m_2 are complex. Since for unitary representations $\vec{A}^+ = \vec{B}$, then $j_1^* = j_2$ and $m_1^* = m_2$. Vectors $\Phi_{m_1 m_2}$ are the eigenvectors of two Hermitian operators $J_3 = A_3 + B_3$ and $K_3 = i(B_3 - A_3)$:

$$J_3\Phi_{m_1 m_2} = m\Phi_{m_1 m_2},\quad K_3\Phi_{m_1 m_2} = \lambda\Phi_{m_1 m_2} \tag{7}$$

and they correspond to the real eigenvalues $m = m_1 + m_2$ and $\lambda = i(m_2 - m_1)$. The basis $\chi_{\lambda m} \equiv \Phi_{m_1 m_2}$ is

$$(\chi_{\lambda m} | \chi_{\lambda' m'}) = \delta_{mm'}\,\delta(\lambda - \lambda'). \tag{8}$$

On the other hand, the infinitesimal operators $\vec{J}$ satisfy the commutation relations of the angular momentum. and in the space of an irreducible representation there exists the basis Ψ_{jm}, on which the operators $J_{\pm} =$

$= J_1 \pm J_2$ and J_3 act by the formulas

$$J_\pm \Psi_{jm} = \left[(j \mp m)(j \pm m + 1)\right]^{1/2} \Psi_{j,m\pm 1}, \quad J_3 \Psi_{jm} = m \Psi_{jm}. \qquad (9)$$

Here j is an integer or half-integer positive number and $m = -j, -j+1, \dots, j-1, j$.

3. Infinitesimaly investigating, the irreducible reprerentations of the Lorentz group SO(3,1) reduces to defining the form of operators $K_\pm = K_1 \pm iK_2$ and K_3 in the basis Ψ_{jm} [4]. For this we expand the vector Ψ_{jm} in eigenfunctions of K_3:

$$\Psi_{jm} = \int_{-\infty}^{\infty} d\lambda \, \langle m_1 m_2 | jm \rangle \, \Phi_{m_1 m_2}. \qquad (10)$$

Applying $J_\pm = A_\pm + B_\pm$ to (10) and taking into account (9), (8) and (6) we obtain a well-known recurrence for the coefficients of expansion (10)

$$\left[(j \mp m)(j \pm m + 1)\right]^{1/2} \langle m_1 m_2 | j \pm 1 \rangle = \left[(j_1 \pm m_1)(j_1 \mp m_1 + 1)\right]^{1/2} \langle m_1 \mp 1, m_2 | jm \rangle + \qquad (11)$$
$$+ \left[(j_2 \pm m_2)(j_2 \mp m_2 + 1)\right]^{1/2} \langle m_1, m_2 \mp 1 | jm \rangle$$

The substitution $\langle m_1 m_2 | jm \rangle = \alpha_\pm C^\pm_{m_1 m_2 m} u^\pm_{jm}(m_1)$, where

$$C^\pm_{m_1 m_2 m} = \left[\frac{\Gamma(j_1 + m_1 + 1)\Gamma(j_2 + m_2 + 1)(j - m)!}{\Gamma(j_1 - m_1 + 1)\Gamma(j_2 - m_2 + 1)(j + m)!}\right]^{\pm 1/2}$$

and $\alpha_+^{-1} = \sin \pi (j_1 - m_1 + 1), \quad \alpha_-^{-1} = \sin \pi (j_2 + m_2 + 1)$

leads to simple difference-recurrence formulas for the functions $u^\mp_{jm}(m_1)$:

$$u^{+}_{j,m+1}(m_1)=\nabla u^{+}_{jm}(m_1), \quad u^{-}_{j,m-1}(m_1)=\Delta u^{-}_{jm}(m_1), \tag{12}$$

where $\Delta f(x) = f(x+a) - f(x)$ and $\nabla f(x) = f(x)-f(x-1)$. From here in exactly the same manner as in deducing the general expression for the Clebsch-Gordan coefficients [2] we find

$$\langle m_1 m_2 | jm \rangle = \frac{(-1)^{j-m}}{\sin\pi(j_1-m_1+1)}\frac{A}{(2j)!}\left[\frac{\Gamma(j_1-m_1+1)\Gamma(j_2-m_2+1)(j+m)!}{\Gamma(j_1+m_1+1)\Gamma(j_2+m_2+1)(j-m)!}\right]^{1/2} \times$$

$$\times\, \Delta^{j-m}_{m_1}\left[\frac{\Gamma(j_1+m_1+1)\Gamma(j_2+j-m_1+1)}{\Gamma(j_1-m_1+1)\Gamma(j_2-j+m_1+1)}\right]. \tag{13}$$

Here A is a constant determined by the normalization condition

$$\|\Psi_{jj}\|^2 = \int_{-\infty}^{\infty} d\lambda \left|\langle m_1 m_2 | jj \rangle\right|^2 = 1. \tag{14}$$

Using the Barns' lemma [5] it is easy to verify that for unitary irreducible representations of the Lorentz group SO(3,1) the condition (14) is satisfied:

a) in the case of the principal series, when $j_1 = j_2^* = \frac{1}{2}(\mu-1+i\gamma)$, $j = |j_1+j_2+1|, |j_1+j_2+1|+1, \ldots$ (μ is an integer or half-integer and γ is an arbitrary real number);

b) for the complementary series, when $j_1 = j_2^* = \frac{1}{2}(\sigma-1)$, $(|\sigma|<1)$, $j = j_1-j_2, j_1-j_2+1, \ldots$.

As in the theory of angular momenta, the coefficients of the expansion (10) are called "complexified Clebsch--Gordan coefficients" .

Using Rodrigues' formula for the Hahn polynomials [2] and with the relations (1), (4) and (13), the complexificated Clebsch-Gordan coefficients can be expressed in terms of the Hahn polynomials of imaginary argument. For the principal series we have

$$\langle m_1 m_2 | jm \rangle = f \frac{[\rho(\lambda)]^{1/2}}{d_{j-m}} p_{j-m}^{(m-\mu, m+\mu)}(\lambda, \gamma) \tag{15}$$

where $\rho(\lambda)$ and d_n are the weight and norm of the polynomials $p_n^{(\alpha,\beta)}(x, \gamma)$ see [1]. In the case of complementary series

$$\langle m_1 m_2 | jm \rangle = f \frac{[\rho(\lambda)]^{1/2}}{d_{j-m}} q_{j-m}^{(m)}(\lambda, \delta). \tag{16}$$

4. The formulas (10),(15) and (16) allow us to derive the matrix elements of the operators $K_\pm$ and K_3 in the basis Ψ_{jm} thus exploiting the above-studied properties of the Hahn polynomials. For instance, using the recurrence relation (2) we get

$$K_3 \Psi_{jm} = a_{jm} \Psi_{j-1,m} + b_{jm} \Psi_{jm} + a_{j+1,m} \Psi_{j+1,m} \tag{17}$$

where for the principal series

$$a_{jm} = \frac{1}{j} \left[\frac{(j^2-m^2)(j^2-\mu^2)(j^2+\gamma^2)}{4j^2-1} \right]^{1/2}, \quad b_{jm} = \frac{m\mu\gamma}{j(j+1)}. \tag{17a}$$

In the case of complementary series set $\mu = 0$ and $\gamma = -i\delta$ in formulas (17) and (17a).

Using (9), (17) and (5) it is easy to determine the action of the operators $K_\pm$ on Ψ_{jm}. Thus, the action of $\vec{J}$ and $\vec{K}$ on Ψ_{jm} for an irreducible representation of SO(3,1) is obtained using the properties of the Hahn polynomials of imaginary argument. Conversely, from the formulas determining the action of $\vec{J}$ and $\vec{K}$ on the basis Ψ_{jm} the group-theoretic interpretation for the main properties of these polynomials arise naturally.

5. The above arguments yield simple expressions for the boost matrix elements of SO(3,1):

$$d^{(\mu,\gamma)}_{jj'm}(t)=(\Psi_{jm}|\, e^{-itK_3}\Psi_{j'm}).$$

In the case of the principal series (μ,γ) formulas (10) and (15) imply

$$d^{(\mu,\gamma)}_{jj'm}(t)=\int_{-\infty}^{\infty} d\lambda\, \bar{p}_{j-m}(\lambda)\, e^{-i\lambda t}\, \bar{p}_{j'-m}(\lambda)\, \rho(\lambda), \tag{18}$$

where $\bar{p}_{j-m}(\lambda) = d^{-1}_{j-m}\, p^{(m-\mu,m+\mu)}_{j-m}(\lambda,\gamma)$. An analogous integral representations for the boost matrix in the case of complementary series is easily derived from (10) and (16).

The integral representation (18) allows us to study the function $d^{(\mu,\gamma)}_{jj'm}(t)$ if we know the properties of the Hahn polynomials of imaginary argument. In particular, owing to (3) we get $d^{(\mu,\gamma)}_{jj'm}(0) = \delta_{jj'}$. The linearization of the product of two Hahn polynomials in (18) leads to the formula

$$d^{(\mu,\gamma)}_{jj'm}(t)=\sum_{j''} c^{jj'}_{j''}\, d^{(\mu,\gamma)}_{j''mm}(t). \tag{19}$$

Aknowledgements. We are grateful to Yu.A.Danilov, V.G.Kadyshevsky, R.M.Mir-Kasimov, A.F.Nikiforov, Ya.A.Smorodinsky and N.Ya.Vilenkin for valuable discussions.

References

1. Atakishiyev,N.M., Suslov,S.K. (1985). J. Phys. A: Math. Gen. 18, 1583.

2. Nikiforov,A.F., Suslov,S.K., Uvarov,V.B. (1985). Classical orthogonal polynomials of a discrete variable. Nauka, Moscow (Russian).

3. Askey,R., Wilson,J. (1982). SIAM J.Math.Anal. 13, 651.

4. Gelfand,I.M., Minlos,R.A., Sapiro,Z.Ya. (1958). Representations of the Rotation Group and of the Lorentz Group, Fizmatgiz, Moscow (Russian).

5. Whittaker,E.T., Watson,G.N. (1927). A course of modern analysis , Cambridge.

$\widetilde{SL}(2,R)$ OVERLAP COEFFICIENTS BETWEEN SO(2) AND SO(1,1) SUBGROUP REDUCTIONS

N.M.Atakishiyev, R.M.Mir-Kasimov[+]
+Joint Institute for Nuclear Research, Dubna 141980

In this paper it is shown that wave functions for a model of the linear oscillator [1-3] in the relativistic configurational x space [4] are the overlap coefficients between the compact and noncompact generators eigenbases for the discrete series of the universal covering group $\widetilde{SL}(2,R)$ representations. Therefore with the help of the explicit form for the wave functions in terms of Pollaczek polynomials [2] it is possible to obtain simple formulas for the overlap coefficients and boost matrix elements corresponding to the discrete series of $\widetilde{SL}(2,R)$ representations.

1. The wave functions for a relativistic linear oscillator [1-3] can be written in momentum representation as

$$\Psi_n(p) = i^n\left(\frac{\hbar}{2mc}\right)^{1/2} \Phi_n^{\nu}(\eta), \tag{1}$$

where $\eta^2 = \frac{2c}{\hbar\omega}p^+$, $p^+ = p_0 + p$ is light-cone variable, $2\nu = 1 + (4\mu^2+1)^{1/2}$, $\mu = \frac{mc^2}{\hbar\omega}$

and the function

$$\Phi_n^{\nu}(\eta) = \left[\frac{2n!}{\Gamma(n+2\nu)}\right]^{1/2} \eta^{2\nu-1/2} e^{-\eta^2/2} L_n^{2\nu-1}(\eta^2) \tag{2}$$

is expressed via the generalized Laguerre polynomials $L_n^{\alpha}(x)$. The phases in (1) have been chosen so that

$\Psi_n(p)$ coincide with the wave functions of the nonrelativistic linear oscillator as the speed of light $c \to \infty$.

As follows from (1), the matrix elements of an arbitrary operator A in momentum representation

$$\int_{-\infty}^{\infty} d\Omega_p \Psi_n^*(p) A \Psi_{n'}(p) = \int_0^{\infty} d\eta\, \Phi_n^{\nu}(\eta) \tilde{A} \Phi_{n'}^{\nu}(\eta)$$

where $d\Omega p = mc\,dp^+/p^+ = 2mc\,d\eta/\eta$ is the invarinat element and $\tilde{A} = \eta^{-1/2} A \eta^{1/2}$. In particular, for the Hamiltonian $H(p) \equiv \hbar\omega K_0$, raising K_+ and lowering K_- operators, which satisfy the commutation relations

$$[K_-, K_+] = 2K_0, \quad [K_0, K_\pm] = \pm K_\pm \qquad (3)$$

and being the dynamical symmetry group $\widetilde{SL}(2,R)$ generators, we obtain the following connection with the realization used in [5]:

$$\tilde{K}_0 = J_0^{\gamma}, \; \tilde{K}_\pm = \pm i(\tilde{K}_1 \pm i \tilde{K}_2), \; \tilde{K}_1 = J_1^{\gamma}, \; \tilde{K}_2 = J_2^{\gamma} \qquad (4)$$

Here $\gamma = (2\nu - 1)^2 - 1/4$. In the considered case of oscillator the values of ν are $[1, \infty[$. However, the formulas (1), (2) and (4) may also be used for the description of $\widetilde{SL}(2,R)$ unitary irreducible representations (UIRs) as $1/2 \leq \nu < 1$, where the lower bound is defined by the requirement that the generators (4) have a spectrum bounded below. (Since UIRs in the negative discrete series $D^-(\nu)$ are obtained from UIRs in the positive discrete series $D^+(\nu)$ via the outer automorphism $K_0 \leftrightarrow -K_0$, $K_1 \leftrightarrow -K_1$ and $K_2 \leftrightarrow K_2$ of $sl(2,R)$ in this paper we consider only $D^+(\nu)$.) The functions $\Psi_n(p)$ form a basis for the positive discrete series $D^+(\nu)$ UIRs of the $\widetilde{SL}(2,R)$ corresponding to the reduction to $SO(2)$ (cf.[5]).

2. In the relativistic configurational x-space wave functions of linear oscillator

$$\psi_n(x)=(2\pi\hbar)^{-1/2}\int d\Omega_p\,\xi(p,x)\,\psi_n(p) \tag{5}$$

are defined decomposing in the matrix elements of representations of the Lobachevski space motion group

$$\xi(p,x)=\left(\frac{p_0-p}{mc}\right)^{-i\tilde{x}}=\left(\frac{\eta^2}{2\mu}\right)^{i\tilde{x}} \tag{6}$$

where $\tilde{x}=x/\lambda$ is a dimensionless variable and $\lambda=\hbar/mc$ is the Compton wavelenght. The functions $\xi(p,x)$ are the eigenfunctions of the noncompact generator K_2, i.e. $K_2\xi(p,x)=\tilde{x}\,\xi(p,x)$, and form Dirac-orthonormalized basis corresponding to the reduction $SL(2,R)\supset SO(1,1)$ [5-8]. They are related to the eigenfunctions $\Phi_{\tilde{x}}(\eta)$ of $\tilde{K}_2$; employed in [5], in the following way:

$$\xi(p,x)=(\pi\eta)^{1/2}(2\mu)^{-i\tilde{x}}\,\Phi_{\tilde{x}}(\eta). \tag{6'}$$

In the configurational x-spacethe difference operators $\left(\nabla x=\frac{d}{dx}\right)$ correspond to the generators (4)

$$\begin{aligned} K_+ &\equiv K_0+K_1=\mu^{-1}[\tilde{x}(\tilde{x}+i)+\mu^2]e^{i\nabla_{\tilde{x}}},\\ K_- &\equiv K_0-K_1=\mu e^{-i\nabla_{\tilde{x}}},\quad K_2=-\tilde{x}, \end{aligned} \tag{7}$$

satisfy the commutation relations

$$[K_-,K_+]=2iK_2,\quad [K_2,K_\pm]=\pm iK_\pm. \tag{3'}$$

Note that as one of the possible examples of realization of $SL(2,R)$ arising in the noncompact subgroup $SO(1,1)$ reduction, a case is given in [8], when

$$J_2 = \lambda\,, \quad J_- = -e^{i\nabla_\lambda}, \quad J_+ = f(\lambda)e^{-i\nabla_\lambda}, \quad f(\lambda) = \lambda(\lambda - i) + j(j+1) \tag{8}$$

It is easy to verify that for $\lambda = -\tilde{x}$ and $j = \nu - 1$ we come to the following relation between the generators (7) and (8): $J_2 = K_2$, $J_\pm = \pm\mu^{\pm 1} K_\pm$.

3. As is known, the overlap coefficients

$$C_n^{(\nu)}(\tilde{x}) \equiv (\Phi_{\tilde{x}}, \Phi_n^\nu) = \int_0^\infty d\eta\, \Phi_{\tilde{x}}^*(\eta)\, \Phi_n^\nu(\eta) \tag{9}$$

relate the coordinates $f(\tilde{x}) \equiv (\Phi_{\tilde{x}}, f)$ of an arbitrary function $f(\eta) \in L^2(R^+)$ in the noncompact generator K_2 eigenbasis to its coordinates $f_n^{(\nu)} \equiv (\Phi_n^\nu, f)$ in the compact generator $\tilde{K}_0$ eigenbasis, i.e.

$$f(\tilde{x}) = \sum_{n=0}^{\infty} f_n^{(\nu)} C_n^{(\nu)}(\tilde{x}), \quad f_n^{(\nu)} = \int_{-\infty}^{\infty} dx\, C_n^{*(\nu)}(x) f(x) \tag{10}$$

Using the explicit form for the wave functions [2]

$$\Psi_n(x) = 2^{\nu - \frac{1}{2}} \left[\frac{n!}{\pi\lambda\,\Gamma(n+2\nu)}\right]^{\frac{1}{2}} \mu^{-i\tilde{x}}\, \Gamma(\nu + i\tilde{x})\, P_n^\nu\!\left(\tilde{x}, \frac{\pi}{2}\right) \tag{5'}$$

through the Pollaczek polynomials $P_n^\nu(x;\varphi)$ and substituting in (5) the relations (1) and (6') it is easy to show that

$$C_n^{(\nu)}(\tilde{x}) = i^n \left[\frac{n!}{2\pi\,\Gamma(n+2\nu)}\right]^{1/2} 2^{\nu - i\tilde{x}}\, \Gamma(\nu - i\tilde{x})\, P_n^\nu\!\left(\tilde{x}; \frac{\pi}{2}\right) \tag{11}$$

The orthogonality of the overlap coefficients (11)

$$\int_{-\infty}^{\infty} dx\, C_n^{*(\nu)}(x)\, C_{n'}^{(\nu)}(x) = \delta_{nn'} \tag{12}$$

is in agreement with the orthogonality relation for the Pollaczek polynomials

$$\int_{-\infty}^{\infty} dx\, P_n^{\nu}(x;\tfrac{\pi}{2})\, P_{n'}^{\nu}(x;\tfrac{\pi}{2})\, \rho^{(\nu)}(x) = \frac{\Gamma(n+2\nu)}{n!}\delta_{nn'} \tag{13}$$

where $\rho^{\nu}(x) = 2^{2\nu-1}\pi^{-1}|\Gamma(\nu+ix)|^2$. Under the action of an arbitrary finite element $g=\begin{pmatrix} a & b \\ c & d\end{pmatrix} \in SL(2,R)$ the coefficients $C_n^{(\nu)}(x)$ transform to

$$[U^{\nu}(g)C_n^{(\nu)}](\tilde{x}) = \left[\frac{a-ib}{a+ib}\right]^{n+\nu}(a^2+b^2)^{-i\tilde{x}}\sum_{k=0}^{\infty}\frac{1}{k!}\left[\frac{1}{2}(ac+bd)\right]^k C_n^{(\nu)}(\tilde{x}+ik) \tag{14}$$

In the cases, when g belongs to the one-parameter compact $\exp(i\gamma\tilde{K}_0)$ and noncompact $\exp(i\beta\tilde{K}_2)$ subgroups, the transformation (14) is simply reduced to multiplication by $\exp[i\gamma(n+\nu)]$ and $\exp(i\beta\tilde{x})$, respectively.

4. As follows from the definition (9) of the overlap coefficients $C_n^{(\nu)}(\tilde{x})$, the matrix elements $D_{nn'}^{\nu}(g) \equiv$ $\equiv(\Phi_n, U^{\nu}(g)\Phi_{n'})$ in the $SO(2)$ basis are related to the matrix elements $D^{\nu}_{\tilde{x}\tilde{x}'}(g) \equiv (\Phi_{\tilde{x}}, U^{\nu}(g)\Phi_{\tilde{x}'})$ in the $SO(1,1)$ basis by

$$D_{nn'}^{\nu}(g) = \int_{-\infty}^{\infty} dx \int_{-\infty}^{\infty} dx'\, C^{*(\nu)}(x)\, D^{\nu}_{xx'}(g)\, C_{n'}^{(\nu)}(x') \tag{15}$$

In particular case, when $g \in \exp(i\beta\tilde{K}_2)$, the $SO(1,1)$ matrix elements satisfy

$$D_{xx'}\begin{pmatrix} \exp(-\beta/2) & 0 \\ 0 & \exp(\beta/2) \end{pmatrix} = e^{i\beta x}\delta(x-x')$$

This leads to an integral representation for the boost matrix elements in the basis

$$D^{\nu}_{nn'}\begin{pmatrix} e^{-\beta/2} & 0 \\ 0 & e^{\beta/2} \end{pmatrix} = (-i)^{n} i^{n'} \frac{n!n'!}{\gamma} \int_{-\infty}^{\infty} dx\, \rho^{(\nu)}(x) P^{\nu}_{n}(x;\tfrac{\pi}{2})\, e^{ix\beta} P^{\nu}_{n'}(x;\tfrac{\pi}{2}) \tag{16}$$

where $\gamma_{nn'}=[n!n'!\Gamma(n+2\nu)\Gamma(n'+2\nu)]^{1/2}$. The integral representation (16) allows one to obtain readily the known analytic expression [9]

$$D^{\nu}_{nn'}\begin{pmatrix} e^{-\frac{\beta}{2}} & 0 \\ 0 & e^{\frac{\beta}{2}} \end{pmatrix} = \frac{(-1)^{n'}}{\gamma_{nn'}} \Gamma(2\nu+n+n') \frac{th^{n+n'}\frac{\beta}{2}}{ch^{2\nu}\frac{\beta}{2}} F(-n,-n';1-2\nu-n-n'; cth^{2}\beta/2) \tag{17}$$

Indeed, as follows from the definition of the weight function $\rho^{(\nu)}(x)$ for the Pollaczek polynomials $P^{\nu}_{n}(x;\pi/2)$, the integral in (16) may be closed when $\beta > 0$ $(\beta < 0)$ in the upper (lower) part of the complex plane $z = x + iy$, thus reducing it to the sum of the integrand residues at the points $z_k = i(\nu + k)$, $k = 0, 1, 2 \ldots$ (when $\beta < 0$ at the points z^{*}_{k}), i.e.

$$D^{\nu}_{nn'}\begin{pmatrix} e^{-\frac{\beta}{2}} & 0 \\ 0 & e^{\frac{\beta}{2}} \end{pmatrix} = i^{n'}(-i)^{n} 2^{2\nu} \frac{n!n'!}{\gamma} \sum_{k=0}^{\infty} \frac{(-1)^{k}}{k!} \Gamma(\nu - iz_k) P^{\nu}_{n}(z_k;\tfrac{\pi}{2})\, e^{i\beta z_k} P^{\nu}_{n'}(z_k;\tfrac{\pi}{2})$$

Now making use of the relation $n!P^{\nu}_{n}(z_k;\frac{\pi}{2}) = i^{n} m^{(2\nu,-1)}_{n}(k)$ where

$$m^{(2\nu,-1)}_{n}(k) = (2\nu)_n F(-n,-k;2\nu;2) = (2\nu)_n \sum_{j=0}^{\min(n,k)} \frac{(-n)_j(-k)_j}{(2\nu)_j} \frac{2^{j}}{j!} \tag{18}$$

are the analytic continuation of the Meixner polynimials $m_n^{(\gamma,\mu)}(z)$ with respect to the parameter μ (see [10]), we get

$$D^{\nu}_{nn'}\begin{pmatrix} e^{-\beta/2} & 0 \\ 0 & e^{\beta/2} \end{pmatrix} = \frac{(-1)^{n'}}{\gamma_{nn'}} 2^{2\nu} e^{-\nu\beta} \sum_{k=0}^{\infty} \frac{(-1)^k}{k!} \Gamma(2\nu+k) e^{-k\beta} m_n^{(2\nu,-1)}(k) m_{n'}^{(2\nu,-1)}(k) \qquad (16')$$

Since (a proof is given in [11])

$$2^{2\nu} e^{-\nu\beta} \sum_{k=0}^{\infty} \frac{(-1)^k}{k!} \Gamma(2\nu+k) e^{-k\beta} m_n^{(2\nu,-1)}(k)\, m_{n'}^{(2\nu,-1)}(k) =$$

$$= \Gamma(2\nu+n+n') th^{n+n'} \frac{\beta}{2} \left[ch\frac{\beta}{2}\right]^{-2\nu} F(-n,-n';1-2\nu-n-n'; cth^2 \beta/2), \qquad (19)$$

we get (17).

We have thus seen that the use of the wave functions of a linear oscillator in the relativistic configurational $\mathcal{X}$-space [1-3] makes it possible to obtain simple formulas for the description of the discrete series $D^{\pm}(\nu)$ UIRs of the universal covering group $\widetilde{SL}(2,R)$.

We are grateful to V.G.Kadyshevsky, W.Lassner, V.I. Man'ko, M.V.Savel'yev and S.K.Suslov for valuable discussions.

References

1. Atakishiyev, N.M., Mir-Kasimov, R.M., Nagiyev, Sh.M. (19 (1980). Theor. Math. Phys., 44, 47 (Russian)

2. Atakishiyev, N.M. (1984). Teor. Math. Phys., 58, 254 (Russian)

3. Atakishiyev, N.M., Mir-Kasimov, R.M. (1985). Preprint E2-85-214, JINP, Dubna, USSR

4. Kadyshevsky, V.G., Mir-Kasimov, R.M., Skachkov, N.B. (1968). Nuovo Cimento, 55A, 233

5. Basu, D., Wolf, K.B. (1982). J.Math. Phys., 23, 189

6. Vilenkin, N.Ya. (1965). Special functions and the theory of group representations, Nauka, Moscow (Russian)

7. Barut, A.O., Phillips, E.C. (1968). Commun. Math.Phys.,

8, 52

8. Kuriyan, J.G., Mukunda, N., Sudarshan, E.C.G. (1968). J.Math. Phys., 9, 2100

9. Bargmann, V. (1947). Ann. Math., 48, 568

10. Atakishiyev, N.M., Suslov, S.K. (1984). Preprint PHE-12, Institut für Hochenergiephysik, Berlin-Zeuthen, GDR

11. Atakishiyev, N.M. (1985). Preprint No 127, Physics Institute, Baku

ASYMPTOTIC REPRESENTATIONS FOR SOME FUNCTIONS AND INTEGRALS CONNECTED WITH AIRY FUNCTION

A.I. Nikishov and V.I. Ritus

P.N. Lebedev Physical Institute, Moscow

Abstract. The asymptotic representations of the functions $Ai_1(z)$, $Gi(z)$, $Gi^1(z)$, $Ai^2(z)$, $Bi^2(z)$, the integral $\int_v^\infty dx\, Ai(x) h(x, v)$ as $v \to -\infty$, and the integrals, differing from it by the change of $Ai(x)$ by $Ai'(x)$ or $Ai_1(x)$, are presented here.

It is convenient to regard the divergent sum of the latter terms of an asymptotic expansion as a symbolic representation of the remainder [1]. In general, this representation does not uniquely determine the remainder but with the help of some additional information of the function the remainder term can, in principle, be uniquely obtained. The presence of several series in an asymptotic representation of a function in a sector of complex variable z-plane reflects linear dependence of this function on other functions, each represented by its own single series. This linear dependence of the functions holds in all the sectors of the z-plane, but outside the considered sector the representation of each function by one series may be unsatisfactory, i.e. the remainder term may not be small.

For the Airy function the Stokes' phenomenon is considered as a consequence of discontinuous behaviour of steepest descent lines over the passes. When z crosses the Stokes ray, the steepest descent line over the higher pass abruptly changes the direction of its asymptotic approach to the steepest descent line over the lower pass for the direction of approach to the opposite end of this line. Therefore when the integration contour, drawn along the steepest descent lines, goes over the higher pass, it starts or stops going over the lower pass while z crosses the Stokes ray, and as a result the recessive series (contribution from the lower pass) discontinuously appears or disappears in the asymptotic representation of a function containing the dominant series.

Introduction. In the theory of quantum processes with particles in a constant or slowly varying external field the Airy function, related functions and integrals with these functions play an important role, see e.g. [2,3]. Here we give asymptotic expansions for some frequency occurring functions. The details see in [4].

1. Asymptotic representations for functions connected with Airy function. We define the Airy function and related functions as in [5,6], but without the factor

$$Ai(x)=\int_0^\infty \cos\left(xt+\frac{t^3}{3}\right)dt;\quad Ai_1(x)=\int_x^\infty Ai(t)\,dt; \tag{1}$$

$$Hi(x)=\int_0^\infty e^{xt-\frac{t^3}{3}}\,dt;\quad Gi(x)=\int_0^\infty \sin\left(xt+\frac{t^3}{3}\right)dt; \tag{2}$$

$$Bi(x)=Hi(x)+Gi(x). \tag{3}$$

The asymptotic expansions for $Ai_1(x)$ are of the form

$$Ai_1(x)\sim\frac{\sqrt{\pi}}{2}\,\frac{e^{-\xi}}{x^{3/4}}\sum_{n=0}^\infty a_n(-\xi)^{-n},\quad x\to\infty, \tag{4}$$

$$\xi=\frac{2}{3}x^{3/2},\ a_0=1,\ a_1=\frac{41}{2^3\cdot 3^2},\ a_2=\frac{9241}{2^7\cdot 3^4},\ a_3=\frac{5^2\cdot 203009}{2^{10}\cdot 3^7},\ldots$$

$$Ai_1(x)\sim\pi-\frac{\sqrt{\pi}}{(-x)^{3/4}}\sum_{k=0}^\infty(-1)^k\Big[a_{2k}\xi^{-2k}\cos\left(\xi+\frac{\pi}{4}\right)+$$

$$+a_{2k+1}\xi^{-2k-1}\sin\left(\xi+\frac{\pi}{4}\right)\Big],\quad \xi=\frac{2}{3}(-x)^{3/2},\ x\to-\infty. \tag{5}$$

Here the coefficients a_n are connected with the coefficients

$$c_n = \frac{(6n-1)!!}{6^{3n} n!\,(2n-1)!!} = \frac{\Gamma(n+\frac{1}{6})\Gamma(n+\frac{5}{6})}{2^{n+1}\,\pi n!}\,, \tag{6}$$

entering the asymptotic expansion for $Ai\,(x)$ [5] by the recurrence

$$a_n = c_n + (n-\tfrac{1}{2})a_{n-1} = \sum_{k=0}^{n} \frac{\Gamma(n+\frac{1}{2})}{\Gamma(k+\frac{1}{2})} c_k = \frac{\Gamma(n+\frac{1}{2})}{2\pi^{3/2}} \sum_{k=0}^{n} 2^k B(k+\tfrac{1}{6}, k+\tfrac{5}{6}). \tag{7}$$

For $|x| \gg 1$ the representations with the remainder are also useful:

$$Ai_1(x) = -\sum_{k=0}^{n}\left[\frac{(3k)!\,Ai'(x)}{(3k)!!!\,x^{3k+1}} + \frac{(3k+1)!\,Ai(x)}{(3k)!!!\,x^{3k+2}}\right] + \frac{(3n+2)!}{(3n)!!!}\int_x^{\infty} t^{-3n-3} Ai(t)\,dt,\ (3n)!!! = 3\cdot 6\cdot 9 \ldots 3n = 3^n n!,\ x>0, \tag{8}$$

$$Ai_1(x) = \pi - \sum_{k=0}^{n}\left[\frac{(3k)!\,Ai'(x)}{(3k)!!!\,x^{3k+1}} + \frac{(3k+1)!\,Ai(x)}{(3k)!!!\,x^{3k+2}}\right] + \frac{(3n+2)!}{(3n)!!!}\int_{-\infty}^{x} t^{-3n-3} Ai(t)\,dt,\ x<0. \tag{9}$$

For $Hi\,(x)$ and $Hi'(x)$ we have [6,4]

$$Hi(z) \sim -\sum_{n=0}^{\infty} \frac{(3n)!}{(3n)!!!\,z^{3n+1}}\,,\ |ph(-z)| < \tfrac{2}{3}\pi; \tag{10}$$

$$Hi(x) \sim \pi^{1/2} x^{-1/4} e^{\xi} \sum_{n=0}^{\infty} c_n \xi^{-n} - \sum_{n=0}^{\infty} \frac{(3n)!}{(3n)!!!\,x^{3n+1}}, \quad \xi = \tfrac{2}{3}x^{3/2},\ x\to\infty; \tag{11}$$

$$Hi'(x) \sim \sum_{n=0}^{\infty} \frac{(3n+1)!}{(3n)!!!\,x^{3n+2}}\,,\ x \to -\infty; \tag{12}$$

$$Hi'(x) \sim \pi^{1/2} x^{\frac{1}{4}} e^{\xi} \sum_{n=0}^{\infty} d_n \xi^{-n} + \sum_{n=0}^{\infty} \frac{(3n+1)!}{(3n)!!!\,x^{3n+2}},\ x\to\infty, \quad d_n = -\frac{6n+1}{6n-1} c_n, \tag{13}$$

c_n are defined in (6).

Substituting asymptotic series for the functions in the relation

$$Ai_1(x) = \pi + Hi(x)Ai'(x) - Hi'(x)Ai(x), \tag{14}$$

or in the Wronskian

$$Bi'(x)Ai(x) - Bi(x)Ai'(x) = \pi \tag{15}$$

gives

$$\sum_{n=0}^{2s} (-1)^n c_n d_{2s-n} = 0, \quad s \geqslant 1, \tag{16}$$

or equivalently

$$\sum_{k=0}^{2s} (-1)^k \frac{\Gamma(k+\frac{1}{6})\Gamma(k+\frac{5}{6})\Gamma(2s-k-\frac{1}{6})\Gamma(2s-k+\frac{7}{6})}{k!(2s-k)!} = 0. \tag{17}$$

Differentiating the known asymptotic series for $Gi(x)$ [6] we have

$$Gi'(x) \sim -\sum_{n=0}^{\infty} \frac{(3n+1)!}{(3n)!!!\, x^{3n+2}}, \quad x \to +\infty \tag{18}$$

We write down also the asymptotic expansions for $Gi(-x)$ and $Gi'(-x)$, $x \to \infty$

$$Gi(-x) \sim \frac{\sqrt{\pi}}{x^{1/4}} \sum_{n=0}^{\infty} (-1)^n \Big[c_{2n} \xi^{-2n} \cos\left(\xi + \frac{\pi}{4}\right) + \tag{19}$$

$$+ c_{2n+1} \xi^{-2n-1} \sin\left(\xi + \frac{\pi}{4}\right)\Big] + \sum_{n=0}^{\infty} \frac{(3n)!}{(3n)!!!\,(-x)^{3n+1}}, \quad \xi = \frac{2}{3} x^{3/2};$$

$$Gi'(-x) \sim \sqrt{\pi}\, x^{1/4} \sum_{n=0}^{\infty} (-1)^n \Big[d_{2n} \xi^{-2n} \sin\left(\xi + \frac{\pi}{4}\right) - \tag{20}$$

$$- d_{2n+1} \xi^{-2n-1} \cos\left(\xi + \frac{\pi}{4}\right)\Big] - \sum_{n=0}^{\infty} \frac{(3n+1)!}{(3n)!!!\,(-x)^{3n+2}}, \quad x \to \infty.$$

Note that the leading terms of the asymptotic series

$$Hi'(x) \sim -\frac{3}{2} x^{-2}, \quad x \to -\infty,$$

$$Gi'(x) \sim \frac{7}{96} x^{-2}, \quad x \to \infty,$$

given in the handbook [5] disagree with the true formulas (12) and (18).

Denoting $W_{1,2}(x)$ the oscillatory and nonoscillatory

(for $x \to -\infty$) parts of $Ai^2(x)$ we find from the known expansions for $Ai(x)$ and $Bi(x)$ [5]

$$w_{1,2}(x) \equiv \frac{1}{2}\left[Ai^2(x) \pm Bi^2(x)\right], \tag{21}$$

$$w_1(x) \sim \frac{\pi}{2} \sum_{n=0}^{\infty} \frac{e_n}{(-x)^{3n+\frac{1}{2}}}, \quad e_n = (-1)^n \frac{(6n-1)!!}{2^{5n} \cdot 3^n \cdot n!}, \tag{22}$$

$$e_0 = 1, \; x \to -\infty ;$$

$$w_2(x) \sim \frac{\pi}{2} \sum_{n=0}^{\infty} \left[\frac{g_{2n}}{(-x)^{3n+\frac{1}{2}}} \sin(2\xi) + \right.$$

$$\left. + \frac{g_{2n+1}}{(-x)^{3n+2}} \cos(2\xi) \right], \quad \xi = \frac{2}{3}(-x)^{3/2}, \; x \to -\infty, \tag{23}$$

$$g_0 = 1, \; g_1 = -\frac{5}{2^3 \cdot 3}, \quad g_2 = -\frac{5 \cdot 41}{2^7 \cdot 3^2},$$

$$g_3 = \frac{5 \cdot 7 \cdot 11 \cdot 59}{2^{10} \cdot 3^4}, \quad g_4 = \frac{5 \cdot 7 \cdot 11 \cdot 12769}{2^{15} \cdot 3^5}, \ldots$$

The coefficients g_n satisfy the recurrence

$$g_n = \frac{(3n-5)(3n-3)(3n-1)}{2^5 \cdot 3 \cdot n} g_{n-2} + (-1)^n \frac{27n^2 - 27n + 5}{2^3 \cdot 3 \cdot n} g_{n-1}. \tag{24}$$

For the integral $\int_v^\infty dx\, Ai(x)\, h(x,v)$ in which the function $h(x,v)$ has no singularities in the integration interval and can be expanded at $x=v$ in the series

$$h(x,v) = \sum_{k=-1}^{\infty} h_k (x-v)^{k/2} = \sum_{k=-1}^{\infty} f_k \left(\frac{x-v}{-v}\right)^{k/2} \tag{25}$$

with coefficients f_k weakly depending on v ($f_k \equiv$ $\equiv h_k(-v)^{k/2} \sim f_0$), we have [3,4] the following asymptotic representation as $v \to -\infty$:

$$\frac{1}{\pi}\int_v^\infty dx\, \Phi(x) h(x) \sim \sum_{k=0}^{\infty} \left\{ \frac{h^{(3k)}(0)}{(3k)!!!} - \left[\frac{t\Phi^2 + \Phi'^2}{\pi\sqrt{-t}} - \frac{w_1''(t)}{2\pi\sqrt{-t}} \right] \right..$$

$$\cdot f_{k,-1} + \left[\frac{\Phi\Phi'}{\pi(-v)^{3/2}} - \frac{w_1'(t)}{2\pi(-v)^{3/2}}\right](f_{k,-1} + f_{k,1}) - \tag{26}$$

$$- \frac{\Phi'(v)}{\pi v} f^+_{k,0} - \frac{\Phi(v)}{\pi v^2}(f^+_{k,0} + f^+_{k,2})\Big\}, \quad t = 2^{-2/3} v.$$

Here for brevity $\Phi(x) \equiv Ai(x)$, $h(x) \equiv h(x, v)$.

The Airy functions in the quadratic combinations depend on $t = 2^{-2/3} v$,

$$f_{k,2m-1} = \frac{(m+\frac{1}{2})(m+\frac{3}{2})}{v^{3k}} \sum_{s=0}^{m+2k} R_{2s-1}(k,m) f_{2s-1}, \tag{27}$$

$$R_{2s-1}(k,m) = \sum_{n_{k-1}=n_{k-1}(s)}^{m+2} (n_{k-1}+\frac{1}{2})(n_{k-1}+\frac{3}{2}) \dots \tag{28}$$

$$\cdot \sum_{n_2=n_2(s)}^{n_3+2} (n_2+\frac{1}{2})(n_2+\frac{3}{2}) \sum_{n_1=n_1(s)}^{n_2+2} (n_1+\frac{1}{2})(n_1+\frac{3}{2}), R_{2s-1}(1,m)=1,$$

$$n_i(s) = \begin{cases} 0, & 0 \leqslant s \leqslant 2i, \\ s-2i, & s > 2i, \end{cases} \tag{29}$$

$$f^+_{k,2m} = \frac{(m+1)(m+2)}{v^{3k}} \sum_{s=0}^{m+2} T_{2s}(k,m) f_{2s}, \tag{30}$$

$$T_{2s}(k,m) = \sum_{n_{k-1}=n_{k-1}(s)}^{m+2} (n_{k-1}+1)(n_{k-1}+2) \dots$$

$$\cdot \sum_{n_2=n_2(s)}^{n_3+2} (n_2+1)(n_2+2) \sum_{n_1=n_1(s)}^{n_2+2} (n_1+1)(n_1+2), \; T_{2s}(1,m) = 1; \tag{31}$$

All the terms in the right-hand side of (26), except for $h^{(3k)}(0)$, are oscillatory functions of v as $v \to -\infty$. For example, from (23) we have

$$\frac{\Phi\Phi'}{\pi(-v)^{3/2}} - \frac{w_1'(t)}{2\pi(-v)^{3/2}} = \frac{w_2'(t)}{2\pi(-v)^{3/2}} = \left[-\frac{1}{2^3 \cdot 3 \cdot v^3} - \frac{5 \cdot 7 \cdot 8 \, 9}{2^8 \cdot 3^4 \cdot v^6} + \dots\right].$$

$$\cdot \sin(2\zeta) + \left[-\frac{1}{2(-v)^{3/2}} - \frac{5\cdot 7}{2^6\cdot 3^3(-v)^{9/2}} + \frac{5\cdot 7\cdot 11\cdot 13\cdot 107}{2^{12}\cdot 3^5(-v)^{15/2}} + \dots\right]\cos(2\zeta), \tag{32}$$

$$\frac{t\Phi^2+\Phi'^2}{\pi\sqrt{-t}} - \frac{w_1''(t)}{2\pi\sqrt{-t}} = \frac{w_2''(t)}{2\pi\sqrt{-t}} = \left[-1 - \frac{1}{2^5\cdot 3^2\cdot v^3} - \frac{5\cdot 7\cdot 3923}{2^{11}\cdot 3^5 v^6} + \dots\right]\cdot$$

$$\cdot \sin(2\zeta) + \left[-\frac{1}{2^3\cdot 3(-v)^{3/2}} - \frac{5\cdot 7\cdot 19}{2^7\cdot 3^4(-v)^{9/2}} + \dots\right]\cos(2\zeta), \tag{33}$$

$$\zeta = \frac{2}{3}(-t)^{3/2} = \frac{1}{3}(-v)^{3/2}.$$

The explicit form of the first few coefficients $f_{k,2m-1}$ and $f^+_{k,2m}$:

$$f_{0,-1} = f_{-1},\quad f_{0,1} = f_1,$$

$$f_{1,-1} = \frac{3}{4v^3}(f_{-1}+f_1+f_3), \tag{34}$$

$$f_{1,1} = \frac{15}{4v^3}(f_{-1}+f_1+f_3+f_5),$$

$$f_{2,-1} = \frac{3}{16v^6}\left[53(f_{-1}+f_1+f_3)+50f_5+35f_7\right],$$

$$f_{2,1} = \frac{15}{16v^6}\left[116(f_{-1}+f_1+f_3)+113f_5+98f_7+63f_9\right],$$

and

$$f^+_{1,0} = 2v^{-3}(f_0+f_2+f_4),$$

$$f^+_{1,2} = 6v^{-3}(f_0+f_2+f_4+f_6),$$

$$f^+_{2,0} = 4v^{-6}\left[10(f_0+f_2+f_4)+9f_6+6f_8\right], \tag{35}$$

$$f^+_{2,2} = 12v^{-6}\left[20(f_0+f_2+f_4)+19f_6+16f_8+10f_{10}\right],$$

$$f^+_{3,0} = 80v^{-9}\left[28(f_0+f_2+f_4)+27f_6+24f_8+18f_{10}+9f_{12}\right],$$

$$f^+_{3,2} = 240v^{-9}\left[84(f_0+f_2+f_4)+82f_6+76f_8+64f_{10}+45f_{12}+21f_{14}\right].$$

In conclusion of this section we note that due to the relations

$$\int_{v}^{\infty} \Phi_1(x) h(x)\, dx = \int_{v}^{\infty} \Phi(x) g(x)\, dx, \tag{36}$$

$$g(x) = \int_{v}^{x} h(x)\, dx = 2h_{-1}(x-v)^{1/2} + h_0(x-v) + \frac{2}{3} h_1 (x-v)^{3/2} + \ldots \tag{37}$$

and

$$\int_{v}^{\infty} \Phi'(x) h(x)\, dx = 2h_{-1}\Phi(t)\Phi'(t) - h_0 \Phi(v) - \int_{v}^{\infty} \Phi(x) \varphi'(x)\, dx, \tag{38}$$

$$h(x) = h_{-1}(x-v)^{-1/2} + \varphi(x) = h_{-1}(x-v)^{-1/2} + h_0 + h_1(x-v)^{1/2} + $$
$$+ h_2(x-v) + \ldots, \qquad t = 2^{-2/3} v, \tag{39}$$

the asymptotic representations for the integrals in the left-hand sides of (36) and (38) are reduced to the considered case, see (25) and (26).

2. <u>Stokes' phenomenon and the choice of saddle points</u>. Let us consider the solution of the Airy equation $w'' = zw$ defined by the integral

$$w(z) = i \int_{C} e^{-i(zt + t^3/3)}\, dt \tag{40}$$

over the contour beginning at infinity in the sector 1 and ending at infinity in the sector 3; see Figs.1-6 which qualitatively indicate the locations of saddle points t_1, t_2 and the character of the steepest descent (SD) lines from the passes in the complex t-plane for the specific values of phase with parameter z which changes in its own complex plane. The arrows mark the integration contour in (4) drawn along the SD lines. This SD contour is convenient for obtaining the asymptotic representation of w(z) by the saddle point method.

For z of the main sector $-4\pi/3 < \mathrm{ph}\, z < 0$ the SD contour goes over one saddle point and we get

$$w(z) = W_n^{(13)}(z) + R_n^{(13)}(z) \sim W_\infty^{(13)}(z), \tag{41}$$

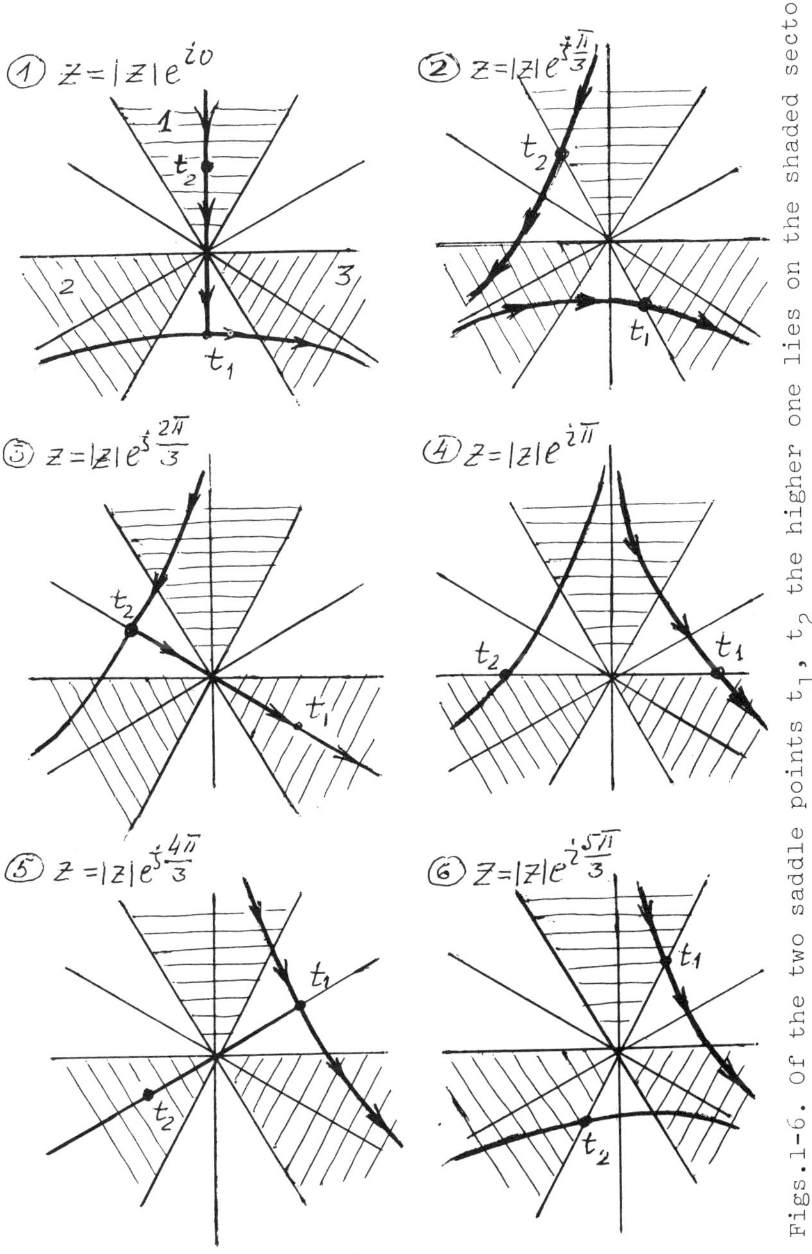

Figs.1-6. Of the two saddle points t_1, t_2 the higher one lies on the shaded sector

where

$$W_n^{(13)}(z) \equiv 2S_n(z) = \frac{\pi^{1/2}}{z^{1/4}} e^{\xi} \sum_{k=0}^{n-1} c_k \xi^{-k}, \qquad (42)$$

$$\xi = \frac{2}{3} z^{3/2}$$

is the sum of the first n terms of the asymptotic series $W_\infty^{(13)}(z)$ (obtained by the saddle point method) and $R_n^{(13)}(z)$ is the remainder term small as compared with $W_n^{(13)}(z)$ as $|z| \to \infty$.

By analytic continuation from the main sector into the complementary one through the Stokes ray phz=o the SD contour going here over the higher pass discontinuously transforms into SD contour C_{12}, also going over the higher pass, plus SD contour C_{23} going over the second, lower pass see Figs. 6, 1, 2. Hence in the complementary sector $0 < ph z < 2\pi/3$ the function w(z) is described by two asymptotic series

$$w(z) = W_n^{(12)}(z) + R_n^{(12)}(z) + W_n^{(23)}(z) + R_n^{(23)}(z) \sim$$
$$\sim W_\infty^{(12)}(z) + W_\infty^{(23)}(z); \qquad (43)$$

the principal sum

$$W_n^{(12)}(z) = W_n^{(13)}(z) = 2S_n(z) \qquad (44)$$

is the analytic continuation of (42) and the remainder jump on the Stokes ray phz=o:

$$R_n^{(13)}(z) - R_n^{(12)}(z) = w_{23}(z) \quad . \qquad (45)$$

(The analytic function $w_{ij}(z)$ is defined by the integral (40) over the contour beginning in the i- and ending in the j-sector at infinity).

On the other hand, by analytic continuation from the main sector into the complementary one through the Stokes ray phz= $-4\pi/3 = 2\pi/3 - 2\pi$ the SD contour C_{13} going here over the higher pass discontinuously transforms itself into the SD contour C_{23} plus the SD contour C_{12} going over the lower pass (Figs. 4, 3, 2). Therefore for the sum $W_n^{(13)}(z)$, principal near $\mathrm{ph}\, z = 2\pi/3$, and $R_n^{(23)}(z)$ of (43) we have

$$W_n^{(23)}(z) = W_n^{(13)}(z e^{-2\pi i}) \equiv 2 S_n(z e^{-2\pi i}) =$$

$$= i \frac{\pi^{1/2}}{z^{1/4}} e^{-\xi} \sum_{k=0}^{n-1} c_k (-\xi)^{-k}, \quad \xi = \frac{2}{3} z^{3/2}, \tag{46}$$

$$w_{12}(z) + R_n^{(23)}(z) = R_n^{(13)}(z e^{-2\pi i}). \tag{47}$$

So in the complementary sector the asymptotic representation (43) is $w(z) = 2S_n(z) + 2S_n(z e^{-2\pi i})$ + remainder (48) of the sum of two analytic continuations of the asymptotic representation from the main sector, continuations corresponding to the right and left detours of the branch point z=o. Thereby the asymptotic representation does not depend on the position of a cut.

References

1. Dingle, R.B. (1973). Asymptotic expansions: their derivations and interpretation. Acad. Press.
2. Aspnes, D.E. (1966). Phys. Rev., 147, 554.
3. Ritus, V.I. (1979). P.N. Lebedev Phys. Inst. Proc.,111, Nauka, Moscow; (1969). ZhETF, 56, 986 (Russian).
4. Nikishov, A.I., Ritus, V.I.(1985). Lebedev Phys. Inst. Preprint N 253, (Russian).
5. Antosiewicz, H.A. (1965). In: Handbook of Mathematical Functions, Abramowitz, M., Stegun, I.A. (ed.), Dover, N.Y.
6. Olver, F.W.J. (1974). Asymptotics and Special functions. Acad. Press.

LORENTZ GROUP AND SOME INTEGRAL REPRESENTATIONS FOR BESSEL FUNCTIONS

R.L. Shapiro, N.Ya. Vilenkin[+]
State Extra-mural Pedagogical Institute, Moscow[+]

1. Introduction. One of important problems of the group representation theory is obtaining new relations for special functions or the group theoretical interpretation of the known relations.

For example some integral relations for Bessel's functions result from consideration of the representations of M(2) [2], M(n) [3] and M(n,C) [4] (the motion groups of the plane, n-dimensional Euclidean and complex spaces respectively). In [5] a realization for the principal series representations of the three-dimensional Lorentz group is suggested, where generators are expressed in terms of Bessel's functions. Group theoretical approach and a method of analytical continuation enable us to obtain integral relations including the infinite-limit integrals with the integrand consisting of Bessel's and exponential functions [1]. Here we derive one of these relations.

2. Representations of SU(1,1). Realization of the Lorentz group representations in [5] is based on homomorphisms of SU(1,1) onto the Lorentz group [2]. We use this realization of representations of SU(1,1). Unlike the known realizations when elements of the compact subgroup are diagonalized, here one of hyperbolic generators, J_2, is in a simple form. Then the second hyperbolic generator J_1 and the generator J_0 of the maximal compact subgroup include Bessel's functions (in the case

of discrete series). We start here from the realization suggested for positive discrete series D_k^+ representation in [5]:

$$[e^{i\xi J_2} f](z) = e^{\xi/2} f(ze^{\xi/2}) \tag{1}$$

$$[e^{i\mu J_0} f](z) = \int_0^\infty z' dz' L^{(k,+)}(z, z'; \mu) f(z') \tag{2}$$

$$[e^{i\nu J_1} f](z) = \int_0^\infty z' dz' M^{(k,+)}(z, z'; \nu) f(z') \tag{3}$$

The kernels $L^{(k,+)}$ and $M^{(k,+)}$ are defined by the formulas

$$L^{(k,+)}(z, z'; \mu) = \\ = (e^{i\pi k}/\sin \tfrac{1}{2}\mu) \exp[-i(z^2+z'^2) \operatorname{ctg} \tfrac{1}{2}\mu] J_{2k-1}(a), \tag{4}$$

$$M^{(k,+)}(z, z'; \nu) = \\ = (e^{i\pi k}/\operatorname{sh} \tfrac{1}{2}\nu) \exp[-\tfrac{1}{2} i(z^2+z'^2) \operatorname{ctg} \tfrac{1}{2}\nu] J_{2k-1}(b), \tag{5}$$

where $a = zz'/\sin\frac{1}{2}\mu$; $b = zz'/\operatorname{sh}\frac{1}{2}\nu$.

3. Some Integral Relations for Bessel's Functions. The integral relations can be obtained with the help of (2) and (4) if the change of the order of integration in the multiple integral is possible (The validity of this assumption will be proved in Sect. 4). A similar relation can be obtained from formulas (3), (5). Alternative ratios for Bessel's functions can be found from the formulas defining continuous series representations. As

$$[e^{i\mu_2 J_0} e^{i\mu_1 J_0} f](z) = [e^{i(\mu_1+\mu_2) J_0} f](z),$$

then changing the order of integrating we get

$$L^{(k,+)}(z, z''; \mu_1 + \mu_2) = \\ = \int_0^\infty z' dz' L^{(k,+)}(z, z'; \mu_2) L^{(k,+)}(z', z''; \mu_1). \tag{6}$$

Substituting the explicit form of $L^{(k,+)}$ we get

$$e^{i\pi k\alpha}\exp[i(x^2+y^2)\alpha]J_{2k-1}(xy\alpha)=$$
$$=\int_0^\infty z\,dz'\exp(-iz'^2/\alpha)J_{2k-1}(yz')J_{2k-1}(xz') \qquad (7)$$

Here $x=(z/\sin\frac{\mu_2}{2})$, $y=(z'/\sin\frac{\mu_1}{2})$, $\alpha=(\sin\frac{\mu_1}{2}\sin\frac{\mu_2}{2}/\sin\frac{\mu_1+\mu_2}{2})$.

If $\alpha \geqslant 0$ then (7) can be transformed into the form close to that given in [1], p. 60:

$$e^{-i\pi k}\exp[i(x^2+y^2)]J_{2k-1}(xy)=$$
$$=\int_0^\infty z\,dz\,\exp(-iz^2)J_{2k-1}(zx)J_{2k-1}(yz). \qquad (8)$$

4. Appendix. Formulas (2), (4) are considered for real μ and the convergence of the integral

$$[e^{i\mu J_0}f](z)=\int_0^\infty \frac{e^{i\pi k}}{\sin\frac{\mu}{2}}\exp[-i(z^2+z'^2)\operatorname{ctg}\frac{\mu}{2}]J_{2k-1}\left(\frac{zz'}{\sin\frac{\mu}{2}}\right)z'dz'$$

is to be justified. For this and to prove the possibility to change the order of integration used to derive formula (8) let us make use of the method of analytical continuation over the parameter μ and set

$$\operatorname{ctg}\frac{1}{2}\mu=\rho+i\nu \qquad \text{for small } \nu$$

$$\sin\frac{1}{2}\mu\approx\frac{1}{\sqrt{1+\rho^2}}\left(1-i\frac{\rho\nu}{1+\rho^2}\right)$$

$$\frac{1}{\sin\frac{1}{2}\mu}\approx\sqrt{1+\rho^2}\left(1+i\frac{\rho\nu}{1+\rho^2}\right)$$

$$\frac{1}{\sqrt{\sin\frac{1}{2}\mu}}\approx\sqrt[4]{1+i\rho^2}\left(1+\frac{\rho\nu}{1+\rho^2}\right)$$

$$z=\frac{zz'}{\sin\frac{1}{2}\mu}\approx zz'\sqrt{1+\rho^2}\left(1+i\frac{\rho\nu}{1+\rho^2}\right),\ \arg z=\arg\operatorname{tg}\frac{\rho\nu}{1+\rho^2}.$$

Since $\arg z$ is small for small ν, the following, asymptotics for Bessel's functions holds (only for the case $m=0$) [3]

$$J_{2k-1}(z)\approx\sqrt{\frac{2}{\pi z}}\cos\left(z-(2k-1)\frac{\pi}{2}-\frac{\pi}{4}\right)-$$

$$-\frac{\Gamma(2k-1+\frac{3}{2})}{\Gamma(2k-1-\frac{1}{2})}\cdot\frac{1}{2z}\sin\left(z-(2k-1)\frac{\pi}{2}-\frac{\pi}{4}\right).$$

Analysing the behaviour of the integral in formula (9) for complex μ , we see that (9) converges monotonously for $\nu<0$, thus proving the validity of (8) as $\nu\to 0$.

References

1. Bateman, H., Erdelyi, A. (1966). Higher Transcendental Functions. 2, Nauka, Moscow (Russian).

2. Vilenkin, N.Ya. (1965). Special Functions and the Theory of group Representations. Nauka, Moscow (Russian).

3. Gradstein, I.S., Ryzhik, I.M. (1962). Tables of Integrals, sums, series and products. M., Physmathgiz, Moscow (Russian).

4. Shapiro, R.L. On the Problems of the Theory of Representations Associated with n-dimensional Unitary Group. - Thesis by Candidate of Phys. - Math. Science M., 1966.

5. Mukunda N., Raghakrishnan B. - J. Math. Phys., 1973, 13, 254.

POISSON TRANSFORMATION FOR HYPERBOLIC SPACE

I.I.Shitikov

Tambov State Pedagogic Institute

The real hyperbolic space is a homogeneous space $X = G/H$, where $G = SO_0(p,q)$, $H = SO_0(p,q-1)$ is the stationary subgroup of the point $x^0 = (0,\dots,0,1)$. Lorentz spaces ($p = 1$) and homogeneous spaces of de Sitter groups ($p = 1$, $q = 4$ and $p = 2$, $q = 3$) are special cases of real hyperbolic spaces. Let S be the maximum boundary of X, see [1], i.e. S is the direct product of two Euclidean spheres of unit radius: $S = S^{p-1} \times S^{q-1}$.

In the space D_ε of infinitely differentiable functions on S of parity ε($\varepsilon = 0$ or 1) the group G acts by the formula

$$T_\chi(g)\varphi(s) = \varphi\left(\frac{sg}{|sg|}\right) |sg|^{\sigma}, \chi = (\sigma, \varepsilon), \quad \sigma \in \mathbb{C}, \text{ where } x = (x_1^2 + \dots + x_p^2)^{\frac{1}{2}}.$$

The group G acts in the space D'_ε dual to D_ε via

$$(T'_\chi(g)F, \varphi) = (F, T_{\bar{\chi}^*}(g^{-1})\varphi),$$

where $F \in D'_\varepsilon$, $\varphi \in D_\varepsilon$, $\sigma^* = 2-n-\sigma$, $\chi^* = (\sigma^*, \varepsilon)$, $\bar{\chi} = (\bar{\sigma}, \varepsilon)$. In [1] the structure of T_χ and T'_χ is discribed. Set

$$\Theta_\chi = \frac{1}{\Gamma(\frac{\sigma+\varepsilon+1}{2})} |s_n|^{\sigma,\varepsilon} = \frac{1}{\Gamma(\frac{\sigma+\varepsilon+1}{2})} |s_n|^{\sigma} \operatorname{sgn}^{\varepsilon} s_n$$

(where s_n is the n-th coordinate of a point of S). This function on S is invariant with respect to H in the representation T'_χ, see [2]. Set

$$P_\chi(x,s) = |[x,s]|^{\sigma,\varepsilon} / \Gamma((\sigma+\varepsilon+1)/2)$$

clearly,

$$P_\chi(x,s) = T'_\chi(g^{-1})\,\theta_\chi(x), \quad x = x^0 g.$$

Call $P_\chi(x,s)$ the <u>Poisson kernel</u>, and the map $P_\chi : D_\varepsilon \to C^\infty(X)$ defined by the formula $(P_\chi\varphi)(x) = \int_S P_\chi(x,s) \times \varphi(s)ds$, the <u>Poisson transformation</u> corresponding to θ_χ. Let Δ be Laplace-Beltrami operator on X. An important property of the Poisson transformation is that its image consists of eigenfunctions of Δ with eigenvalue $\sigma^*\sigma$:

$$\Delta f = \sigma^*\sigma f. \tag{1}$$

Using our scheme we can get the classical Poisson integral, see e.g. [3]. In this case $G = SO_0(1, 2)$, $H = SO_0(2)$, $\theta = 1$, and the Poisson transformation is represented

$$(P\varphi)(r,\alpha) = \frac{1}{2\pi}\int_0^{2\pi} \frac{1 - r^2}{1-2r\cos(\alpha-\beta) + r^2}\,\varphi(\beta)\,d\beta.$$

Here (r, α) are polar coordinates in the unit disk.

This scheme can be transferred without difficulty on any pseudo-Riemannian symmetric space, embracing, as is known, Riemannian symmetric spaces and semisimple Lie groups. In [4] the Poisson transformation for pseudo-Riemannian symmetric spaces is also constructed but for the general representation . We consider Poisson transformation at integral points.

The Poisson **kernel** $P_\chi(x,s)$ is a generalized coherent state of quantum mechanic system, see [5]. Here G stands for the group of coherent states of the system, function θ_χ is the initial state of the system. Note that $P_\chi(x,s$ is a function of a pair of points: a point of X and a point on its boundary S. It is known, see [5], that with the help of coherent states it is possible to map X to S (for example a hyperboloid to a cone). It is possible to rise the inverse problem: to map S to X, i.e. to reconstruct the function on X by its values on S. This map is realized with the help of the Poisson transformation.

List the main properties of the Poisson transformation for a real hyperbolic space X.

1. Let H_χ be the subspace in $C^\infty(X)$, consisting of functions satisfying (1). The topology in H_χ is that of uniform convergence with all the derivatives on every compact. Let H_z be the eigensubspace of Laplace-Beltrami operator on S with eigenvalue λ_z, $z = (l, m)$, l and m - integers, see [1]. Then the Poisson transformation may be represented as follows

$$(P_\chi \varphi)(t,s) = \sum_z C_{\chi,z} R_{\chi,z}(t) \varphi_z(s)$$

where the series converges in topology of H_χ and (t,s) are polar coordinates on X, $\varphi_z \in H_z$, $R_{\chi,z}(t)$ a s lution of the radial (depending on t) part of equation (1) (it is a hypergeometric function whose explicit form we skip) Numbers $c_{\chi,z}$ are called <u>Poisson coefficients</u>. Their explicit expression is

$$C_{\chi,z} = (-1)^m 2^{2-\sigma} \cdot \pi^{n/2} \frac{\Gamma(\sigma+1)}{\Gamma\left(\frac{\sigma+\varepsilon+1}{2}\right) \Gamma\left(\frac{\sigma+q-l+m}{2}\right) \Gamma\left(\frac{2+\sigma-l-m}{2}\right)}$$

In other words we may say, that the Poisson transformation in the "basis" $R_{\chi,z}(t)$, $\varphi_z(s)$ is diagonal with a Poisson coefficients as diagonal elements.

2. If σ is non integer, then the image of the Poisson transformation is dense in H_χ . If σ is integer, we may get then the image's charactetistics with the help of G--invariant closed subspaces in H_χ. Description of such subspaces presents is of interest itself. We can describe them, but we will not do that here.

In [6] for p = 1 and q = 2 full characterization of the

image of the Poisson transformation for harmonic functions is given.

3. Let B_χ be an operator in D_ε intertwining representations T_χ and $T_{\chi*}$, see [1], $M_\chi : H_\chi \to D_\varepsilon$, defined for $\mathrm{Re}\,\sigma > \frac{1}{2}(2-n)$ by the formula

$$(M_\chi f)(s) = \lim_{t\to+\infty} \mathrm{ch}^{-\sigma} t \cdot f(t,s)$$

Then $M_\chi P_\chi = B_{\chi*} / \Gamma(\frac{\sigma+\varepsilon+1}{2})$. This fact establishes a connection of the Poisson transformation with intertwining operators. From this we get the behaviour of the Poisson transformation at infinity, and consequently on the boundary S of X.

4. Point out the connection of the Poisson transformation with the discrete series of G: The image of the Poisson transformation for $\sigma \in \mathbb{Z}$ and $\sigma < \frac{1}{2}(2-n)$ belongs to $L^2(X)$. Thus we get all the discrete series of unitary representations of the group of X.

For complex and quaternion hyperbolic spaces, i.e. for homogeneous spaces $U(p,q)/U(p,q-1) \times U(1)$ and $Sp(p,q)/Sp(p,q-1)\,Sp(1)$, the Poisson transformation is similarly constructed. The essential information about these homogeneous spaces, their boundary, representations of the pseudo-unitary and symplectic groups may be found in [7] and [8]. In this case H-invariants are expressed by the formula $\Theta_\sigma = |s_n|^\sigma / (\frac{\sigma+d}{2})$ where $\sigma \in \mathbb{C}$, $d = 2$ or 4 for complex or quaternion case respectively. The Poisson kernel (coherent state) $P_\sigma(x,s) = |[x,s]|^\sigma / \Gamma(\frac{\sigma+d}{2})$ cor esponds to this function. The Poisson transformation is the integral

$$(P_\sigma \varphi)(x) = \frac{1}{(\frac{\sigma+d}{2})} \int_S |[x,s]|^\sigma \varphi(s)ds.$$

Here S is the boundary of the hyperbolic space, see [7]. The main properties of the Poisson transformation are preserved. So, the Poisson coefficients are expressed as follows:

$$c_{\sigma,z} = (-1)^m 4\pi^{nd} \frac{\Gamma(\frac{\sigma+d}{2})}{\Gamma(\frac{1+pd}{2})\,\Gamma(\frac{2+\sigma-1-m}{2})\,\Gamma(\frac{\sigma+qd-1+m}{2})}$$

Connection with intertwining operators takes place, the formula is similar. Property 4 is transferred here almost without changes but instead of $\sigma \in \mathbb{Z}$ and $\sigma < \frac{1}{2}(2-n)$; Let $\sigma \in 2\mathbb{Z}$ and $\sigma < \frac{1}{2}(2-nd)$.

References

1. Molchanov,V.F. (1970). Representations of pseudo-orthogonal group associated with a cone, Math. USSR Sbornik, 10, 333-347. (Russian).

2. Molchanov,V.F. (1976). Spherical functions on hyperboloids, Math.USSRSbornik, 28, 119-138. (Russian).

3. Titchmarsh,E. (1939). The theory of functions. Oxford University press.

4. Oshima,T. (1979). Poisson transformation on symmetric space, Proc; Japan.Acad., 55A, 323-327.

5. Perelomov,A.M. (1977). The generalized coherent states and applications, Usp.Fis.Nauk (Russian Phys.Surveys) 123, 23-55. (Russian).

6. Shitikov,I.I. (1982). Poisson transformation for imaginery Lobochevskian space. Functional Analysis, Ulianovsk, 19, 162-171. (Russian).

7. Molchanov,V.F. (1984). Representations of pseudo-unitary group associated with cone, Functional Analysis, Ulianovsk, 22, 55-66. (Russian).

8. Faraut,J. (1979). Distributions spheriques sur les espaces hyperboliques. - S.Math; pures et appl., 58, 369-444.

ON THE CONTINUOUS SUBGROUPS OF THE GENERALISED POINCARE GROUP P(1,n)

L.F.Barannik, A.F.Barannik, W.I.Fushchich [+]

V.G.Corolenko Poltava pedagogical Institute
[+] Mathematical Institute of the Ukrainian SSR Academy of Sciences, Kiev

1. Introduction. A number of problems of theoretical and mathematical physics needs a description of the subgroup structure of the generalised Poincaré group: investigation of physical systems with variable mass and spin [1-3], construction of exact solutions of some nonlinear differential equations [4], reduction of the representations of P(1,n) to its subgroups ([5]).

The systematic study of the continuous subgroups of the inhomogeneous groups of transformations of the quantum mechanics was started in [6], where the general method was suggested for classification of subalgebras of finite dimensional Lie algebras L with nontrivial abelian ideal N as a semidirect summand: $L = N \oplus E$ up to conjugation by a given subgroup. Subalgebras of Lie algebras of P(1,3) ([6]), P(1,4) ([7,8]) and Euclidean groups E(3) 9 , E(4) 10 were classified. But because of its very general character this method cannot be always effectively used.

In this paper we develope Patera-Winternitz-Zassenhaus method for groups P(1,n) which gives a possibility to reduce the problem of classification of the subalgebras of P(1,n) up to P(1,n)-conjugation to the description of the irreducible subalgebras of the AO(1,k) up to O(1,k)-conjugation and to the description of the irreducible subalgebras of AO(k) (k = 2,...,n) up to O(k)-

conjugation.

2. Maximal subalgebras. Let $\langle X_1,\dots,X_s\rangle$ be a vector space spanned by $X_1,\dots,X_s$ over $\mathbb{R}$, $U = U^{n+1}$ the $(n+1)$-dimensional pseudoeuclidean space with the scalar product

$$(X,Y)=x_0y_0-x_1y_1-\dots-x_ny_n; \qquad (1)$$

$O(1,n)$ the group of linear transformations of U^{n+1} preserving $(,)$. Suppose that $O(1,n)$ is realized by real $(n+1)\times(n+1)$ matrices.

The Poincaré group $P(1,n)$ is the multiplicative group of matrices of the form $\begin{pmatrix}\Delta & y\\ 0 & 1\end{pmatrix}$, where $\Delta\in O(1,n)$, $y\in\mathbb{R}^{n+1}$.

Denote by AG the Lie algebra of a Lie group G. Then $AP(1,n)$ is determined by the following commutation relations:

$$[J_{\alpha\beta},J_{\gamma\delta}]=g_{\alpha\delta}J_{\beta\gamma}+g_{\beta\gamma}J_{\alpha\delta}-g_{\alpha\gamma}J_{\beta\delta}-g_{\beta\delta}J_{\alpha\gamma},$$
$$[P_\alpha,J_{\beta\gamma}]=g_{\alpha\beta}P_\gamma-g_{\alpha\gamma}P_\beta,\ J_{\beta\alpha}=-J_{\alpha\beta},\ [P_\alpha,P_\beta]=0, \qquad (2)$$

where $g_{00}=-g_{11}=\dots=-g_{nn}=1$, $g_{\alpha\beta}=0$ for $\alpha\neq\beta$ $(\alpha,\beta=0,1,\dots,n)$.

Generators of rotations $J_{\alpha\beta}$ generate $AO(1,n)$, these of translations P_α generate the commutative ideal N and $AP(1,n) = N \oplus AO(1,n)$.

Let C be a $(n+2)\times(n+2)$ matrix such that $\varphi_C: X\to CXC^{-1}$ is an automorphism of $AP(1,n)$. If $C\in G$, $G\subset P(1,n)$ then φ_C is called a G-automorphism. If $C=\mathrm{diag}(\lambda E, 1)$ $(\lambda\in R,\ \lambda>0)$ then φ_C is called homothety.

Subalgebras K and K' of $AP(1,n)$ are called $P(1,n)$-conjugate if there exists a $P(1,n)$-automorphism φ_C of $AP(1,n)$ such that $\varphi_C(K) = K'$.

Let us identify N with the space U of $(n+1)$-dimensional columns: $P_i\to(0,\dots,0,1,0,\dots,0)$ 1 on the i-th place. We consider elements of $AO(1,n)$ as $(n+1)\times(n+1)$ matrices. Then $[X,Y] = X\cdot Y$ for every $X \in AO(1,n)$, $Y \in N$.

Let $W \subset U$ be a nondegenerate subspace. A subalgebra $F \subset AO(W)$ is called irreducible if the identity representation of F in $AO(W)$ is irreducible and fully reducible if so is its identity representation.

Theorem 1. The maximal subalgebras of $AO(1,n)$ are exhausted up to $O(1,n)$-conjugation by the maximal irreducible subalgebras and by the following algebras: $AO(n)$;

$$AO(1,k) \oplus AO'(n-k),$$

where

$$AO'(n-k) = \langle J_{ab} \,/\, a,b = k+1,\dots,n\rangle,\ k = 2,\dots,n-1;$$

$$\langle G_1,\dots,G_{n-1}\rangle \dashv\!\!\!\!\oplus (AO(n-1) \oplus \langle J_{on}\rangle),$$

where $G_a = J_{oa} - J_{an}$ $(a = 1,2,\dots,n-1)$.

Proof. Let F be a subalgebra of $AO(1,n)$, $U_1 \subset U$ an F invariant subspace. If U_1 is degenerate then it contains a 1-dimensional F-invariant isotropical subspace W conjugate to $\langle P_o+P_n\rangle$. In this case F is a subalgebra of the algebra $L = \{X \in AO(1,n) : XY \in W$ for any $Y \in W\}$. Quite clearly $L = \langle G_1,\dots,G_{n-1}\rangle \dashv\!\!\!\!\oplus (AO(n-1) \oplus \langle J_{on}\rangle)$. If U_1 is a nondegenerate space then it is pseudoeuclidean or euclidean.

According to Witt's theorem the normalizer U_1 in $AO(1,n)$ is conjugate to $AO(n)$ or $AO(1,k) \oplus AO'(n-k)$. Theorem is prosed.

Let $AE(n) = \langle P_1,\dots,P_n\rangle \dashv\!\!\!\!\oplus AO(n)$, $AE'(n-k) = \langle P_{k+1},\dots,P_n\rangle \dashv\!\!\!\!\oplus AO'(n-k)$, $A\tilde{G}(n-1)$ be the extended Galilei algebra with the basis $M = P_o+P_n$, $P_o, P_1,\dots,P_{n-1}$, $G_1,\dots,G_{n-1}$, $J_{ab}(a,b = 1,\dots,n-1)$. By Theorem 1 the description of subalgebras of $AP(1,n)$ is reduced to the description of subalgebras of $AE(n)$, $AP(1,k) \oplus AE'(n-k)$, $A\tilde{G}(n-1) \dashv\!\!\!\!\oplus \langle J_{on}\rangle$ up to $P(1,n)$ conjugation and also algebras $U \dashv\!\!\!\!\oplus F$ where F is an irreducible subalgebra of

AO(1,n).

Let π be a projection of AP(1,n) on AO(1,n), F a subalgebra of AO(1,n) and $\hat{F}$ a subalgebra of AP(1,n) such that $\pi(\hat{F})=F$. If $\hat{F}$ is P(1,n)-conjugate to $W \oplus F$, where W is an F-invariant subspace of U, then F is called splitting in AP(1,n). When every $\hat{F} \subset AP(1,n)$ is splitting we say that F possesses splitting extensions in AP(1,n) .Splitting subalgebras of other algebras of inhomogeneous transformations are similarly defined.

<u>Statement 1</u>. Let $L = \langle G_1,\dots,G_{n-1}\rangle \oplus (AO(n-1) \oplus \oplus \langle J_{on}\rangle)$, where $G_a = J_{oa}-J_{an}$ $(a=1,2,\dots,n-1)$. A subalgebra $F \subset AO(n-1) \oplus \langle J_{on}\rangle$ possesses only splitting extensions in L if and only if F is semisimple or F is not conjugate to a subalgebra of AO(n-2).

It is not difficult to establish that AO(1,n) has only one maximal solvable subalgebra $\langle G_1 \dots G_{n-1},\dots,J_{on}\rangle$ $(2k=n-1)$ up to O(1,n)-conjugation if n is odd.

If n is even then AO(1,n) has two maximal solvable subalgebras $\langle J_{12},J_{34},\dots,J_{n-1,n}\rangle$ and $\langle G_1,\dots,G_{n-1}, J_{12},J_{34},\dots,J_{n-3,n-2},J_{on}\rangle$.

This and Statement 1 implies the description of the maximal abelian subalgebras of AO(1,n).

<u>Statement 2.</u> The maximal abelian subalgebras of AO(1, n) are exhausted up to O(1,n)-conjugation by the following algebras:

1) $n=2k+1$: $\langle J_{12}, J_{34},\dots, J_{n-2,n-1}, J_{on}\rangle$; $\langle G_1,G_2,\dots,G_{n-1}\rangle$; $\langle G_1,G_2,\dots,G_{2m}, J_{2m+1,2m+2},\dots,J_{n-2,n-1}\rangle$ $(m=1,\dots,k-1)$;

2) $n=2k$; $\langle J_{12},J_{34},\dots,J_{n-1,n}\rangle$; $\langle J_{12},J_{34},\dots,J_{n-3,n-2},J_{on}\rangle$; $\langle J_{12},J_{34},\dots,J_{n-3,n-2},G_{n-1}\rangle$; $\langle G_1,G_2,\dots,G_{n-1}\rangle$; $\langle G_1,\dots,G_{2m},G_{n-1},J_{2m+1,\, 2m+2},\dots,J_{n-3,n-2}\rangle$ $(m=1,\dots,k-2)$.

2. Fully reducible subalgebras.

Theorem 2. If $n \geqslant 2$ then an irreducible subalgebra of $AO(1,n)$ is swmisimple and compact.

Proof. Let F be an irreducible subalgebra of $AO(1,n)$. Then $F = 2(F) \oplus Q$, where $Z(F)$ is the centre and Q is the Levi factor. If F is absolutely irreducible then from Schur's lemma we get $Z(F)=0$. If F is not absolutely irreducible and $Z(F)\neq 0$ then $\dim Z(F)=1$ and a squared nonzero matrix of $Z(F)$ coincides with $(-\lambda^2)E$, where $\lambda \in R$, $\lambda \neq 0$. The latter contradicts Statement 2. So F is semisimple.

If F is compact then there exists symmetric matrix $C \in GL(n+1, \mathbb{R})$ a such that $C^{-1}FC \subset AO(n+1)$. Since $\exp(C^{-1}FC) = C^{-1}\exp FC$, then an irreducible group in $O(n+1)$ preserves $x_0^2+\ldots+x_n^2$ and $\lambda_0^2 x_0^2 - \lambda_1^2 x_1^2 - \ldots$ $\ldots - \lambda_n^2 x_n^2$ $(\lambda_0, \lambda_1, \ldots, \lambda_n$ are zero real numbers). The contradiction proves the second part of the theorem.

Statement 3. A reducible subalgebra F of $AO(1,n)$ is fully reducible if and only if it is conjugate to a subalgebra of one of the following algebras: $AO(n)$, $AO(n-1) \oplus \langle J_{on} \rangle$ or to $L_1 \oplus L_2$, where L_1 is an irreducible subalgebra of $AO(1,k)$ $(k > 1)$ and L_2 is a subalgebra of $AO'(n-k)$.

Statement 3 is a consequence of Statement 1, Theorem 2 and the fact that the G_a-action is not fully reducible on the space $\langle P_o+P_n, P_a \rangle$.

Statement 4. A fully reducible subalgebra F of the algebra $AO(1,n)$ possesses only splitting extensions in $AP(1,n)$ if and only if F is a semisimple algebra or it is not conjugate to a subalgebra of one the following algebras: $AO(n)$, $AO(1,n-1)$.

Let Γ be an identity representation of a fully reducible subalgebra $F \subset AO(1,n)$ not conjugate to a subalgebra of $AO(n-1) \oplus \langle J_{on} \rangle$. Then $\Gamma = \Gamma_1 + \ldots + \Gamma_m$, where Γ_i is an irreducible representation of F in $AO(W_i)$

$(i = 1,\dots,m)$. Set $F_i = \{\mathrm{diag}(0,\dots,\Gamma_i(X),\dots,0) \mid X \in F\}$. Then F_i is an irreducible subalgebra of $AO(W_i)$. If $F_i \neq 0$ then F_i is called an irreducible part of F. If Γ_i and Γ_j are equivalent representations then suppose that $\Gamma_i(X) = \Gamma_j(X)$ for every $X \in F$. Having united equivalent non-zero representations we obtain non-zero disjunctive primary subrepresentation of the representation Γ. Subalgebras of $AO(1,n)$ corresponding to them and constructed according to the same rule as irreducible parts F_i are called primary parts of algebra F.

Theorem 3. Let $K_1, K_2, \dots, K_q$ be primary parts of a non-zero fully reducible subalgebra F of $AO(1,n)$, $V \subset U$ F-invariant subspace. Then $V = V_1 \oplus \dots \oplus V_q \oplus \tilde{V}$, where $V_i = [K_i, V_i] = [K_i, V]$, $[K_j, V_i] = 0$ for $j \neq i$, $V = \{X \in V \mid [F,X] = 0\}$. If a primary algebra K is a semidirect sum of irreducible subalgebras $S_1, S_2, \dots, S_r$ of $AO(W_1)$, $AO(W_2), \dots, AO(W_r)$ respectively then up to $O(1,n)$ conjugation non-zero subspaces W of U such that $[K,W] = W$ are exhausted by the subspaces $W_1, W_1 \oplus W_2, \dots, W_1 \oplus W_2 \oplus \dots \oplus W_r$.

Proof. The expansion $V = V_1 \oplus \dots \oplus V_q \oplus \tilde{V}$ is a consequence of Goursat's theorem on subalgebras of the direct sum of Lie algebras. Let K be a primary part of F. By Statement 3 we may assume that K is a primary subalgebra of $AO(n)$. It follows directly from Schur's lemma that if Q is an irreducible subalgebra of $AO(m)$ then the group of automorphisms of the algebra $\langle P_1, \dots, P_m \rangle \oplus Q$ expands into a direct product of $E(m)$ and the homothety group. Hence the second statement of Theorem 3 follows.

By Theorem 3 the description of subalgebras $\hat{F} \subset AP(1,n)$ for which $\pi(\hat{F})$ is a fully reducible algebra not conjugate to a subalgebra of $AO(n-1) \oplus \langle J_{0n} \rangle$ reduces to the description of irreducible representations of $AO(1,k)$

and $AO(k)$ $(k = 2,3,\ldots,n)$. The remaining cases can be reduced to that of $A\tilde{G}(n-1) \dotplus \langle J_{on}\rangle$.

3. Subalgebras of Galilei algebra. The Galilei algebra $AG(n)$ is determined by the following commutation relations:

$$[J_{ab},J_{cd}]=\delta_{ad}J_{bc}+\delta_{bc}J_{ad}-\delta_{ac}J_{bd}-\delta_{bd}J_{ac};$$

$$[P_a,J_{bc}]=\delta_{ab}P_c-\delta_{ac}P_b;\ [P_a,P_b]=0;$$

$$[G_a,J_{bc}]=\delta_{ab}G_c-\delta_{ac}G_b;\ [G_a,G_b]=0;$$

$$[P_a,G_b]=0;\ [P_0,J_{ab}]=[P_0,P_a]=0;\ [G_a,P_0]=P_a$$

$$(a,b,c,d=1,2,\ldots,n).$$

This algebra is a quotient of $A\tilde{G}(n)$ modulo its centre $\langle M\rangle$. We denote by $A\bar{G}(n)$ the isochronous Galilei algebra obtained of $AG(n)$ modulo $\langle P_0\rangle$.

Statement 5. A subalgebra $F\subset AO(n)$ possesses only splitting extensions in $A\bar{G}(n)$ if and only if F is semisimple and not conjugate to a subalgebra of $AO(n-1)$.

The algebra $A\bar{G}(n)$ is the semidirect sum of two euclidean algebras $\langle P_1,\ldots,P_n\rangle \dotplus AO(n)$ and $\langle G_1,\ldots,G_n\rangle \dotplus AO(n)$. Therefore Statement 5 follows from Statement 1.

Statement 6. Let $m=[n/2]$, $\delta=0$, for $n=2m$, $\delta=1$ for $n=2m+1$. The maximal abelian subalgebras of $AG(n)$ are exausted up to $G(n)$-conjugation by the following algebras:

$$\langle P_0,P_1,\ldots,P_n\rangle;\ \langle G_1+\alpha P_0,P_1,P_2,\ldots,P_n\rangle\ (\alpha>0);$$

$$\langle G_1,\ldots,G_n,P_1,\ldots,P_n\rangle;\ \langle P_0,\delta P_n,J_{12},J_{34},\ldots,J_{2m-1,2m}\rangle;$$

$$\langle G_n,P_n,J_{12},J_{34},\ldots,J_{2m-1,2m}\rangle\quad (n=2m+1);$$

$$\langle J_{12},J_{34},\ldots,J_{2a-1,2a},P_0,P_{2a+1},\ldots,P_n\rangle\ (a=1,\ldots,m-1);$$

$$\langle J_{12}, J_{34}, \ldots, J_{2a-1,2a}, P_{2a+1}, \ldots, P_n, G_{2a+1}, \ldots, G_n \rangle \ (a=1,\ldots,m-1);$$

$$\langle J_{12}, J_{34}, \ldots, J_{2a-1,2a}, G_{2a+1} + \alpha P_0, P_{2a+1}, \ldots, P_n \rangle \ (\alpha > 0;\ a=1,\ldots,m-1)$$

Proof. AG(n) has only one maximal solvable subalgebra up to G(n)-conjugation:

$$K = \langle P_0, P_1, \ldots, P_n, G_1, \ldots, G_n, J_{12}, J_{34}, \ldots, J_{2m-1,2m} \rangle.$$

Let L be a maximal abelian subalgebra of AG(n). Then $L \subset K$. If the projection of L onto $\langle P_0 \rangle$ does not vanish then $P_0 \in L$ and the projection of L onto $\langle G_1, \ldots, G_n \rangle$ vanishes or the projection of L onto $\langle P_0, G_1, \ldots, G_n \rangle$ coincids with $\langle G_{2s+1} + \alpha P_0 \rangle$. It is not difficult that the projection of L onto $\langle J_{2d-1,2d}, P_{2d-1}, P_{2d}, G_{2d-1}, G_{2d} \rangle$ coincides with $\langle J_{2d-1,2d} \rangle$ or with a subalgebra of $\langle P_{2d-1}, P_{2d}, G_{2d-1}, G_{2d} \rangle$. It means that L is conjugate to one of the algebras listed in the Statement.

Every subalgebra F of AO(n) is a fully reducible Lie algebra of linear transformations of $\langle P_1, \ldots, P_n, G_1, \ldots, G_n \rangle$. In the same way as in Theorem 3 we see that the description of subalgebras of AG(n) can be reduced to the description of irreducible subalgebras of AO(m) (m = 2,...,n) and to the classification of spaces obtained by uniting subspaces irreducible with respect to a given algebra F AO(m) accordingly to $\langle P_1, \ldots, P_n \rangle$, $\langle G_1, \ldots, G_n \rangle$.

Note that in [11] all the subalgebras of AG(3) up to G(3)-conjugation are listed improving the classification given in [12].

We write $F: S_1, \ldots, S_m$ instead of $S_1 \niplus F, \ldots, S_m \niplus F$. Let $V_a^b = \langle P_a, \ldots, P_b \rangle$, $W_a^b = \langle G_a, \ldots, G_b \rangle$, $N_{a,b} = \langle G_a + \alpha P_b, G_b - \alpha P_a \rangle$.

Theorem 4. Let F be the projection of $\hat{F} \subset AG(4)$ on AO(4). Subalgebras $\hat{F}$ of AG(4) such that $\dim F \geqslant 3$ are exhausted by subalgebras of AG(3) and the following algebras up to G(3)-conjugation:

$$AO(4): 0, \langle P_0\rangle, V_0^4, V_1^4, W_1^4, V_1^4+W_1^4, V_0^4+W_1^4;$$

$$\langle J_{12}+J_{34}, J_{13}-J_{24}, J_{23}+J_{14}, J_{12}-J_{34}\rangle: 0, \langle P_0\rangle,$$

$$V_0^4, V_1^4, W_1^4, N_{1,2}^{\alpha}+N_{3,4}^{-\alpha}, V_1^4+W_1^4, V_0^4+W_1^4 \ (\alpha \neq 0);$$

$$\langle J_{12}+J_{34}, J_{13}-J_{24}, J_{23}+J_{14}, J_{12}-J_{34}+\alpha P_0\rangle: 0, V_1^4,$$

$$V_1^4+W_1^4 \ (\alpha \neq 0);$$

$$\langle J_{12}+J_{34}, J_{13}-J_{24}, J_{23}+J_{14}\rangle: 0, \langle P_0\rangle, V_0^4, V_1^4, W_1^4,$$

$$N_{1,2}^{\alpha}+N_{3,4}^{-\alpha}, V_1^4+W_1^4, V_0^4+W_1^4 \ (\alpha \neq 0);$$

$$AO(3) \oplus \langle G_1, G_2, G_3, G_4+\lambda P_4\rangle \ (\lambda \neq 0);$$

$$AO(3): S_j, V_0^3+S_j \ (j=1,2,6), V_1^3+S_j \ (j \neq 4,7),$$

$$W_1^3+S_j \ (j=1,2,6), V_1^3+W_1^3+S_j,$$

where S_j coincides with one of the spaces:

$$\langle P_4\rangle, \langle G_4\rangle, \langle P_0+\alpha G_4\rangle, \langle P_0, P_4\rangle, \langle P_0+\alpha G_4, P_4\rangle,$$

$$\langle P_4, G_4\rangle, \langle P_0, P_4, G_4\rangle \ (\alpha > 0).$$

References

1. Fushchich, W.I., Krivsky, I.Yu. (1968). Nucl. Phys. B7, 79.
2. Fushchich, W.I., Krivsky, I.Yu. (1969). Nucl. Phys. B14, 573.
3. Fushchich, W.I. (1974). Lett. Nuovo Cim. 10, 163.
4. Fushchich, W.I. (1983). In: Algebra-theoretical methods in problems of mathematical physics. Math.Inst. the Ukr.SSR Acad. of Sci., p. 4, Kiev (Russian).

5. Nikitin,A.G., Fushchich,W.I., Yurik,I.I. (1976).Theor. Math. Phys. 26, 206 (Ruaaian).

6. Patera,J., Winternitz,P., Zassenhaus,H. (1975). J. Math. Phys. 8, 1597.

7. Fedorchuk,V.M. (1979). Ukr. Mat. zhurn. 31, 717.

8. Fedorchuk, V.M. (1981). Ukr. Mat. zhurn. 33, 696.

9. Beckers,J., Patera,J., Perroud,M. , Winternitz,P. (1977). J. Math. Phys. 18,72 .

10. Barannik,A.F., Barannik,L.F., Moskalenko,Yu.D.(1983). In: Algebra-theoretical methods in problems of mathematical physics. Math. Inst. the Ukr. SSR Acad. Sci. p. 119, Kiev (Russian).

11. Fushchich,W.I., Barannik,A.F., Barannik, L.F. (1985). Prepr. 19, 85. Math. Inst. the Ukr.SSR Acad. Sci., Kiev.

12. Sorba,P. (1976). J. Math. Phys. 17, 941.

Part VI
QUANTUM AND CLASSICAL MECHANICS

PERTURBATION THEORY AT LARGE ORDER

R.Damburg, R.Propin, V.Martyshchenko[+]

[+]Institute of Electronic and Computer Sciences, Riga

1. Introduction. Perturbation theory (PT) was introduced in quantum mechanics by Schrödinger in 1926. The Reyleigh-Schrödinger perturbation theory which has its roots in classical mechanics and mathematics is the most general technique solving quantum-mechanical eigenvalue problems.

For many problems of quantum mechanics it is easy to calculate the first two or three terms in perturbation series (PS). However, with growth of the order of the term the difficulties very rapidly increase. In the early days quantum mechanics calculating a considerable number of coefficients in PS was impossible because computers were not available. Perhaps that is why large order perturbation theory (LOPT) was not studied long ago.

At the end of 60-s (PT) went into a more sophisticated stage.

Note that PS are usually (but not always, as we will show) asymptotic and have zero radius of convergence. Therefore, to obtain valuable information PS should be sheared. (We do not discuss here the summation procedure for asymptotically diverging series using Pade or Borel methods which allow, in principle, to determine eigenvalues exactly [12].)

It is known that in the cases where the terms in asymptotic series are sign-changing, the highest accurasy is achieved if the series is sheared just before the smallest in magnitude term. Then, the error of asymptotic calculation does not exceed the first discarded term. But in our cases the terms are not sign-changing. The situation here is more complicated and this question has not been discussed in mathematical literature.

We will not consider the whole history of the LOPT but only mention some important papers. Bender and Wu [1] and Simon [2] were the first to study LOPT in the problem of anharmonic oscillator with positive and negative anharmonicity αx^4. The analysis was extended by many people including us [3].

There are close connections between the LOPT anharmonic oscillator, Stark's effect for hydrogen and the so-called two- and three-well problems of quantum mechanics [3-13]. In all the cases when eigenvalues are calculated asymptotically in addition to PS, some exponentially small terms (in the large parameter of the problem) should be taken into consideration. Exponentially small terms, which are imaginary, allow one to determine asymptotical behaviour of the coefficients of PS for large orders. Consequently, the possibility to calculate analytically the term which is the smallest in magnitude in PS for the chosen value of the large parameter of the problem, arises. This term determines the value of the intrinsic error for the whole asymptotic expansion for the eigenvalue. The highest accuracy for asymptotic expansion is achieved if PS is sheared just before the smallest term but all exponentially small corrections greater than the smallest term in PS are taken into account.

A comprehensive review of the LOPT was given by Simon in [2] who summarized rigorous results and listed the open problems. Some of those are now solved at least partially [9, 11]. We present two examples not considered by Simon.

2. Coulomb angular spheroidal equation (CASE). CASE results after separating variables in the Schrödinger equation for $Z_1 e Z_2$ ion $(-1 \leqslant \eta \leqslant 1)$:

$$\frac{d}{d\eta}\left[(1-\eta^2)\frac{dY}{d\eta}\right]+\left[-A-p^2(1-\eta^2)+2pX\eta-\frac{m^2}{1-\eta^2}\right]Y=0. \quad (1)$$

Consider p as a large parameter, as constant. Set

$$\beta=(2p(X-m-1)-A)/4p. \quad (2)$$

The spectrum of (1) is infinite discrete and nondegenerate. Set $Y(\eta)=\exp(p\eta)(1-\eta^2)^{\frac{1}{2}m}\Phi(\eta)$ and $Y=2p(1-\eta)$. Then (1) takes the form

$$Y\frac{d^2\Phi}{dY^2}+(m+1-Y)\frac{d\Phi}{dY}+\beta\Phi=$$

$$=\frac{1}{4p}\left\{Y^2\frac{d^2\Phi}{dY^2}+(2m+2-Y)Y\frac{d\Phi}{dY}+[m(m+1)+(X-m-1)Y]\Phi\right\}. \quad (3)$$

Using standard PS we can find β_{PT} in the form

$$\beta_{PT}=\sum_{k\geq 0}\frac{\beta_k}{(4p)^k}, \quad \beta_0=n \quad (4)$$

where n is a positive integer or zero. The coefficients β_k can be determined from recurrence relations up to la large k for any X, n, m.

To analyse the results, we consider the behaviour of β_k for large k [14]

$$\beta_k=-\frac{\Gamma(k+4n+2m+2-2X)}{\Gamma(n+1)\Gamma(n+m+1)\Gamma(n+1-X)\Gamma(n+m+1-X)}\Big[1+ \quad (5)$$

$$+\frac{\rho_1}{k+4n+2m+1-2X}+\frac{\rho_2}{(k+4n+2m+1-2X)(k+4n+2m-2X)}+\dots\Big]$$

Formulas for ρ_1 and ρ_2 are given in [14].

In Table I we compare the results for β_{40} with the asymptotical values obtained by formula (5) for noninteger X.

Table I. Accuracy of the asymptotic formula (5) for β_{40}.

X	$\beta_{40,\text{numer.}}$
0.25	-3.11234(48)
0.50	-2.43173(47)
1.75	-3.79927(42)

Formula (5)		
1 term of expansion	2 terms of expansion	3 terms of expansion
-3.46853(48)	-3.11525(48)	-3.11317(48)
-2.59714(47)	-2.43482(47)	-2.43211(47)
-3.61877(42)	-3.79970(42)	-3.79964(42)

When X is integer, there are three different possibilities: a) $n+1-X>0$; b) $n+1-X\leq 0$ but $n+m+1-X>0$; and c) $n+m+1-X\leq 0$.

In the case (a) two different eigenvalues of (1) have the same PS. The degeneracy is removed by exponentially small terms. In cases (b) and (c) according to (5) $\beta_k=0$ for all k. It means that the corresponding β_k grow with k essentially slower than formula (5) predicts.

The case (c) is trivial: (4) converges to the exact value of β if $\rho > \rho_0$, where ρ_0 depends on χ, n, m. The exact values of β for the case (c) can be also obtained solving an algebraic equation of order $\chi - m$.

The case (b) is much more unexpected. Here the series (4) also converge. However, β_{PT} does not coincide with the exact value of β. For example, for $\chi = 2$, $n = 1$, $m = 2$ the correct asymptotic expression for β is

$$\beta_{as} = \frac{1}{2} + \frac{1}{4\rho} + \frac{1}{2}\sqrt{1 - \frac{2}{\rho} + \frac{1}{4\rho^2}} + \frac{1}{6}(4\rho)^6 \exp(-4\rho)\left[1 - \frac{26}{4\rho} + \frac{222}{(4\rho)^2} + \dots\right], \tag{6}$$

whereas

$$\beta_{PT} = \frac{1}{2} + \frac{1}{4\rho} + \frac{1}{2}\sqrt{1 - \frac{2}{\rho} + \frac{1}{4\rho^2}}. \tag{7}$$

States with $m = 1$, $\chi = n + 1$ represent the simplest examples of the case (b). We have $\beta_{PT} = n$ here. Exponentially small terms here can be formed from comparatively simple recurrences, e.g. for $n = 0$

$$\beta_{as} = (4\rho)^2 \exp(-4\rho) \sum_{k=0} \frac{1}{(4\rho)^k} \sum_{s=0}^{k} x_s x_{k-s} \tag{8}$$

where $x_0 = 1$, $x_s = -\sum_{1 \le \ell \le s} (\ell+1)!\, x_{s-\ell}$, $s \ge 1$.

In [14] the asymptotic solutions of (1) are discussed in more detail.

3. The O(2) anharmonic oscillator and the double-well potential. Consider asymptotic solutions $(R \to \infty)$ for the Scrödinger equations for the $O(2)$ anharmonic oscillator with negative anharmonicity

$$\left(-Y\frac{d^2}{dY^2}-\frac{1}{4Y}+\frac{1}{4}Y-\frac{1}{8R^2}Y^2-E^{osc}\right)\Psi(Y)=0,$$
$$0\leqslant y<\infty,\quad E^{osc}=\operatorname{Re}E^{osc}\pm\frac{1}{2}i\gamma \tag{9}$$

and for the double-well potential

$$\left[-\frac{1}{2}\frac{d^2}{dX^2}+\frac{1}{8R^2}(R^2-X^2)^2-E_{g.u}\right]\Psi_{g.u}(X)=0,$$
$$-\infty<X<\infty,\quad E_{g.u}=E_c\mp\frac{1}{2}\Delta E. \tag{10}$$

We see that

$$E^{osc}_{PT}=(\operatorname{Re}E^{osc})_{PT}=(E_{g.u})_{PT}=(E_c)_{PT}=\sum_{k=0}\frac{E_k}{R^{2k}},\ E_0=n+\frac{1}{2}. \tag{11}$$

This was first noted by Seznec and Zinn-Justin [15] who calculated numerically 60 coefficients E_k for both cases and found that they coincided with accuracy of at least 12 places and to all orders computed. This observation was extended by Avron and Seiler [16] who showed that the agreement holds analytically for all the states up to 11 orders at least. A rigorous proof of (11) is given in [11].

However, from (11) the identity of the eigenvalues of (9) and (10) does not follow.

We have calculated [11] the intrinsic error ω for PS (11). For $n=0$

$$\omega=\left|\frac{E_k}{R^{2k}}\right|_{min}=\sqrt{\frac{3}{2\pi}}\,4R\exp\left(-\frac{4}{3}R^2\right)\left[1-\frac{31}{12R^2}+O(R^{-4})\right]. \tag{12}$$

Calculations show [11] that the asymptotic expansion for

$\operatorname{Re} E^{osc}$ does not contain real exponentially small terms which are greater than ω. Therefore, the asymptotic expansion for $\operatorname{Re} E^{osc}$ simply coincides with (11).

We have also studied the function

$$f(R) = E_c(R) - \operatorname{Re} E^{osc}(R) \tag{13}$$

and found

$$f(R) = \frac{\gamma}{2\pi}\left[2\ln R + 3\ln 2 - \Psi(1) + O(R^{-2})\right] \tag{14}$$

where

$$\gamma = 8R^2 \exp\left(-\frac{4}{3}R^2\right)\left[1 + O(R^{-2})\right]. \tag{15}$$

Comparing (14), (15) and (12), we see that $f(R) > \omega$. Therefore, calculating E_c asymptotically it is necessary to take into account $f(R)$, i.e. $E_c = E_{PT} + f(R)$. Then, the error in the value of E_c does not exceed ω. IN [11] we presented the comparisons of the results obtained by the asymptotic formulas for E_c, $\operatorname{Re} E^{osc}$, $f(R)$, ΔE, γ, E_k, etc. with numerical data for different values of R.

4. Conclusion. It is known that the asymptotic expansions are used for practical calculations even more often than the converging ones.

But the above two examples show that in this area there still remain many interesting open problems.

The examined problems can be also solved numerically, but numerical solution requires some "input" data for eigenvalues, and here again the asymptotic values are very desireable, especially if the highly excited states are involved.

References

1. Bender, C., Wu, T.T., (1969). Phys.Rev., 184, 1231; (1973). D7, 1620.

2. Simon, B., (1970). Ann.Phys., NY, 58, 76; (1982). I.J.Q.C., 21, 3.

3. Damburg, R., Propin, R., (1983). J.Phys.B, 16, 3741.

4. Damburg, R., Propin, R. (1968). J.Reine Angew. Math., 233, 28.

5. Damburg, R., Propin, R. (1968). J.Phys.B, 1, 681.

6. Damburg, R., Propin, R. (1971). J.Chem.Phys., 55, 612.

7. Damburg, R., Propin, R. (1982). Preprint LAFI-050, Salaspils.

8. Damburg, R., Propin, R. (1982). Izv.AN Latv.SSR. Ser. Phys., 4,3 (Russian).

9. Damburg, R., Propin, R. (1982). Int.J.Quant.Chem., XXI, 191.

10. Damburg, R., Propin, R. (1982). J.Phys.A, 15, 3481; (1984). 17, 329.

11. Damburg, R., Propin, R. Martyshchenko, V. (1984). J. Phys.A, 17, 3493.

12. Damburg, R., Propin, R., Graffi, S., Grechi V., Harrell, E., Cizek, J., Paldus, J., Silverstone, H. (1984). Phys. Rev. Lett., 52, 1112.

13. Damburg, R., Kolosov, V. (1983). Ridberg States of Atoms and Molecules, Stebbings, R., Dunning, F. ed. CUP Cambridge, 31.

14. Silverstone, G., Damburg, R., Propin, R., Martyshchenko, V. (1985). Preprint LAFI-75, Salaspils.

15. Seznec, R., Zinn-Justin, J. (1979). J.Math. Phys., 20, 1398.

16. Avron, J., Seiler, R. (1981). Phys.Rev.D, 23, 1316.

PHASE SPACE EIGENFUNCTIONS OF MULTIDIMENSIONAL QUADRATIC HAMILTONIANS

V.V.Dodonov, V.I.Man'ko

P.N.Lebedev Physical Institute, Leninsky Prospect 53, 1179124 Moscow

Since Wigner [1] introduced in quantum mechanics the function nowdays called by his name this construction has been studied and used in a number of papers, see e.g. recent reviews [2-4]. Therefore it is suprising that so far complete sets of explicit solutions of the Schroedinger equation in the Wigner representation have been known only for few cases: for a harmonic oscillator [5-8] and a particle in a uniform constant field [9]. In [10] we have shown that using the method of quantum time-dependent integrals of motion one can easily obtain the equilibrium Wigner functions (i.e. the Weyl symbols [11] of the operator $\exp(-\beta \hat{H})$, where $\hat{H}$ is the Hamiltonian of the system) for quite arbitrary multidimensional quadratic Hamiltonians (i.e. nonuniform quadratic forms of the operators of Cartesian coordinates and conjugate to them momenta). In this paper we apply the same method to obtain explicit expressions for the "phase space eigenfunctions" (PSE), i.e. the functions of the form [12]

$$W_{nm}(x,p,t)=\int\Psi_n\left(x+\frac{\xi}{2},t\right)\Psi_m^*\left(x-\frac{\xi}{2},t\right)\exp(-ip\xi/\hbar)d\xi \qquad (1)$$

(the Wigner functions is a special case of eq. (1) when m = n) the functions $\Psi_n(x,t)$, $\Psi_m(x,t)$ being the solutions in the coordinate space of the Schroedinger equation with an arbitrary time-dependent multidimensional quadratic Hamiltonian.

The functions $\psi_n(x,t)$ were obtained for the first time in [13] for uniform quadratic Hamiltonians and in [14] in the most general case. However, using eq. (1) directly is rather difficult since the functions $\psi_n(x,t)$ are expressed in terms of the multidimensional Hermite polynomials of several variables, and analytical expressions for the corresponding integrals were unknown. Therefore we prefer another approach.

First recall that for any quantum system with N degrees of freedom described by the Hamiltonian

$$\hat{H} = \frac{1}{2}\hat{q}_\alpha B_{\alpha\beta}(t)\hat{q}_\beta + C_\alpha(t)\hat{q}_\alpha \equiv \frac{1}{2}\hat{q}B(t)\hat{q} + C(t)\hat{q} \tag{2}$$

where $\alpha, \beta = 1,2,\ldots, 2N$, there exist exactly N non-hermitian operators — integrals of motion, which are linear in $\hat{q}_\alpha$ and satisfy Bose commutation relations [14]:

$$\hat{A}_n(t) = \Lambda_{n\alpha}(t)\hat{q}_\alpha + \delta_n(t), \quad n = 1,2,\ldots,N; \alpha = 1,2,\ldots 2N \tag{3}$$

$$[\hat{A}_n(t), \hat{A}^+_m(t)] = \delta_{nm}, \quad [\hat{A}_n(t), \hat{A}_m(t)] = 0. \tag{4}$$

Here the operators $\hat{q}_\alpha$, $\alpha = 1,2,\ldots,2N$, are combined into the 2N - dimensional vector $\hat{q}$ with the components $\hat{q}_n = \hat{p}_n$, $\hat{q}_{N+n} = \hat{x}_n$, $n = 1,2,\ldots, N$. We assume that the Greek indices run 1 to 2N, while the italic ones 1 to N. The coefficients $C_\alpha(t)$ and $B_{\alpha\beta}(t) = B_{\beta\alpha}(t)$ are real functions of time. These coefficients are naturally combined into the 2N-dimensional vector $C(t)$ and the symmetric 2N x 2N matrix $B(t)$. Similarly, complex coefficients $\Lambda_{n\alpha}(t)$ and $\delta_n(t)$ constitute the rectangular N x 2N matrix $\Lambda(t)$ and the N-dimensional vector $\delta(t)$ such that

$$\frac{d\Lambda}{dt} = \Lambda\Sigma B, \quad \frac{d\delta}{dt} = \Lambda\Sigma C, \quad \Sigma = \begin{Vmatrix} 0 & E_N \\ -E_N & 0 \end{Vmatrix} \tag{5}$$

(E_N is the N x N unit matrix). These equations are consequences of (3) and the equation

$$i\hbar\partial\hat{A}/\partial t=[\hat{H},\hat{A}] \tag{6}$$

which every integral of motion must satisfy. The conditions (4) impose the following requirements on the coefficients $\Lambda_{n\alpha}$ (t):

$$\Lambda_{n\alpha}(t)\Sigma_{\alpha\beta}\Lambda^*_{m\beta}(t)=\frac{i}{\hbar}\delta_{nm},\ \Lambda_{n\alpha}(t)\Sigma_{\alpha\beta}\Lambda_{m\beta}(t)=0 \tag{7a}$$

or in the matrix form:

$$\Lambda(t)\Sigma\Lambda^+(t)=\frac{i}{\hbar}E_N,\ \Lambda(t)\Sigma\tilde{\Lambda}(t)=0 \tag{7b}$$

(the asterisk means the complex conjugation, the cross the Hermitian conjugation, and the tilde the transposition). One can verify that relations (7) hold at any moment thanks to (5).

Using operators $\hat{A}_n$ and $\hat{A}^+_n$ one can construct N commuting Hermitian operators : integrals of motion, which are nothing but the number of quasiparticles operators

$$\hat{K}^{(n)}=\frac{1}{2}(\hat{A}^+_n\hat{A}_n+\hat{A}_n\hat{A}^+_n)=\frac{1}{2}\hat{q}D^{(n)}\hat{q}+F^{(n)}\hat{q}+|\delta_n|^2,\ n=1,2,\dots N \tag{8}$$

(there is no summation over n in general, we will imply the summation over repeated Greek indices only.) The elements of the symmetric 2N x 2N matrix D^n and 2N-dimensional vector $F^{(n)}$ are as follows:

$$D^{(n)}_{\alpha\beta}=\Lambda^*_{n\alpha}\Lambda_{n\beta}+\Lambda_{n\alpha}\Lambda^*_{n\beta} \tag{9}$$

$$F^{(n)}_{\alpha}-\Lambda^*_{n\alpha}\delta_n+\Lambda_{n\alpha}\delta^*_n\ . \tag{10}$$

Our first aim is to obtain the phase space eigenfunctions of the operators $\hat{K}^{(n)}$, i.e. the functions $W_{\lambda\rho}(q,t)$ such that

$$\hat{K}^{(n)} W_{\lambda\rho} = (\lambda_n + \tfrac{1}{2}) W_{\lambda\rho}, \quad \lambda_n = 0,1,2,\ldots; \quad \lambda = (\lambda_1, \lambda_2, \ldots \lambda_N) \tag{11}$$

$$W_{\lambda\rho} \hat{K}^{(n)} = (\rho_n + \tfrac{1}{2}) W_{\lambda\rho}, \quad \rho_n = 0,1,2,\ldots; \quad \rho = (\rho_1, \rho_2, \ldots \rho_N). \tag{12}$$

In these equations left and right operators are implied. Let us recall the correspondence between $\hat{q}\hat{\rho}$, $\hat{\rho}\hat{q}$ (where $\hat{\rho} = |\psi_n\rangle\langle\psi_m|$) and their Weyl symbols:

$$\hat{q}W = \left(q - \frac{i\hbar}{2}\Sigma\frac{\partial}{\partial q}\right)W(q), \quad W\hat{q} = \left(q + \frac{i\hbar}{2}\Sigma\frac{\partial}{\partial q}\right)W(q) \tag{13}$$

(Σ is defined in (5)).

The most direct way to obtain an expression for $W_{\lambda\rho}$ is to calculate at first the PSE of the operators $\hat{A}_n(t)$, i.e. the functions $W_{\alpha\beta}$ corresponding to the coherent states [14,15] of the operators $\hat{A}_n$:

$$\hat{A}(t) W_{\alpha\beta} = \alpha W_{\alpha\beta}, \quad W_{\alpha\beta} \hat{A}^{+} = \beta^{*} W_{\alpha\beta}. \tag{14}$$

Here α and β are arbitrary complex N-dimensional vectors, whose components are eigenvalues of $\hat{A}_n(t)$.

The evolution equation for any PSE (following from the Schroedinger equation) in the case of the Hamiltonian (2) coincides with the classical Liouville equation for the distribution function:

$$\frac{\partial W}{\partial t} = \frac{\partial}{\partial q_\gamma}\left([\Sigma B q + \Sigma C]_\gamma W\right). \tag{15}$$

Consequently, any PSE depends in fact not on 2N+1 variables q_α, $\alpha = 1,2,\ldots,2N$, and t, but on 2N variables: first integrals of the classical system with the same Hamiltonian (see also refs. [6,7]):

$$z_n=\Lambda_{n\alpha}q_\alpha+\delta_n\,,\quad z_n^*=\Lambda^*_{n\alpha}q_\alpha+\delta^*_n\,;\ n=1,2,\ldots N$$
$$W(q_1,q_2,\ldots,q_{2N},t)=W(z_1,\ldots,z_N,z_1^*,\ldots,z_N^*). \tag{16}$$

This means that PSE depends on time only through the time dependences of the functions $\Lambda_{n\alpha}(t)$ and $\delta_n(t)$.

If one takes (3), (13) into account and makes the transformation to the new variables z_n, z_n^* in accordance with eqs. (16), then the following equations arise instead of (14):

$$\left(z_n+\frac{1}{2}\frac{\partial}{\partial z_n^*}\right)W_{\alpha\beta}=\alpha_n W_{\alpha\beta};\ \left(z_n^*+\frac{1}{2}\frac{\partial}{\partial z_n}\right)W_{\alpha\beta}=\beta_n^* W_{\alpha\beta} \tag{17}$$

where $n = 1,2,\ldots,N$. The normalized solution of these equations is as follows:

$$W_{\alpha\beta}=2^N\exp\left\{-2zz^*+2\alpha z^*+2\beta^* z-\alpha\beta^*-\frac{1}{2}\left(|\alpha|^2+|\beta|^2\right)\right\} \tag{18}$$

$$\int W_{\alpha\beta}(q,t)\,\frac{dq}{(2\pi\hbar)^N}=\exp\left\{\alpha\beta^*-\frac{1}{2}\left(|\alpha|^2+|\beta|^2\right)\right\}. \tag{19}$$

Eq. (19) is the consequence of the connection between the functions $W_{\alpha\beta}$ and $W_{\lambda\rho}$:

$$W_{\alpha\beta}=\exp\left\{-\frac{1}{2}\left(|\alpha|^2+|\beta|^2\right)\right\}\sum_{\lambda,\rho=0}^{\infty}\ \prod_{i=1}^{N}\frac{\alpha_i^{\lambda_i}(\beta_i^*)^{\rho_i}}{(\lambda_i!\,\rho_i!)^{1/2}}W_{\lambda\rho} \tag{20}$$

and the normalization condition [12]

$$\int W_{\lambda\varrho}(q,t)\,\frac{dq}{(2\pi\hbar)^N} = \delta_{\lambda\varrho} \tag{21}$$

Eq. (20) is, in turn, a consequence of the usual connection between the coherent state and Fock's states [15]:

$$|\alpha\rangle = \exp\left(-\frac{1}{2}|\alpha|^2\right)\sum_{n=0}^{\infty}\frac{\alpha^n}{\sqrt{n!}}|n\rangle .$$

To calculate the integral (19) it is convenient to introduce the new variables

$$u=\sqrt{\frac{\hbar}{2}}(z+z^*),\quad v=i\sqrt{\frac{\hbar}{2}}\,(z^*-z). \tag{22}$$

The transformation matrix between the variables (p,x) and (v,u) is symplectic due to eqs. (7), (16):

$$T\Sigma\tilde{T}=\Sigma,\quad \begin{pmatrix} v\\ u\end{pmatrix}=T\begin{pmatrix} p\\ x\end{pmatrix}.$$

Consequently, det T = 1. Now calculating (19) has become trivial.

Now one can easily obtain an explicit expressions for the functions $W_{\lambda\varrho}$: it suffices to use the expansion (20) and the following relations (see also [16]):

$$\exp(a_1x_1+a_2x_2-a_1a_2)=\sum_{n=0}^{\infty}\frac{(a_1x_1)^n}{n!}\left(1-\frac{a_2}{x_1}\right)^n\exp\left(\frac{a_2}{x_1}x_1x_2\right)=$$

$$=\sum_{n=0}^{\infty}\sum_{m=0}^{\infty}\frac{a_1^n a_2^m}{n!}(-1)^m x_1^{n-m}L_m^{n-m}(x_1x_2).$$

We have used the known generating function for the associated Laguerre polynomials L_m^{d-m} from [17]. We finally get:

$$W_{\lambda\varrho} = 2^N \prod_{n=1}^{N} 2^{\mu_n} (-1)^{\varrho_n} \left(\frac{\varrho_n!}{\lambda_n!}\right)^{1/2} (z_n^*)^{\mu_n} e^{-2\varkappa_n} L_{\varrho_n}^{\mu_n}(4\varkappa_n) = \tag{23}$$

$$= 2^N \prod_{n=1}^{N} 2^{-\mu_n} (-1)^{\lambda_n} \left(\frac{\lambda_n!}{\varrho_n!}\right)^{1/2} z_n^{-\mu_n} e^{-2\varkappa_n} L_{\lambda_n}^{-\mu_n}(4\varkappa_n);$$

where $\varkappa_n = z_n z_n^*$, $\mu_n = \lambda_n - \varrho_n$. Formula (23) is a generalization of a similar formula for the PSE of the one-dimensional harmonic oscillator with the constant frequency [5-8] to the case of arbitrary quadratic non-stationary multidimensional Hamiltonian (2). By the way we have proved a remarkable result: for quite an arbitrary quadratic Hamiltonian there exists a complete set of completely factorized PSE. To emphasize the significance of this result it is worth mentioning that such a factorization of wave functions in the coordinate representation takes place under very special conditions only: it is necessary that the quadratic form $qB(t)q$ be diagonalizable by a time-independent linear transformation of coordinates $x_1, \ldots, x_N$ (and induced transformation of momenta); clearly, for an arbitrary matrix $B(t)$ this is imposiible. Moreover, even if $B=\text{const}$, then at least positive definiteness of B is needed otherwise, the form qBq cannot be in general diagonalized by a linear canonical transformation [18,19].

As an example let us consider an oscillator with a time-dependent frequency:

$$H = p^2/2m + m\omega^2(t)x^2/2 - f(t)x. \tag{24}$$

The integral of motion $\hat{A}(t)$ was found in [20]:

$$\hat{A}(t)=\frac{i}{\sqrt{2m\hbar}}\{\varepsilon(t)\hat{p}-m\dot{\varepsilon}(t)\hat{x}-\gamma(t)\} \tag{25}$$

where the complex function $\varepsilon(t)$ satisfies the equation

$$\ddot{\varepsilon}+\omega^{2}(t)\varepsilon=0 \tag{26}$$

and the condition $\operatorname{Im}(\dot{\varepsilon}\varepsilon^{*})=1$ and

$$\gamma(t)=\int\varepsilon(\tau)f(\tau)\,d\tau .$$

The number of quasiparticles operator is (see also [21])

$$\hat{K}(t)=(2m\hbar)^{-1}\{|\varepsilon|^{2}\hat{p}^{2}+m^{2}|\dot{\varepsilon}|^{2}\hat{x}^{2}-m\operatorname{Re}(\dot{\varepsilon}\varepsilon^{*})(\hat{x}\hat{p}+\hat{p}\hat{x})-$$
$$-2\hat{p}\operatorname{Re}(\varepsilon\gamma^{*})+2\hat{x}\operatorname{Re}(\dot{\varepsilon}\gamma^{*})+|\gamma|^{2}\}. \tag{27}$$

The PSE is of the form (23), the variables $z(x,p,t)$ and $\mathscr{K}(x,p,t)$ are given by (25), (27), where one should replace the operators $\hat{x},\hat{p}$ by c-numbers. If the frequency and the force do not depend on time, then the function $\varepsilon(t)$ can be chosen in the form $\omega^{-1/2}\exp(i\omega t)$, and one gets the following expression;

$$W_{\lambda\rho}(x,p,t)=(-1)^{\rho}2^{\lambda-\rho+1}\left(\frac{\rho!}{\lambda!}\right)^{1/2}\left\{\frac{e^{-i\omega t}}{\sqrt{2m\hbar\omega}}(m\omega x_{*}-ip)\right\}^{\lambda-\rho}\times$$
$$\times e^{-2H_{*}}L_{\rho}^{\lambda-\rho}(4H_{*});\; x_{*}=x-f/m\omega^{2};\; H_{*}=\frac{p^{2}+m^{2}\omega^{2}x_{*}^{2}}{2m\hbar\omega}; \tag{28}$$

$$\hat{H}_{*}W_{\lambda\rho}=(\lambda+\tfrac{1}{2})W_{\lambda\rho},\quad W_{\lambda\rho}\hat{H}_{*}=(\rho+\tfrac{1}{2})W_{\lambda\rho}.$$

Another example is charged particle in a uniform but time-dependent magnetic field. In the coordinate representation this problem was studied in [21-23]. Let the Hamiltonian be of the form (we consider the motion across the field)

$$H=\frac{1}{2m}\left\{\left(p_1+\frac{1}{2}m\omega(t)x_2\right)^2+\left(p_2-\frac{1}{2}m\omega(t)x_1\right)^2\right\} \tag{29}$$

where $\omega(t)$ is the cyclotron frequency. It is convenient to choose the annihilation operators (3) as follows [22],

$$\hat{A}_{\pm}(t)=\frac{i}{\sqrt{2m\hbar}}\left\{\varepsilon(t)(\hat{p}_1\pm i\hat{p}_2)-m\dot{\varepsilon}(t)(\hat{x}_1\pm i\hat{x}_2)\right\}e^{\pm i\Omega(t)}. \tag{30}$$

The function $\varepsilon(t)$ satisfies, as before, eq. (26) (with the replacement $\omega\to\omega/2$); and $\Omega(t) = \frac{1}{2}\int\omega(\tau)d\tau$. Then the PSE of the operators $\hat{A}^+_{\pm}\hat{A}_{\pm}$ has the following form (the meaning of quantum numbers, we hope, is clear):

$$W_{\lambda_+\lambda_-\varrho_+\varrho_-}=4(-1)^{\lambda_++\lambda_-}\left(\frac{\lambda_+!\lambda_-!}{\varrho_+!\varrho_-!}\right)^{1/2}\left(\frac{i}{\sqrt{m\hbar}}\right)^{\varrho_++\varrho_--\lambda_+-\lambda_-}\times$$

$$\times\left\{\varepsilon(p_1+ip_2)-m\dot{\varepsilon}(x_1+ix_2)\right\}^{\varrho_+-\lambda_+}\left\{\varepsilon(p_1-ip_2)-m\dot{\varepsilon}(x_1-ix_2)\right\}^{\varrho_--\lambda_-}\times \tag{31}$$

$$\times\exp\left\{-2\varkappa_+-2\varkappa_-+i\Omega(t)(\varrho_+-\lambda_+-\varrho_-+\lambda_-)\right\}L_{\lambda_+}^{\varrho_+-\lambda_+}(4\varkappa_+)L_{\lambda_-}^{\varrho_--\lambda_-}(4\varkappa_-);$$

$$\varkappa_{\pm}=\left\{|\varepsilon|^2(p_1^2+p_2^2)+m^2|\dot{\varepsilon}|^2(x_1^2+x_2^2)-\right.$$

$$\left.-2m\,\mathrm{Re}(\dot{\varepsilon}\varepsilon^*)(x_1p_1+p_2x_2)\pm 2m(p_1x_2-p_2x_1)\right\}.$$

Let us draw attention to the fact that the function (31) is simultaneously the eigenfunction of the operator $\hat{A}_-^+\hat{A}_- - \hat{A}_+^+\hat{A}_+$ coinciding up to Planck's constant with the angular momentum projection, while the function $\chi(x,p,t)$ and the corresponding operator $\hat{A}_+^+A_+ + 1/2$) is the integral of motion generalizing the energy (if ω = const, then χ_+ coincides with the classical Hamiltonian). However, the factorization of the phase space eigenfunction is achieved not in the energy-angular momentum variables, but in the variables χ_+, χ_- corresponding to the integrals of motion with discrete equidistant spectra, which are bound from below.

We would like to emphasize once again that solutions like (23) can be constructed for quite an arbitrary quadratic Hamiltonian, no matter whether the spectrum of the Hamiltonian is discrete or not. For example, choosing the function $\varepsilon(t) = a + bt$, $\mathrm{Im}\,(ba^*) = 1$, one gets from (23) PSE with discrete spectrum corresponding to the free motion.

Now let us consider an example of PSE with the continuous spectrum, namely, the eigenfunction of the commuting operators

$$\hat{F}_n = \frac{1}{2}\hat{P}_n^2(t) - f_n\hat{X}_n(t), \quad f_n = \mathrm{const}, \; n = 1,2,\dots N \qquad (32)$$

$$\hat{X}_n(t) = \sqrt{\frac{\hbar}{2}}\,(\hat{A}_n + \hat{A}_n^+), \quad \hat{P}_n(t) = i\sqrt{\frac{\hbar}{2}}\,(\hat{A}_n^+ - \hat{A}_n). \qquad (33)$$

To obtain the solutions of the equations

$$\hat{F}_n W_{\xi\eta} = \xi_n W_{\xi\eta}, \quad W_{\xi\eta}\hat{F}_n = \eta_n W_{\xi\eta}, \qquad (34)$$

$$n = 1,2,\dots N; \quad \mathrm{Im}\,\xi = \mathrm{Im}\,\eta = 0$$

it is natural to introduce the new variables u,v (22). With these variables eqs. (34) assumes the following form

$$\left(\frac{1}{2}v_n^2 - f_n u_n\right)W \pm \frac{i\hbar}{2}\left(v_n \frac{\partial W}{\partial u_n} + f_n \frac{\partial W}{\partial v_n}\right) - \frac{\hbar^2}{8}\frac{\partial^2 W}{\partial u_n^2} = \chi_n^{(\pm)} W,$$
$$\chi_n^{(+)} = \eta_n,\ \chi_n^{(-)} = \xi_n,\ n = 1,2,\dots N. \tag{35}$$

The PSE is again completely factorized: $W = \prod_{n=1}^{N} W^{(n)}(u_n, v_n)$ The consequences of eqs. (35) are the equations

$$v_n \frac{\partial W}{\partial u_n} + f_n \frac{\partial W}{\partial v_n} = W(\eta_n - \xi_n)/i\hbar,\ n = 1,2,\dots N.$$

Therefore,

$$W^{(n)}(u_n, v_n) = \exp\{i(\xi_n - \eta_n)v_n/\hbar f_n\}\tilde{W}^{(n)}(\varphi_n);\ \varphi_n = v_n^2/2 - f_n u_n.$$

The function $\tilde{W}^{(n)}(\varphi_n)$ obeys the equation

$$\varphi\tilde{W}(\varphi) - \frac{\hbar^2}{8}f^2\tilde{W}''(\varphi) = \frac{1}{2}(\xi + \eta)\tilde{W}(\varphi).$$

Finally, the normalized PSE has the following form (we assume $f_n > 0$):

$$W_{\xi\eta} = \left(2\pi^{-1/2}\hbar^{-2/3}\right)^N \prod_{n=1}^{N} f_n^{-2/3} \exp\{i(\xi_n - \eta_n)v_n/\hbar f_n\} \times$$
$$\times \mathrm{Ai}\left(2(\hbar f_n)^{-2/3}\{\varphi_n(q,t) - (\xi_n + \eta_n)/2\}\right), \tag{36}$$

where Ai(x) is the Airy function:

$$\mathrm{Ai}(x) = \frac{1}{2\sqrt{\pi}}\int_{-\infty}^{\infty} \exp(it^3/3 + ixt)\,dt.$$

The function (36) satisfies the condition

$$\int W_{\xi\eta}(q,t)\frac{dq}{(2\pi\hbar)^N} = \delta(\xi - \eta).$$

By the way we have obtained in the phase space representation the solution of the problem of quantum particle motion under the influence of the time-dependent force f(t): the function (36) is the PSE of the qudratic integral of motion coinciding at the initial moment with the energy of the particle, if one chooses $f_n = f(0)$ (we consider the $N = 1$ case) and

$$u(x,p,t)= x-pt+\gamma_1(t), \quad v(p,t)=p-\gamma_0(t),$$

$$\gamma_n(t)=\int_0^t \tau^n f(\tau)\,d\tau, \quad n=0,1.$$

The special cases of the solution (36) foz $\xi = 2$ were found previously in [9] (f = const) and [24] (an arbitraty time-dependence f(t)).

Reference

1. Wigner,E. (1932). Phys. Rev. 40, 749.
2. Tatarskii,V.I. (1983). Sov. Phys. Uspekhi 26,311.
3. Balazs,N,L., Jennings,B.K. (1984). Phys. Rep. 104,347.
4. Hillery,M., O'Connell, R.F., Scully,M.O., Wigner,E.P. (1984). Phys. Rep 106, 121.
5. Groenewold,H.J. (1946). Physica 12, 405.
6. Bartlett, M.S., Moyal,J.E. (1949). Proc. Cambridge Phil. Soc. 45, 545.
7. Takabayasi,T. (1954). Progr. Theor. Phys. 11,341.
8. Kruger,J.G., Poffin,A. (1977). Physica, 87A, 132.
9. Balazs,N.L. (1980). Physica 102A, 236.
10. Akhundova,E.A., Dodonov,V.V., Man'ko,V.I. (1982). Physica 115A, 215.
11. Weyl,H. (1928). Zs. Physik 46, 1.
12. Moyal,J.E. (1949). Proc. Cambridge Phil. Soc. 45, 99.
13. Chernikov,N.A. (1967). Zhurn. Eksp. Teor. Fiz. 53, 1006.
14. Malkin,I.A., Man'ko,V.I., Trifonov, D.A. (1973). J. Math. Phys. 14, 576.
15. Glauber,R.J. (1963). Phys. Rev. 131, 2766.
16. Dodonov,V.V., Man'ko,V.I., Semjonov,V.V. (1984). Nu-

ovo Cimento 83B, 145.

17. Bateman Manuscript Project: Higher Transcendental Functions edited by Erdely (McGrew Hill, N.Y., 1953), vol. II, sect. 10,12.

18. Williamson,J. (1936). Amer. J. Math. 58, 141.

19. Broadbridge,P., Hurst, C.A. (1981). Physica 108A,39.

20. Malkin,I.A., Man'ko,V.I. (1970). Phys. Lett. 32A,243.

21. Lewis,H.R., Riesenfeld,W.B.,Jr. (1969). J. Math. Phys. 10, 1458.

22. Malkin,I.A., Man'ko,V.I., Trifonov,D.A. (1970). Phys. Rev. 2D, 1371.

23. Dodonov,V.V., Malkin,I.A., Man'ko,V.I. (1972). Physica 59, 241.

24. Dodonov,V.V., Man'ko,V.I., Shakhmistova,O.V. (1984). Phys. Lett. 102A, 295.

PATH INTEGRALS ON COMPACT KÄHLER MANIFOLDS

A.V.Gorokhov

Kuibyshev State University

1. Introduction. Quantum-mechanical path integrals and the approximation to which they have led have plaied an essential role in the physics in recent years. It is well known that the hamiltonian methods are significant in the path integral formulation of the quantum theory. Unlike the lagrangian methods this approach allows us to avoid some difficulties and obtain correct results for the transition amplitude, if we use some convention about the operator ordering [1 - 4]. Let us consider, as usual, the quantum-mechanical problem with the Hamiltonian $\hat{H} = \hat{H}(\hat{p}, \hat{x})$, wehre $\hat{p}$ and $\hat{x}$ are the canonical momentum and position operators ($[\hat{x}_\alpha, \hat{p}_\beta] = i\hbar \hat{I} \cdot \delta_{\alpha\beta}$, $\alpha, \beta = \overline{1, n}$; $\hat{I}$ is the identity operator). The most simple and carefull way leading here to the correct path integral construction is connected with the Weyl-symbols language and (or) with the Glauber coherent states [2,5-6].

A remarkable property of the Glauber coherent states (CS) $|P, Q\rangle$ is their close relation to the quamtum-mechanical Weyl-Heisenberg group W :

$$|P,Q\rangle = \exp(-i\lambda/\hbar)\,\hat{T}(P,Q;\lambda)\,|0\rangle, \qquad (1)$$

where

$$\hat{T}(P,Q;\lambda) = \exp(i\lambda\hat{I}/\hbar)\exp(iP\cdot\hat{x}/\hbar)\exp(-iQ\cdot\hat{p}/\hbar)$$

is the representation operator of W , and $|0\rangle$ is the vacuum vector in a Hilbert space.

The coherent states $|P,Q\rangle$ can be represented as a square integrable function:

$$\langle x|P,Q\rangle = (\pi\hbar)^{-n/4} \exp\{[iP\cdot x - \tfrac{1}{2}(x-Q)^2]/\hbar\}. \tag{2}$$

The variables P and Q label the point in classical phase space $\mathbb{R}_{2n}$, which can be considered as an orbit of W [5, 6].

The Glauber CS is a very usefull system for calculating matrix elements of the time-evolution operator $\hat{U}(t) = \exp(-it\hat{H}/\hbar)$. This calculations lead to the coherent-state path integrals. Such form of quantum-mechanical path integral was first discussed by Klauder [9], see also [2,7, 8].

The coherent-state path integrals is also related to W and W can be treated as the dynamical group of the quantum mechanical problem meaning that the Hamiltonian $\hat{H} = \hat{H}(\hat{p}, \hat{x})$ is function of generators of W.

But, in physical applications we have a wide class of quantum systems for which there are no momentum or position operators. The well-known examples are spin systems, N-level molecule in problems of quantum optics, model-hamiltonians of fermions in nuclear and solid state physics, etc.

The euclidian phase space $\mathbb{R}_{2n}$ is the simplest example of a symplectic manifold, i.e. the phase spaces of a classical hamiltonian system [10-12]. The quantization of these manifolds with the use of path integrals is an old problem. IN [13] Berezin suggested a general definition of quantization of arbitrary mechanical systems (without path integrals). This definition was illustrated by the systems with one degree of freedom for which the phase space is Lobachevsky plane or a sphere.

In [14-22] path integrals in coherent state representation of the dynamic group of the quantum problem have been introduced. This approach is convenient when the Hamiltonian can be expressed as a polynomial in generators of a Lie algebra in the space of a representation of a group G. Then, the coherent states on G can be build generalizing states (1) for the dynamical groups other than W [23].

Such approach enables one to quantize Kahler manifolds which naturally arrize here.

The main purpose of this paper is the further study of the path integral quantization of the Kahler manifolds connected with the orbits of classical Lie groups SU(N) and SO(N). We generalize our previous results [18, 20] and conclude that the path integral constructions of [16-22] are only quasiclassical, i.e. they are suitable for deriving quasiclassical asymptotics only.

<u>2. Coherent states, covariant symbols of operators and path integrals.</u> If we know the dynamical group G of the quantum-mechanical problem, then the unitary irreducible representation of G is realized in the states space of the system and the Hamiltonian $\hat{H}$ is an operator-valued polynomial

$$\hat{H} = \sum_{n_1,\dots,n_r} \omega_{n_1 \dots n_r} \hat{A}_1^{n_1} \dots \hat{A}_r^{n_r}, \tag{3}$$

where $\hat{A}_1, \dots, \hat{A}_r$ are generators of an irreducible representation $\hat{T}(g)$ of G. For example, Hamiltonians of type (3) arise in the manybody problems.

According to the quantum-mechanical principles information concerning dynamics of the system is contained in matrix elements of the evolution operator

$$\hat{U}(t,t_0) = T\exp\left[-\frac{i}{\hbar}\int_{t_0}^{t}\hat{H}(\tau)d\tau\right]. \tag{4}$$

To calculate the matrix elements of $\hat{U}(t, t_0)$ introduce CS system on G [23, 24]. This system is defined by an orbit of any vector $|0\rangle$ in the representation $\hat{T}(g)$ with stationary subgroup $G_0 \subset G$, i.e.

$$G/G_0 = X \ni g_\xi \rightarrow |\xi\rangle = \hat{T}(g_\xi)|0\rangle. \tag{5}$$

The vector $|0\rangle$ is taken so that G_0 would be maximal. The CS are labelled by the points of X.

It is known [11, 12] that we can introduce on X local complex coordinates, i.e. $z = (z^1, \ldots, z^n)$, $2n = \dim(G/G_0)$. Then it is convenient to realize the Hilbert space as the space of holomorphic functions on X with the scalar product:

$$\langle\psi_1|\psi_2\rangle = \int_X \bar{\psi}_1(z)\psi_2(z)\exp[-\rho(z,\bar{z})]d\mu(z,\bar{z}),$$

where $|\psi\rangle \rightarrow \psi(z) = \langle z|\psi\rangle / \langle z|0\rangle$, $\rho(z,\bar{z}) = 2\ln|\langle z|0\rangle|$, $|z\rangle \equiv |\xi(z^1,\ldots,z^n)\rangle$ and $d\mu(z,\bar{z})$ is an invariant measure on X normalized by the condition

$$\int_X \exp[-\rho(z,\bar{z})]d\mu(z,\bar{z}) = 1.$$

It is easy to verify that for any holomorphic functions $\psi(z)$ we have

$$\psi(z) = \int_X K(z,\bar{w})\exp[-\rho(w,\bar{w})]\psi(w)d\mu(w,\bar{w}), \tag{6}$$

where $K(z,\bar{w}) = \langle z | w \rangle / \langle z | 0 \rangle \langle 0 | w \rangle$ is the reproducing kernal [23]. The holomorphic in z and anti-holomorphic in $\bar{w}$ function $K(z,\bar{w})$ is called Bergmann kernal [24, 25].

Thanks to the existence of the kernel K(z,z) one can introduce the G-invariant symplectic 2-form

$$\omega^2 = \frac{i}{2} \sum_{\alpha=1}^{n} \sum_{\beta=1}^{n} \left[\partial^2 \ln K(z,\bar{z}) / \partial z^\alpha \partial \bar{z}^\beta\right] dz^\alpha \wedge d\bar{z}^\beta, \quad (7)$$

in X which makes X into a Kähler manifold [11, 24, 25].

Let every operator F defined on the space of a representation of G be assigned to the covariant (in Berezin's terms) symbol $\mathfrak{F}(z,\bar{w})$ [7, 8]:

$$\mathfrak{F}(z,\bar{w}) = \langle z | \hat{F} | w \rangle / \langle z | w \rangle \quad (8)$$

and

$$(\hat{F}\psi)(z) = \int_X \mathfrak{F}(z,\bar{w}) \frac{K(z,\bar{w})}{K(w,\bar{w})} d\mu(w,\bar{w}). \quad (9)$$

If $\hat{F} = \hat{F}_1 \cdot \hat{F}_2$ then

$$\hat{F}_1 \cdot \hat{F}_2 \rightarrow \mathfrak{F}(z,\bar{z}) = \int_X \mathfrak{F}_1(z,\bar{w}) \mathfrak{F}_2(w,\bar{z}) \frac{K(z,\bar{w}) K(w,\bar{z})}{K(z,\bar{z}) K(w,\bar{w})} d\mu(w,\bar{w}). \quad (10)$$

Thanks to ω^2 we can introduce the Poisson bracket of the symbols:

$$\{\mathcal{F}_1, \mathcal{F}_2\}(z,\bar{z}) = \frac{i}{\hbar}\sum_{\alpha,\beta} g^{\alpha\bar{\beta}}(z,\bar{z})\left(\frac{\partial \mathcal{F}_1}{\partial z^\alpha}\frac{\partial \mathcal{F}_2}{\partial \bar{z}^\beta} - \frac{\partial \mathcal{F}_1}{\partial \bar{z}^\beta}\frac{\partial \mathcal{F}_2}{\partial z^\alpha}\right), \tag{11}$$

(here $\sum_\beta g_{\alpha\bar{\beta}}\, g^{\lambda\bar{\beta}} = \delta^\lambda_\alpha$, $\sum_\alpha g_{\alpha\bar{\beta}}\, g^{\alpha\bar{\gamma}} = \delta^\gamma_\beta$,

$g_{\alpha\bar{\beta}}(z,\bar{z}) \equiv \partial^2 \ln K(z,\bar{z}) / \partial z^\alpha \partial \bar{z}^\beta$)

and study the corresponding classical dynamics, i.e. the flows on X.

Taking into account formulas (8)-(10) we can show that the covariant symbol of the evolution operator (4) is a path integral over the Kähler manifold [16, 18-22].

Consider the evolution operator (4) on a small time integral $\Delta t = t - t_0$. Since $|\Delta t| \ll 1$, we see that the covariant symbol of the operator $\hat{U}(t_0+\Delta t, t_0)$ is of the form

$$\underline{\mathcal{U}}(z,\bar{z}\,|\,t_0+\Delta t, t_0) = \exp\left[-\frac{i}{\hbar}\int_{t_0}^{t_0+\Delta t} \underline{\mathcal{H}}(z,\bar{z}\,|\,\tau)\,d\tau\right], \tag{12}$$

where $\underline{\mathcal{H}}(z,\bar{z}\,|\,\tau)$ is the covariant symbol of the Hamiltonian (3) which in general explicitly depends on time.

If the interval $t - t_0$ is not small, then dividing it into m small parts $\Delta t = (t-t_0)/m$ and using the well-known group property of the evolution operator, we approximate the covariant symbol $\underline{\mathcal{U}}(z,\bar{z}/t, t_0)$ by m-th convolution of the symbols of the form (12):

$$\underline{\mathcal{U}}^{(m)}(z,\bar{z}\,|\,t,t_0) = \int_X \cdots \int_X \exp\left[i S^{(m)}/\hbar\right] \prod_{\ell=1}^{m-1} d\mu(z_\ell, \bar{z}_\ell), \tag{13}$$

where

$$S^{(m)} = -i\hbar\Big\{\sum_{\ell=1}^{m}\big[\ln K(z_{\ell-1},\bar z_\ell) - \frac{1}{2}\ln K(z_{\ell-1},\bar z_{\ell-1}) - \frac{1}{2}\ln K(z_\ell,\bar z_\ell)\big] +$$

$$+\frac{1}{2}\ln\big[K(z_m,\bar z_m)/K(z_0,\bar z_0)\big]\Big\} - \sum_{\ell=1}^{m}\int_{t_{\ell-1}}^{t_\ell}\mathcal{H}(z_{\ell-1},\bar z_\ell|\tau)\,d\tau. \tag{14}$$

Suppose that the path integral representation of the symbol of the evolution operator is a limit (in any sense) of this m-time (lattice) approximation, see e.g. [7].

To investigate the limit $m\to\infty$, $\Delta t\to 0$ in (13) make the substitution $z(t_\ell) = z(t_{\ell-1}) + \Delta z_{\ell-1}$ $(z(t_\ell) = (z^1(t_\ell),\dots,z^n(t_\ell)) \equiv z_\ell)$ and use the expansion of the terms in $S^{(m)}$ into the power series in Δz and $\Delta\bar z$. We then derive the covariant symbol in the form

$$\mathcal{U}(z,\bar z|t,t_0) = \lim_{\epsilon\to+0}\int \exp(iS_\epsilon/\hbar)\prod_{t_0<\tau<t} d\mu(z(\tau),\bar z(\tau)) \tag{15}$$

where $S_\epsilon = S + \Delta S(\epsilon)$ (16)

$$S = \int_{t_0}^{t}\Big\{i\hbar\sum_{\alpha=1}^{n}\big[\bar Z_\alpha(\tau+0)\dot z^\alpha(\tau) - Z_\alpha(\tau)\dot{\bar z}^\alpha(\tau)\big] - \mathcal{H}(z(\tau),\bar z(\tau+0)|\tau)\Big\}d\tau -$$

$$-\frac{i\hbar}{2}\ln\big[K(z(t),\bar z(t))/K(z(t_0),\bar z(t_0))\big]. \tag{17}$$

The term $\sum_{\alpha=1}^{n}\big[\bar Z_\alpha(\tau+0)\dot z^\alpha(\tau) - Z_\alpha(\tau)\dot{\bar z}^\alpha(\tau)\big]$ in (17) is the part linear in Δz and $\Delta\bar z$ (of $S^{(m)}$). In [18, 20] we only considered these terms (as it is usual in physical applications). Similar results are obtained in

[16, 19, 21, 22] by another methods. In (17) [18, 20]:

$$\bar{Z}_\alpha(\tau+0)=\frac{1}{2}\left.\frac{\partial \ln K(w,\bar{w})}{\partial w^\alpha}\right|_{w=z(\tau+0),\,\bar{w}=\bar{z}(\tau+0)}$$

$$Z_\alpha(\tau)=\frac{1}{2}\left.\frac{\partial \ln K(w,\bar{w})}{\partial \bar{w}^\alpha}\right|_{w=z(\tau),\,\bar{w}=\bar{z}(\tau)},$$

while

$$\Delta S(\epsilon)=-i\hbar\sum_{L\equiv\ell_1+\ldots+\ell_n=2}^{\infty}\frac{\epsilon^{L-1}}{L!}\int_{t_0}^{t}\Big\{B_{\ell_1\ldots\ell_n}(\tau)\cdot[\dot{\bar{z}}^1(\tau)]^{\ell_1}\ldots[\dot{\bar{z}}^n(\tau)]^{\ell_n}+$$

$$+(-1)^L B^*_{\ell_1\ldots\ell_n}(\tau+0)\cdot[\dot{z}^1(\tau)]^{\ell_1}\ldots[\dot{z}^n(\tau)]^{\ell_n}\Big\}d\tau, \tag{18}$$

where $B_{\ell_1\ldots\ell_n}(\tau)=\frac{1}{2}\left.\frac{\partial^L \ln K(z,\bar{z})}{(\partial\bar{z}^1)^{\ell_1}\ldots(\partial\bar{z}^n)^{\ell_n}}\right|_{z=z(\tau),\,\bar{z}=\bar{z}(\tau)}$,

$$B^*_{\ell_1\ldots\ell_n}(\tau+0)=\frac{1}{2}\left.\frac{\partial^L \ln K(z,\bar{z})}{(\partial z^1)^{\ell_1}\ldots(\partial z^n)^{\ell_n}}\right|_{z=z(\tau+0),\,\bar{z}=\bar{z}(\tau+0)}.$$

The shifted arguments in (17),(18) correspond to the definition of the prelimit expression (14).

The non-standard terms in action (16) are due to the natural requirement that the path integral (15) should agree with its prelimit expression (13). If we neglect these terms in (15) and (16) than the reconstruction of prelimit expression from (15) (without $\Delta S(\epsilon)$) gives

$$\underline{\mathcal{U}}'^{(m)}(z,\bar{z}|t,t_0)=\int\limits_X\cdots\int\limits_X \exp\left[iS'^{(m)}/\hbar\right]\prod_{\ell=1}^{m-1}d\mu(z_\ell,\bar{z}_\ell), \qquad (19)$$

where

$$S'^{(m)}=-i\hbar\sum_{\ell=1}^{m}\left\{\bar{\underline{Z}}_\alpha(t_\ell)\left[z^\alpha(t_\ell)-z^\alpha(t_{\ell-1})\right]-\underline{Z}_\alpha(t_{\ell-1})\left[\bar{z}^\alpha(t_\ell)-\right.\right.$$

$$\left.\left.-\bar{z}^\alpha(t_{\ell-1})\right]\right\}-\frac{i\hbar}{2}\ln\left[K(z_m,\bar{z}_m)/K(z_0,\bar{z}_0)\right]-\sum_{\ell=1}^{m}\int_{t_{\ell-1}}^{t_\ell}\mathcal{H}(z_{\ell-1},\bar{z}_\ell|\tau)d\tau. \qquad (20)$$

Clearly the first terms in (20) and (14) do not coincide. Moreover, we believe that the approximation (19) leads to incorrect results, because the integrals in (19) are divergent after substituting into (20) the explicit expressions for $\bar{Z}_\alpha(t_\ell)$ and $Z_\alpha(t_{\ell-1})$. (It can be easily verified e.g. for the SU(2) CS path integral.) To test the path integral representation (15), recall that quantum system with Hamiltonian (3), linear in generators of the time-evolution operator, takes the form of representation operators of the dynamical group G, and the covariant symbol $\mathcal{U}(z,\bar{z}|t,t_0)$ can be evaluated exactly. We used (15) for the SU(2)-spin system $(\hat{H}=-\omega\hat{J}_3)$ and got the right result for the propagator, which coincides with the direct evaluation of the matrix element $\langle z|\exp[-i(t-t_0)\hat{H}/\hbar]\,|z\rangle$, where $|z\rangle$ is the SU(2) CS [20, 23].

Note that for the usual (Glauber) CS path integrals $\Delta S(\epsilon)=0$, because then $\ln K(z,\bar{w})$ is a bilinear function. In general, the non-standard terms are due to the curvature of the Kähler manifolds.

<u>3. Quasiclassical approximations.</u> Consider quasiclassical approximation for the path integral (15). It is known that for its determination we can use the stationary phase method. This method gives us quasiclassical propagator as

$$\mathcal{U}_{cl}(z,\bar{z}|t,t_0)=\sum\{\tilde{\mathcal{U}}\exp(iS_{cl}/\hbar)\} \tag{21}$$

sum over all classical trajectories $z=z(t)$ determined from the Hamiltonian equations

$$\dot{z}^{\alpha}=\{z^{\alpha},\mathcal{H}\},\quad \dot{\bar{z}}^{\alpha}=\{\bar{z}^{\alpha},\mathcal{H}\} \tag{22}$$

with the boundary values $z(t_0)=z$, $\bar{z}(t)=\bar{z}$. These equations follow from the stationarity of the action $\delta S=0$.

Many authors [14, 17-22, 26, 27] applyed such equations to nuclear physics and quantum optics.

Note that $\Delta S(\epsilon)$ vanishes in the action computed over smooth paths and gives no contribution the classical equations of motion and the quasiclassical propagator.

Suppose for simplicity that the classical path obeying (22) and the boundary values is unique.

Using the symbol (21) and the method of [28] we can derive the quasiclassical quantization condition for the periodic solutions of equations (22) [16, 19, 20]:

$$2\int_{\Sigma_E}\omega^2=\left(2n+\frac{\nu}{2}\right)\pi,\qquad n=0,1,2,\dots, \tag{23}$$

where

$$2\int_{\Sigma_E}\omega^2\equiv i\int_{\Sigma_E}\sum_{\alpha,\beta=1}^{n}\frac{\partial^2\ln K(z,\bar{z})}{\partial z^{\alpha}\partial\bar{z}^{\beta}}dz^{\alpha}\wedge d\bar{z}^{\beta}=$$

$$= i \sum_{\alpha=1}^{n} \oint_{\Gamma_E} (\bar{Z}_\alpha dz^\alpha - Z_\alpha d\bar{z}^\alpha).$$

Here Γ_E is a closed (periodic) classical orbit with the energy E, Σ_E the surface with the boundary Γ_E.

In (23) ν is the Maslov-Morse index of Γ_E [10].

For a compact dynamical group G "the area" of the surface in (23) cannot surpass some finite value, this limites the energy spectra of the system. (This agrres with the well known fact that all the irreducible unitary representations of a compact group are finite dimensional. Examples of applications of (22) and explicit evaluations of $\tilde{\mathcal{U}}$ are considered in [20].

4. Conclusion. We have proposed a new path integral construction (15). We saw that the action in this path integral has an unusual form. Note that the naive quantization of the classical systems on a Kähler manifold via path integral performed by replacing the usual expression $\sum P_\alpha dQ_\alpha$ in the path integral over phase space by its Kähler analogue, i.e. the 1-form $i\sum(\bar{Z}_\alpha dz^\alpha - Z_\alpha d\bar{z}^\alpha)$, does not give the right result.

In the quasiclassical treatment the non-standard terms are not essential. But these terms give an assential contribution quantum corrections to quasiclassics.

In his exelent work [7] Berezin noticed that non-smooth paths give the main contribution the Feynman path integral. From this point of view the non-standard terms are still more important. This and related problems and further details will be discussed elsewhere.

References

1. Popov, V.N. (1976). Path Integrals in Quantum Field Theory and Statistical Physics. Atomizdat, Moscow (Russian).

2. Klauder, J.R. (1980). Acta Phys.Austriaca Suppl., XXII, 3.

3. Prokhorov, L.V. (1982). Particles & Nuclei, 13, 1094 (Russian).

4. Ramond, P. (1981). Field Theory. A Modern Primer. (Benjamin/Cummings, London.

5. Klauder, J.R., Sudarshan, E.C.G. (1968). Fundamentals of Quantum Optics. Benjamin, N.Y.

6. Malkin, I.A., Man'ko, V.I. (1978). Dynamical Symmetries and Coherent States of Quantum Systems. Nauka, Moskow, (Russian).

7. Berezin, F.A. (1980). Russian Phys. Surveys, 132, 587 (Russian).

8. Berezin, F.A., Shubin, M.A. (1983). Schrodinger Equation. Moscow State Univ., Moskow (Russian).

9. Klauder, J.R. (1960). Ann. Phys., N.Y., 11, 123.

10. Arnol'd, V.I. (1974). Mathematical Methods of Classical Mechanics, Nauka, Moskow (Russian).

11. Dubrovin, B.A., Novikov, S.P., Fomenko, A.T. (1979). Modern Geometry, Nauka, Moskow (Russian).

12. Fomenko, A.T. (1983). Differential Geometry and Topology. Supplementary Capitals. Moscow State Univ., Moskow (Russian).

13. Berezin, F.A. (1975). Commun. Math. Phys., 40, 153.

14. Klauder, J.R. (1979). Phys. Rev. 19D, 2349.

15. Novikov, L.F. (1977). Theor. Math. Phys., 30, 218; (1983). 54, 193 (Russian).

16. Kuratsuji, H., Suzuki, T. (1980). J. Math. Phys., 21, 472;

Kuratsuji, H., Mizobuchi, Y. (1981). J.Math. Phys., 22, 757.

17. Gorokhov, A.V. (1980). In: Group Theoretical Methods in Physics, Nauka, Moskow, p.246 (Russian).

18. Gorokhov, A.V. (1983). In: Group Theoretical Methods in Physics, II, Nauka, Moskow, p.200 (Russian).

19. Kuratsuji, H., Suzuki, T. (1983). Suppl. of Progr. Theor. Phys., 74&75, 209.

20. Gorokhov, A.V. (1983). Group Theoretical Methods in Problem of Quantum Physics, Part III. Kuibyshev (Russian).

21. Yaffe, L.G. (1982). Rev. Mod. Phys., 54, 407.

22. Gerry, Ch.C., Silverman, S. (1982). J.Math.Phys., 23, 1995.

23. Perelomov, A.M. (1972). Commun. Math. Phys., 26, 222.

24. Onofri, E. (1975). J.Math.Phys., 16, 1087.

25. Hua, L.K. (1963). Harmonic Analysis of Functions of Several Complex Variables in Classical Domains, Am. Math.Soc., Providence, R.I.

26. Feng, D.H., Gilmore, R. (1980). Phys. Lett., 90B, 327.

27. Kramer, P., Saraceno, M. (1981). LN in Phys., 140, 1.

28. Keller, J.B. (1958). Ann.Phys. (N.Y.), 4, 180.

ON RELATIVISTIC AND NON-RELATIVISTIC COULOMB PROBLEMS FOR A PARTICLE OF ARBITRARY SPIN

A.G. Nikitin

The Ukrainian Academy of Sciences, Institute of Mathematics, Kiev

To describe the motion of a charged particle in central field is one of the main problems of quantum mechanics. For an electron moving in the Coulomb field this means to describe the atom of hydrogen.

At present the Coulomb problem for particles of higher spins becomes more and more pressing because of the experimental search to exotic atoms with orbital particle of spin $s > \frac{1}{2}$ [1]. In this note we present a solution of this problem for any spin s starting from relativistic motion equations as well as from Galilei invariant ones.

To describe a charged particle of any spin in external electromagnetic field the following covariant wave equations can be used

$$(\pi_\mu \pi^\mu - m^2 - \frac{e}{2s} S_{\mu\nu} F^{\mu\nu}) \Psi(x_0, \bar{x}) = 0 \qquad (1)$$

where $\pi_\mu = i \frac{\partial}{\partial x^\mu} - e A_\mu$, A_μ and $F^{\mu\nu}$ are the vector-potential and tensor of the electromagnetic field, $S_{\mu\nu}$ are the $(2s+1)\times(2s+1)$ matrices which realize the representation $D(s\ 0)$ of the Lie algebra of the homogeneous Lorentz group. These matrices can be written in the form

$$S_{ab} = \varepsilon_{abc} S_c, \quad S_{0a} = -i S_a, \quad a, b = 1,2,3 \qquad (2)$$

where S_a are generators of the irreducible representation $D(s)$ of $O(3)$ satisfying the relations

$$[S_a, S_b] = i\varepsilon_{abc} S_c, \quad S_a S_a = s(s+1). \tag{3}$$

The second-order equations (1) can be obtained from a number of covariant first-order relativistic wave equations, in particular from the wave equations considered in [2-6]. For s = ½ formulas (1)-(3) give the Zaitsev-Feynman-Gell-Mann equation [7] which follows from the Dirac one.

For the Coulomb field A_μ and $F^{\mu\nu}$ can be chosen in the form

$$A_0 = -\frac{Ze}{x}, \quad A_a = 0, \quad F_{ab} = 0, \quad F_{0a} = -\frac{Zex_a}{x^3}. \tag{4}$$

Substituting (4) into (1) and considering stationary states $\Psi(x_0, \bar{x}) = \exp(-i\varepsilon x_0)\Psi(\bar{x})$ one gets the following system

$$\left[\left(\varepsilon + \frac{\alpha}{x}\right)^2 + \Delta - m^2 - \frac{i\alpha}{s}\frac{\bar{S}\cdot\bar{x}}{x^3}\right]\Psi(\bar{x}) = 0 \tag{5}$$

where Δ is the Laplace operator.

Equations (5) are manifestly invariant under O(3) and so can be solved by separating variables. Actually the operator in brackets commutes with the total angular momentum operator $\underline{J} = \underline{x} \times \underline{p} + \underline{S} \equiv \underline{L} + \underline{S}$ so it is convenient to represent $\Psi(\underline{x})$ as a linear combination of spherical spinors

$$\Psi(\underline{x}) = \varphi_\lambda(x)\mathfrak{R}^s_{jj-\lambda m}, \quad \lambda = -s, -s+1, \ldots, -s+2n_{sj}, \quad n_{sj} = \min(s, j) \tag{6}$$

where $\{\mathfrak{R}^s_{jj-\lambda\, m}\}$ is the complete set of eigenfunctions of the commuting operators $\bar{J}^2$, J_3 and $\bar{L}^2$ corresponding to the eigenvalues $j(j+1)$, m and $(j-\lambda)(j-\lambda+1)$. Substituting (6) into (5) one obtains by separating angular variables the following system of ordinary differential equations

$$D\varphi_\lambda = x^{-2} b_{\lambda\lambda'} \varphi_{\lambda'} \tag{7}$$

where

$$D=\left(\varepsilon+\frac{\alpha}{x}\right)^2-m^2+\frac{d^2}{dx^2}+\frac{2}{x}\frac{d}{dx}-\frac{j(j+1)}{x^2},$$
$$b_{\lambda\lambda'}=(\lambda^2-\lambda d_j)\delta_{\lambda\lambda'}-\frac{i\alpha}{s}a^{sj}_{\lambda\lambda'},\quad d_j=2j+1, \tag{8}$$

and $a^{sj}_{\lambda\lambda'}$ are the coefficients of the following expansion

$$\frac{\underline{S}\cdot\underline{x}}{x}\mathfrak{R}^{s}_{jj-\lambda m}=a^{sj}_{\lambda'\lambda}\,\mathfrak{R}^{s}_{jj-\lambda' m}. \tag{9}$$

The values of $a^{sj}_{\lambda'\lambda}$ for s = ½, 1 are well-known and given e.g. in [8]. We succeeded in calculating $a^{sj}_{\lambda'\lambda}$ for any s and j [9]

$$a^{sj}_{\lambda'\lambda}=-\frac{1}{2}\left(\delta_{\lambda\lambda'+1}\,a^{sj}_{\lambda+s}+\delta_{\lambda\lambda'-1}\,a^{sj}_{\lambda+s+1}\right), \tag{10}$$

where

$$a^{sj}_{\mu}=\frac{\mu(d_j-\mu)(d_s-\mu)(d_{js}-\mu)}{(d_{js}-2\mu-1)(d_{js}-2\mu+1)},$$
$$d_s=2s+1,\ d_{js}=d_j+d_s,\ \mu=0,1,\dots,2n_{sj}. \tag{11}$$

Eqs. (7), (8) with given matrices $a^{sj}_{\lambda\lambda'}$ can be solved exactly. The matrix $\|b_{\lambda\lambda'}\|$ commutes with the operator D and is diagonalizable, so (8) reduces to a subsequence of noncoupled equations of the kind

$$D\varphi=x^{-2}b^{sj}\varphi \tag{12}$$

where b^{sj} are the eigenvalues of the matrix $\|b_{\lambda\lambda'}\|$. Each of the equations (12) in its turn reduces to an equation for a degenerate hypergeometric function

$$z\frac{d^2F}{dz^2}+(k+1)\frac{dF}{dz}+\left(\beta-\frac{k+1}{2}\right)F=0 \tag{13}$$

where

$$F=z^{-\frac{k}{2}}\exp\left(\frac{z}{2}\right)\left(\frac{z}{m^2-\varepsilon^2}\right)^{\frac{1}{2}}\varphi,\quad z=2(m^2-\varepsilon^2)^{\frac{1}{2}}\,x,$$

$$\beta = \frac{\varepsilon\alpha}{(m^2-\varepsilon^2)^{\frac{1}{2}}}, \quad k^2 = d_j^2 + 4(b^{sj})^2 - 4\alpha^2.$$

Consider the solutions corresponding to the coupled states $(\varepsilon^2 < m^2)$. These solutions can be written in such a form:

$$F = C\mathcal{F}\left(\frac{k+1}{2} - \beta,\ k+1,\ z\right) \tag{14}$$

where $\mathcal{F}$ is a degenerate hypergeometric function, C an arbitrary constant. From the finiteness of F as $z \to \infty$ one obtains $\beta = (k+1)/2 + n',\ n' = 0, 1, \dots,$ or

$$\varepsilon = m\left[1 + \alpha^2/(n' + \tfrac{1}{2} + [(j+\tfrac{1}{2})^2 - \alpha^2 - b^{sj}]^{\frac{1}{2}})^2\right]^{-1/2}. \tag{15}$$

Formula (15) gives the energy levels of any spin particle in the Coulomb field and coincides for s = ½ with the Zommerfeld field for the electron. The values of b^{sj} coincide with the roots of the matrix characteristic equation

$$\det \left\| (\lambda^2 - \lambda d_j - b^{sj})\delta_{\lambda\lambda'} - \frac{i\alpha}{s} a^{sj}_{\lambda\lambda'} \right\| = 0 \tag{16}$$

where $a^{sj}_{\lambda\lambda'}$ are coefficients (10).

To analyse the spectrum (15) it suffices to consider the approximate solutions of (16) which can be expressed in the form

$$b^{sj} = \lambda^2 - \lambda d_j + b^{sj}_{\lambda}\alpha^2 + o(\alpha^4). \tag{17}$$

One obtains from (16), (17) the values of the coefficients

$$b^{sj}_{\lambda} = \frac{1}{8s^2}\left(\frac{a^{sj}_{s+\lambda}}{j-\lambda+1} - \frac{a^{sj}_{s+\lambda+1}}{j-\lambda}\right) \tag{18}$$

where a^{sj}_{μ} $(\mu = s+\lambda \text{ or } \mu = s+\lambda+1)$ are given in (11). In the end one gets the following approximate form for the energy levels

$$\varepsilon = m\left(1 - \frac{\alpha^2}{2n^2} + \frac{\alpha^4 (b_\lambda^{sj} - 1)}{n^3(2\ell+1)} + \frac{3}{8}\frac{\alpha^4}{n^4}\right),$$
$$n = n' + j + \lambda = 1, 2, \ldots, \quad \ell = j - \lambda = 0, 1, \ldots, n-1. \tag{19}$$

Relation (19) gives the fine structure of the energy spectrum of an arbitrary spin particle in the Coulomb field. Besides a nonessential constant term, formula (19) includes the Balmer term $-\frac{m\alpha^2}{2n^2}$ and the additional ones which are proportional to α^4 and represent the contribution of spin-orbit coupling. These additional terms lead to splitting the energy levels corresponding to a fixed value of the main quantum number n. It is not difficult to calculate that the number of the fine structure sublevels for a given n is

$$N_n = n(2s+1) - \left[s + \frac{1}{2}\right], \; n \geq s+1; \; N_n = n^2, \; n \leq s, \tag{20}$$

where is the integer part of the number. The exception is the case $s = \frac{1}{2}$ when $b_\lambda^{sj} = \frac{2\lambda}{d_j}$ and the number of the fine structure sublevels is n^2 (any level for $j \neq n - \frac{1}{2}$ is twice degenerate). A more detailed discussion of the spectrum (19) is given in [9].

So the relativistic wave equations (1) make it possible to solve the Coulomb problem immediately for any spin.

It has been demonstrated in [10-12] that the motion of a spinning particle can be described in terms of equations invariant with respect to the Galilei group. Starting from the Galilei-invariant equations proposed in [11] one can obtain the following equations for the wave function of an arbitrary spin particle in the Coulomb field

$$\left(\hat{\varepsilon} + \frac{\alpha}{x} + \frac{\Delta}{2m} - \frac{\alpha g}{2sm}\frac{\mathbf{S}\,\mathbf{x}}{x^3}\right)\Psi(\mathbf{x}) = 0 \tag{21}$$

where g is a parameter.

The solutions of (21) can be found by analogy with the relativistic equations below. As a result one obtains the exact and approximate formulas for the coupled states energy levels in the form [12, 13]:

$$\hat{\varepsilon} = -\frac{m\alpha^2}{\left(\sqrt{(j+\frac{1}{2})^2 + b^{sj}} + n' + \frac{1}{2}\right)^2} \approx \tag{22}$$
$$\approx -\frac{m\alpha^2}{2n^2} + \frac{mg^2\alpha^4 b^{sj}_{\lambda}}{n^3(2l+1)} + o(\alpha^6).$$

Comparing (22) and (19) one concludes that for $q^2 = -1$

$$\hat{\varepsilon} = \varepsilon - m - \left(\frac{3}{8} - \frac{n}{2l+1}\right)\frac{m\alpha^4}{n^4} = \varepsilon - m - \left\langle -\frac{p^4}{8m^3}\right\rangle, \; p^4 = \Delta^2$$

where averaging is made with the Schrodinger wave functions. In other words formula (22) (obtained using Galilei-invariant wave equations) takes into account all the "relativistic" corrections to the Balmer energy levels (caused by the spin-orbit, quadruple and Darwin interactions), neglecting the mass term and relativistic correction $\left\langle -\frac{p^4}{8m^3}\right\rangle$ to the kinetic energy. It proves once more that the spin-orbit coupling should not be considered as a relativistic effect but can be successfully described in the frames of the Galilei-invariance approach [10, 11].

The author thanks Professor W.I. Fushchich for discussing the paper and useful comments.

References

1. Betty, S.J. (1982). Phys. Elem. Particles and Atom. Nucl. 13, 164. (Russian).
2. Lomont, I.S., Moses, H.E. (1960). Phys. Rev. 118, 337.
3. Dowker, I.S. (1967). Proc. Roy. Soc., A293, 35.
4. Hurley, W.J. (1971). Phys. Rev. D15, 327.
5. Nikitin, A.G., Fushchich, W.I. (1978). Theor. Math. Phys. 34, 319;(1978). Phys. Elem.Particles and Atom. Nucl., 9, 501. (Russian).

6. Gaseau, J.P. (1980). J. Phys. G: Nucl. Phys. 6, 1459.

7. Zaitsev, G.A. (1955). ZhETPh, 28, 530. (Russian); Feynman, R.P., Gell-Mann, M. (1958). Phys. Rev. 109, 193.

8. Warshalovich, D.A., Moskalev, A.N., Hersonsky, V.K. (1975). Quantum theory of angular momentum. Nauka, Leningrad.(Russian).

9. Nikitin, A.G. (1983). Theor. Math. Phys. 57, 257.

10. Fushchich, W.I., Nikitin, A.G. (1976). Lett. Nuovo Cim. 16, 81; Nikitin, A.G., Fushchich, W.I. (1980). Theor. Math. Phys. 44, 34. (Russian).

11. Fushchich, W.I., Nikitin, A.G. (1980). Phys. Elem. Particles and Atom. Nucl. 12, 1157; Nikitin, A.G. (1981). In: Algebraic-Theoretical Studies in Mathematical Physics. Naukova Dumka, Kiev, p. 90 (Russian).

12. Fushchich, W.I., Nikitin, A.G. (1984). Nuovo Cim., 81A, 644.

13. Nikitin, A.G. (1985). Acta Phys. Pol. B16, 3.

PLANE-WAVE ASYMPTOTIC FORM FOR A SYSTEM OF CHARGED PARTICLES

A.K.Liepinsh, R.K.Peterkop[+]

[+]Institute of Physics, Latvian SSR Academy of Sciences, Riga-Salaspils

1.Introduction. The asymptotic form of plane type wave function for a system of three or more charged particles has been obtained by Redmond (see [1]). For applications it is interesting to know the asymptotic form for partial waves corresponding to some total orbital momentum L and projection M [2]. It is difficult to obtain the partial wave asymptotic form from the Redmond asymptotic form. Here, the asymptotic form for partial wave with L = 0 is obtained using the rules for constructing the general asymptotic form [3, 4]. We consider the case of two electrons in the field of a nucleus with infinite mass and charge Z = 1. Atomic units are used. The results can be generalized for more complicated systems of charged particles.

2. General asymptotic form. In the general case, the asymptotic form of wave function describing two electrons in the field of a nucleus can be expressed as

$$\psi \sim A_0(\mathfrak{R})\, r^{-n_0} \exp(iS_f + i\gamma), \qquad (1)$$

where S_f satisfies the Hamilton-Jacobi equation for free (non-interacting) particles

$$(\nabla_1 S_f)^2 + (\nabla_2 S_f)^2 = 2E \qquad (2)$$

and reduces to the form

$$S_f = \varkappa r \mu(\Omega), \tag{3}$$

where $\varkappa = (2E)^{1/2}$, E is the energy of the system, $r = (r_1^2 + r_2^2)^{1/2}$, $\vec{r}_1$ and $\vec{r}_2$ are the position vectors of electrons, Ω the set of 5 angular variables for the six-dimensional configuration space.

The amplitude $A_o(\Omega)$ and n_o are determined by the continuity equation for the classical flow of free particles

$$\nabla_1(A_o^2 r^{-2n_o} \nabla_1 S_f) + \nabla_2(A_o^2 r^{-2n_o} \nabla_2 S_f) = 0. \tag{4}$$

The logarithmic phase γ is determined by the equation

$$\nabla_1 S_f \cdot \nabla_1 \gamma + \nabla_2 S_f \cdot \nabla_2 \gamma = -V, \tag{5}$$

where V is the Coulomb potential

$$V = -\frac{1}{r_1} - \frac{1}{r_2} + \frac{1}{r_{12}}. \tag{6}$$

It follows from equations (4) and (5) that the asymptotic form is determined by the choice of the classical action S_f of free particles.

The solution of (5) in the general case is of the form analogous to the plane-wave case

$$\gamma = \frac{1}{k_1} \ln(k_1 r_1 + \vec{k}_1 \vec{r}_1) + \frac{1}{k_2} \ln(k_2 r_2 + \vec{k}_2 \vec{r}_2) - \frac{1}{k_{12}} \ln(k_{12} r_{12} + \vec{k}_{12} \vec{r}_{12}). \tag{7}$$

Unlike the plane-wave case, $\vec{k}_1$ and $\vec{k}_2$ are not necessarily constants. In the general case they can depend on $\mathfrak{R}$ and are defined as

$$\vec{k}_1 = \nabla_1 S_f, \quad \vec{k}_2 = \nabla_2 S_f. \tag{8}$$

Each term in the logarithmic phase (7) corresponds to a definite term in the potential (6) and the equation (5) is valid separately for them. Using this it is not difficult to generalize (7) for the general case of system of charged particles. The logarithmic phase in the form (7) corresponds to the asymptotic form "plane + ingoing wave". The asymptotic form "plane + outgoing wave" can be obtained if we change the sign of S_f and consider the complex conjugate wave function.

To construct the asymptotic expansion it is essential to reduce γ to the form

$$\gamma = W \ln r + \beta, \tag{9}$$

where W abd β do not depend on r.

3. Partial wave with orbital momentum L = 0. A plane-wave-type classical action for two free particles with total orbital momentum L = 0 is

$$S_f = \left[k_1^2 r_1^2 + k_2^2 r_2^2 + 2k_1 k_2 r_1 r_2 \cos(\theta - \theta_k)\right]^{1/2}, \tag{10}$$

where θ and θ_k are the angles between $\vec{r}_1$, $\vec{r}_2$ and $\vec{k}_1$, $\vec{k}_2$. The component of the six-dimensional plane wave corresponding to L = 0 is

$$\Phi_f = \sum_\ell (2\ell+1)(-1)^\ell j_\ell(k_1 r_1) j_\ell(k_2 r_2) P_\ell(\cos\theta) P_\ell(\cos\theta_k), \tag{11}$$

where $j_\ell(x)$ is a spherical Bessel function and $P_\ell(x)$ is a Legendre polynomial.

We seek a solution of the Schrödinger equation for interacting particles as asymptotic expansion

$$\Phi^{+}=r^{-n_0}\exp(iS_f+i\gamma)\sum_{n\geq 0}A_n(\alpha,\theta)(2i\varkappa r)^{-n}, \tag{12}$$

where $\alpha = \arctan(r_2/r_1)$. Substituting it in the Schrödinger equation leads to a system whose first and second equations are

$$[D+(\tfrac{5}{2}-n_0)\mu+\tfrac{1}{2}(\Delta^*\mu)]A_0=0, \tag{13}$$

$$[D+(\tfrac{3}{2}-n_0)\mu+\tfrac{1}{2}(\Delta^*\mu)]A_1=$$

$$=-\exp(-i\beta)\Delta^*[A_0\exp(i\beta)]-(iW+4-n_0)(iW-n_0)A_0, \tag{14}$$

where

$$D=\frac{\partial\mu}{\partial\alpha}\frac{\partial}{\partial\alpha}+\frac{4}{\sin^2 2\alpha}\frac{\partial\mu}{\partial\theta}\frac{\partial}{\partial\theta} \tag{15}$$

and Δ^* is the angular part of the Laplaceian for two particles

$$\Delta_1+\Delta_2=\frac{1}{r^5}\frac{\partial}{\partial r}r^5\frac{\partial}{\partial r}+\frac{\Delta^*}{r^2}. \tag{16}$$

Eq. (13) describes free particles and is equivalent to (4). Eq. (14) takes into account the Coulomb interaction. The function $\mu(\alpha,\theta)$ has a stationary point at $\alpha=\alpha_k=\arctan(k_2/k_1)$ and $\theta=\theta_k$, where $\mu=1$ and $D=0$. It follows from (13) that

$$n_0=\frac{5}{2}+\frac{1}{2}\Delta^*\mu(\alpha_k,\theta_k). \tag{17}$$

Eq. (13) has analytic solutions. If $\sin 2\alpha_k \sin\theta_k = 0$, then $n_0=1$ and

$$A_0(\alpha,\theta)=\frac{1}{2i\varkappa\rho^{M}(\alpha,\ \theta)}\,. \tag{18}$$

If $\sin 2\alpha_k \sin\ \theta_k\ \neq\ 0$, then $n_0=3/2$ and

$$A_0(\alpha,\theta)=-\ \frac{\exp(i\pi/4)}{(2\pi\varkappa^3\sin\ 2\alpha_k\ \sin\theta_k\rho^{M}(\alpha,\theta)\sin\ 2\alpha\sin\theta)^{1/2}}\,. \tag{19}$$

The normalization factors in (18) and (19) are chosen in agreement with the asymptotic form of (11). Eq. (14) was solved by numerical integration over the classical trajectories, where D is an ordinary differential operator

$$D=\frac{r}{\varkappa}\ \frac{d}{dt}=\frac{\partial\rho^{M}}{\partial\alpha}\ \frac{d}{d\alpha}=\frac{4}{\sin^2 2\alpha}\ \frac{\partial\rho^{M}}{\partial\theta}\ \frac{\partial}{\partial\theta}\,, \tag{20}$$

and t is the time.

In the Cartesian coordinates, the trajectories are straight lines

$$\vec{r}_1=\vec{k}_1t+\vec{c}_1,\quad \vec{r}_2=\vec{k}_2t+\vec{c}_2\,, \tag{21}$$

where $\vec{c}_1$ and $\vec{c}_2$ satisfy

$$\vec{k}_1\times\vec{c}_1+\vec{k}_2\times\vec{c}_2=0. \tag{22}$$

The stationary point corresponds to t = ∞ .

In the α , θ -coordinates, trajectories are determined by the equation

$$\cot 2\alpha=\cot 2\alpha_k\cos(\theta-\theta_k)-\frac{2C}{\sin^2 2\alpha_k}\ \sin(\theta-\theta_k)\,, \tag{23}$$

where $C=d\alpha/d\theta$ at $\theta=\theta_k$. For fixed α_k and θ_k determining a stationary point (the direction of the plane wave) changing C we get a set of trajectories which cross the stationary point.

Numerical calculations of the function $A_1(\alpha,\ \theta)$

were performed for energy E = 1 and several values of parameters α_k and θ_k. Except the boundary region, where $\sin 2\alpha \sin\theta \approx 0$, we have $1 \leqslant |A_1/2A_0\varkappa| \leqslant 30$. These numbers characterise the convergence and the region of applicability of the asymptotic expansion (12). If the energy decreases, then $|A_1/A_0|$ increases. At boundary points, where $\sin 2\alpha \sin\theta = 0$, A_0 and A_1 have singularities.

4. Comparison with the Altick asymptotic form. Altick [5, 6] has suggected an asymptotic form where instead of (10) another solution of the Hamilton-Jacobi equation is used

$$S_{fA} = k_1 r_1 + k_2 r_2 = \varkappa r \cos(\alpha - \alpha_k). \tag{24}$$

This simple solution describes trajectories $\theta = const = \theta_k$. In Altick's procedure the asymptotic form is expanded in Legendre polynomials $P_\ell(\cos\theta)$ and the problem is reduced to solving an infinite set of differential equations with the variable α. However, it is unnecessary, since formulas (4)-(8) can be used. It follows that if (24) is applied, then

$$n_0 = 2, \quad A_0 = F(\theta)/\sin\alpha\cos\alpha, \tag{25}$$

where $F(\theta)$ is an arbitrary function. It can be shown that in Altick's procedure $F(\theta) = const \cdot \delta(\theta - \theta_k)$. Since (24) equals (10) at $\theta = \theta_k$, which is the stationary point for all values of α, Altick's asymptotic form [5] is an expression for our asymptotic form.

In [6], Altick has suggested an asymptotic form, where (24) is also used but the term corresponding to the electron interaction is neglected in the logarithmic phase. This leads to a branching point in the amplitude A_0 at

$\alpha = \alpha_k$, i.e. for the stationary direction which is not acceptable from the physical point of view.

References

1. Rosenberg, L. (1973). Phys.Rev., D8, 1833.

2. Peterkop, R., Liepinsh, A. (1981). J.Phys., B14, 4125; (1985). B18, 2763.

3. Peterkop, R., Rabik, L. (1977). Theor. Math. Phys., 31, 348. (Russian).

4. Peterkop, R. (1977). Phys. Lett., 62A, 81.

5. Altick, P.L. (1982). Phys. Rev., A25, 128.

6. Altick, P.L. (1983). J.Phys., B16, 3543.

SINGLE-VALUEDNESS OF WAVE FUNCTIONS AND SCATTERING IN MULTIPLY CONNECTED SPACES

G.N. Afanasiev
Joint Institute for Nuclear Research, Dubna

1. Usually it is believed that the following three conditions suffice for the Aharonov-Bohm (AB) effect: 1) the space accessible for the incident particles is multiply connected; 2) the curlless magnetic vector potential (MVP) in this multiply connected region is nonzero; 3) the wave functions (wf) are singlevalued. We prove that actually they do not suffice. Exactly, a counterexample is given for which in the same multiply connected space with nonzero VMP and single-valued wf the AB effect may or may not exist depending on specific configuration of the magnetic field. We study which of these peculiarities are responsible for the appearance of the AB effect.

2. The renewed interest in the AB effect is due to the series of papers by Bocchieri et al. [1]. These authors claim that the AB effect does not exist and is nothing but mathematical fiction. I disagree with that though appreciate attracting the physicists' interest to this intricate problem. The whole discussion on the AB effect rests on the fact that in multiply connected spaces the momentum operator has inequivalent irreducible representations and there are no reliable criteria for choosing one of them [2,3]. Different representations lead to different physical results. It is known that the AB effect arises only if one uses IR corresponding to the single-valued wf [4-6]. On the other hand, the real physical space is simply connected. In fact, any real cylindrical solenoic has a finite length and the repulsive potential

inside it is not infinitely high. So, there are no inaccessible regions, any contour can be shrinked into a point. There is no AB scattering in a simply connected region as one gets rid of MVP using a suitable gradient transformation. In a multiply connected space this procedure fails. For an infinite cylindrical solenoid one obtains after a transformation an additional singular magnetic field on the solenoid axis [7]. This leads to the problem whose physics is not equivalent to the original one. Therefore it was suggested [8] that the observed nonzero AB scattering is due to the magnetic field tails. Later Roy [9] obtained the following important result: if MVP satisfies certain conditions, the scattering is determined solely by the fields accessible for incident particles. For a finite impenetrable solenoid (no matter how long) these conditions are satisfied. This means that the AB scattering is due to the leakages of the magnetic field. The subsequent discussion [10-12] did not lead to any definite conclusion. The approach adopted here is somewhat intermediate between these two extremes. Usually making calculations with a cylindrical solenoid one disregards the field tails. In spite of their smallness they, however, are of fundamental importance as they indicate simple-connectedness of the real physical space. One may postulate that an ideal multiply connected space is to be obtained from the simply connected one passing to the limit as the solenoid's length and the repulsion inside it tend to infinity. Since wf are single-valued in a simply connected space, they should be the same in the limit multiply connected one.

3. Is the origin of the AB effect the space accessibl for the incident particles in multiply connectedness. The multiple connectedness, in turn, is caused by truncating the part of the space where magnetic field is nonzero. This can be achieved by switching inside that regions

the infinite repulsive potential or simply putting $\psi = 0$ on the boundaries and inside the truncated regions. We consider first the multiply connected space obtained by excluding the infinite cylindric tube C_0 of radius b and axis z from the whole space S. Charged particles coming from the negative x direction are scattered on C_0 and do not penetrate inside. Let ψ_0 be the wf and f_0 the scattering amplitude. Let C_0' be the cylindric solenoid coaxial to C_0 of a radius $a < b$. Outside C_0' only one component of the MVP is nonzero: $A_\varphi = \frac{Ha^2}{2\rho} = \frac{\Phi}{2\pi\rho}$ (H and Φ are the magnetic field and its flow). Let ψ be the wf and f the scattering amplitude. Clearly, $f \neq f_0$ and this is the cause of the AB effect. On the other hand, the Schroedinger equation is formally satisfied by

$$\psi = \psi_0 \exp\left(\frac{ie\chi}{\hbar c}\right), \quad \chi = \frac{\Phi\varphi}{2\pi} \tag{3.1}$$

where ψ_0 is the wf in the absence of magnetic field. If ψ_0 in (3.1) is a single-valued solution then unitary of (3.1) guarantees the conservation of the particle density and current, so $f = f_0$ and the magnetic field does not contribute to the scattering (no AB effect). Now, require single-valuedness of ψ in (3.1). Then, ψ_0 should satisfy the following boundary value:

$$\psi_0(2\pi) = \psi_0(0)\exp\left(-\frac{ie\Phi}{\hbar c}\right). \tag{3.2}$$

It is easy to verify that (3.1) coincides with the single-valued solution of the Schroedinger equation obtained by the direct method (without using (3.1)) if ψ_0 in (3.1) satisfies (3.2). Now note that MVP outside the solenoid can be presented as a gradient of the function χ of (3.1). In what follows χ is called the generating function. It changes Φ when the contour crosses the half plane $\varphi = 0,\ x > a$. (This guarantees that $\oint A_e dl = \chi(2\pi) - \chi(0) = \Phi$ as it should be.)

Now insert into C_0 (Fig.1) a toroidal solenoid instead of cylindric one). The generating function for

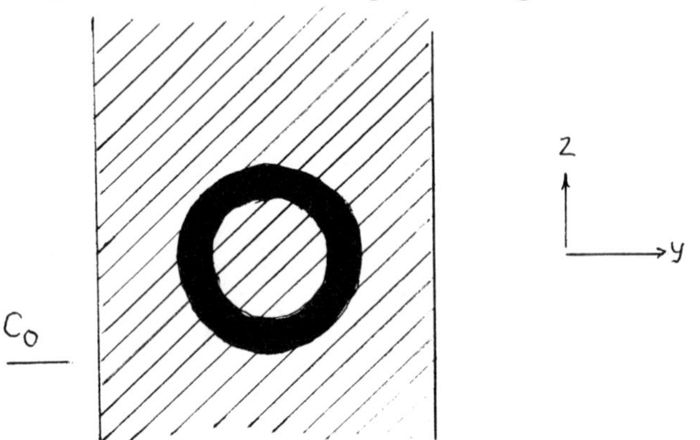

Fig.1. The absence of the AB effect in multiconnected space with nonzero MVP. The toroidal solenoid is embedded into infinite impenetrable cylinder.

toroidal solenoid is explicitly given in [13]. The following two features of χ_T will be needed later. First, χ_T decreases as $1/z^2$ at large distances. Second, χ_T suffers a finite jump (equal to Φ) when one crosses the part $\rho < d - R$ of the equatorial $z = 0$ plane. (Clearly, the equation of the solenoid's surface is $(\rho - d)^2 + z^2 = R^2$.) In other words: for a toroidal solenoid the discontinuities of the generating function are located on the equatorial circle of the radius $d - R$. Since all the singularities of the toroidal generating function χ_T lie inside the cylindrical tube C_0 , then (3.1) (where Ψ_0 is the wf in the absence of magnetic field and χ the generating function for the toroidal solenoid) is everywhere continuous and single-valued. Due to the unitarity of the transformation (3.1) the magnetic field gives no contribution to the scattering amplitude.

Another example of the multiply connected space with

nonzero MVP is shown on Fig.2, where the toroidal solenoid is embedded into one "arm" of the impenetrable torus.

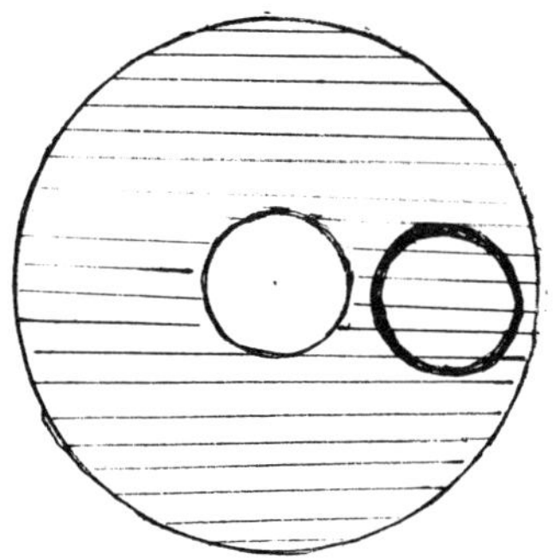

Fig.2. The same as in Fig.1 but for a toroidal solenoid embedded into the one arm of the impenetrable torus.

As in the previous case there is no AB scattering. The discontinuities of the generating function are not accessible for the incident particles and (3.1) (where ψ_0 is the scattering wf for the impenetrable torus in the absence of magnetic field) is everywhere continuous and single-valued.

One more example with the zero AB effect is illustrated on Fig.3, where the solenoid is embedded into the impenetrable sphere. The cause is the simple connectedness of the accessible space.

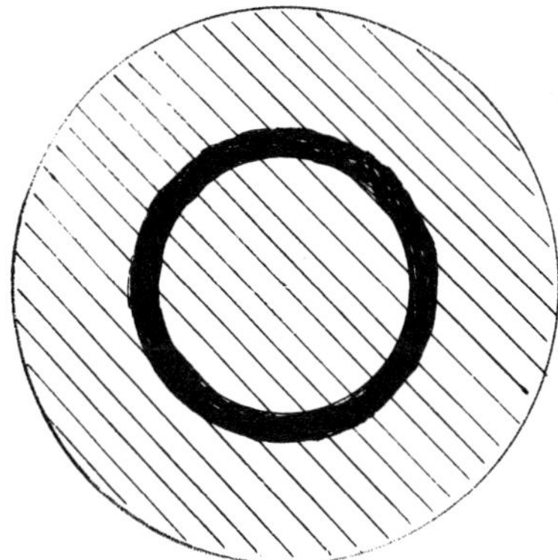

Fig.3. The toroidal solenoid is surrounded with the impenetrable sphere.

We see that the existence of the AB effect is not determined solely by the multiple connectedness of the accessible space, but depends also on specific features of the magnetic field. What are these features? Perhaps, the accurate formulation is as follows [14]. The AB effect is absent when it is possible to hide discontinuities of the generating function inside the inaccessible region. Due to the continuity and the single-valuedness of the MVP this also holds for any other choice of the discontinuity region.

The usual statement "the origin of the AB effect is due to the phase difference acquired by the wave function" is not entirely correct because of the singlevaluedness of this function. This shows a typical nonzero AB effect is illustrated on Fig.4 a) and scattering in a simply connected space on Fig.4 b). The inaccessible regions are shaded. In both cases there are paths 1 and 2 along which the phase changes by $\pm \frac{e\Phi}{2\hbar c}$. In spite of this similarity an AB scattering takes place in the first case and is absent in the second one.

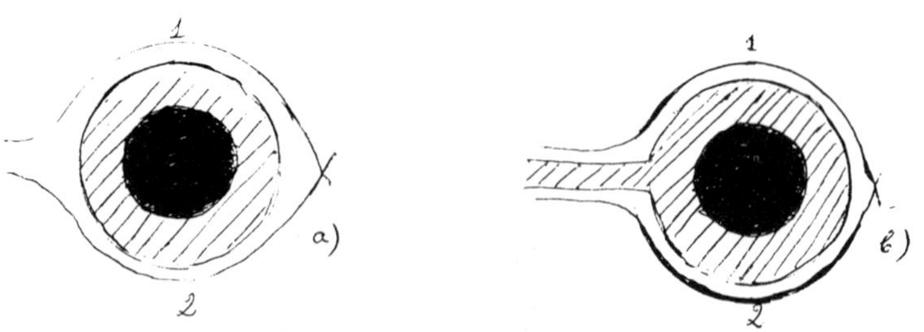

Fig.4. The inaccessible regions are shaded. The cylindric solenoid is shown as a black circle. On Fig. a) the accessible space is multiply connected, on Fig. b) it is simply connected.

4. We want to emphasize here that the whole discussion on the existence of the AB effect is of somewhat scholastic nature. In fact, the problem is to find the single-valued continuous solutions of the Schroedinger equation with single-valued, finite and continuous electromagnetic potentials. In this sense the problem does not differ from any other quantum-mechanical problem. Consider e.g. the Schroedinger equation for a 2-dimensional hydrogen atom. Its finite square integrable solutions are:

$$\Psi_{h,m} \sim \exp\left(-\frac{y}{2}\right) y^{m+\gamma} \cdot L_h^{2m+2\gamma}(y) \cdot \exp[i(m+\gamma)\vartheta], \; y = 2k\varrho.$$

The solutions corresponding to different discontinuity parameters are unitary inequivalent. The physical characteristics (such as differences of energy levels, probability density, moments of radius $\langle \varrho^k \rangle$, transition rates, etc.) depend explicitly on γ. The experiments (in the 3-dimensional case) clearly show that only single-valued wf $(\gamma = 0)$ are realized. The observed series (Layman, Pashen, Bahlmer etc.) of the spectral lines correspond to them. The theoretical uncertainties are mainly due to the simple fact that one can always kill the MVP in the curl-free region with relations similar to (3.1) in which Ψ and Ψ_0 are solutions of the Schroedinger equation with and without MVP resp. The unitary character of the transformation (4.4) guarantees existence or absence of the AB effect for both Ψ and Ψ_0 simultaneously. One has either single-valued Ψ many-valued Ψ_0 or single-valued Ψ_0 and many-valued On the other hand there is no unitary equivalence between the single-valued functions Ψ and Ψ_0, and this is the reason for the appearance of the AB effect. The simplest way to verify (theoretically) its existence is to find the single-valued solution of an unabridged (i.e.

with MVP) Schroedinger equation. The relations similar to (3.1) can be used to kill the potentials of the Schroedinger equation completely or partially, but in that case one should mind the boundary values which the transformed wave function should satisfy. One should keep in mind that to solve the free Schroedinger equation with a discontinuous boundary value is not easier than to find single-valued solution of the unabridged initial equation. These considerations are rather trivial but their unclear exposition had given rise in the past to numerous misunderstandings. We note also that the present treatment casts a doubt that multiple connectedness of the accessible space is the main condition for existence of the AB effect. Together with nontriviality of MVP and single-valuedness of wf it is necessary but not sufficient.

References

1. Bocchieri, P., Loinger, A. (1978). Nuovo Cim. A47, 1; (1980). A59, 121; (1981). A66, 164; (1979). Lett Nuovo Cim. 25, 476; (1982). 35, 469; (1984). 39, 148; Boc-Chieri, P., Loinger, A., Siragusa, G. (1979). Nuovo Cim. A51, 1; (1980). A56, 55.
2. Goldin, G.A., Menikoff, R., Sharp, D.H. (1981). J. Math. Phys. 22, 1614; Henneberger, W.C. (1981). J. Math Phys. 22, 116; Roy, S.M., Singh, V. (1984). Nuovo Cim. A79, 391; Bawin, M., Burnel, A. (1983). J. Phys., A16, 2173.
3. Ruijsenaars, S.N.M. (1983). Ann. Phys. (N.Y.) 146, 1.
4. Kretzschmar, M.Z. (1985). Z. Phys. 185, 1965; Merzbacher, E. (1962). Am. J. Phys. 30, 237; Yang, C.N.(1983) In: Proc. Int. Symp. Foundations cf Quantum Mechanics, Tokyo, pp.5-9; Aharanov, Y. ibid. pp. 10-19, Bohm, D., Hiley, D.J. (1979). Nuovo Cim. A52, 295.
5. Rothe, M.J. (1981). Nuovo Cim. A62, 54.
6. Berry, M.V. (1980). Eur. J. Phys. 1, 240.
7. Mignaco, J.A., Novaes, C.A. (1979). Lett. Nuovo Cim. 26, 453.
8. Strocchi, F., Wightman, A.S. (1974). J. Math. Phys. 15, 2198; Casati,G.,Guarneri,I.(1979).Phys.Rev.Lett.42,1579

9. Roy, S.M. (1980). Phys. Rev. Lett. 44, 111.

10. Greenberger, D.M. (1981). Phys. Rev. D23, 1460; Klein, U., ibid., 1463; Lipkin, H.J., ibid., 1466.

11. Babiker, M., Loudon, R. (1984). J. Phys. A17, 2973; Bawin, M., Burnel. A. (1984). Aharonov-Bohm effect and multivalued vector potentials, Preprint Liege Univ.

12. Home, D., Sengupta, S. (1983). Amer. J. Phys. 51, 942.

13. Afanasiev, G.N. (1983). JINR, E2-83-339, Dubna; (1984). JINR, E4-84-65, Dubna.

14. Afanasiev, G.N. (1985). JINR Rapid Communications, N 6-85, p. 17-25, Dubna.

INTEGRALS OF MOTION AND TRANSITION AMPLITUDES FOR QUADRATIC SYSTEM WITH NON-DIAGONALIZED HAMILTONIAN

E.V.Ivanova

S.Ordgonikidse Moscow aviation institute

The integrals of motion and connected with them coherent states methods are convenient in the study of quadratic and some non-quadratic quantum systems [1] .

Let us consider the N-dimensional system with Hamiltonian

$$H = \sum_{\alpha,\beta=1}^{2N} Q_\alpha B_{\alpha\beta} Q_\beta \equiv \vec{Q} B \vec{Q}, \tag{1}$$

where the Hermitian matrix B(t) is an arbitrary function of time, and components of $\vec{Q}$ satisfy the following commutation relations (here after c = h = 1)

$$[Q_\alpha, Q_\beta] = i \mathcal{J}_{\alpha\beta}; \qquad \mathcal{J} = \begin{Vmatrix} O_N & -E_N \\ E_N & O_N \end{Vmatrix}, \tag{2}$$

where E_N is the N x N unit matrix, and $Q_j = p_j = -i \dfrac{\partial}{\partial q_j}$; $Q_{N+j} = q_j$; $1 \leqslant j \leqslant N$. For the quantum systems (1) in [1] 2N Hermitian integrals of motion are constructed:

$$i \frac{\partial}{\partial t} \vec{I} = [H, \vec{I}], \quad \vec{I}(t=0) = \begin{Vmatrix} \vec{p} \\ \vec{q} \end{Vmatrix}, \tag{3}$$

$$\vec{I} = \Lambda(t) \vec{Q}, \quad [\vec{I}_\alpha(t), \vec{I}_\beta(t)] = i \mathcal{J}_{\alpha\beta}, \tag{4}$$

where Λ (t) is defined from

$$\dot{\Lambda} = -2\Lambda \mathcal{J} B ; \quad \Lambda(0) = E_{2N} . \tag{5}$$

The commutation relations (4) are invariant under the transformation

$$\vec{I}' = S\vec{I} + \vec{d}, \tag{6}$$

where S is a real symplectic 2N x 2N matrix and $\vec{d}$ is a real vector. Therefore we may select operators $\vec{I}'$ as independent integrals of motions. They correspond to another set of initial values (see (3)).

Using the invariants (4), the annihilation and creation Bose operators $\vec{A}$, $\vec{A}^+$ can be constructed:

$$\left\| \begin{matrix} \vec{A} \\ \vec{A}^+ \end{matrix} \right\| = T\vec{I} ; \qquad T = \frac{1}{\sqrt{2}} \left\| \begin{matrix} iE_N & E_N \\ -iE_N & E_N \end{matrix} \right\| \tag{7}$$

Resting on [3,4] , in [2] it was demonstrated that the sufficient condition for diagonalizability of the Bose Hermition quadratic Hamiltonian (1) with the help of real linear canonical transformation (6) is its non-negative definiteness, i.e., in this case the Hamiltonian (1) may be represented in the form of N independent one-dimensional Hamiltonians:

$$H = \sum_{j=1}^{N} H_j = \sum_{j=1}^{N} (\mu_j p_j^2 + \nu_j q_j^2) = \tag{8}$$
$$= \sum_{j=1}^{N} (\omega_j A_j^+ A_j + \gamma_j A_j^2 + \delta_j A_j^{+2})$$

In [2] it was demonstrated that the Hamiltonian

$$H = p_x^2 + xy, \tag{9}$$

can not be transformed to the type (8) by any time-independent linear canonical transformation (6).

Note that when Bose stationary Hamiltonian of type (1) is diagonalizable, its diagonalization, i.e., its representation in terms of the creation-annihilation operators in the form (8), allows us to find easily stationary and coherent states and the spectrum of the system. This method of considering stationary quantum quadratic systems is given in detail in [5].

Now consider the non-diagonalized Hamiltonian (9). It is easy to verify that the integrals of motion (3)-(5) for the Hamiltonian (9) are following

$$I_1 = p_x + ty; \tag{10}$$

$$I_2 = -t^3 p_x + p_y + tx - \frac{t^3}{3} y; \tag{11}$$

$$I_3 = -2t p_x + x - t^2 y; \tag{12}$$

$$I_4 = y; \tag{13}$$

$$i \frac{\partial}{\partial t} I_\alpha = [H, I_\alpha]; \qquad \alpha = 1, 2, 3, 4. \tag{14}$$

The Bose integrals of motion (7) for the Hamiltonian are

$$A_1 = \frac{1}{\sqrt{2}} (I_3 + i I_1) = \tag{15}$$

$$= \frac{1}{\sqrt{2}} \left\{ (-2t + i) p_x + x + t(-t + i) y \right\};$$

$$A_2 = \frac{1}{\sqrt{2}} (I_4 + i I_2) = \tag{16}$$

$$= \frac{1}{\sqrt{2}} \left\{ -i t^2 p_x + i p_y + i t x + (1 \;\; i \frac{t^3}{3}) y \right\};$$

$$[A_1, A_1^+] = [A_2, A_2^+] = 1; \quad [A_1, A_2] = [A_1, A_2^+] = 0. \qquad (17)$$

Let us consider the classical Hamiltonian corresponding to the quantum Hamiltonian (9). Doing so we substitute the classical generalized momentum $\mathcal{P}_x = p_x + e\mathcal{A}_x$ ($c = h = 1$) for the momentum operator $p_x = -i\frac{\partial}{\partial x}$ in (9). Here $\vec{\mathcal{A}}$ is the value of the vector-potential at the point in which the particle with charge e is located [6].

The classical trajectory of the particle in the phase space is

$$x = -y_0 t^2 + 2\mathcal{P}_{0x} t + x_0; \quad y = y_0;$$
$$\mathcal{P}_x = -y_0 t + \mathcal{P}_{0x}; \quad \mathcal{P}_y = \frac{1}{3} y_0 t^3 - \mathcal{P}_{0x} t^2 - x_0 t + \mathcal{P}_{0y}, \qquad (18)$$

where x_0, y_0, $\mathcal{P}_{0x}$, $\mathcal{P}_{0y}$ are initial values of coordinates and generalized momenta of the particle respectively.

Thus, the motion is performed along the X-axis with a constant acceleration depending on the initial value of the coordinate y.

Note that the simplest case of particle motion with constant acceleration is motion in constant homogeneous electric field. The quantum analogue of such a motion is considered in [6, § 24].

It is easy to see that the energy spectrum E for the Hamiltonian (9) is continuous and orthonormalized eigenstates of the Hamiltonian can be written as

$$\Psi(x,y) = \frac{1}{\sqrt{\pi \sqrt[3]{y_0}}}\, \Phi\left(x - \frac{E}{y_0}\right) \delta(y - y_0),$$

where $\Phi(\xi)$ is the Airy function (see [6,7]).

Now let us obtain the expression for the coherent states of the system (9), i.e. for the eigenstates of the A_1 and A_2:

$$A_j \,|\vec{\alpha};t\rangle = \alpha_j \,|\vec{\alpha};t\rangle, \quad j=1,2, \quad |\alpha_1,\alpha_2;t\rangle \equiv |\vec{\alpha};t\rangle,$$

$$i\frac{\partial}{\partial t}|\vec{\alpha};t\rangle = H\,|\vec{\alpha};t\rangle; \quad \frac{1}{\pi^2}\int d^2\vec{\alpha}\,|\vec{\alpha};t\rangle\langle\vec{\alpha};t| = 1; \tag{19}$$

$$d^2\alpha_j = d(\mathrm{Re}\,\alpha_j)\, d(\mathrm{Im}\,\alpha_j),$$

where α_1 and α_2 are complex numbers.

The coherent states can be expressed as

$$|\vec{\alpha};t\rangle = |\vec{0};t\rangle \exp\Big\{-\frac{1}{2}\Big(|\alpha_1|^2+|\alpha_2|^2+\frac{1-2ti}{1+2ti}\alpha_1^2+\alpha_2^2\Big) - \frac{\sqrt{2}\,i}{1+2ti}\Big[i\alpha_1 x + it^2\alpha_1 y + (-2t+i)\alpha_2 y\Big]\Big\}. \tag{20}$$

where the state $|\vec{0};t\rangle$ is defined in accordance with the requirements

$$A_1|\vec{0};t\rangle = A_2|\vec{0};t\rangle = 0; \quad |\vec{0};t\rangle \equiv |0,0;t\rangle;$$
$$i\frac{\partial}{\partial t}|\vec{0};t\rangle = H\,|\vec{0};t\rangle; \quad \langle\vec{0};t|\vec{0};t\rangle = 1. \tag{21}$$

The solution $|\vec{0};t\rangle$ for the Hamiltonian (9) is given by

$$|\vec{0};t\rangle = \frac{1}{i\sqrt{\pi(1+2ti)}}\exp\Big\{\frac{1}{2(2t-i)}\Big[ix^2 - 2t(1+it)xy + \Big(i-2t-\frac{2}{3}t^3-\frac{1}{3}t^4\Big)y^2\Big]\Big\}. \tag{22}$$

The coherent states $|\vec{\alpha};t\rangle$ are generating functions for the Fock states $|n_1,n_2;t\rangle$ (see [1]):

$$|\alpha_1, \alpha_2; t\rangle = \exp\left[-\frac{1}{2}(|\alpha_1|^2 + |\alpha_2|^2)\right] \cdot \sum_{n_1 n_2 = 0}^{\infty} \frac{\alpha_1^{n_1} \alpha_2^{n_2}}{\sqrt{n_1! n_2!}} |n_1, n_2; t\rangle \quad (23)$$

The Fock states of the system (9) satisfy the following relations:

$$A_1^+ A_1 |\vec{n}; t\rangle = n_1 |\vec{n}; t\rangle; \quad A_2^+ A_2 |\vec{n}; t\rangle = n_2 |\vec{n}; t\rangle;$$
$$i \frac{\partial}{\partial t} |\vec{n}; t\rangle = H |\vec{n}; t\rangle; \quad (24)$$

$$\langle \vec{m}; t | \vec{n}; t\rangle = \delta_{m_1 n_1} \delta_{m_2 n_2}; \quad n_{1,2}, m_{1,2} = 0, 1, 2, \ldots$$

Using (20) and (23) we can express $|\vec{n}; t\rangle$ in terms of Hermite polynomials of two variables [7] (see also [1])

$$|n_1, n_2; t\rangle = \frac{|\vec{0}; t\rangle}{\sqrt{n_1! n_2!}} H_{n_1, n_2}\left(\sqrt{2} \frac{x + t^2 y}{1 - 2ti}, y\right). \quad (25)$$

Using the common scheme for calculating the radiation power corresponding to transitions between coherent and Fock states [8], it is easy to calculate the dipole radiation of the system (9). To do this, we express the interaction Hamiltonian in terms of Bose integrals of motion (15) and (16):

$$H_{int} = -\frac{e}{m} \vec{A}_{rad} \vec{P}; \quad (26)$$

$$\vec{A}_{rad} = \sum_{\lambda, \rho} \sqrt{\frac{2\pi}{V \omega_\lambda}} \vec{e}_{\lambda\rho} (c_{\lambda\rho} + c_{\lambda\rho}^+); \quad (27)$$

$$p_x = -\frac{1}{\sqrt{2}} \left[(A_1 + A_1^+) i + t (A_2 + A_2^+) \right] \quad (28)$$

$$p_y = \frac{1}{\sqrt{2}}\left[-t\,(A_1+A_1^+) + i\,t^2 (A_1 - A_1^+) + \frac{t^3}{3}(A_2+A_2^+) - i\,(A_2 - A_2^+)\right] \quad (29)$$

Here m is the mass of the charge e, and $\vec{\mathcal{A}}_{rad}$ is the vector potential of the radiation field. Consider the case $e = 1$; $m = \frac{1}{2}$. We suppose also that the polarization vector $\vec{e}_{\lambda\rho}$ lies in the xy-plane.

Thus, summing over the polarization vectors and integrating with respect to the solid angle , for the transitions between the mixed states of the system $|\alpha, n; t\rangle \rightarrow |\beta, n; t\rangle$ we obtain the following expression for the spontaneous dipole radiation:

$$W(\omega_\lambda) = \frac{3}{4\pi}\left[\left(2|\alpha_0|^2 + 1 - \alpha_0^2 + \alpha_0^{*2}\right)\frac{\delta(\omega_\lambda)}{\omega_\lambda^2} - \left(\alpha_0^2 - \alpha_0^{*2}\right)\frac{\delta(\omega_\lambda)}{\omega_\lambda}\right], \quad (30)$$

Here we assume that the initial and final states of the partical are described by the density matrices ρ_{in} and ρ_f respectively which in the P-Glauber representation (see [1,5]) are

$$\rho_{in,f} = \int d^2\alpha\, P_{in,f}(\alpha)\ |\alpha; t\rangle\langle\alpha; t| ;$$
$$P_{in} = \delta^{(2)}(\alpha - \alpha_0) ; \qquad P_f = \frac{1}{\pi}. \quad (31)$$

Noted that the problem of the radiation of a charge moving with a constant acceleration is considered in detail in [9] and references therein.

The author is greteful to V.I.Man'ko for encouragement and helpful discussions.

References

1. Malkin,I.A., Man'ko,V.I. (1979). Dynamic Symmetries and Coherent States of Quantum Systems. Nauka, Moscow (Russian).

2. Dodonov,V.V., Man'ko,V.I. (1982). Trudy PhIAN, 152, 145 (Russian).

3. Bogolyubov,N.N., Tyablikov,S.V. (1949). ZhTPh 19,256 (Russian).

4. Williamson,J. (1936). Amer. J. Math. 58 , 141

5. Ivanova,E.V., Malkin,I.A., Man'ko,V.I. (1975). Nucl. Phys., 21, 664.

6. Landau,L.D., Lifshitz,E.M. (1974). Quantum Mechanics. Nauka, Moscow, pp. 98-100 (Russian).

7. Erde'lyi,A. et al. (1953). Higher Transcendental Functions. McGraw-Hill Book Co., N.Y. 2.

8. Ivanova,E.V., Malkin,I.A., Man'ko,V.I. (1977). Int. J. Theor. Phys. 16, 503.

9. Ginzburg,V.L. (1975). Theoretical Physics and Astrophysics. Nauka, Moscow (Russian).

AN ALGEBRAIC APPROACH TO COUPLED OSCILLATORS

S.V. Prants
Pacific Oceanology Institute, Vladivostok

A model of two coupled harmonic oscillators is widely used in quantum electronics and quantum optics to describe such processes as parametric generation and amplification, parametric up- and down-frequency-conversion [1]. Under the assumption of a strong pumping field, the model is suitable for investigating three- and four-wave mixing, Raman and Brillouin scattering. It may also be useful in quantum acoustics to describe the parametric process of backward-wave echo generation [2].

The most general bilinear model Hamiltonian is given as

$$\mathcal{H}(t)=\sum_{1}^{2}\omega_{ij}(t)a_i^+a_j+\left[b_{ij}(t)a_ia_j+d_i(t)a_i\right]+\left[H.c.\right], \quad (1)$$

where a_i^+ and a_j are the field mode (electromagnetic or acoustic) creation and annihilation operators with the usual commutation relation $[a_i, a_j^+]=\delta_{ij}$, $\omega(t)$, $b(t)$ and $d(t)$ are arbitrary complex valued functions of time.

The purpose of this paper is twofold. First, we apply the Lie algebraic approach to find an exact solution for the evolution equation with the Hamiltonian (1). Then, we use this general solution to treat the parametric process of backward-wave phonon echo generation.

If a model Hamiltonian $\mathcal{H}(t)=\sum_{j=1}^{m}h_j(t)H_j$, where H_j are constant operators and $h_j(t)$ are complex valued time-dependent functions, generates a finite-dimensional Lie algebra $L_n:\{H_1, H_2, \ldots, H_m, \ldots, H_n\}, n \geq m$, then the time-dependent Schrodinger equation

$$i\hbar\, dU(t,t_0)/dt = \mathcal{H}(t)\, U(t,t_0) \tag{2}$$

with the initial value $U(t,t_0) = I$ has an exact solution in the form

$$U(t,t_0) = \prod_{j=1}^{n} \exp\{g_j(t,t_0) H_j\}. \tag{3}$$

The complex valued functions of time $g_j(t,t_0)$ obey the equation [3]

$$(i\hbar)^{-1} \sum_{j=1}^{n} h_j(t) H_j = \sum_{j=1}^{n} \dot{g}_j(t,t_0) \left\{ \prod_{k=1}^{j-1} \exp[g_k(t,t_0)\, ad\, H_k] \right\} H_j, \tag{4}$$

which, in general, is equivalent to a system of n coupled nonlinear differential equations for $g_j(t,t_0)$. If is solvable, all the solutions are easily obtained by simple integration. If L_n is not solvable and is not simple, it decomposes into the semidirect sum of the semisimple subalgebra S and the radical R, i.e. $L = S \oplus R$. Such a decomposition gives rise to the corresponding decomposition $\mathcal{H} = \mathcal{H}_S + \mathcal{H}_R$ or $\mathcal{H}$. In this case the initial system of equations reduces to independent sets of equations of smaller dimensions. The time-evolution operator factors into the product of two unitary operators $U = U_S\, U_R$ wuch that

$$i\hbar\, dU_S/dt = \mathcal{H}_S(t) U_S, \quad i\hbar\, dU_R/dt = U_S^{+} \mathcal{H}_R(t) U_S U_R. \tag{5}$$

The dynamic algebra of two coupled oscillators is the semidirect sum $sp(4,R) \oplus n(2)$, where $u(2)$ is the Heisenberg-Weyl algebra, constructed from $\{a_i a_j, a_i^+ a_j, a_i^+ a_j^+\}$ and $\{a_i^+, a_j, I\}, i,j = 1,2$ respectively [4]. In the model Hamiltonian (1) we may "decouple" the parametric-amplifier part generating the dynamical algebra $u(1,1)$: $\{a_1^+ a_1 + 1/2,$

$a_2^+ a_2 + 1/2, \quad a_1^+ a_2^+, \quad a_1 a_2 \},$
and the frequency-converter part generating $u(2)$:

$$\{a_1^+ a_1 + 1/2, a_2^+ a_2 + 1/2, a_1^+ a_2, a_1 a_2^+\}.$$

Our algebraic approach reduces the solution of the Schrodinger equation (2) with any hermitian bilinear Hamiltonian (1) to the solution of eq.(2) with the Hamiltonian that generates the semidirect sum u(1,1) ⊕ n(2). The solution is valid for any process with the dynamic algebra u(2) because of its connection with u(1,1) via the Weyl unitary trick $a \to a$, $a^+ \to a^+$, $b \to -ib^+$, $b^+ \to -ib$.

The hermitian Hamiltonian which generates u(1,1)⊕ n(2) is

$$\mathcal{H} = \sum_{j=1}^{2} \left[\hbar\omega_j a_j^+ a_j + d_j(t) a_j + d_j^*(t) a_j^+\right] - \sum_{i<j=1}^{2} \left(\hbar \mathfrak{R}(t) a_i a_j + \text{H.c.}\right) \tag{6}$$

where $\mathfrak{R}(t) = \mathfrak{R} \exp\left[i(\omega_1 + \omega_2)t\right]$, $d_j(t)$ are arbitrary functions. Using (3) and (5), we may represent the time-evolution operator as a product of two exponential operators $U = U_{u(1,1)} U_{n(2)}$, where

$$U_{u(1,1)} = \exp g_1 (a_1^+ a_1 + 1/2) \exp g_2 (a_2^+ a_2 + 1/2) \exp g_3 a_1 a_2 \exp g_4 a_1^+ a_2^+, \tag{7}$$

$$U_{n(2)} = \exp(f_1 a_1) \exp(f_2 a_1^+) \exp(f_3 a_2) \exp(f_4 a_2^+) \exp(f_5 I). \tag{8}$$

The time-dependent functions g obey eq.(4) which, in the case u(1,1), is equivalent to a set of four coupled differential equations

$$i\omega_1 = \dot{g}_1 + \dot{g}_4 g_3 \tag{9a}$$

$$i\omega_2 = -\dot{g}_2 + \dot{g}_4 g_3 \tag{9b}$$

$$i\Omega \exp i(\omega_1+\omega_2)t=(\dot g_3+\dot g_4 g_3^2)\exp(-g_1-g_2) \tag{9c}$$

$$i\Omega \exp i(-\omega_1+\omega_2)t=\dot g_4 \exp(g_1+g_2). \tag{9d}$$

Adding eqs. (9a) and (9b), we find formal solutions for g_1 and g_2 with the accuracy up to integrating constants : $g_1=-i\omega_1 t+G(t)$ and $g_2=-i\omega_2 t+G(t)$, where $G(t)$ is an arbitrary function of time. Substituting $\dot g_4$ from eq.(9d) and g_3 from (9a) into (9c) we see that G(t) satisfies $\ddot G+\dot G^2-\Omega^2=0$ which has the solution $G(t)=\ln\cosh\Omega t$. Now, the solution of eqs. (9) is:

$$g_1=-i\omega_1 t+\ln\cosh\Omega t,\quad g_2=-i\omega_2 t+\ln\cosh\Omega t,$$
$$g_3=(i/2)\sinh 2\Omega t,\qquad g_4=i\tanh\Omega t. \tag{10}$$

In the case u(1,1) it is necessary to solve a set of four nonlinear differential equations. The problem is simplified for n(2). The quantities $f_1,\dots,f_5$ are found from eq.(4) solved in quadratures:

$$f_1=i\hbar\int_{t_0}^{t}(d_1e^{g_1}-g_3d_2^*e^{-g_2})dt,\quad f_3=i\hbar\int_{t_0}^{t}(d_2e^{g_2}-g_3d_1^*e^{-g_1})dt,$$

$$f_2=i\hbar\int_{t_0}^{t}(d_1^*e^{-g_1}+d_2g_4e^{g_2}-g_3g_4d_1^*e^{-g_1})dt, \tag{11}$$

$$f_4=i\hbar\int_{t_0}^{t}(d_2^*e^{-g_2}+d_1g_4e^{g_1}-g_3g_4d_2^*e^{-g_2})dt,$$

$$f_5=-\int_{t_0}^{t}(\dot f_2 f_1-\dot f_4 f_3)dt.$$

Thus, we have found a general exact solution (7), (8), (10), (11) for the temporal evolution of two coupled oscillators. To illustrate the Lie algebraic method let us develop a quantum theory of backward-wave phonon echoes in dielectrics and compare it with the observation of the exponential growth of the echo amplitude [5] . The backward-wave phonon echo is generated as a consequence of the annihilation of a photon of frequency 2ω and wave vector $\vec{k}_e$ and creation of two phonons $(\omega, \vec{k})$ and $(\omega, -\vec{k})$ propagating in opposite directions. Treating the electromagnetic field $(2\omega, \vec{k}_e)$ classically, the total Hamiltonian of this quadratic parametric process is given by the expression (6) with $d_j(t) = 0$ and $\Omega(t) = \gamma\omega \cdot E_c \times \exp[-i(\omega+\omega-2\omega)t + i(\vec{k}-\vec{k}-\vec{k}_e)\vec{r}]$, γ is the nonlinear piezoelectric coefficient. In this case the time-evolution operator (7) is an element of the dynamic group U(1,1) with parameters $g_1, \dots, g_4$ given by (10). Using (7) and (10), we can calculate the temporal evolution of the creation a_1^+ and annihilation a_2 operators for the forward- and backward-travelling phonons, respectively

$$a_1^+(t) = U^+_{u(1,1)} a^+(t=0) U_{u(1,1)} = e^{i\omega t} e^{-t/T_1} (a_1^+ \cosh \gamma E_o \omega t -$$
$$- i a_2 \sinh \gamma E_o \omega t), \tag{12}$$
$$a_2(t) = e^{-i\omega t} e^{-t/T_1} (a_2 \cosh \gamma E_o \omega t + i a_1^+ \sinh \gamma E_o \omega t),$$

where phonon relaxation time T_1 has been introduced in (12) phenomenologically. To find the threshold of generation we take $d a_1^+/dt \geq 0$ and $d a_2/dt \geq 0$. The result is $\gamma E_o \omega \geq 1/T_1$. Since we know the magnitude of E_o and the sound attenuation T_1, this condition provides an estimate of the nonlinear piezoelectric coefficient γ and, vice versa, if knowing the magnitudes of E_o and γ we can estimate T_1.

The expectation value of the number of backward-traveling phonons can be calculated in the usual way.

$$a_2^+ a_2 = \mathrm{tr}\, U^+_{u(1,1)} a_2^+(t=0) a_2(t=0) U_{u(1,1)} \rho_o =$$
$$= (\bar{n}_{a_1} + 1) e^{-2t/T_1} \sinh^2 \gamma E_o \omega t, \tag{13}$$

where $\bar{n}_{a_1}$ is the average number of phonons a_1 in the forward-traveling mode at t=0, ρ_o is the initial equilibrium density matrix. Thus we see that the expectation value of the number of phonons $\langle a_2^+ a_2 \rangle$ is an exponentially increasing function of time for long times. In a recent $\omega - 2\omega$ experiment [5] on LiNb O_3 the observed variation of the backward-wave echo amplitude has agreed well with the $\sinh^2 \Theta$, where Θ is proportional to the amplitude of the pump electric field, E_0.

Our algebraic approach is general enough meaning that various quadratic parametric processes in the framework of the model of the two coupled oscillators can be described, grouped and classified by their dynamic algebras. The algebraic concept is useful in predicting new physical effects involving electromagnetic and acoustic resonance and scattering phenomena which manifest the same dynamic algebra.

References

1. Louisell, W.H. (1964). Radiation and Noise in Quantum Electronics. New York: McGraw Hill.
2. Kopvillem, U.Kh., Prants, S.V. (1985). Polarisation Echo. Nauka, Moscow. (Russian).
3. Wei, J., Norman, E. (1963). J. Math. Phys., 4, 575.
4. Malkin, I.A., Man'ko, V.I. (1979). Dynamic Symmetries and Coherent States of Quantum systems. Nauka, Moscow. (Russian).
5. Meredith, D.J., Miyasato, T., Wigmore, J.K. - Phys.Rev., 1984, 30B, 7366.

EVOLUTION EQUATIONS FOR STATISTICAL OPERATORS OF QUANTUM LINEAR OPEN SYSTEMS

V.V. Dodonov, V.I. Man'ko

P.N. Lebedev Physical Institute, Moscow

The aim of this paper is to investigate open quantum systems obeying evolution equations of the type

$$\dot{\hat{\varrho}} = \hat{z} L \hat{z} \hat{\varrho} + \hat{\varrho} \hat{z} R \hat{z} + \hat{z} K \hat{\varrho} \hat{z}. \qquad (1)$$

Here $\hat{\varrho}$ is the statistical operator (density matrix) of the system, L, R, K complex $2N \times 2N$ matrices, $\hat{z}$ an $2N$ -dimensional vector operator whose components satisfy the commutation relations

$$\hat{z}_\alpha \hat{z}_\beta - \hat{z}_\beta \hat{z}_\alpha = \Sigma_{\alpha\beta}, \qquad (2)$$

where $\Sigma_{\alpha\beta}$ are c-numbers constituting $2N \times 2N$ antisymmetric matrix Σ . In concrete applications the vector $\hat{z}$ usually takes either the form $\hat{z} = (\hat{p}, \hat{q})$, where $\hat{q}$, $\hat{p}$ are N-dimensional vectors of Cartesian coordinates and conjugate to them momenta, or the form $\hat{z} = (\hat{a}, \hat{a}^+)$, where $\hat{a}$, $\hat{a}^+$ are the annihilation and creation boson operators. The interest in eq.(1) has two reasons: clearly (1) is the most general equation leading to linear equations of motion for average values $\langle \hat{z}_\alpha \rangle = \mathrm{Tr}(\hat{\varrho} \hat{z}_\alpha)$:

$$d\langle \hat{z} \rangle / dt = A \langle \hat{z} \rangle \qquad (3)$$

(A is a $2N \times 2N$ matrix determined by matrices R, L, K); besides, equations like (1) are often used to describe quantum dissipative systems. Nonetheless, great number of papers is devoted to the systems dealing with various special cases of eq.(1) only, while the general properties of this equation, especially the conditions which must be imposed on matrices L, R, K so that eq.(1) would determine a positive semidefinite quantum dynamic semigroup [1], have not been studied in detail. Our paper fills this gap. First, note that eq.(1) preserves the hermiticity of $\hat{\varrho}$ if the matrices L, R, K satisfy the conditions

$$L=\tilde{S}R^{+}S, \quad R=\tilde{S}L^{+}S, \quad K=\tilde{S}K^{+}S \tag{4}$$

where the matrix S connects the operators $\hat{z}$ and $\hat{z}^{+}$:

$$\hat{z}^{+}=S\hat{z}, \quad S^{*}S=I_{2N}, \quad \Sigma^{+}=S\Sigma\tilde{S} \tag{5}$$

Hereafter the tilde means the matrix transposition, the cross the hermitian conjugation, and the asterisk the complex conjugation, I_N is the $N\times N$ unit matrix. Another condition ensures the possibility of deriving eq.(3) from eq.(1):

$$L+R+\tilde{K}=0 \tag{6}$$

(it guarantees also that $\mathrm{Tr}\,\hat{\varrho}$ does not change in time). Then

$$A=\Sigma(L-\tilde{R}). \tag{7}$$

Eqs. (4), (6) allow us to determine the matrices L and R provided K is given. It is essential, however, that the matrix K cannot be arbitrary. Indeed, eq.(1) ought to preserve the most important property of the statistical operator - its positive semidefiniteness. To get the corresponding restriction on K we use the known expression [1-3] for the most general linear evolution equation determining a quantum dynamic semigroup:

$$\dot{\hat{\varrho}}=-\frac{i}{\hbar}[\hat{H},\hat{\varrho}]+\sum_{j}(2\hat{\Phi}_j\hat{\varrho}\hat{\Phi}_j^{+}-\hat{\Phi}_j^{+}\hat{\Phi}_j\hat{\varrho}-\hat{\varrho}\hat{\Phi}_j^{+}\hat{\Phi}_j). \tag{8}$$

Here $\hat{H}$ is a hermitian operator, while $\hat{\Phi}_j$ can be arbitrary linear operator (the number of different operators $\hat{\Phi}_j$ in eq.(8) may be arbitrary). In our case it is natural to choose the operator $\hat{\Phi}_j$ as a linear combination of operators $\hat{z}_\alpha$: $\hat{\Phi}_j = \varphi_{j\alpha}\hat{z}_\alpha$ (hereafter the summing over the repeated Greek subscripts is assumed), and the operator $\hat{H}$ as a quadratic form of the same operators: $\hat{H} = \frac{1}{2}\hat{z}B\hat{z}$, $B = \tilde{B}$. Then comparing (1) and (8) we get

$$K = 2\tilde{\Psi}S, \quad L = -\frac{i}{2\hbar}B - \frac{1}{2}\tilde{K}, \quad R = \frac{i}{2\hbar}B - \frac{1}{2}\tilde{K} \tag{9}$$

where $\Psi_{\alpha\beta} = \sum_j \varphi^*_{j\alpha}\varphi_{j\beta}$. In this case the conditions (4) and (6) are automatically satisfied, provided $B = \tilde{S}B^+S$ ensuring the hermiticity of $\hat{H}$. Eq.(7) can be now rewritten as follows

$$A = \Sigma\left[\frac{1}{2}(K - \tilde{K}) - \frac{i}{\hbar}B\right]. \tag{7a}$$

The matrix Ψ is obviously hermitian and (this is its most important feature) nonnegative definite: for any complex vector x we have

$$x^*_\alpha \Psi_{\alpha\beta} x_\beta = \sum_j |\varphi_{j\alpha}x_\alpha|^2 \geqslant 0.$$

Consequently, $Q = S^+\tilde{K}$ is hermitian and nonnegative definite. Therefore any quantum dynamical semigroup generated by eq.(1) is determined by the two independent matrices: $Q = Q^+ \geqslant 0$ and $B = \tilde{B} = \tilde{S}B^+S$.

Now suppose that the matrix A in (3) is given, and the problem is to construct an equation similar to (1) having (3) as its consequence. If $\Sigma^{-1}A$ is symmetric, then the system is closed with the Hamiltonian determined by the matrix $B = i\hbar\Sigma^{-1}A$, and one can assume that $K = 0$. But if $\Sigma^{-1}A$ has a nonzero antisymmetrical part, then one has to assume the general form of (1) with $K \neq 0$. However, the matrix

A (or (3)) determines only the antisymmetric part of the matrix K. The symmetric part of K should be chosen so as $S^+\tilde{K}$ would be nonnegative definite. The significance of the symmetric part of K becomes clear if one considers an equation for the covariance matrix following from (1):

$$\dot{M} = AM + M\tilde{A} + \Sigma\tilde{K}\Sigma, \quad M = \|M_{\alpha\beta}\|, \quad M_{\alpha\beta} = \langle\hat{z}_\alpha\hat{z}_\beta\rangle = \mathrm{Tr}(\hat{\rho}\hat{z}_\alpha\hat{z}_\beta). \tag{10}$$

The matrix M must satisfy certain inequalities generalizing the well-known uncertainty relations $\Delta x\Delta p \geq \hbar/2$ [4]. Therefore the condition $S^+\tilde{K} \geq 0$ is equivalent to the preservation of these inequalities in time [5].

Eq.(1) corresponds to the description of the quantum system in the Schrodinger picture. The equivalent description in the Heisenberg picture consists in replacing eq.(1) by the operator Langevin equation [6,7]

$$d\hat{z}/dt = A\hat{z} + \hat{\chi}(t), \quad \hat{\chi} = (\hat{\chi}_1, \hat{\chi}_2, \ldots, \hat{\chi}_{2N}) \tag{11}$$

where

$$\langle\hat{\chi}(t)\rangle = 0, \quad \langle\hat{\chi}_\alpha(t)\hat{\chi}_\beta(t')\rangle = \delta(t-t')X_{\alpha\beta}. \tag{12}$$

The solution of eq. (11) is

$$\hat{z}(t+\Delta t) = \hat{z}(t) + A(t)\hat{z}(t)\Delta t + \int_t^{t+\Delta t}\hat{\chi}(\tau)\,d\tau \quad \text{as} \quad \Delta t \to 0.$$

Then with the same accuracy we get

$$M_{\alpha\beta}(t+\Delta t) = M_{\alpha\beta}(t) + \Delta t\left[A(t)M(t) + M(t)\tilde{A}(t)\right]_{\alpha\beta} + \Delta t X_{\alpha\beta}(t).$$

Comparing this expression with (10) we get the following relation

$$X = \Sigma\tilde{K}\Sigma = S^*\Sigma^+Q\Sigma, \quad X^+ = SX\tilde{S}. \tag{13}$$

Thus SX is hermitian and nonnegative definite. It

is often convenient to solve not the operator equation (1) but the equivalent equations for the density matrix or its analogues in some representations. For example, for the density matrix in the coordinate representation $\rho(x', x'')$ one gets from eqs.(1), (7a), (9), (13) that (here $\hat{z} = (\hat{p}, \hat{q}), y = (x', x'')$, so that $S = I_{2N}$):

$$\frac{\partial \rho(y)}{\partial t} = \left\{ \frac{1}{4} \frac{\partial}{\partial y} \left[\tilde{T}_+ D T_+ + i\hbar (\tilde{T}_+ A T_- + \tilde{T}_- \tilde{A} T_+) \right] \frac{\partial}{\partial y} - \right.$$
$$- \frac{i}{2\hbar} y \left[\tilde{T}_- D T_+ + i\hbar (\tilde{T}_- A T_- - \tilde{T}_+ A T_+) \right] \frac{\partial}{\partial y} - \tag{14}$$
$$\left. - \frac{1}{4\hbar^2} y \left[\tilde{T}_- D T_- - i\hbar (\tilde{T}_- A T_+ + \tilde{T}_+ \tilde{A} T_-) \right] y - \frac{1}{2} \operatorname{Tr} A \right\} \rho(y)$$

where

$$T_+ = \begin{Vmatrix} 0 & 0 \\ I_N & I_N \end{Vmatrix}^N, \quad T_- = \begin{Vmatrix} I_N & -I_N \\ 0 & 0 \end{Vmatrix}_N .$$

The hermitian matrix $D = X + \tilde{X}$ should satisfy the conditions

$$D_* \equiv D + i\hbar (A \Sigma_* + \Sigma_* \tilde{A}) = D_*^+ \geqslant 0, \quad \Sigma_* = \begin{Vmatrix} 0 & I_N \\ -I_N & 0 \end{Vmatrix} . \tag{15}$$

Eq.(14) is rather cumbersome. The most simple equation can be obtained for the Wigner function [8]:

$$W(p,q) = \int \rho(q + \xi/2, q - \xi/2) \exp(-ip\xi/\hbar) d\xi = W(y), \quad y = (p,q)$$

$$\frac{\partial W}{\partial t} = -\frac{\partial}{\partial y} (A y W) + \frac{1}{4} \frac{\partial}{\partial y} D \frac{\partial}{\partial y} W . \tag{16}$$

For example, for systems with one degree of freedom conditions (15) are reduced to the inequalities

$$D_p \geqslant 0, \quad D_q \geqslant 0, \qquad D = \begin{Vmatrix} D_p & D_{pq} \\ D_{pq} & D_q \end{Vmatrix} . \tag{17}$$
$$D_p D_q - D_{pq}^2 = \det D \geqslant (\hbar \operatorname{Tr} A)^2,$$

In the special case of a quantum one-dimensional damped oscillator relations (15), (17) are obtained in [9-12]. In the general case (for an arbitrary number of degrees of freedom and an arbitrary A) conditions (15) are derived in [5].

If the quantum system is described in terms of the boson annihilation and creation operators, i.e. $\hat{z} = (\hat{a}, \hat{a}^+)$, then it is convenient to use Glauber's P-representation [13]

$$\hat{\rho} = \pi^{-1}\int P(\alpha,\alpha^*)|\alpha\rangle\langle\alpha|\, d^2\alpha \tag{18}$$

where $P(\alpha,\alpha^*)$ is a real function of complex argument α, and $|\alpha\rangle$ means Glauber's coherent state. The function $P(y)$, where $y=(\alpha,\alpha^*)$ satisfies, as Wigner's function, an equation of the Fokker-Planck type:

$$\frac{\partial P}{\partial t} = -\frac{\partial}{\partial y}(AyP) + \frac{1}{4}\frac{\partial}{\partial y}G\frac{\partial}{\partial y}P, \tag{19}$$

$$G = D + AS + S\tilde{A}, \quad D = X + \tilde{X}, \quad S = \begin{Vmatrix} 0 & I_N \\ I_N & 0 \end{Vmatrix}.$$

The matrix D in this case satisfies the conditions

$$D_* \equiv S(D - A\Sigma_* - \Sigma_*\tilde{A}) = D_*^+ \geq 0. \tag{20}$$

The matrix Σ_* is the same as in (15). For the systems with one degree of freedom we have

$$A = \begin{Vmatrix} \xi & \eta \\ \eta^* & \xi^* \end{Vmatrix}, \quad D = \begin{Vmatrix} D_a & D_n \\ D_n & D_a^* \end{Vmatrix}, \quad G = \begin{Vmatrix} D_a + 2\eta & D_n + 2\mathrm{Re}\,\xi \\ D_n + 2\mathrm{Re}\,\xi & D_a^* + 2\eta^* \end{Vmatrix}$$

$$D_n = D_n^*,$$

and the elements of D obey the inequalities

$$D_n \geq 2|\mathrm{Re}\,\xi|, \quad D_n^2 - |D_a|^2 \geq (2\mathrm{Re}\,\xi)^2 = (\mathrm{Tr}\,A)^2. \tag{21}$$

Besides the usual P-representation, there exist various "generalized P-representations" [14], in which instead of the "diagonal" expansion (18) "non-diagonal" expansions are used:

$$\hat{\varrho} = \int d\mu(\alpha,\beta) P(\alpha,\beta) |\alpha\rangle\langle\beta^*| \left(\langle\beta^*|\alpha\rangle\right)^{-1}.$$

Here α and β are independent complex vectors. The integration measure $d\mu(\alpha, \beta)$ and the domain of integration may be determined in different ways. For all such generalizations eqs. (19), (20) hold as well, where $y = (\alpha, \beta)$. Note that in such a case the function $P(y)$ depends on $4N$ real arguments. The P-representation does not exist for an arbitrary statistic operator. But for an arbitrary statistic operator $\hat{\varrho}$ there exists the R-representation [13], i.e. the analytic function $R(\alpha^*, \beta) = \exp\left(\frac{1}{2}|\alpha|^2 + \frac{1}{2}|\beta|^2\right)\langle\alpha|\hat{\varrho}|\beta\rangle$. The function $R(y)$, where $y = (\alpha^*, \beta)$, satisfies the equation

$$\frac{\partial R}{\partial t} = \Big\{ \frac{1}{4} y (D + S\tilde{A} + AS) y + \frac{1}{2} y (A - S\tilde{A}S - DS) \frac{\partial}{\partial y} + \frac{1}{4} \frac{\partial}{\partial y} (SDS - SA - \tilde{A}S) \frac{\partial}{\partial y} - \frac{1}{4} \mathrm{Tr}\,(2A + SD) \Big\} R(y). \tag{22}$$

The matrix D again satisfies the conditions (20).

References

1. Davies, E.B. (1976). Quantum Theory of Open Systems. Academic Press, London.

2. Belavin, A.A., Perelomov, A.M., Popov, V.S., Zel'dovich, B.Ya. (1969). ZhETPh, 56, 264. (Russian)

3. Spohn, H. (1980). Rev. Mod. Phys. 52, 569.

4. Dodonov, V.V., Man'ko, V.I. (1985). In: Group Theoretical Methods in Physics, Markov, M.A. et al. (Eds.). Harwood Academic Publishers, Chur-London-Paris-N.Y., vol.1, p.591.

5. Dodonov, V.V., Man'ko, O.V. (1985). Physica 130A, 353.

6. Lax, M. (1968). Fluctuation and Coherence Phenomena in Classical and Quantum Physics. Gordon and Breach, N.Y.

7. Haken, H. (1975). Rev. Mod. Phys. 47, 67.

8. Wigner, E.P. (1932). Phys. Rev. 40, 749.

9. Dodonov, V.V., Man'ko, V.I. (1985). In: Group Theoretical Methods in Physics, Proceedings of the International Seminar, Zvenigorod, 24-26 November 1982, Markov, M.A. et al. (Eds.). Harwood Academic Publishers, Chur-London-Paris-N.Y., vol.1, p.705.

10. Barchielli, A. (1983). Nuovo Cimento 74B, 113.

11. Dekker, H., Valsakumar, M.C. (1984). Phys. Lett. A104, 67.

12. Sandulescu, A. and Scutaru, H. (1985). JINR Rapid Communications, JINR, Dubna, N10-85, p.5.

13. Glauber, R.J. (1963). Phys. Rev. 131, 2766.

14. Drummond, P.D., Gardiner, C.W. (1980). J. Phys. A13, 2353.

NORMAL FORMS OF THREE-DIMENSIONAL RELAXATIONAL DYNAMICAL SYSTEMS

B.A.Malomed, M.I.Monastyrsky

Institute of Theoretical and Experimental Physics, Moscow

1. Introduction. Normal forms of differential equations of motion is a significant part of the modern dynamical systems theory [1-3]. Investigation of non-hamiltonian dynamical systems with the phase of dimension $D \geqslant 3$, that typically show stochastic behaviour first found in the Lorenz model [4] (other examples are described e.g. in the survey [5]), gives rise to new problems concerning description of normal forms. The investigation of stochastic properties of particular systems is, as a rule, a complicated problem essentially based on numerical methods. That is why relaxational systems, i.e. those admitting separation of fast and slow motions, for which the stochasticity can be investigated analytically [6,7], are of special interest. A relaxational system with 3-dimensional phase space is described by a system of differential equations

$$\dot{x} = f(x,y,z),\ \dot{y} = g(x,y,z),\ \epsilon\dot{z} = h(x,y,z), \tag{1}$$

where ϵ is a small parameter, z the fast variable, and x,y the slow ones. A trajectory in the 3-dimensional phase space consists of pieces corresponding to the slow motion along the two-dimensional surface $h(x,y,z) = 0$ (called the fast motion null-cline), and of pieces corresponding to the fast motion along the lines parallel to the z-axis. The slow motion turns into the fast one when a trajectory intersects the shooting line (called the stall line in [7]) lying on the null-cline and determined by the equations

$$h(x,y,z)=0, \quad \partial h/\partial z=0. \tag{2}$$

In particular, for important class of relaxational systems with z-like null-clines (Fig. 1) the existence

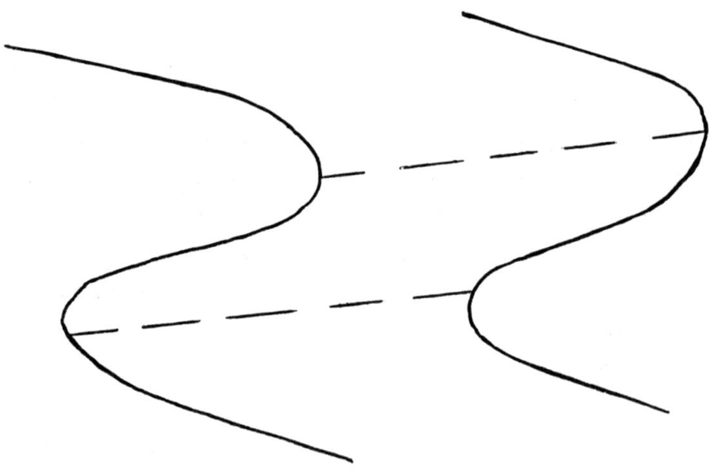

Fig. 1. The fast motion null-cline in the three-dimensional phase space

of the stochasticity can be often proved analytically as $\epsilon \to 0$ [6,7].

Stochastic properties of the considered systems detected as $\epsilon \to 0$ should persist in general at ϵ small but finite. Indeed, the system's entropy (which is a quantitative characteristic of the stochasticity), being finite for an infinitely small ϵ [7], in general cannot suddenly vanish since ϵ is finite. At the same time , a new phenomena may occur for finite ϵ . In particular, the shooting line turns into a shooting strip of a finite width. Further, as $\epsilon \to 0$, on the stable part of the null-cline there lies a separatrix, i.e. a slow motion trajectory tangent to the shooting line. It separates the trajectories which fall down off the null-cline and those which remain on it (Fig. 2). An

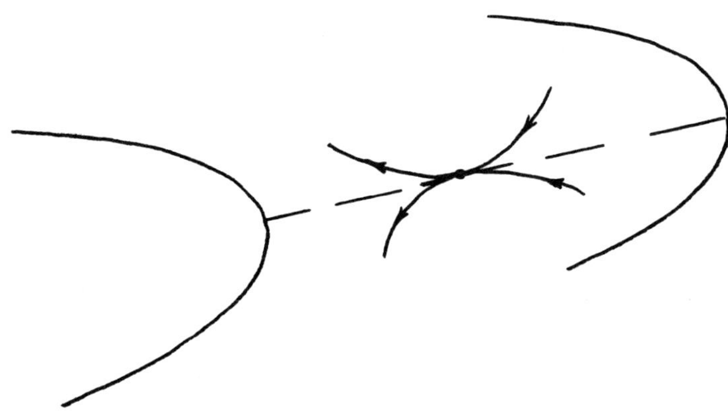

Fig. 2. The two one-dimensional separatrices in the case $\epsilon \to 0$.

analogous separatrix, tangent to the shooting line at the same point, lies on the unstable part of the null-cline (beneath the shooting line, Fig. 3). For finite ϵ these one-dimensional separatrices turn into a 2-dimensional separatrix surface which may have a complicated form.

In sect. 2 of this paper we derive a universal nonlinear 2 order system containing one continuous and one discrete (sign) parameters to describe trajectories of the system (1) for a finite ϵ in a vicinity of the shooting line (2). We perform an analytical investigation of that system for some particular cases.

In general - the system can be solved only numerically (this will be considered elsewhere). In the end of sect. 2 we discuss a feasible form of the separatrix surface determined by the universal system. The form of that surface, as well as the system itself should be universal.

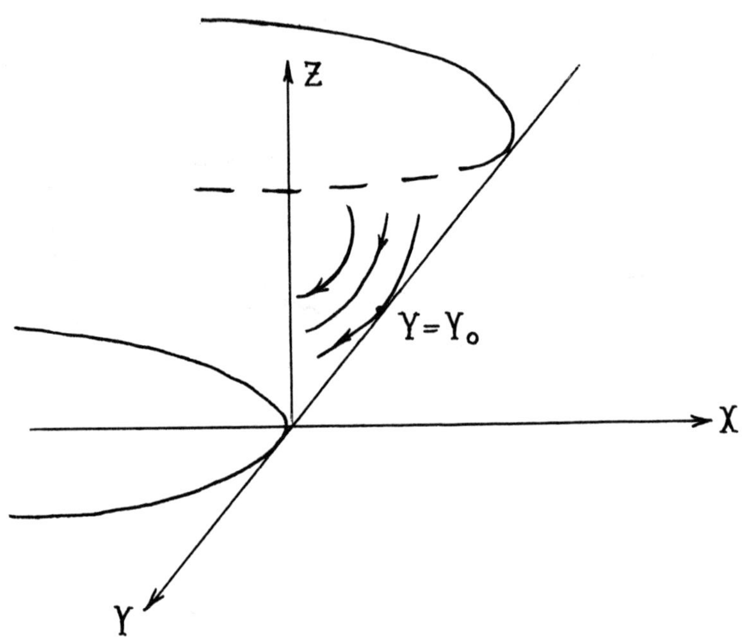

Fig. 3. The trajectories not falling down from the null-cline

As it is known from Whitney's theory [8,2] , a 2-dimensional surface imbedded into 3-dimensional space may possess, parallel with a fold (the null-cline of Fig. 1 has two folds), a singularity of the cusp type. In sect. 3 we obtain a universal system describing trajectories of the system (1) in vicinity of the cusp. This system admits explicit analytic investigation yielding that in the cusp's vicinity trajectories do not intersect the shooting line, i.e. this vicinity does not contribute to stochastic properties of the system.

2. Fold-type systems

2.1. General Investigation. According to Whitney's theorem [8,2] there are three distinct structurally stable types of mappings of the plane (x,z) into itself. A map belonging to any of these types can be brought into a certain canonical form by means of a smooth coordinate transformation. The first type is equivalent to the identity map. The second type is the fold:

$$x' = x,\ z' = z^2. \tag{3}$$

The third type is the cusp:

$$x' = x,\ z' = z^3 - xz. \tag{4}$$

According to (3), in the fold's vicinity the dynamical system (1) can be turned in appropriate coordinates into the form

$$\dot{x} = F(y) + G(y)x + H(y)z, \tag{5}$$

$$\dot{y} = 1, \tag{6}$$

$$\epsilon \dot{z} = -(x + z^2), \tag{7}$$

where F(y), G(y) and H(y) are arbitrary regular functions. In these coordinates the shooting line (2) takes the form x=0, z=0 (Figs 3, 4) while y may be identified with the time due to (6).

Intersecting the shooting line by a slow motion trajectory (which in the limit $\epsilon \to 0$ lies on the null-cline) plays the key role in stochastic motion. Denote the moment of intersection y_0 and expand

$$F(y) = F_0 + F_1(y - y_0) + \dots, \tag{8}$$

$$G(y) = G_0 + G_1(y - y_0) + \dots,\quad H(y) = H_0 + H_1(y - y_0) + \dots\ . \tag{9}$$

First consider the case $F_0 \neq 0$. As we will see in this case trajectories in a vicinity of y_0 behave as shown an Fig. 2. It is convenient to introduce new local coordinates

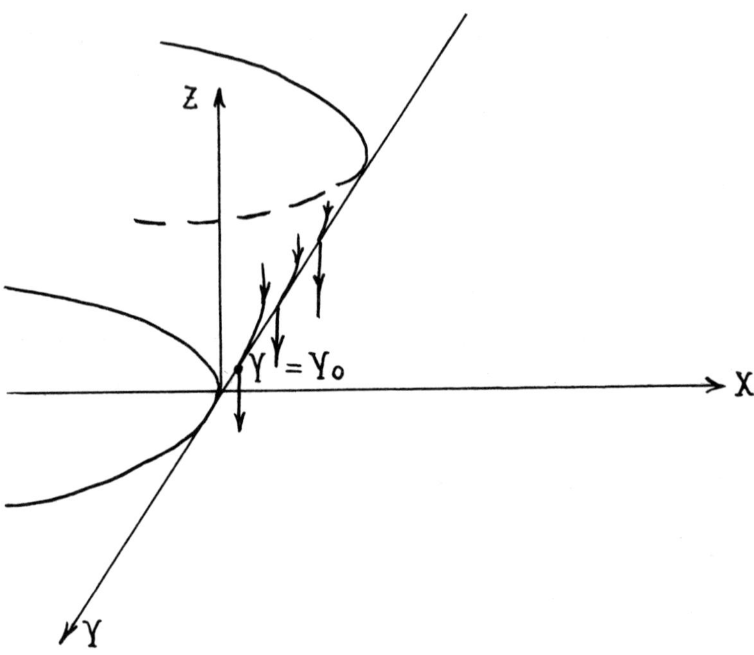

Fig. 4. The trajectories falling down from the null-cline

$$\zeta = (F_0\epsilon)^{-1/3} z, \quad \xi = (F_0\epsilon)^{-2/3} x. \tag{10}$$

After eliminating the time, the system (5)-(7) in the coordinates (10) reduces, with (8), (9), to the Riccati equation determining the form of a trajectory:

$$\frac{d\zeta}{d\xi} = -\xi - \zeta^2. \tag{11}$$

A general solution of (11) can be readily found:

$$\zeta = u^{-1}\frac{du}{d\xi}, \quad u = Ai(-\xi) + cBi(-\xi), \tag{12}$$

where Ai and Bi are two standard Airy functions, and c an arbitrary constant. As is clear from (12), $\zeta \to \infty$ as $u \to 0$. According to (10) and (17), infinite ζ implies shooting (falling down) of the trajectory off the null-cline (Fig. 4). Since first zeroes of the functions $Ai(-\xi)$ and $Bi(-\xi)$ are points ~ 1, it is clear that, according to (10) and (12), the shooting line $x=0$, $z=0$ (existing as $\epsilon \to 0$) at finite ϵ turns into a shooting strip of width $\sim \epsilon^{2/3}$ in the x-direction.

Now let us proceed with the case $F_0 = 0$. At infinitely small ϵ this case corresponds to the separatrix trajectory of the slow motion tangent to the shooting line at $Y = Y_0^{(0)}$ (Fig. 3). Inserting the expansions (8),(9) into r.h.s. of (5) and denoting $Y - Y_0 \equiv t$, one obtains from (5)-(7) the following system:

$$\frac{dx}{dt} = F_1 t + G_0 x + H_0 z, \tag{13}$$

$$\epsilon \frac{dz}{dt} = -(x + z^2). \tag{14}$$

With the same accuracy that (13), (14) were derived from (5)-(7) , it is easy to eliminate the term $G_0 x$ from (13) getting the limit system

$$\frac{d\xi}{d\tau} = f\tau + (\mathrm{sgn}\, H_0)\zeta, \tag{15}$$

$$\frac{d\zeta}{d\tau} = -(\xi + \zeta^2). \tag{16}$$

Introduce new local coordinates

$$\xi = (\epsilon |H_0|)^{-1} x, \quad \zeta = (\epsilon |H_0|)^{-1/2} z, \quad \tau = (|H_0|/\epsilon)^{1/2} t \tag{17}$$

and the notation

$$f = F_1 / H_o^2 . \tag{18}$$

The system (15), (16) is just a universal system describing the behaviour of trajectories in a vicinity of $Y_o^{(o)}$ (V.I.Arnold noted that the system (15),(16) is rather general. It is important for investigating equations of the form $F(x,y, dy/dx) = 0$ arising e.g. in pursuit problem and in investigating asymptotic lines on surfaces of constant curvature.) Excluding ξ turns this system into

$$\frac{d^2\zeta}{d\tau^2} + 2\zeta \frac{d\zeta}{d\tau} + f\tau \pm \zeta = 0.$$

2.2. <u>Special cases</u>. The special case $H_o=0$ when (15), (17) and (18) are not defined needs special consideration. In this case excluding $G_o x$ turns (13) into $dx/dt = F_1 t$, i.e.

$$x = \frac{1}{2} F_1 t^2 + k \quad \text{, where } k \text{ is a constant.} \tag{19}$$

Fig. 3 depicts (19) for $F_1 < 0$. Therefore (14) in the new local coordinates

$$\zeta = \left(-\frac{1}{2}\epsilon^2 F_1\right)^{-1/4} z, \quad \xi = \left(-2\epsilon^2 / F_1\right)^{-1/4} \tag{20}$$

takes the form of the Riccati equation which determines the form of a trajectory:

$$\frac{d\zeta}{d\xi} - \xi^2 + \zeta^2 + \gamma = 0, \tag{21}$$

where

$$\gamma = (k/\epsilon)\sqrt{-F_1/2} . \tag{22}$$

Similarly to (12), a general solution of (21) can be represented in the form

$$\xi = u^{-1}\frac{du}{d\xi}, \quad u = D_{\gamma/2-1/2}(\sqrt{2}\,\xi) + c D_{\gamma/2-1/2}(-\sqrt{2}\,\xi), \tag{23}$$

where $D_{\gamma/2-1/2}$ is a parabolic cylinder function and c an arbitrary constant. It is known [9] that at $\gamma<0$ the function $D_{\gamma/2-1/2}(\sqrt{2}\,\xi)$ is positive for all real ξ, and it possesses the asymptotic form

$$D_{\gamma/2-1/2}(\sqrt{2}\,\xi) \approx (2\pi)^{1/4}\left(\Gamma(1/2-\gamma/2)\right)^{-1/2} \times |\sqrt{2}\,\xi|^{\frac{1}{2}(1-\gamma\,\mathrm{sgn}\,\xi)} \exp\left(-\frac{1}{2}\xi|\xi|\right). \tag{24}$$

It is clear from (23) and (24), that $u(\xi)$ has no zeros if $\gamma<0$ and $c>0$, i.e. the trajectories described by (24) do not fall down off the null-cline. They remain on it and perform one more cycle of slow motion. If $\gamma<0$ and $c<0$ then, as follows from (23) and (24), $u \to \mp\infty$ as $\xi \to \pm\infty$, i.e. in this case $u(\xi)$ has at least one zero., and the corresponding trajectory falls down. Thus, for fixed negative γ the trajectory with $c = c_0 \equiv 0$ is the separatrix separating trajectories that fall down and those that do not. The same situation takes place for $\gamma = 0$. In this case the parabolic cylinder functions reduce to the modified Bessel functions [9], and the function $u(\xi)$ in (23) takes the form

$$u = |\xi|^{1/2}\left[I_{-1/4}(\xi^2/2) + \tilde{c}\, I_{1/4}(\xi^2/2)\right] \tag{25}$$

(the constants c in (23) (for $\gamma = 0$) and $\tilde{c}$ in (25) do not coincide). The function (25) has no zero if $\tilde{c} > -1$, i.e. for $\gamma = 0$ the separatrix trajectory corresponds to $\tilde{c} = c_0 \equiv -1$.

For $\gamma > 0$ there are two possibilities. If γ is sufficiently small then $D_{\gamma/2-1/2}(\sqrt{2}\,\xi)$ has, as before, no zero, and if γ is sufficiently large then it has one or more zeroes. Therefore there should exist a positive constant $\gamma_0 \sim 1$ such that for $\gamma < \gamma_0$ there is a separatrix trajectory corresponding to some $c = c_0(\gamma)$, while for $\gamma > \gamma_0$ all the trajectories fall down. This inference becomes quite clear if one recalls that γ is proportional to k (see (22)) and, turning to (19), at sufficiently large k trajectories get far into the range $x > 0$, i.e. far beyond the shooting line (Fig. 3).

To conclude this subsection let us note that the original system (13), (14) can be explicitly solved in one more particular case, $F_1 = 0$. According to (18), in this case $f = 0$, and (15), (16) yield the linear equation for ρ^2:

$$\frac{1}{2}(\operatorname{sgn} H_0)\frac{d(\rho^2)}{d\xi} + \rho^2 + \xi = 0.$$

Its solution is evident:

$$\rho^2 = -(\operatorname{sgn} H_0)\xi^2 + \mathrm{const}\cdot\exp\left(-2(\operatorname{sgn} H_0)\xi\right). \tag{26}$$

Since (26) remains finite at finite ξ we encounter a trivial situation when the trajectories do not fall down off the null-cline.

2.3. Feasible Structure of the Separatrix Surface. It is clear that, as well as in the particular case (19)-(24), the variety of trajectories, i.e. solutions of the second order system (15), (16), depend on two parameters which we as before, denote γ and c. In general case one may expect that some condition $c = c_0(\gamma)$ singles out, as in the above particular case, separatrix trajectories spanning a two-dimensional separatrix surface (γ and ξ may be viewed as coordinates parametrizing this surface). To determine the surface's form one should numerically solve the equations (15),(16). Even in the explicitly studied particular case the analysis of the form of the surface defined by the solution (26) is not very simple and also needs engaging some numerical methods. In particular, one may assume the surface to have sufficiently complicated "plicate" form, i.e. instead of one $c_o(\gamma)$ there may exist some values $c_o^{(1)}(\gamma),\ldots, c_o^{(n)}(\gamma)$ such that trajectories remain on the null-cline provided

$$c_o^{(1)}(\gamma)<c<c_o^{(2)}(\gamma);\ \ldots\ ;\ c_o^{(n-2)}(\gamma)<c<c_o^{(n-1)}(\gamma);$$

$$c>c_o^{(n)}(\gamma),$$

and they fall down off it if

$$c<c_o^{(1)}(\gamma);\ c_o^{(2)}(\gamma)<c<c_o^{(3)}(\gamma);\ \ldots\ ;\ c_o^{(n-1)}(\gamma)<c<$$

$$<c_o^{(n)}(\gamma).$$

In principle, n may equal infinity.

3. Cusp type systems. When the null-cline has a cusp the equations of motion can be brought, as follows from (4), instead of (5)-(7) the following form:

$$\dot{x}=a+a_1x+a_2y+a_3z,\ \dot{y}=b+b_1x+b_2y+ \tag{27}$$

$$+b_3z,\ \epsilon\dot{z}=z^3-yz-z,$$

This is a universal system describing trajectories in vicinity of the cusp. The corresponding null-cline is determined as

$$z^3-yz-x=0, \tag{28}$$

and the shooting line is

$$y = 3z^2, \quad x = -2z^3. \tag{29}$$

The cusp is located at $x=y=z=0$. The shooting line (29) and the null-cline (28) are depicted in Fig. 5.

In this section we are not concerned with effects stipulated by finiteness of ϵ. Instead, we concentrate on studying trajectories in the cusp's vicinity as $\epsilon \to 0$. Then slow motion trajectories lie on the surface (28). As follows from (29) the shooting line is close to the parabola $y=3z^2$ lying in the plane $x=0$. Therefore we consider a slow motion trajectory going through some point . A which lies on the null-cline to the right from the shooting line (Fig.6). If $b<0$ in (27), the situation is evident: all the trajectories go to the right and do not intersect the shooting line. Since stochastic properties of the original system are stipulated just by fast motion after intersecting the shooting line, we see that for $b < 0$ the cusp's vicinity does not contribute to stochastization of dynamics of the system.

If $b > 0$ the trajectory approaches the shooting line. Considering the trajectory in this case, one should single out two regions: $|X| \sim |Y|$ and $|X| \ll |Y|$. In the former region (28) yields

$$z = x^{1/3} + \frac{1}{3} y x^{-1/3}. \tag{30}$$

Differentiating (30) gives

$$\dot{z} = \frac{1}{3} x^{-2/3} \dot{x} + \frac{1}{3} x^{-1/3} \dot{y} - \frac{1}{9} y x^{-4/3} \dot{x}. \tag{31}$$

Then we substitute (27) into (31) to obtain

$$\dot{z} = \frac{1}{3} a x^{-2/3}. \tag{32}$$

As is seen from (32), z is large as compared with x and y, therefore the trajectory goes almost vertically, i.e. almost parallel to the shooting line, and does not intersect it.

In the region $|X| \ll |Y|$ (28) gives

$$z = x/y. \tag{33}$$

Differentiating (33) and with (27) we get

$$\dot{z} = a/y + a_2. \tag{34}$$

One sees from (31) that z does not become infinite when $x \to 0$ (as it would follow from (32)); nevertheless, z remains large as compared with x and y, i.e. the trajectory indeed goes almost vertically. Therefore in the case $b \geqslant 0$, $a \neq 0$ the cusp's vicinity does not generate stochastic motion either.

As last, in the degenerate case $b > 0$, $a = 0$, when (32) vanishes, more accurate calculating (31) in the region $|X| \sim |Y|$ yields

$$\dot{z} = \frac{1}{3}(a_3 + b)x^{-1/3}. \tag{35}$$

In the narrow region $|X| \ll |Y|$ a more thorough calculation of $\dot{z}$ shows that $\dot{z} \sim \dot{x}, \dot{y}$, but this does not alter the general form of the trajectory.

Thus, the developed analysis of the dynamical system (1) in the cusp's vicinity yields that this vicinity does not contribute to stochastic properties of the system.

In conclusion note that it would be quite interesting to derive and investigate universal (normal) forms of differential equations analogous to (15)-(16), (21) and (27) describing singularities of null-clines of relaxational dynamical systems with manydimensional phase space where the singularity theory of mappings is sufficiently advance.

The authors are indebted to V.I.Arnold and A.S.Pikovsky for useful discussions.

References

1. Arnold,V.I. (1974). Mathematical Methods of Classical Mechanics. Nauka, Moscow (Russian).

2. Arnold,V.I., Varchenko,A.N., Gusein-zade,S.M. (1982). Singularities of Differentiable Maps. 1. Nauka, Moscow (Russian).

3. Takens,F. (1976). In: L.N. in Math. 525,p. 143.

4, Lorenz,E. (1963). J. Atmosp. Sci. 20, 130.

5, Helleman, R.H.G. (1980). In: Fundamental Problems in Statistical Mechanics, vol. 5. North Holland.

6. Pikovsky,A.S., Rabinovich,M.I. (1978). Doklady AN SSSR 239, 301 (Russian).

7. Malomed,B.A. (1983). Physica D8, 343.

8. Whitney,H. (1955). Ann. Math. 62, 374.

9. Handbook of Mathematical Functions. (1964).. Abramowitz M., Stegun I.A. (ed.). National Bureau of Standards.

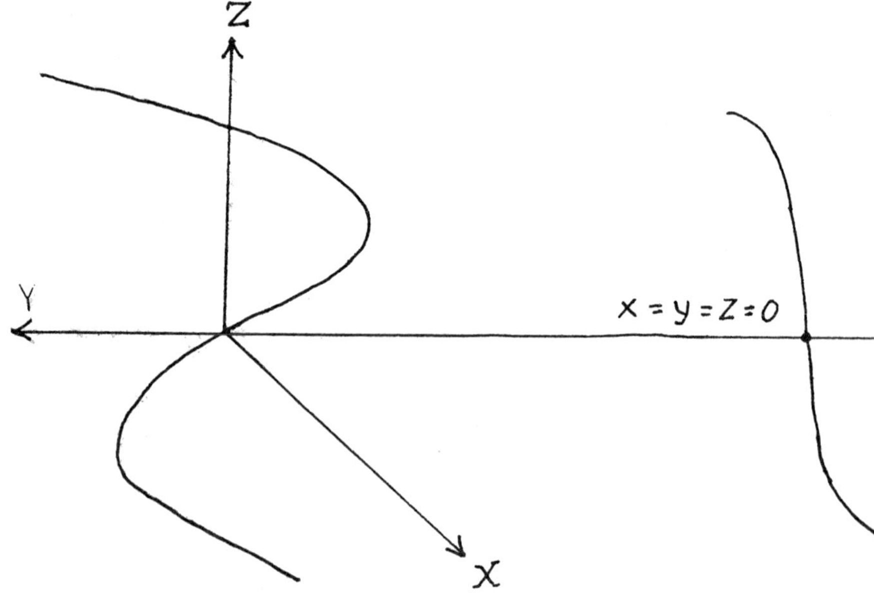

Fig.5. The nullcline of the cusp type.

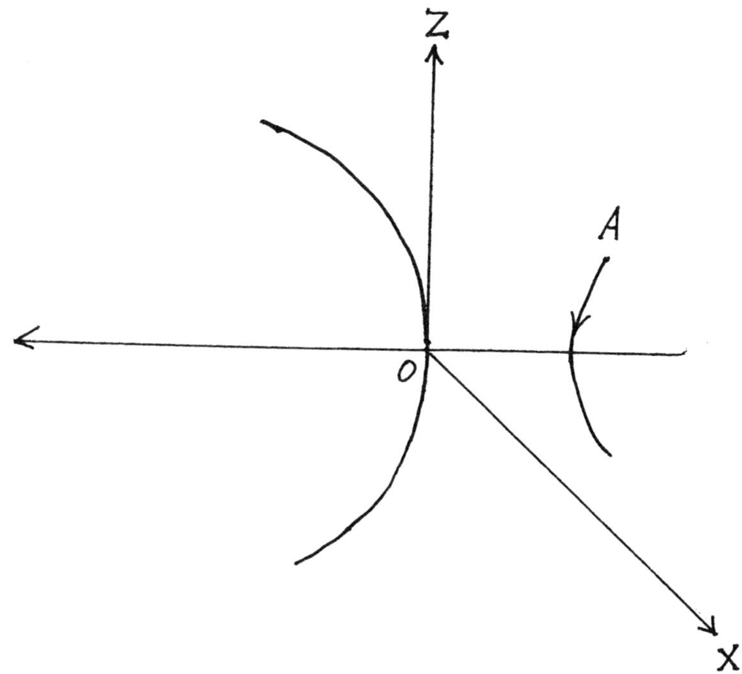

Fig.6. The tragectories in the vicinity of the cusp.

ONE EXACT SOLUTION OF THE RELATIVISTIC TWO-BODY PROBLEM IN THE POINCARE APPROACH

G.N. Afanasiev, R.A. Asanov
Joint Institute for Nuclear Research, Dubna, Moscow Region, 141980

In 1906 H. Poincare [1] made a relativistic generalization of the Newtonian equations for two gravitating point-like masses taking into account the finite propagation velocity (equal to the speed of C) of the retarded interaction. His equations do not exactly fix the form of the interaction. Oversimplifying the form of the interaction Poincare and his successors have failed in the description of astronomic effects, so-called "three crucial tests" of General Relativity. Only in 1943-1944 with a similar approach G. Birkhoff and collaborators [2] have shown that "three crucial tests" can be explained. In 1981 the interaction form was exactly fixed [3] by requiring the Poincare equations to coincide with the equations of GR. From this the correct description of "three crucial tests" and time delay of radar signals immediately follow.

Now it is natural to try to search e.g. for particular exact solutions of the two-body relativistic problem with retarded interaction. It is known [4] that in the (retarded) electrodynamics for two charged particles a uniform circular motion does not exist. For this case in 1972 J. Singh [5] succeeded in introducing additional forces such that circular motion becomes possible. (It is also known that there exists a circular motion in the case of the half-sum of retarded and advanced interactions and in the case when the first particle produces a retarded force and the second particle an advanced one [4,6]).

Here the (retarded) Poincare approach is treated (details see in [7]), and some conditions on the form of the interaction are established for existing of a uniform circular relativistic motion of two particles with equal masses. A particular case is treated, where in the nonrelativistic limit equations of motion coincide with the gravitational Newtonian ones and a connection between the radius and angular velocity is obtained in that limit: the third Kepler law. There is given an exact solution describing motion along a straight line. Here the chosen interaction reduces in the nonrelativistic limit to the interaction with a constant force. Note, that until now there has been no physical prescription for fixing the form of the interaction.

Following Poincare write the equations of motion for two interacting particles:

$$\frac{d^2 x_{1\mu}}{d\tau_1^2} = f_1 x_\mu + f_2 \frac{dx_{1\mu}}{d\tau_1} + f_3 \frac{dx_{2\mu}}{d\tau_2} + f_4 \frac{d^2 x_{2\mu}}{d\tau_2^2} \tag{1a}$$

$$\frac{d^2 x_{2\mu}}{d\tau_2^2} = -g_1 x_\mu + g_2 \frac{dx_{2\mu}}{d\tau_2} + g_3 \frac{dx_{1\mu}}{d\tau_1} + g_4 \frac{d^2 x_{1\mu}}{d\tau_1^2}, \tag{1b}$$

($\mu = 0,1,2,3$). Here τ_1 and τ_2 are proper times of particles. ($d\tau = \sqrt{dt^2 - c^{-2}(d\vec{x})^2}$); $x_{1\mu}$ and $x_{2\mu}$ are particle's coordinates; $x_\mu = x_{1\mu} - x_{2\mu}$. The masses of particles are absorbed into the functions f and g. Eqs. (1a) and (1b) should pass one into another under the change of indices $1 \rightleftarrows 2$. This gives $g_i(1,2) = f_i(2,1)$, where f and g are arbitrary functions invariant w.r.t. the Poincare group. Equations treated by Poincare did not contain 4-accelerations (in the r.h.s., i.e. $f_4 = g_4 = 0$). We need them however, to include the electromagnetic two-body problem into this scheme.

Eqs. (2) are not sufficiently determined from the physical point of view since they contain two coordinates and two proper times. So, one must somehow connect t_1 and t_2 . Poincare suggested that the time t of "attracted" body is related to the time $\tilde{t}$ of "attracting" one by the following relativistically invariant condition:

$$\tilde{t} = t - r/c, \text{ here } r = \Big(\sum_{i=1}^{3} x_i^2\Big)^{1/2}, x_i = z_i(t) - \tilde{z}_i(t - r/c), \quad (2)$$

$z_i(t)$ and $\tilde{z}_i(\tilde{t})$ are coordinates of the attracted and attracting bodies. Eq. (2) means that the interaction (propagating with the speed of light c) "leaves" the attracting body before it "reaches" the attracted one (causality). In Eqs. (1a) it is assumed that particle 1 is attracted and particle 2 is attracting one. It is convenient to divide Eqs. (1a) into the spatial and time parts:

$$\frac{d^2 x_{1i}}{d\tau_1^2} = f_1 x_i + f_2 \frac{dx_{1i}}{d\tau_1} + f_3 \frac{dx_{2i}}{d\tau_2} + f_4 \frac{d^2 x_{2i}}{d\tau_2^2}, i = 1,2,3, \quad (3)$$

$$\frac{d^2 t_1}{d\tau_1^2} = f_1 \frac{r}{c} + f_2 \frac{dt_1}{d\tau_1} + f_3 \frac{dt_2}{d\tau_2} + f_4 \frac{d^2 t_2}{d\tau_2^2} .$$

The orthogonality of the 4-velocity to the 4-acceleration gives

$$f_1 c A + c^2 f_2 + c^2 C f_3 + f_4 D = 0. \quad (4)$$

A similar relation holds for g . Thanks to these relations and the symmetry $1 \rightleftarrows 2$ we have only three arbitrary functions, say f_1, f_3, f_4 . The quantities which enter (4) are the following invariant combinations of the 4-coordinates, velocities and accelerations:

$$A=\frac{x_\mu}{c}\frac{dx_{1\mu}}{d\tau_1}=(1-\beta_1^2)^{-1/2}\left(\tau-\frac{\vec{z}\vec{v}_1}{c}\right), C=\frac{1}{c^2}\frac{dx_{1\mu}}{d\tau_1}\frac{dx_{2\mu}}{d\tau_2}=$$

$$=\frac{1}{\sqrt{1-\beta_1^2}}\frac{1}{\sqrt{1-\beta_2^2}}\left(1-\frac{v_1v_2}{c^2}\right), D=\frac{dx_{1\mu}}{d\tau_1}\frac{d^2x_{2\mu}}{d\tau_2^2}= \tag{5a}$$

$$=\frac{1}{\sqrt{1-\beta_1^2}}\frac{1}{1-\beta_2^2}\left[\frac{\vec{v}_2\vec{w}_2^{\,2}}{1-\beta_2^2}\left(1-\frac{\vec{v}_1\vec{v}_2}{c^2}\right)-\vec{v}_1\vec{w}_2\right].$$

Here $\vec{v}_1, \vec{v}_2, \vec{w}_1, \vec{w}_2$ are the usual 3-velocities and accelerations: $\beta_1=v_1/c, \beta_2=v_2/c$. Below we need three more invariants:

$$B=\frac{1}{c}x_\mu\frac{dx_{2\mu}}{d\tau_2}=\frac{1}{\sqrt{1-\beta_2^2}}\left(z-\frac{\vec{z}\vec{v}_2}{c}\right), E=-x_\mu\frac{d^2x_{2\mu}}{d\tau_2^2}=\frac{\vec{z}\vec{w}_2}{1-\beta_2^2}-$$

$$-\frac{\vec{v}_2\vec{w}_2}{(1-\beta_2^2)^{3/2}}\frac{B}{c}, F=-\left(\frac{d^2x_{2\mu}}{d\tau_2^2}\right)^2=\frac{1}{(1-\beta_2^2)^{3/2}}\left[\vec{w}_2^{\,2}+\frac{\vec{v}_2\vec{w}_2}{c^2(1-\beta_2^2)}\right]. \tag{5b}$$

Note that Poincare used only the invariants A,B,C which do not contain accelerations. It is convenient to change in the first three Eqs. (3) the proper time by the coordinate one. We get

$$\frac{w_{1i}}{1-\beta_1^2}=f_1x_i-v_{1i}\left[\frac{z}{c}f_1+\frac{f_3}{\sqrt{1-\beta_2^2}}+f_4\frac{\vec{w}_2\vec{v}_2}{c^2(1-\beta_2^2)^2}\right]+$$

$$+v_{2i}\left[\frac{f_3}{\sqrt{1-\beta_2^2}}+f_4\frac{\vec{w}_2\vec{v}_2}{c^2(1-\beta_2^2)^2}\right]+\frac{w_{2i}}{1-\beta_2^2}f_4. \tag{6}$$

Due to the invariance with respect to particle's permutation it suffices to consider only (6). Following Poincare suppose that f is constructed from A,...,F. The explicit form of f depends on the physical nature of the problem (gravity, electromagnetism, nuclear forces), desired form of solutions, etc.

Now, find the conditions under which Eqs.(6) have solutions corresponding to a stationary circular motion. Let the masses of particles be equal. Then they move along the same circle of radius a with a constant angular velocity

$$x_{1,2} = \pm a \cos \omega t , \quad y = \pm a \sin \omega t . \tag{7}$$

Next, substitute (7) into (6). All the quantities referring to particle 2 should be taken at some earlier moment of time:

$$\begin{aligned} x_2 &= -a \cos[\omega(t-t_0)] = -a \cos(\omega t - \varphi_0), \\ y_2 &= -a \sin[\omega(t-t_0)] = -a \sin(\omega t - \varphi_0). \end{aligned} \tag{8}$$

The motion is stationary, so t_0 = const, $\varphi_0 = \omega t_0$ = = const. The angle φ_0 of the delay is determined as follows: an interaction which exits particle 2 at the time $t-t_0$ should reach particle 1 at the time t. This gives:

$$\beta \equiv \frac{\omega a}{c} = \frac{\varphi_0}{\sqrt{2(1+\cos\varphi_0)}} = \frac{\varphi_0/2}{\cos(\varphi_0/2)} . \tag{9}$$

It follows that $\varphi_0 = 0$ for $\beta = 0$ and takes maximum value ($\sim 85^0$) for $\beta = 1$. The motion is determined by two constants a and ω , while two others (β and φ_0) are defined from (9). In what follows we need the values of invariants A,...,F for this circular motion; denote them as $A_0,\dots,F_0$. Inserting (7) and (8) into (6) we get the following relations:

$$\frac{f_3^0}{\omega\sqrt{1-\beta^2}} = \left(\frac{f_1^0}{\omega^2} + \frac{1}{1-\beta^2}\right)\frac{\sin\varphi_0}{1+\cos\varphi_0} - \frac{f_1^0\,\beta\sqrt{2}\cos\varphi_0}{\omega^2\sqrt{1+\cos\varphi_0}}\,,$$

$$f_4^0(1-\beta^2)^{-1} = -2f_1^0\omega^{-2} - (1-\beta^2)^{-1} - \frac{f_1^0\beta\sqrt{2}\sin\varphi_0}{\omega^2\sqrt{1-\cos\varphi_0}}\,. \qquad (10)$$

Here $f_i^0 = f_i(A,\ldots,F)\big|_{A=A_0,\ldots,F=F_0}$. We may consider (10) as testing equations for the existence of circular motion in this relativistic problem.

With (10) it is easy to see that there are no circular motions (if $v < c$) in the retarded electrodynamics. The same holds for the simplified version of the Poincare equations defined by: $f_2 = f_4 = 0$, $f_1 = \gamma/B^3$, $f_3 = -\gamma A/cB^3C$, $\gamma = const$. Synge interaction [5] satisfies condition (10) if $m_1 = m_2$, $e_1 = -e_2$.

Now, the invariants $A_0,\ldots,F_0$ depend only on a and ω. Vice versa, expressing a and ω via $A_0,\ldots,F_0$ and substituting them into (10) we get equations of the type:

$$f_3^0 = \Psi_3^0(f_1^0, A_0,\ldots,F_0),\; f_4^0 = \Psi_4^0(f_1^0, A_0,\ldots,F_0). \qquad (11)$$

It is clear that great majority of interactions of the form

$$f_3 = \Psi_3(f_1, A,\ldots,F),\; f_4 = \Psi_4(f_1, A,\ldots,F) \qquad (12)$$

with arbitrary f_1 (and various Ψ_3 and Ψ_4) admit the circular motion with arbitrary a and ω. Let us try to reduce this arbitrariness demanding that (6) in the nonrelativistic limit differ from the Newtonian gravitational equations in terms of order . Finally we fix the interaction (not uniquely actually):

$$f_1 = \gamma/B^3,\; f_4 = 0,$$

$$f_3 = -f_1 \frac{A}{c} \frac{\sqrt{2x} + y\sqrt{s}}{y\sqrt{x} + \sqrt{2s}} \frac{1}{\sqrt{xs}} , \quad x = \frac{\varepsilon}{c^2} , \quad y = \frac{D}{\sqrt{\varepsilon F}} ,$$
$$s = x - y^2 + 2 . \tag{13}$$

In this case the parameters a and ω of the circular motion are related by the equation

$$2 - \frac{\omega^2 a^3}{\gamma (1-\beta^2)^{5/2}} \left[\beta \sin \varphi_0 + \sqrt{2(1+\cos\varphi_0)} \right]^3 +$$
$$+ \sqrt{2}\, \beta \sin \varphi_0 = 0 , \tag{14}$$

whose nonrelativistic limit is the third Kepler law $\gamma / 4\omega^2 = a^3$.

One can find particular explicit straight-line motion solutions of the main Eqs.(6) if one makes the following choice of the functions:

$$f_1 = \gamma_1 / A, \ f_3 = f_4 = 0, \ g_1 = \gamma_2 / \tilde{A}, \ g_3 = g_4 = 0, \ \tilde{A} = \frac{\tilde{\varepsilon} - \tilde{\vec{\varepsilon}}\, \vec{v}_2 / c}{\sqrt{1 - \beta_2^2}} .$$

Then (6) take the form

$$w_{1x} = \gamma_1 \frac{x_1 - \tilde{x}_2}{|x_1 - \tilde{x}_2|} (1-\beta_1^2)^{3/2} , \quad w_{2x} = \gamma_2 \frac{x_2 - \tilde{x}_1}{|x_2 - \tilde{x}_1|} (1-\beta_2^2)^{3/2} . \tag{15}$$

The quantities with tilde are taken at the times $t - \varepsilon/c$ or $t - \tilde{\varepsilon}/c$. The nonrelativistic form of these equations was studied thoroughly by P. Appel [8]. Since the relativistic case does not differ essentially from the nonrelativistic one (the particle's speed cannot exceed that of light in the former case), we consider it very briefly. For $x_1 > \tilde{x}_2$ the solution of (15) looks like

$$x_1(t) = \frac{c^2}{\gamma_1}\left[1+\left(\frac{\gamma_1}{c}t + \frac{\beta_{10}}{\sqrt{1-\beta_{10}^2}}\right)^2\right]^{1/2} - \frac{c^2}{\gamma_1}\frac{1}{\sqrt{1-\beta_{10}^2}} + x_1^0,$$

$$\beta_1(t) = \left(\frac{\gamma_1}{c}t + \frac{\beta_{10}}{\sqrt{1-\beta_{10}^2}}\right)\left[1+\left(\frac{\gamma_1}{c}t + \frac{\beta_{10}}{\sqrt{1-\beta_{10}^2}}\right)^2\right]^{-1/2},$$

$$x_2(t) = -\frac{c^2}{\gamma_2}(1+N^2)^{1/2} - \frac{c^2}{\gamma_2}(1-\beta_{20}^2)^{-1/2} + x_2^0,$$

$$\beta_2(t) = N(1+N^2)^{-1/2}, \quad N = -\frac{\gamma_2}{c}t + \beta_{20}(1-\beta_{20}^2)^{-1/2}.$$

Here $x_1^0, x_2^0, v_1^0, v_2^0, x_1, x_2, v_1, v_2$ are positions and velocities of particles at the initial (t) and given moment of time, $\beta_i = v^i/c$, $\beta_{i0} = v_i^0/c$. For $x_2 > \tilde{x}_1$ similar equations hold. For simplicity, let the particle's masses be equal

$(\gamma_1 = \gamma_2 = \gamma)$.

First consider the repulsion case $(\gamma < 0)$. Let at the initial moment the particles be at rest and situated symmetrically with respect to the origin $(x_1^0 = -x_2^0 = x_0 > 0)$. Then the solution for $t > 0$ is

$$x_1(t) = -x_2(t) = x_0 + \frac{c^2}{\gamma}\sqrt{1+\gamma^2 t^2 c^{-2}} - c^2/\gamma,$$

i.e. particles fly away with increasing speed tending to c as $t \to \infty$. Now let the initial particle velocities be of the same magnitude and directed towards the origin $(v_2^0 = -v_1^0 = v_0 > 0)$. Then, for $t > 0$ speed of the particles decrease. If the initial speed is not very large $v_0 < v_0^c = \dfrac{\sqrt{\gamma x_0 + (2+\gamma x_0)/c^2}}{1+\gamma x_0/c^2}$, then the particles approach each other, stop at the moment $t = \frac{v_0}{\gamma}\frac{1}{\sqrt{1-\beta_0^2}}$ (at the distance $\pm\left[x_0 - \frac{c^2}{\gamma}(1-\beta_0^2)^{-1/2} + \frac{c^2}{\gamma}\right]$ from the origin) and then fly away as in the previous case. If $v_0 > v_0^c$, the particles meet at the origin, pass through each other and then fly away as in the first case.

The situation looks more interesting if $\gamma > 0$ (attraction). Let particles initially be at rest $(x_1^0 = -x_2^0 = x_0 > 0)$. Later particles move towards each other and meet at the origin at time $t_0 = \left[\frac{2x_0}{|\gamma|}\left(1 + \frac{x_0|\gamma|}{2c^2}\right)\right]^{1/2}$. After that $x_1(t) - x_2(t) < 0$, i.e. particle order is reversed. The particles decelerate and stop at moment $2t_0$ (at distance $\mp x_0$ from the origin). The process is repeated. In short, one has a periodic motion with period $4t_0$ and amplitude x_0.

We are very thankful to Prof. N.A. Chernikov for useful discussions.

References

1. Poincare, H. (1906). Rend. Circ. Mat. Palermo, 21, 129; (1956). Oeuvres, Gauthier-Villars, Paris, IX, p.494-550.
2. Birkhoff, G.D. (1943). Proc. Nat. Acad. Sci. USA, 29, 231. Barajas, A. (1944). Proc. Nat. Acad. Sci. USA, 30, 54. Barajas, A. et al. (1944). Phys. Rev., 66, 138.
3. Afanasiev, G.N., Asanov, R.A. (1981). Ann. Phys. (Leipzig), 38, 169.
4. Dettman, J.W., Schild, A. (1954). Phys. Rev., 95, 1057. Schild, A. (1963). Phys. Rev., 131, 2763. Anderson,C.M., Bayer, H.C. (1970). Ann. Phys. (N.Y.), 60, 67. Anderson, J.L. (1967). Principles of Relativity Physics, AP. N.Y.
5. Synge, J.L. (1972). In: Magic without Magic, Freeman, S.F., p.117-133.
6. Shavokhina, N.S. (1982). Dokl. Akad. Nauk USSR, 265, 852.(Russian).
7. Asanov, R.A., Afanasiev, G.N. (1985). JINR Prepr. P2-85-357. (Russian).
8. Appel, P. (1953). Traite de mecanique rationelle, 1 (ch.XI, ex.18 at ch.XIII) Gauthier-Villars, Paris.

TOROID MOMENTS AS GENERAL PHYSICAL PARAMETERS

V.M.Dubovik

Joint Institute for Nuclear Research, Dubna

Searches for effects caused by the toroid moment has become more popular. Their simple definitions and some applications are given here.

The investigations we made in the early seventies of the multipole expansion structure of current in the framework of the Maxwell-Lorentz electrodynamics allow us to reveal the third moment family in addition to the charge and magnetic ones: we called it toroid moments T_{lm} [1,2]. That provides completeness of the multipole expansion and the geometric representation of arbitrary electromagnetic sources (moving charges as well as macroscopic currents). An ideal model of the first generic term of the new family is a conventional "point-like" toroidal coil (with an even number of windings). Such a system is the one that cannot interact with static uniform fields and teacts to (local) displacement or conductivity currents [1,2,3].

We have studied in detail properties of T_{lm} and their classical and quantum manifestation. An exact solution was given to the problem of multipole emission in vacuum (in a general form [1,2] and for particular cases [4]). There it is shown that reconstructing multipole structure of a source is possible only when the angular analysis is complemented with spectral analysis of emission multipoles [1].

<u>In atomic physics</u> it is pointed out that in unhomogeneous and/or nonstationary fields Stark and Zeeman's effects are accompanied by the toroid effects with level-splitting

energy $\Delta W \sim \vec{T}(\text{rot}\vec{H} + \dot{\vec{D}})$ [1]. It is also shown that precisely T_l are responsible for the violation of the Siegert theorem and give corrections to sum rules of the Thomas-Reihe-Kuhn type [5].

In nuclear physics they demonstrate that contributions to the cross-section of electron scattering on nuclei ^{9}Be can mask form-factors M_2 [6]. As an example all the form-factors have been calculated and a complete experiment has been planned for analysing the transition $\frac{3}{2}^- \to \frac{1}{2}^+$ in the nucleus ^{9}Be [6]. In [7] it is indicated that electric spin flip E1-transitions result from the spin part of T_1, and there is calculated the so-called hindered transitions occurring due to that part in heavy nuclei (^{175}Lu, ^{181}Ta). Criteria for observations of such transitions are also given.

For processes of internal conversion on nuclei it has recently [8] been established that the so-called parameters of penetration of electronic current into a nucleus under study measured independently, should in total reduce to a single physically reasonable quantity, the transitional toroid dipole T_1 (or moment T_l with $l>1$)[9]. The values of T_1 were extracted for 15 nuclei [9] from experiments. One should also note some interesting possibilities of studying giant resonances calculating which by the collective-coordinate method in the model of nuclear hydrodynamics is performed in [10].

For elementary particles we have derived formulas for the toroid operator of E1-transition of a composite system [11], and as an example, estimates of T_1 contributions have been carried out in quasinuclear [11] and quarkonian systems [4].

For subparticles, leptons and quarks the toroid dipole (whose spin part is sometimes called anapole [3]) has became not more exotic than, say, the electron anomalous magnetic moment $\alpha/2\pi$. Indeed, in electroweak models for the value of α responsible is the first correction to virtual production of W or Z-boson in the $e^+e^-\gamma$ -vertex that represent, as we know, physical realities. First calculations of T_1 were made in [12] and in a diploma work of my student in 1975. Toroid moments of leptons and quarks were also estimated on the basis of polarization experiments on nucleons and nuclei [13]. In papers [14,15] a remarkable fact is indicated that a Majorana massive neutrino possesses only two internal characteristics, spin and anapole (actually, toroidal dipole, since in the context of the problem studied in [15] anapole is not a physical quantity, see [1]). Recently, proceeding directly from the CPT selection rules [2], it has been proved that electromagnetic properties of any majoron ($s \geqslant 1/2$) can be described by toroid moments only [16].

In recent years the existence of toroid current configurations has led to interesting results in the theory of condensed matter. Great toroid effects were discovered theoretically [17] (and possibly, verified experimentally on CuCl [18] and on CdS [19]). Apparently, a new collective phenomenon. the superdiamagnetism, is discovered. It is determined by a cooperation of electron-hole pairs in the exciton liquid (by analogy with the Cuper parameters of superconductivity). The experimentally observed magnetic susceptibility is close to the ideal one (10^{-2}) even at a liquid nitrogen temperature, which is explained by a spontaneous production and aligning of toroid moments at phase transitions in the substances we have said above. We assume the discovered phenomenon will be of wide practical use.

A fundamental application of the toroid dipole as a structure unit would be a consistent construction of current electrostatics [20] and more complicated quantum gauge theories. An important problem is derivation of soliton-like solutions of nonlinear equations having toroid (more precisely poloidal) structure. In particular, toroid structures can appear in the fragmentation of magnetic fields in superconductors of second kind. Solutions of the Landau-Ginzburg equation having axial structure in the case of such fragmentation are known for a long time, and the corresponding objects are called Abrikosov vortices. In principle, magnetic field lines may not escape through the walls of a sample into vacuum, but they get closed inside it. Then, electromagnetic macrostructure of such a vortex may exactly represent a model of a toroid dipole: a closed (toroid) bunch of magnetic field lines contracted by (poloidal) superconductivity currents circulating around it. Derivation of solutions with a similar symmetry would be of interest not only for superconductivity theory but also for modern gauge models dessribing the structures and interactions of fundamental particles and bearing certain similarity with the superconductivity theory.

In gauge models, SO(10) and others, where a nontrivial structure of vacuum arises owing to the introducing the sector of Higgs particles, solutions analogous to Abrikosov vartices have recently [21] been constructed. A closed vortex obtained in [21] possesses remarkable properties, in particular, it may provide an intensive nucleonsynthesis when intersecting with galactic magnetic fields.

2. Let us give here the most simple derivation of the toroid moment definitions and relations between various definition forms.

A formal reason of the toroid moment availability in the multipole expansion of the cureent density $\vec{j}$ $(\vec{r},t)$ is that its transverse part (div $\vec{j}_\perp = 0$) may be expressed through the two scalar functions [2]

$$\vec{j}_\perp(r) \equiv \vec{L}\Psi + \vec{K}\chi, \quad \vec{L} \equiv \vec{r}\times\vec{\nabla}, \quad \vec{K} \equiv \vec{\nabla}\times\vec{L}. \qquad (1)$$

Here $\Psi(r)$ and $\chi(r)$ are pseudoscalar and scalar functions, respectively, representing the two-component vector via two orthogonal manifolds, i.e.

$$\int_V \vec{L}\Psi \cdot \vec{K}\chi\, d^3r = \int_S (\Psi(\vec{r})\vec{r}) \times (\vec{K}\chi)\hat{r}\, d\vec{\sigma} = 0 \qquad (1a)$$

The representation of a transverse field (flow) in the form (1a) was possibly used by F.Neumann and by Debye for description of transverse fields. The operators $\vec{L}$ and $\vec{K}$ satisfy the relations

$$\vec{\nabla}\cdot\vec{L} = 0, \quad \vec{r}\cdot\vec{L} = 0, \quad \vec{\nabla}\cdot\vec{K} = 0, \quad \vec{r}\cdot\vec{K} = -\vec{L}\cdot\vec{L}, \quad \vec{L}\cdot\vec{K} = 0. \qquad (2)$$

Therefore, Ψ and χ may be extracted from (1) by the following projection operations

$$\vec{L}\cdot\vec{j}_\perp = L^2\Psi, \quad \vec{r}\cdot\vec{j}_\perp = -L^2\chi. \qquad (3)$$

Consequently, as in the usual multipole expansion Ψ and χ may expand in complete set of nonsingular solutions of the scalar Helmholtz equation

$$\vec{L}\cdot\vec{j}_\perp = L^2 \sum_{l,m} \int k^2 dk \alpha_l M_{lm}(k^2) f_l(kr) Y_{lm}(\hat{r}) = -\sum_{lmk} \alpha_l M_{lm} f_l\, l(l+1) Y_{lm}, \qquad (4)$$

$$\vec{r}\cdot\vec{j}_\perp = -L^2 \sum_{l,m} \int k^2 dk (-\alpha_l) T_{lm}(k^2) f_l(kr) Y_{lm}(\hat{\vec{r}}) = \sum_{lmk} (-\alpha_l) T_{lm} f_l\, l(l+1) Y_{lm}. \qquad (5)$$

Therefore, magnetic M_{lm} and toroid T_{lm} multipole moments in the useful normalization are defined by:

$$M_{lm} = \frac{1}{l+1} \sqrt{\frac{4\pi}{2l+1}} \int \vec{L} \cdot \vec{j}_\perp \, r^l Y^*_{lm} d^3r, \tag{6}$$

$$T_{lm} = \frac{1}{l+1} \sqrt{\frac{4\pi}{2l+1}} \int \vec{r} \cdot \vec{j}_\perp \, r^l Y^*_{lm} d^3r. \tag{7}$$

The set of values (6) and (7) is not the full parametrization $\vec{j}_\perp$ because the spherical radial harmonic systems are not closed in R_3. The full parametrization includes for each moment of the range l the whole series (in general, an infinite one) of parameters of the same angular part but of another radial dependence ($r^l \to r^{l+2n}$, n = = 0,1,...), the so-called mean 2n-power radii.

In Cartesian coordinates T_{1m}, T_{2m} and T_{3m} are expressed in the form

$$\vec{T} = \frac{1}{2}\int \vec{r}(\vec{r}\vec{j})d^3r, \quad T_{ik} = \frac{1}{2}\int (r_i r_k - \frac{r^2}{2}\delta_{ik})(\vec{r}\vec{j}_\perp)d^3r,$$

$$T_{ikl} = \frac{1}{6}\int \left[r_i r_k r_l - \frac{1}{5}r^2(r_i\delta_{kl} + r_l\delta_{ik} + r_k\delta_{il})\right](\vec{r}\vec{j}_\perp)d^3r. \tag{8}$$

For the not transverse current, i.e. if div $\vec{j} = -\dot{\rho}$, formulas more general than (6) and (7) have been found expanding in the vector harmonics composing a closed set in R_3 [1,2]. Then, toroid moment of rank l is

$$T_{lm} = \frac{i}{2(l+1)(2l+3)} \sqrt{\frac{4}{2l+1}} \int \vec{j} \cdot \mathrm{rot}\left[r^{l+2}\vec{L}Y^*_{lm}\right], \tag{9}$$

For l = 1 it reduces to:

$$\vec{T} = \frac{1}{10} \int \left[\vec{r}(\vec{r}\vec{j}) - 2r^2\vec{j}\right]d^3r. \tag{10}$$

It may be shown that if a current configuration possesses only toroid dipole and all lowest moments $\dot{\vec{d}}$, $\dot{Q}_{ik}$, μ_i vanish, (10) is translation-invariant.

The condition of current transversality in terms of moments of their radii in dimension 3 is:

$$\dot{\vec{r}}_d^{\,2} = \int [2\vec{r}(\vec{j}\vec{r}) + r^2\vec{j}]\, d^3r = 0. \qquad (11)$$

Using the relation $2\int \vec{r}(\vec{r}\vec{j})d^3r = -\int r^2\vec{j}_\perp d^3r$, we get one more definition of $\vec{T}$ (see, e.g.[17] for transverse current)

$$\vec{T} = -\frac{1}{4}\int r^2\vec{j}_\perp d^3r. \qquad (12)$$

It is clear that toroid dipole can be called an r.m.s. radius of the transverse current, and what is a geometrical image of this dipole. In the simplest case of axial symmetry of $\vec{j}(\vec{r})$ (12) is a characteristic of a mean double cylindric layer. When a system has no moment $\dot{\vec{d}}$ ($\dot{\vec{d}} = \int \vec{j} d^3r = 0$) and current lines are closed ($\mathrm{div}\,\vec{j}_\perp = 0$), this layer in a limited volume transforms into a torus with poloidal currents in it. These geometric properties of model $\vec{T}$ were predicted by the type of symmetry of K in (1). Note that unlike T_{1m}, higher toroid moments T_{lm} are neither mean 2n-power radii (the series of the latter compose, by definition, a toroid dipole formfactor), nor radii of longitudinal charge moments (see[2], p.510 and [22]).

3. A toroid dipole may only interact with external currents including the displacement current, that runs through the point of its location [1]:

$$W = \vec{T}(\vec{r} = 0)[\dot{\vec{D}}(0) + \vec{I}^{ext}(0)] \,. \qquad (13)$$

Note that current on torus can be treated only as equivalent of the magnetization $\vec{M}$ inside the torus. Then, the interior magnetic field may be considered an internal field of a closed circular chain of magnetic dipoles while the dipole of the system $\vec{T}$ is $\frac{1}{2}\int[\vec{r}\times\vec{M}]d^3r$. The interaction energy of this toroid dipole with current (external magnetic field) equals

$$W = \vec{T}^{(M)} \cdot \vec{I}^{ext} = \vec{T}(\mathrm{rot}\vec{B} - \dot{\vec{D}}). \qquad (14)$$

In the system of crossed fields toroid dipole may appear as the polarizability and may nonlinearly interact with the fields

$$W = \vec{T}(\vec{E} \times \vec{B}). \qquad (15)$$

In electrodynamics, on the basis of the Larmor-Pistolkors duality principle describing symmetry of the electrodynamics equations with respect to the electric and magnetic characteristics, one may transform (13) via the change $\vec{D} \to -\vec{B}$ ($e \to -g$, where g is magnetic charge, a pseudoscalar) and introduce toroid dipole that now is a pseudovector [23]. It is interesting that despite the absence of magnetic charges effects dual to (13) are imitated by a toroid axial moment, e.g. of a closed chain of electric dipoles $\vec{G}^{(d)} = \frac{1}{2}\int[\vec{r} \times \vec{d}]d^3z$ that interacts with fields as follows

$$W = -\vec{G}\dot{\vec{B}} = \vec{G}' \,\mathrm{rot}\vec{E}. \qquad (16)$$

When matter is anisotropic the nonlinear responses with the interaction energy are also possible:

$$W = \vec{G}\,(\vec{D} \times \vec{E}) \quad \text{or} \quad W = \vec{G}'\,(\vec{M} \times \vec{H}).$$

Note that $\vec{G}$ realises the last of possible symmetries of electromagnetic dipole moments in the three-dimensional world [23]:

	$\vec{d}$	$\vec{\mu}$	$\vec{T}$	$\vec{G}$
P ($\vec{r} \to -\vec{r}$)	-	+	-	+
T ($t \to -t$)	+	-	-	+

Dipoles $\vec{T}$ and $\vec{G}$ (as well as moments of higher dimensions) being introduced into the theory of condensed matter as order parameters allow a uniform classification of some known effects and prediction of new phase transitions (see [23] and references therein).

References

1. Dubovik,V.M., Cheshkov,A.A. (1974). Sov.J.Part. Nucl. 5, 318-341.
2. Dubovik,V.M., Tosunyan,L.A. (1983). Sov.J.Part. Nucl. 14, 504-519.
3. Zeldovich,YA.B. (1958). Sov. Phys. JETP 6, 1184-1189.
4. Dubovik,V.M., Cheshkov,A.A. (1970). Preprints R2-5823, R2-5284, JINR, Dubna (Russian).
5. Dubovik,V.M. (1975). Preprint E2-9262, Dubna.
6. Bluvstein,R.E., Dubovik,V.M., Cheshkov,A.A. (1978). Sov.J.Nucl. Phys. 15, 59.
7. Dubovik,V.M., Eramzhyan,R.A., Tosunyan,L.A. (1976). Preprint E4-9979, JINR, Dubna.
8. Band,I.M., Listengarten,M.A., Feresin,A.P. (1976). Anomalies in Gamma-Ray Conversion Coefficients, Nauka, M.-L. (Russian).
9. Listengarten,M.A., Feresin,A.P., Grigor'ev,V.N. (1981). Abstracts of Papers at the 31st Symposium on Nucl. Spectroscopy and Nucl.Structure, Izv.Akad.Nauk SSSR, ser.phys. 45, 2038.
10. Semenko,S.F. (1981), Sov.J.Nucl.Phys. 34, 356.
11. Dubovik,V.M., Tosunyan,L.A. (1981). JETP Lett. 31,65; Dubovik,V.M., Tosunyan,L.A., Zenkin,S.V. (1980). Preprint E2-80-671, JINR, Dubna.
12. Audin,Z.Z., Baran,S.A., Barut,A.O. (1973). Nucl.Phys., 1355, 101; (1973). B62, 333.
13. Rekalo,M.P. (1978). Sov.J.Nucl.Phys. 28, 852.
14. Kobzarev,I.Yu., Okun',L.B. (1972). In: Problems of Theor. Phys. (memorial volume to Igor' E.Tamm) Nauka, Moscow, 219 (Russian).
15. Kayser,B. (1982). Phys.Rev. D26, 1662; (1984). D30, 1023,Nieves, J. ibid. (1982). D26, 3152.
16. Radescu,E. (1985). Phys.Rev. D32, 1266.
17. Ginzburg,V.L. et al. (1984). Sol.State Comm. 50, 334.
18. Brand,N.B. et al. (1978). JETP Lett. 27, 37 (Russian).
19. Cote et al. (1983). Phys.Rev. B28, 5041.

20. Miller,M.A. (1984). Usp.Fis.Nauk (Russian Phys.Surveys) 142, 142 (Russian).

21. Witten,E. (1984). Preprint Princeton Univ.

22. Khriplovich,I.B. (1981). Parity Nonconservation in Atomic Phenomena, Nauka, Moscow. Appendix.

23. Dubovik,V.M., Tosunyan,L.A., Tugushev,V.V. (1985). Preprint E17-85-623, JINR, Dubna.

SET THEORETICAL STRUCTURE OF SINERGETICS

U.H. Kopvillem

Pacific Oceanology Institute, Radio Str. 7,
Vladivostok 690032

Sinergetics is a field of science with not quite definite open set of problems. It seeks for set-theoretical structures with interesting (for theory or practice) concrete realizations. It is kin, I believe, to Goedel's incompleteness theorem meaning that it assures the existence of true statements (in the usual ambiguous language) impossible to formulate in a formal language. Search for anologies is the source of results in sinergetics. In [2] this was applied to predict new phenomena in coherent and correlated processes. The obtained algorithms turned out to be fit also in biology and informatics [3,4].

Note that all 7 Thom's catastrophes stem from generalised Bloch-type equations for simple spin-systems. It is remarkable. It shows that the actual nonformal reasoning in physics is applicable to solving nonphysical problems such as those of biology [5-7].

Give some examples how to extrapolate notions of quantum systems to nonphysical systems. In quantum mechanics we know two different states of a system of N identical particles: a Bloch state and a Dicke state. In the first case the particles are independent, coherently oscillate and radiate with an intensity proportional to N^2. In the Dicke case the particles are firmly bound into one macroscopical body. In this case the system oscillations are macroscopic movements in real space. The Dicke radiation is not generally coherent. A bloch state can transform into a Dicke state and vice versa, a state of the system may be a superpositional Bloch-Dicke state. A living subject is something like one macromolecule and so is apparently in a complicated Dicke state. A subradiant Dicke state can explain the comportment of a sperm not involved in a metabolism or in a state of deep lethargy. A system of computers may also be in a Bloch state. In this case the number of operations in unit time must be proportional to N^2, where N is the number of identical computers in the system. This may be true in the following two situations. 1) The computers can operate only in pairs. The number of pairs $M = \frac{1}{2}N(N-1)$. For $N \gg 1$ we have $M \sim N^2$. 2) The computers work by means of spin echo dynamic memory. The speed of the operation is proportional to the phase relaxation time T_2. If T_2 is the superradiation time then $T_2^{-1} \propto N^2$. In a Dicke-like state the system of computers operates at a speed proportional to $N^{\pm}(L)^2$, the + sign for the operation in clustering by $(L)^2$ computers and if the clusters by L^2 subcompute (perturb one another) the number of clusters R_F containing F particles is $R_F=(1/F) \prod_{\varkappa=0}^{F-1} (N-\varkappa)$. It seems that the case 1) is also a Dicke-like state and only case 2) is a real Bloch state.

Let us discuss properties of operators of observable quantities Q in the Heisenberg representation of non-perturbed energy [2]. Let at time t=0 the density matrix S_0 has only diagonal elements. At $t>0$ S_0 does not change, while $Q(t)=L^{-1}(t)QL(t)$. If $\langle a|Q(t)|a\rangle=0$ we can measure some physical quantity at time t. If $\langle a|Q|b\rangle \neq 0$ then it is possible in principle to get oscillating signals at time t. This is really so if L(t) transforms $\langle a|Q|b\rangle \neq 0$ into diagonal $\langle a|Q(t)|a\rangle \neq 0$. We see that the condition $\langle a|Q|b\rangle \neq 0$ means that the system possesses nonequilibrium radiating and coherent states. The superpositional principle allows us to perform transformation $\langle a|Q|b\rangle \rightarrow \langle c|Q(t)|c\rangle$. The transformation from energy states (Fock states) to coherent superpositional states (Glauber states for oscillators and Bloch states for spins) is a transformation from an equilibrium state to a macroscopic state of rotation. The Dicke states are also equilibrium states. They can radiate only in the framework of quantum mechanics and only spontaneously and uncoherently. A superpositional state of two Dicke states is a coherent state and ratiates classically. We propose to generate such states in non-physical systems by analogy. The algorithm "superposition of two equilibrium states gives a nonequilibrium state works very well in psychology: a superposition of rage and fear gives a non-equilibrium nonpredictable behaviour. We may call a nonequilibrium or superpositional state also an active state. It is extremely interesting that the nature itself spontaneously generates active states (Bloch states). Such a process is called a boson avalance [6]. We end our discussion with some remarks about avalanches.

All the general features of a boson avalanche are present for a system of N two-level particles [8]. The dynamics of the evolution of the intensity of radiation I(t) is described by the equation:

$$\ddot{I} + \delta \dot{I} + \alpha e^{I} = \Omega, \quad \alpha = 8N, \quad \delta = \tau / T_1$$
$$\Omega = -[2N(1 - \tau/T_2) + 2\tau^2/T_1 T_2] \qquad (1)$$

where δ , α and τ are numerical scaling dimensionless parameters: τ the spontaneous relaxation time of an isolated radiator j (= 1,2,...), T_1 the energy relaxation time, δ the friction coefficient, αe^{I} the exponential restoring force and Ω a constant force. For $T_1 = \infty$ and the initial value $I(0) = 2(N - \tau/T_2)$ the solution of (1) is:

$$I(t) = (\pi\tau/2)(1/T_2 - N/\tau)^2 \operatorname{sech}^2[(1/T_2 - N/\tau)(t - t_d)] \qquad (2)$$

$$t_\alpha = (1/\tau_R - 1/T_2)^{-1} \ln[((T_2 - \tau_R)/T_2)\sqrt{N}]^2, \quad \tau_R = \tau/N$$

Here τ_R is the superrelaxation time, t_α the delay time and æ the energy of the radiated quantum. Very important is the condition of avalanche launching: at $t=0$ the system must possess a macroscopical superpositional angle $\theta_0 = 2/\sqrt{N-2}$, very small for $N \sim 10^{23}$. It is a universal measure of global unstability. Nevertheless the physical origin of θ_0 is not known in general. The equation (1) is Lienard-type [9]: $\ddot{x} + f(x)\dot{x} + g(x) = 0$.

Variating control parameters of (1) we can get as a solution equilibrium states, periodic motion, multiply periodic motion and turbulence (chaos). In an extended system delay times of propagation and launching cause nonsimultaneous generation of avalanches in different parts of the system. The interplay of these fields causes chaos.

Boson avalanches are frequently encountered in nature. Every signal may considered as a sum of Fourier components or sum of solitons. A sum of solitons is a sum of avalanche signals because an avalance generates a soliton. Avalanches are classical processes and may be used in various non-physical situations to get insight into avalanche-like occurrences. An avalanche is possible to generate in nonlinear systems where the behaviour drastically changes with the increase of excitation. From (2) we know that it is tremendously important to estimate the quantities θ_0, N and t_α, even approximately. Equidistant unbounded energy spectra are not suited to generate avalanches. Again it is troublesome to search for the analogy that permits to use the information on avalanche dynamics. As we know, algebras generate spectra so we see the connection of avalanches with the structure of algebras. Again sinergetics is structure-dependent. Processes with dynamic memory like echo are also useful.

Their structure is described by nonsolvable Lie algebras [10,11]. It is possible to get an idea how to restore the memory of a living subject by analysing the ways to restore a physical echo with the same structure.

The development of sinergetics will help us to perform new ingeneous powerful computers of intelligent informators. To every question the informator answers finding the sinergetical structure of the question, performing a quick physical process with the same structure and translating into the real language.

References

1. Godel, K. (1931). The undecidability theorem. Monatsh. Math. Phys., 38, 173.
2. Kopvillem, U.H. (1962). Moment excitation of a system of weakly interacting particles. ZhETPh, 42, 5, 1333-1343. (Russian).
3. Kopvillem, U.H. (1964). Doctor's dissertation. Kazan Univ., Kazan. (Russian); Asadullin, Ya.Ya., Kopvillem, U.H. (1967). Solid State Phys. 9, 2737-2739.(Russian).
4. Kopvillem, U.H. (1981). In: Dynamic processes in ocean and atmosphere. FESC Acad. Sci., Vladivostok, p.97-113 (Russian).
5. Kopvillem, U.H. (1971). In: Paramagnetic resonants. Nauka, Moscow, p.218-223; VINITI-Dep.1093-69.(Russian).
6. Nagibarov, V.R., Kopvillem, U.H. (1968). ZhETPh, 54,1, 312-317. (Russian).
7. Kopvillem, U.H., (1984). Zero point oscillations and biorythms. 8th Int. Congress of biophysics, Bristol, U.K. Abstracts, 152.
8. Kopvillem, U.H., Samodurov, A.A., Chudnovsky, V.M. (1983). In: Avalanche-like energy processes in ocean. All-Union Conf. on ocean's energetics. FESC Acad. Sci., Vladivostok. (Russian).
9. Kitano, M. (1984). Spin-related optical bistability and tristability. Kyoto University, Uji, Kyoto 611.
10. Kopvillem, U.H., Prants, S.V. (1985). Polarizational echo. Nauka, Moscow, (Russian).
11. Kopvillem, U.H., Saburova, R.V. (1982). Paraelectric resonance, Nauka, Moscow (Russian).

CLOSED ANALYTIC EXPRESSIONS FOR SOME USEFUL SUMS AND INTEGRALS INVOLVING LEGENDRE FUNCTIONS

G.N. Afanasiev
Joint Institute for Nuclear Research, Dubna

1. The [1] we have obtained in three different ways the components of the vector magnetic potential (VMP) for the toroidal solenoid and their scattering amplitude. As they satisfy the same equations $(\Delta \vec{A} = -\frac{4\pi}{c}\vec{j}, \operatorname{div} \vec{A} = 0)$ with the same boundary values, they should coincide (see e.g. [2]). By comparing these equations one can derive simple closed analytic expressions for some integrals and sums involving Legendre functions, not included in mathematical textbooks, treatises or original publications [3-11]. We study the convergence of the mentioned series in the most suspicious cases. New expressions found for the sums and integrals may turn useful for computing magnetic fields with configurations close to thc toroidal ones.

2. The VMP of the toroidal solenoid $(\rho - d)^2 + z^2 = R^2$ may be viewed as a superposition of VMP's for separate coils. At the point (ρ, z, φ) we get [1]:

$$A_z(\rho, z) = \int_0^{2\pi} (d - \rho\cos\varphi) \cdot F(\rho, z, \varphi) \cdot d\varphi,$$
$$A_\rho(\rho, z) = z \int_0^{2\pi} \cos\varphi \cdot F(\rho, z, \varphi)\, d\varphi \tag{2.1}$$

(due to the axial symmetry A_z and A_ρ do not depend on φ and $A_\varphi = 0$ $\quad j = const$ set

$$F(\rho, z, \varphi) = \frac{\sqrt{R}\,j}{2\pi} \frac{1}{[(\rho\cos\varphi - d)^2 + z^2]^{3/4}} \cdot Q_{\frac{1}{2}}\left\{\frac{\rho^2 + z^2 + d^2 + R^2 - 2d\rho\cos\varphi}{2R[(\rho\cos\varphi - d)^2 + z^2]^{1/2}}\right\}$$

A_z and A_ρ may be found from the Poisson equation $\Delta\vec{A} = -\frac{4\pi}{c}\vec{j}$ (in which variables are separated in the toroidal coordinates). We get [1]:

$$A_\rho = \frac{2\sqrt{2}g}{\pi}\sqrt{ch\mu - \cos\vartheta}\sum_{n=1}^{\infty} R_n^1(\mu)\cdot \sin n\vartheta,\quad A_z = \frac{2\sqrt{2}g}{\pi}\cdot\sqrt{ch\mu-\cos\vartheta}\sum_{n\geq 0} R_n^0(\mu)\cos n\vartheta \tag{2.2}$$

where

$$R_n^1 = -Q_{n-\frac{1}{2}}(ch\mu_0)\left[P_{n-\frac{1}{2}}(ch\mu_0) - P_{n-\frac{3}{2}}(ch\mu_0)\right]\cdot Q^1_{n-\frac{1}{2}}$$

$$R_n^0 = A_n(\mu_0)\cdot P_{n-\frac{1}{2}}(ch\mu_0)\cdot Q_{n-\frac{1}{2}} \tag{2.3}$$

inside of the solenoid $(\mu > \mu_0)$ and

$$R_n^1 = -Q_{n-\frac{1}{2}}(ch\mu_0)\left[Q_{n+\frac{1}{2}}(ch\mu_0) - Q_{n-\frac{3}{2}}(ch\mu_0)\right]\cdot P^1_{n-1/2},$$

$$R_n^0 = A_n(\mu_0)\cdot Q_{n-\frac{1}{2}}(ch\mu_0)\cdot P_{n-1/2} \tag{2.4}$$

outside of it $(\mu < \mu_0)$. Here $ch\mu_0 = \frac{d}{R}$, $A_n(\mu_0) = \frac{C_n}{1+\delta_{n,0}}$, $C_n = (n+\frac{1}{2})\cdot Q_{n+\frac{1}{2}}(ch\mu_0) - (n-\frac{1}{2})Q_{n-\frac{3}{2}}(ch\mu_0)$.
Hereafter the argument of the Legendre function is μ_0 or μ unless otherwise stated. (2.1) and (2.2) satisfy the same equations and boundary values (e.g. it is easy to check that $div\,\vec{A} = 0$, while $H_\varphi = (rot\,\vec{A})_\varphi = \frac{g}{\rho}$ inside of the solenoid (1) and zero outside it). However, a direct comparison of (2.1) and (2.2) is inconvenient due to their complexity. So, consider particular cases. Set $\rho = 0$ in (2.1). Then one has on the z-axis:

$$A_z(\rho=0,z) = \frac{\sqrt{R}\,g d}{(d^2+z^2)^{3/4}}\,Q_{\frac{1}{2}}\left(\frac{d^2+z^2+R^2}{2R\sqrt{d^2+z^2}}\right). \tag{2.5}$$

Put $\mu = 0$ in (2.2) (this corresponds to the z axis). Looking at the behaviour of $P^m_\nu(ch\mu)$ as $\mu\to 0$ and comparing (2.5) and (2.2) we get:

$$\sqrt{1-\cos\vartheta}\cdot\sum_{n\geq 0} A_n Q_{n-\frac{1}{2}}(ch\mu_0)\cos n\vartheta = \frac{\pi}{2\sqrt{2}}\frac{ch\mu_0}{\left(1+2\frac{sh^2\mu_0}{1-\cos\vartheta}\right)^{3/4}}\,Q_{\frac{1}{2}}\left(\frac{1+\frac{sh^2\mu_0}{1-\cos\vartheta}}{\sqrt{1+\frac{2sh^2\mu_0}{1-\cos\vartheta}}}\right). \tag{2.6}$$

For the particular values of ϑ we derive from (2.6) closed expressions for infinite sums involving Legendre functions. Set $\vartheta = \pi$. Then

$$\sum_{n\geq 0} (-1)^n (2n+1) Q_{n-\frac{1}{2}} \cdot Q_{n+\frac{1}{2}} = \frac{\pi}{4} \frac{1}{\sqrt{ch\mu}} Q_{\frac{1}{2}}\left(\frac{1+ch^2\mu}{2ch\mu}\right). \tag{2.7}$$

In (2.7) we changed $\mu_0 \to \mu$. Both sides of (2.6) are equivalent to ϑ^3 as $\vartheta \to 0$. Equating the coefficients of ϑ^3 we get

$$\sum_{n\geq 0} (2n+1)^2 \cdot Q_{n-\frac{1}{2}} \cdot Q_{n+\frac{1}{2}} = \frac{\pi^2}{8} \frac{ch\mu}{sh^3\mu}. \tag{2.8}$$

Not so nice looks (2.6) for $\vartheta = \frac{\pi}{2}$:

$$Q_{-\frac{1}{2}} \cdot Q_{\frac{1}{2}} + \sum_{n=1}^{\infty} (-1)^n Q_{2n-\frac{1}{2}} \left[(4n+1) Q_{2n+\frac{1}{2}} - (4n-1) Q_{2n-\frac{3}{2}}\right] =$$

$$= \frac{\pi}{\sqrt{2}} \frac{ch\mu}{(ch2\mu)^{3/4}} \cdot Q_{\frac{1}{2}}\left(\frac{1+ch2\mu}{2\sqrt{ch2\mu}}\right). \tag{2.9}$$

The sum (2.8) seems most suspicious from the convergence point of view (due to the factor $(2n+1)^2$). Prove that (2.8) converges. For fixed μ and $\nu \gg 1$ one has [3] :

$$Q_\nu(ch\mu) \sim \sqrt{\frac{\pi}{2sh\mu}} \cdot \exp(-\mu\nu) \text{ and}$$

$$\left(n+\tfrac{1}{2}\right)^2 Q_{n-\frac{1}{2}} \cdot Q_{n+\frac{1}{2}} \sim \frac{\pi}{2sh\mu}\left[n \cdot \exp(-\mu n)\right]^2 =$$

$$= \frac{\pi}{2sh\mu} \exp\left[2(\ln n - \mu n)\right] \quad \text{as } n \to \infty. \tag{2.10}$$

It follows from (2.10) that the ratio of the two successive terms in (2.8) equals $\exp(-\mu)$ as $n \to \infty$, so (2.8) converges for any finite $\mu > 0$ as a geometric progression. From (2.6) we get the following closed expression for the integral

$$\int_0^{2\pi} \frac{d\vartheta \cdot \cos n\vartheta}{\sqrt{1-\cos\vartheta}} \frac{Q_{\frac{1}{2}}(x)}{\left(1+\frac{2\,sh^2\mu}{1-\cos\vartheta}\right)^{3/4}} = \frac{2\sqrt{2}}{ch\mu}\left[(n+\tfrac{1}{2})Q_{n+\frac{1}{2}}-(n-\tfrac{1}{2})\cdot Q_{n-\frac{3}{2}}\right]\cdot Q_{n-\frac{1}{2}}$$

$$x=\left(1+\frac{sh^2\mu}{1-\cos\vartheta}\right)\left(1+\frac{2\,sh^2\mu}{1-\cos\vartheta}\right)^{-1/2}. \quad (2.11)$$

Now consider the VMP component $\mathring{A}_\varsigma$. As $\varsigma\to 0$ it decreases as the first power of μ. Equating the coefficients of μ in (2.1) and (2.2) we get

$$\sum_{n=1}^{\infty}(4n^2-1)Q_{n-\frac{1}{2}}\left(Q_{n+\frac{1}{2}}-Q_{n-\frac{3}{2}}\right)\cdot\sin n\vartheta = \frac{\pi}{\sqrt{2}}\, ch\mu\cdot sh^2\mu\,\frac{\sin\vartheta}{(1-\cos\vartheta)^{5/2}}\cdot$$

$$\cdot\left(1+\frac{2\,sh^2\mu}{1-\cos\vartheta}\right)^{-7/4}\cdot\left[2\cdot Q^1_{\frac{1}{2}}(x)-3\cdot Q_{1/2}(x)\right].$$

(Here again we have made the change $\mu_0\to\mu$). As earlier, consider specializations of ϑ. As $\vartheta\to 0$ we get (2.8), nothing new. At $\vartheta=\pi$ we get

$$\sum_{n=1}^{\infty}(-1)^n n\cdot(n+1)\cdot(2n+1)\cdot Q_{n-\frac{1}{2}}\cdot Q_{n+\frac{1}{2}} = -\frac{\pi}{32}\,\frac{1}{(ch\mu)^{5/2}}\cdot$$

$$\cdot\left[3(1+ch^2\mu)\cdot Q_{\frac{1}{2}}(x_0)+2\,sh^2\mu\cdot Q^1_{\frac{1}{2}}(x_0)\right],\quad x_0=\frac{1+ch^2\mu}{2ch\mu}. \quad (2.13)$$

The convergence of (2.13) is proved along the same lines as that of (2.8). Finally, for $\vartheta=\frac{\pi}{2}$:

$$\sum_{n=0}^{\infty}(-1)^n(4n+1)\cdot(4n+3)\cdot Q_{2n+\frac{1}{2}}\cdot\left(Q_{2n+\frac{3}{2}}-Q_{2n-\frac{1}{2}}\right)=$$

$$=\frac{\pi}{\sqrt{2}}\,\frac{ch\mu\cdot sh^2\mu}{(ch\,2\mu)^{7/4}}\cdot\left[2\cdot Q^1_{\frac{1}{2}}(y)-3Q_{\frac{1}{2}}(y)\right],\quad y=\frac{ch^2\mu}{\sqrt{ch\,2\mu}}. \quad (2.14)$$

3. New relations may be obtained if one equates the integral $\oint A_\ell\, d\ell$ along the closed contour passing through the solenoid hole to the magnetic field flow $\Phi=2\pi g(d-\sqrt{d^2-R^2})=2\pi g a(cth\,\mu_0-1)$. For definiteness choose a contour with fixed $\mu\,(<\mu_0)$ and φ. Then

$$\oint A_\ell d\ell = a\int_{-\pi}^{\pi} d\vartheta \; \frac{A_\theta}{ch\mu - \cos\vartheta}\;, \quad a=\sqrt{d^2-R^2} \tag{3.1}$$

Here $A_\theta = -\left[sh\mu\cdot\sin\vartheta\cdot A_\varrho + (1-ch\mu\cdot\cos\vartheta)A_z\right](ch\mu-\cos\vartheta)^{-1}$ is the component of $\vec{A}$ tangent to the contour. Integrating (3.1) and equating it to Φ we find:

$$\sum_{n\geq 0} Q_{n-\frac{1}{2}}\, Q_{n+\frac{1}{2}} = \frac{\pi^2}{4}\cdot(cth\mu - 1)\,. \tag{3.2}$$

This relation may be proved without recourse to the physical aspects of the problem. For this consider the integral $\int_0^{2\pi}\frac{\cos n\vartheta}{ch\mu-\cos\vartheta}\,d\vartheta$. Direct integration gives: $2\pi\cdot\frac{1}{sh\mu\cdot\exp(\mu n)}$. On the other hand, the integrand may be viewed as a product of two factors: $\frac{\cos n\vartheta}{(ch\mu-\cos\vartheta)^{\frac{1}{2}+\lambda}}\cdot\frac{1}{(ch\mu-\cos\vartheta)^{\frac{1}{2}-\lambda}}$ $\left(|\lambda|<\frac{1}{2}\right)$. Expanding both of them in $\cos n\vartheta$, integrating and comparing two integration procedures one get:

$$Q^{\lambda}_{n-\frac{1}{2}}\cdot Q^{-\lambda}_{-\frac{1}{2}} + \sum_{m\geq 1}\left(Q^{\lambda}_{n+m-\frac{1}{2}} + Q^{\lambda}_{m-n-\frac{1}{2}}\right)\cdot Q^{-\lambda}_{m-\frac{1}{2}} = \frac{\pi^2}{2}\,\frac{1}{\exp(\mu n)sh\mu\cdot\cos\pi\lambda}\cdot \tag{3.3}$$

For $\lambda=0,\ n=1$ we get (3.2). For $\lambda=0, n=0$ we get $(Q_{-\frac{1}{2}})^2 + 2\sum_{m=1}^{\infty}(Q_{m-\frac{1}{2}})^2 = \frac{\pi^2}{2sh\mu}$. The integral $\int_{-\infty}^{\infty} A_z(\varrho=0,z)\,dz$ should also be equal to Φ. One may confirm this by putting $\mu=0$ in (2.2) for A_z and integrating w.r.t. ϑ $\left(dz = -\frac{d\vartheta}{1-\cos\vartheta}\right)$. On the other hand, one may substitute A_z given by (2.5). So

$$\sqrt{R}\,d\int_0^{\infty}\frac{dz}{(d^2+z^2)^{3/4}}\cdot Q_{\frac{1}{2}}\left(\frac{d^2+z^2+R^2}{2R\sqrt{d^2+z^2}}\right) = \pi\left(d-\sqrt{d^2-R^2}\right)$$

or in dimensionless variables

$$\int_0^{\infty}\frac{dx}{(1+x^2)^{3/4}}\cdot Q_{\frac{1}{2}}\left(\frac{1+z^2+x^2}{2z\sqrt{1+x^2}}\right) = \frac{\pi}{\sqrt{z}}\left(1-\sqrt{1-z^2}\right), \quad z\leqslant 1. \tag{3.4}$$

4. The following analytic formulas are due to a simple fact that outside the solenoid (where $\vec{H} = \mathrm{rot}\,\vec{A} = 0$) the VMP may be viewed as a gradient of a (generating) function χ. As $\oint A_\ell\, d\ell \neq 0$ along some closed paths χ is a discontinuous function. For the thin solenoid $\left(\frac{R}{d} \ll 1\right)$ it is determined by the relations following from (2.2):

$$\frac{A_\Theta}{\mathrm{ch}\,\mu - \cos\vartheta} = \frac{g\pi}{\sqrt{2}} \exp(-2\mu_0) \frac{P_{-\frac{1}{2}} \cos\vartheta - P_{\frac{1}{2}}}{\sqrt{\mathrm{ch}\,\mu - \cos\vartheta}} = \frac{1}{a} \frac{\partial \chi_0}{\partial \vartheta}$$

$$\frac{A_\mu}{\mathrm{ch}\,\mu - \cos\vartheta} = \sqrt{2}\pi g \cdot \exp(-2\mu_0) \frac{\sin\vartheta}{\sqrt{\mathrm{ch}\,\mu - \cos\vartheta}} \cdot P^1_{-\frac{1}{2}} = \frac{1}{a} \frac{\partial \chi_0}{\partial \mu} . \tag{4.1}$$

Now integrate (4.1) w.r.t. ϑ and μ and equate the results:

$$\chi_0 = ga \cdot \exp(-2\mu_0) \cdot \Big\{ -2\vartheta + \sum_{n \geq 1} \frac{\sin n\vartheta}{n} \Big[P_{-\frac{1}{2}} (Q_{n+\frac{1}{2}} + Q_{n-\frac{3}{2}}) - $$

$$ - 2 P_{\frac{1}{2}} \cdot Q_{n-\frac{1}{2}} \Big] \Big\} = -2ga \cdot \exp(-2\mu_0) \Big[\sum_{n \geq 1} \sin n\vartheta \cdot \int_0^\mu P^1_{-\frac{1}{2}} (Q_{n+\frac{1}{2}} - $$

$$ - Q_{n-\frac{3}{2}}) d\mu + \pi \cdot \left(1 - \cos\frac{\vartheta}{2}\right) \cdot \mathrm{sgn}\,\vartheta \Big] . \tag{4.2}$$

Taking into account [6] that

$$\sum_{n \geq 1} \frac{1}{n} \frac{\sin n\vartheta}{n^2 - \frac{1}{4}} = 2\vartheta - 2\pi \left(1 - \cos\frac{\vartheta}{2}\right) \cdot \mathrm{sgn}\,\vartheta \qquad (-\pi < \vartheta < \pi)$$

and comparing the coefficients of $\sin n\vartheta$ we get for $n \geq 1$:

$$\int_0^\mu P^1_{-\frac{1}{2}} Q^1_{n-\frac{1}{2}} \mathrm{sh}\,\mu \cdot d\mu =$$

$$ = \frac{1}{4} \left(\frac{1}{4n^2} - 1 \right) \cdot \Big[P_{-\frac{1}{2}} \cdot (Q_{n+\frac{1}{2}} + Q_{n-\frac{3}{2}}) - 2 P_{\frac{1}{2}} Q_{n-\frac{1}{2}} \Big] + \frac{1}{4n^2} . \tag{4.3}$$

As $\mu \to \infty$ eq. (4.3) turns into:

$$\int_0^\infty P^1_{-\frac{1}{2}} \cdot Q^1_{n-\frac{1}{2}} \, sh\mu \, d\mu = \frac{1}{4n^2} . \qquad (4.4)$$

Applying the same procedure to a finite solenoid, we get [1] the system of finite-difference equations for the integrals

$$F_n = \int_0^{\mu} (Q_{n-\frac{1}{2}})^2 \, sh\mu \, d\mu \quad \text{and} \quad C_n = \int_0^{\mu} P_{n-\frac{1}{2}} \cdot Q_{n-\frac{1}{2}} \, sh\mu \, d\mu .$$

The consistency condition defines these integrals:

$$n \cdot F_n = sh\mu \cdot \sum_{k=0}^{n-1} \frac{1}{2k+1} \left(Q_{k+\frac{1}{2}} \cdot Q^1_{k+\frac{1}{2}} - Q_{k-\frac{1}{2}} \cdot Q^1_{k-\frac{1}{2}} \right) - \frac{1}{2} \sum_{k=0}^{n-1} \frac{1}{(k+\frac{1}{2})^2} ,$$

$$n C_n = sh\mu \sum_{k=0}^{n-1} \frac{1}{2k+1} \left(Q_{k+\frac{1}{2}} \cdot P^1_{k+\frac{1}{2}} - Q_{k-\frac{1}{2}} P^1_{k-\frac{1}{2}} \right) . \qquad (4.5)$$

As $\mu \to \infty$ the first of these relations turns into a known one [3]

$$2n \int_0^\infty (Q_{n-\frac{1}{2}})^2 \, sh\mu \, d\mu = \frac{\pi^2}{2} - \sum_{k=0}^{n-1} \left(k + \frac{1}{2} \right)^{-2} \qquad (4.6)$$

whereas the second one diverges as $\mu/2$. As $\mu_0 \to \infty$ the generating function for the finite solenoid turns into χ_0 (given by (4.2)). Many other useful formulas result from this condition [12]. For lack of space we give here only one of them:

$$\sum_{k=1}^{n} \frac{\Gamma(k-\frac{1}{2}) \Gamma(n-k+\frac{1}{2})}{\Gamma(k+1) \Gamma(h-k+1)} = 2\sqrt{\pi} \; \frac{\Gamma(h+\frac{1}{2})}{\Gamma(h+1)} \qquad (4.7)$$

which although looks like the Dougoll formula [3] does not reduce to it.

Thus we gave closed analytic expressions for the infinite sums (2.7)-(2.9), (2.13), (2.14), (3.3), (4.7) and the integrals (2.11), (3.4), (4.3), (4.4), (4.5) that we could not find in the literature and hope are new.

References

1. Afanasiev, G.N. (1983). JINR, E2-83-339, Dubna; (1984). JINR, E4-84-65, Dubna.
2. Stratton, J.A. (1941). Electromagnetic theory, New York.
3. Bateman, H. Erdelyi, A. (1953). Higher transcendental functions, 1, McGraw-Hill Co., N.Y. et al.
4. Hobson, E.W. (1931). Theory of spherical and ellipsoidal harmonics, Cambridge Univ. Press.
5. Robin, L. (1957-1959). Fonctions spherique de Legendre of fonctions spheroidales, 1-3, Gauthier-Villars, Paris.
6. Gradschteyn, I.S., Ryzik, I.M., Tables of integrals, series and products (1965). Acad. Press, N.Y.
7. Prudnikov, A.P., Brychkow, Yu.A., Marichev, O.I. (1981) Integrals and series, 1; (1983). 2, Nauka, Moscow. (Russian).
8. Magnus, W., Oberhettinger, F., Soni, R.P. (1966). Formulas and theorems for the special functions of mathematical physics, Springer, Berlin.
9. Hansen, E.R. (1975). A table of series and products, Prentice-Hall, N.Y.
10. Abramowitz, M., Stegun, I.A. (1965). Handbook of mathematical functions, Dover, N.Y.
11. Haeringen, H. van, Kuk, L.P. (1981). J. Math. Phys., 22, 2482; Haeringen, H. van. (1982). J. Math. Phys., 23, 1964; (1983). 24, 1054; Gervois, A., Navelef, H. (1985). J. Math. Phys., 26, 633; Beig, R. (1985). J. Math. Phys., 26, 769.
12. Afanasiev, G.N. (1985). HINR, P5-85-118, Dubna.